APPLIED MINERAL
EXPLORATION
WITH SPECIAL REFERENCE TO

URANIUM

APPLIED MINERAL
EXPLORATION
WITH SPECIAL REFERENCE TO

URANIUM

ROBERT V. BAILEY
MILTON O. CHILDERS

WESTVIEW PRESS
BOULDER, COLORADO

Published in 1977 in the United States of America by
   Westview Press, Inc.
   1898 Flatiron Court
   Boulder, Colorado 80301
   Frederick A. Praeger, Publisher and Editorial Director

Library of Congress Cataloging in Publication Data

Bailey, Robert V.
   Applied mineral exploration with special reference to uranium.

   1. Prospecting. 2. Uranium. I. Childers, Milton O., joint author. II. Title.
TN271.U7B27   622'.18'493   76-30874
ISBN: 0-89158-210-X

Contents

List of Figures

List of Tables

Preface

The purpose of this book is to provide a good explanation of the fundamentals of organizing, operating, and concluding an exploration program, particularly uranium exploration. To our knowledge, no text currently exists which adequately discusses practical aspects of exploration, and this book represents a step in the direction of providing a definitive guide and instruction manual. The need for a book of this type became apparent to us as a result of our dealings with various companies and individuals involved in some way with mineral exploration, including independent explorationists, mining company personnel, utility company personnel, state land commissioners, representatives of Indian tribes, fee mineral owners, and others. We have also observed that many explorationists were insufficiently knowledgeable in this area to carry out effective and efficient exploration programs. These observations apply to many of the large companies involved in mineral exploration and/or production as well as to smaller companies and individuals.

Our own familiarity with the literature suggested that no book existed which dealt with the practical aspects of present-day exploration, and further searches revealed no good complete references on the subject. The geologic literature abounds with descriptions of various types of mineral deposits, mineralogy, theories of origin, studies of component minerals, paragenesis, etc., but never has anyone

published information dealing with the geology of the deposits and a full-spectrum treatise on how to carry out the search for them from a practical standpoint.

The scope of this book is limited to considerations pertaining to exploration, especially uranium exploration. Further, most of the text involves exploration practices currently being used in the industry, along with a review of some new development or research projects which are being studied or which show promise. Although much of the text, such as the chapters on mining law, legal and accounting aspects, etc., is applicable to exploration for most mineral deposits, it was necessary to focus specific exploration considerations on one type of deposit, and our subject is uranium. In fact, we first planned to write a text dealing with exploration for uranium in sedimentary rocks. It evolved to a discussion of exploration in general with an emphasis on uranium when we realized that a large part of the information pertained to exploration for any mineral. We make special reference to uranium primarily because that is where our expertise lies, and secondarily because of the strong and increasing interest in uranium exploration.

The book is directed to those who will benefit from an understanding of the way an exploration program is conceived and carried out. It should be of general interest to certain landowners, lawyers, accountants, landmen, securities analysts, and governmental agencies; it should be of specific interest to mineral explorationists, those teaching mineral exploration, and those financing or otherwise participating in mineral exploration programs.

To best accomplish our purpose, the book has been organized in the general order most exploration programs follow, as explained further in chapter 1. First we describe the personnel, and then the geology of known deposits; this is followed by a discussion of preliminary investigation. Later, the land considerations are discussed, and finally, the steps leading to production.

We have not included certain tables or figures which are available from other sources. For example, we have not included a table of uranium minerals and their formulas and characteristics. This has been done by others, especially the U.S. Geological Survey in publications such as professional papers describing known uranium mining areas. In addition, we see little practical value in listing the minerals. After all, arguments are still going on about the mineral composition of the ore being produced from the Grants district of New Mexico, the largest uranium-producing area in the United States. Further, uranium occurs in many diverse minerals which are the result of the environment in which the minerals were deposited and the chemical character of the depositional medium. We considered preparing tables to show uranium minerals which have been identified in the various districts or deposits, but the idea was abandoned because we seemed to be accomplishing no worthwhile purpose. Most uranium explorationists are looking for uranium regardless of the mineral assemblage in which it occurs. The environment is the most important exploration guide, not any specific mineral.

As we began to formulate our thoughts

about what should be included in this publication, the list of important items grew at a rapid rate. Efficiently operating an exploration program is a complex task which requires experience, hard work, long hours, and working knowledge of subject matter ranging from mining law to geochemistry. Our intent here is to present information which will be of practical value, but at the same time to present important theoretical aspects. We have tried to cite significant references whenever possible. The references are selected. We may have omitted some good and pertinent references, but we have tried to include most of those useful to an explorationist.

As the text began to take form, we were somewhat amazed at the knowledge of law needed by the mineral explorationist, how many legal pitfalls exist, and how much time the explorationist must spend dealing with or considering legal aspects. Of the thirteen chapters in this book, three have specific orientation toward laws dealing with exploration and/or development, and five chapters deal to a significant extent with legal aspects. Some individuals, particularly those in the academic field, who reviewed an outline of the text have commented that certain chapters do not seem sufficiently related to exploration to be included in a book of this type. They wonder if, for example, mining law, land ownership and leasing, selling a deal, and legal and accounting aspects should be discussed at all. Our answer is that in order to carry out an exploration program, the explorationist and his associates *must* deal with these problems. Admittedly, in the course of their careers, some geologists and engineers will have little to do with the above topics (for example, individuals who work for the federal government or universities). On the other hand, those practicing in industry, particularly those involved with applied exploration, must deal with these problems adequately if their programs are to be successful. Too often, those in academia fail to grasp the significance of certain aspects of exploration because they have never had any practical experience in the industry. There is no way one may become a "well-rounded" explorationist without dealing successfully within a broad spectrum of disciplines, including mining law, land, and salesmanship.

We have not gone into great detail on mining law, but we believe some discussion of basic principles should be included here because other sources are often difficult for an explorationist to acquire. Mining law can be expected to undergo some dramatic changes in the coming years, just as drilling and other exploration techniques will change. Nevertheless, such anticipated changes do not mean that discussions of current laws and techniques are futile. For example, we have found in the examination of old prospects that a good understanding of the prospector's motive often includes consideration of the legal and financial climate at the time of prospecting.

We discuss land and its acquisition at considerable length in chapter 5 and elsewhere. Without land, a prospect is essentially valueless to the economic geologist. We have also devoted one chapter (8) to the subject of selling a deal or project because that function is such an important part of the mineral exploration industry. It is one thing to develop a concept for an exploration program or project, but quite another to obtain financing for the program or persuade others that they should spend money on it. This aspect appears to have been inadequately discussed in many other texts dealing with mining exploration.

As authors, perhaps we should comment on our credentials for writing this book. Our combined experience is 36 years in mineral exploration, mostly for uranium

in sedimentary rocks. We are both practicing geologists, actively engaged in formulating, selling, and carrying out exploration programs. We have both been employed at times by major mining corporations and have also performed services as geological consultants in the mineral industry. We have successfully operated independently, assembling and selling deals. In 1973, we formed Power Resources Corporation (Powerco) as the executive officers; in 1974, Powerco completed a public stock offering of 400,670 shares at $1.50 per share (stock now traded on the over-the-counter market), which netted the company approximately $490,000. By using this as "seed" money, in less than three years we had assembled and sold deals wherein our partners had agreed to spend more than $25 million to earn an interest in the various projects, provided the partners carried exploration through all phases. Powerco retained carried interests ranging mostly from 20 to 25 percent. Powerco's investment in these deals is zero, because the deals were structured so that Powerco got more than the initial investment returned as a payment at the outset by the financial partner. In addition, a coal deal was sold with a commitment of $5 million if all phases are completed. Thus, this book was written during a period when the company we manage successfully competed in the uranium exploration field with other exploration groups, large and small, many of which were better financed than Powerco. In addition, we are pleased to state that our exploration work during those three years has resulted in some new uranium ore discoveries.

The most difficult part of this book was finding the time to complete the writing and rewriting in a one-year period and, at the same time, to assemble and sell deals and carry on the other functions necessary to run a successful small exploration company. The result is that most of the information herein derives from first-hand experience. Because of the broad spectrum of disciplines involved in a mineral exploration program, along with continuing development of sophisticated exploration techniques, it is becoming imperative to have an exploration team composed of several specialists, the most important of which is the explorationist himself.

In this text we have used metric measurements (often followed by English) wherever possible; in some situations, it was necessary to use English measurements because of continuing usage in the United States. For example, we have used many existing maps in our illustrations. These were drawn to English scale and it is not reasonable to convert them for use here.

We wish to acknowledge the increasingly important role played by women in the exploration industry, and we do not intend to slight anyone by using the masculine pronoun in almost all cases. It was simply too cumbersome to write "he or she" and "his or her" throughout the book.

Acknowledgments

We wish to acknowledge the time spent in interviews or discussions by the following persons: Bob Schafer, John Wold, and Cotter Ferguson in Casper, Wyoming; Dave Wentworth, Gil Griswold, Dean Clark, and our friend who believes humates may do it all, John Motica, in Albuquerque, New Mexico; Dick Bassham, landman in Denver, Colorado, who was very helpful with chapter 5; Hal Bloomenthal, our friend and legal counsel from whom we have learned the little we know about law, and who commented on chapters 4 and 9; Rick Austermann, now a lawyer with the firm of Roath and Brega in Denver, who drafted most of chapter 4; Maureen Harris-Taylor, writer, whose talents were of great value in editing many of the chapters; Bruce Smith, geophysicist, who commented upon and gave important suggestions on some of the discussions concerning geophysics; Mike Reimer, geochemist, who gave important suggestions concerning radon and helium detection; Terry Offield, geophysicist, who reviewed chapter 3 and make several improvements; Robert Raforth, geologist, who helped assemble some materials for the book; Sue Nies of Altair Drafting, who drafted the maps and many diagrams; Rhoda Bailey, who helped with research and acquisition of materials for the book; Rick Neville, president of Scinti-Log in Casper, Wyoming, who helped gather material on logging methods and equipment; Bob Rodriguez of Wyoming Mineral Corp., who gave several suggestions on developments in geophysics; Stanley Hallman of Elmer Fox, Westheimer, and Co., Certified Public Accountants, who provided a summary of accounting aspects included in chapter 9.

1

The Mineral Explorationist
and the Role He Plays

INTRODUCTION

In the preparation of this book, this first chapter proved to be one of the more difficult and challenging to write, because it attempts to define the role of an explorationist and how he or she is motivated. Very little has been written on the subject in other geology or mining texts, and virtually no pertinent references exist in the literature. Consequently, many of the thoughts expressed here reflect our personal opinions.

We also attempt to review in this first chapter the general activities of an explorationist, as well as his education, experience, individual capabilities, and motivation. In addition, we comment on certain other aspects of exploration, such as the role (or nonrole) of some companies. We are pleased to see that more articles are appearing in magazines and journals about current exploration problems, such as the diminishing number of areas in which exploration may be carried out.

We shall begin by explaining why we use the term "explorationist" rather than geologist, prospector, or any other term. An explorationist is one who carries out exploration for valuable mineral deposits through the use of specialized knowledge of geology or ore habits. The term "prospector" implies an elementary form of search for mineral deposits. "Geologist" is too inclusive on the one hand and too restrictive on the other: many geologists are working in fields unrelated to exploration, such as construction, environmental, and academic areas, whereas it is also true that an explorationist might have a degree in engineering or no degree at all. There are a few explorationists who have learned enough about the geology of specific areas or the characteristics of certain types of mineral deposits to do a competent job without the benefit of formal geologic education. These individuals are rare, however, and it is true that most competent explorationists are geologists.

However, the term "explorationist" is

not appropriate for positions where advanced technical training and experience are required—for example, carrying out detailed surface or underground mapping, ore reserve calculations, interpretation of drilling results, and other activities which require the technical expertise of a geologist. Thus, we use the term "explorationist" in some instances and "geologist" in others.

WHAT AN EXPLORATIONIST DOES

Work Related to Geology

We wish to emphasize that the major efforts of a mineral explorationist should be directed toward the discovery of commercial mineral deposits. Elementary though this may seem, we have observed that the efforts of those supposedly engaged in exploration often are diverted to activities such as detailed petrographic work, accounting or bookkeeping, and (especially in large corporations) shuffling paper in so-called administrative functions. We do not wish to imply that these activities are unnecessary in support of an exploration program; indeed, they are often very important, but they can provide a convenient screen behind which persons who are incompetent, poorly motivated, or otherwise incapable of carrying out good exploration programs may hide. Such screens are not conducive to ore discovery, and not one ton of ore will be found behind them.

Figure 1-1 illustrates the chain of events involved in mineral exploration from initial concept to production. The chain consists of ten critical links, from the original geologic idea, or concept, to the production of minerals. Studying each link in chronological order should clarify what the explorationist does, each step of the way, in progressing from an idea to mineral production.

Link 1 is the *geologic idea or concept;*
or a favorable observation from field work. The explorationist must first develop an idea, or perhaps devise a theory, which justifies additional work and which might eventually lead to a discovery. To initiate such concepts, the explorationist usually must have a thorough background in, and an understanding of, the geology of known ore deposits (this is discussed in further detail in this chapter under "Characteristics of a Good Explorationist").

Link 2 is *further study; geology appears favorable.* This link directly follows Link 1 because the next logical step after an idea or concept has been developed is to carry out additional study to determine whether the idea is sound when placed under close scrutiny. The effective explorationist should be familiar with the various methods of regional or reconnaissance exploration, such as airborne radiometrics, remote sensing, aerial photography, and airborne geologic surveys, to carry out this link of the exploration chain.

Link 3 is *favorably located land available; no insurmountable environmental or political problems.* After the studies carried out in Link 2 appear favorable, the land situation must be examined to determine whether the land is available so that a play might be put together. If the land is not available, then the chain stops here and can progress no further. Environmental or political problems can preclude exploration, as pointed out in chapter 5; indeed, in many areas, land problems are insurmountable.

Link 4 is *favorable land costs; funds available.* Once you have discovered that favorably located land is available, the next logical step is to examine the land costs. Once determined, adequate funds must be available to put together the land play. Other potential hazards must be considered, such as unreasonable royalties which may be demanded by landowners. Where a

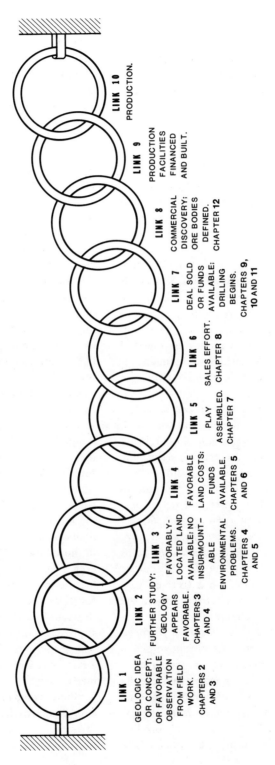

CHAIN-OF-EVENTS DIAGRAM
FROM IDEA TO PRODUCTION

LINK 1
GEOLOGIC IDEA
OR CONCEPT;
OR FAVORABLE
OBSERVATION
FROM FIELD
WORK.
CHAPTERS 2
AND 3

LINK 2
FURTHER STUDY:
GEOLOGY
APPEARS
FAVORABLE.
CHAPTERS 3
AND 4

LINK 3
FAVORABLY-
LOCATED LAND
AVAILABLE. NO
INSURMOUNT-
ABLE
ENVIRONMENTAL
PROBLEMS.
CHAPTERS 4
AND 5

LINK 4
FAVORABLE
LAND COSTS;
FUNDS
AVAILABLE.
CHAPTERS 5
AND 6

LINK 5
PLAY
ASSEMBLED.
CHAPTER 7

LINK 6
SALES EFFORT.
CHAPTER 8

LINK 7
DEAL SOLD
OR FUNDS
AVAILABLE;
DRILLING
BEGINS.
CHAPTERS 9,
10 AND 11

LINK 8
COMMERCIAL
DISCOVERY;
ORE BODIES
DEFINED.
CHAPTER 12

LINK 9
PRODUCTION
FACILITIES
FINANCED
AND BUILT.

LINK 10
PRODUCTION.

Figure 1-1. Chain-of-events diagram.

company is involved, the geologist must now convince (promote, sell) corporate management that land acquisition costs are justified. An individual explorationist must satisfy *himself* that land costs are justified. Again, if the first three links are present, but Link 4 is not available, then the chain ends with three links.

Link 5 is the *play assembled.* At this point, most of the critical land usually has been acquired, the basic geologic concept has been refined, and the prospect begins to form a package which may appear attractive to corporate management or to an investor.

Link 6 is the *sales effort.* The next step, after the play is assembled as a package, is an effort to interest other parties in the value and feasibility of the project, stressing good geology and a favorable land situation, and urging that money be invested to test the idea. (Where a company is involved, this may have been accomplished in Link 4.)

Link 7 is *deal sold or funds available; drilling begins.* Once the deal has been sold, drilling or other evaluation should begin promptly on the project to determine whether the geologic basis is valid. Failure to move promptly at this juncture can increase land holding costs and may waste valuable time of exploration personnel.

Link 8 is *commercial discovery; ore bodies defined.* After the first seven steps are accomplished, a commercial discovery must be made if the chain is to remain intact. If there is no discovery, then the chain ends with seven links.

Link 9 is *production facilities financed and built.* This step is usually beyond the realm of the explorationist, as we have defined the role. This function falls to the developers—such as the mining engineers, who take over the property from the explorationist, build the necessary production facilities, and bring the property into

production, which becomes Link 10. As a general rule, the explorationist is involved only in Links 1 through 8. Of course, geologists are frequently involved in the design and operation of mining facilities; however, we are discussing here the exploration functions only, which involve Links 1 through 8.

Frequently, when a beginning geologist is hired by a company or by an individual, he will be involved only in Links 7 through 10. In this capacity, he will probably be required to work on exploration drilling programs, the definition of ore bodies, or the calculation of reserve estimates, in close association with experienced personnel. He might also be involved in ore control in the mine during the projection phase or might work on development drilling in or near the mine as mining progresses. The beginning geologist is rarely capable of carrying out unassisted the activities involved in Links 1 through 7.

In describing some typical duties of the mineral explorationist, we should list literature studies, log interpretations, logging of samples from drill holes, staking of drilling locations, and field work—examining prospective areas, general preparation of miscellaneous maps (such as contour and isopach maps), and similar office and field duties. As an explorationist gains more experience, he can usually become involved in more links of the chain. Generally, however, the considerable experience of a mature explorationist is required to put together the total picture from Link 1 through Link 8. Often the most efficient approach is through a team effort involving one or more geologists, a land department, a lawyer or legal department, and company management.

Some companies believe that mineral explorationists fall into two categories: the "frontier" type, who carries out exploration in remote areas, and the explorationist

Figure 1-2. Explorationists examine an outcrop of sandstone in an attempt to determine whether it appears to be a favorable host rock for uranium. Evidence suggesting alteration is also sought.

(geologist), who conducts exploration in the general vicinity of known ore deposits. The latter often works near existing mines, attempting to increase the available reserves in a particular mine (this is often referred to as development work). This type of classification may suffice for a personnel manager, but we shall attempt to give a more thorough definition of the job, and also the character and motivation of the person filling it. Generally, a geologist who performs well in development drilling will also perform well in exploration, and vice versa, provided that his experience is comparable.

Evaluation

Another duty typically performed by the explorationist is the evaluation of properties—his own, those of another party, or properties available for acquisition. Some companies rely heavily on submittals in order to acquire new properties for exploration and production. Thus the explorationist may sometimes become involved in evaluating submitted properties, carrying out field work (often accompanied by the party submitting the properties), conducting research, and interviewing numerous people to learn all he can about the submitted properties (fig. 1-2).

The explorationist employed by a company also may be required to evaluate prospects belonging to that company (either those generated by another explorationist or from some other source) and make recommendations based on his findings. If the explorationist recommends proceeding with a prospect, a proposed exploration program and a budget for exploring the property must be developed.

Evaluation is an extremely important part of the explorationist's job, and good evaluation requires in-depth knowledge of mineral habits and known occurrences. The competent explorationist must be able to handle this responsibility without delay and without error.

Many "old-school" exploration geologists have said that the major sin of an explorationist is to turn down the submittal of an unproven major deposit, or to examine a prospect on behalf of his company and declare it to be unworthy of pursuit, only to learn later that it yielded a major discovery. However, explorationists are required to examine numerous prospects in many different geologic, political, and economic climates during the course of a career, and we believe it is inevitable that a major deposit will be passed over occasionally. No two situations are alike, and each environment differs significantly. A geologist may be operating under specific budget restrictions or may have certain other directions from the corporate management concerning the political situation. There are many reasons why he may not be able to carry out an optimum investigation in a given situation, and even if a study is carried out with favorable results, a recommendation not to proceed may result because of environmental or other considerations. However, we are not suggesting convenient excuses for the ultraconservative explorationist to use in turning down every deal.

We believe that any good explorationist, given adequate time and proper financial backing, will ultimately make an ore discovery, either near an existing district, or perhaps in an entirely new district. Persistance, determination, and the desire to win are essential characteristics in the explorationists being sought by corporate management—those who are capable of leading the way to new discoveries, and who will also enjoy personal and financial success.

Promotion

Whether an explorationist is working alone or is employed by a small or large company, he is likely to be involved in some promotion as an integral part of his job. As we define the term, "promotion" means selling an idea to another party—convincing someone of the merit of one's concepts. Sometimes such promotion can lead to the sale of a project or property.

Thus, promoters sell ideas, projects, or properties; if they are successful, they have convinced others that their concepts are correct or at least have merit. The chief executives of many mineral exploration companies have been, and usually continue to be, successful promoters who have advanced in a company. Many of them have been able to promote mineral exploration projects in the course of their careers. The same holds true for many successful independent explorationists; they have been able to promote their concepts or plays and have enjoyed success as a result.

The term "promoter" occasionally has undesirable or negative connotations; indeed, an occasional promoter in the mineral industry is unscrupulous, incompetent, or even totally dishonest. These individuals can do damage to the industry and to investors before they are weeded out through their own failure or by honest competition. The unusual dishonest promoter attracts much more attention than does the typical honest promoter, and many people associate the term "promoter" almost unconsciously with dishonesty. We are convinced, however, that the term should be preserved and used to refer to people who are engaged in the necessary and difficult job of organizing and gaining support for potentially productive projects. Promotion is a widely accepted and very necessary function within the exploration industry, and the role has become well respected. We encourage students and novice explorationists to study and refine promotional skills; the importance of this function in an exploration program cannot be overemphasized. If the university offers basic and advanced courses in selling or salesmanship, some of them should be taken as important electives.

To illustrate how promotion might function within a company, let's take as an example a company geologist assigned to a project which entails carrying out exploration in the Colorado Plateau region for uranium-ore deposits of a certain minimum size. The explorationist may review the literature thoroughly, become acquainted with residents in the area, begin conducting field work, map outcrops, perhaps do some subsurface work, and, finally, present recommendations to company management. Now, enter the promoter: company management must be convinced (1) that the ideas are sound, (2) that pursuit of the ideas will be a worthwhile investment, and (3) that the chances for accomplishing the goals of finding the desired reserves are realistic. If the explorationist is able to do this, then he is a successful promoter, whether he is aware of this or not.

Let's take as another example a district geologist in a larger company who has perhaps eight or ten geologists working under his supervision in various capacities. These geologists, either working in the field or conducting research, may approach the district geologist with new ideas; they *should* be attempting to promote these ideas to their supervisor. If the district geologist believes that the ideas have merit, justifying additional expenditures, then such ideas must be promoted to the management of the company at a higher level. The district geologist, too, in this case, assumes the role of promoter, attempting to convince corporate management that the ideas are good ones. Good management *expects* its geologists to promote projects with merit. The same is true with an individual prospector who has acquired some leases on properties, or who may have staked some claims. If he wishes to sell these to a company—either joint venturing the properties or making some other deal— he must also become a promoter, convincing the prospective purchaser that he has a property which is of value. The college professor or head of a department who approaches the Energy Research and Development Administration, the National Science Foundation, or the U.S. Geological Survey to obtain a grant for a special study becomes a promoter. If he is a successful promoter, the grant may be obtained.

Consequently, every step of the way, almost everyone involved in exploration programs must participate in some form of promotional activities. We describe Link 6 in the chain-of-events diagram as sales or promotional efforts which, if successful, will allow the program to be carried on to Link 7. If the promotional effort is unsuccessful, then the project is generally abandoned, or it remains untested. Sometimes the properties involved are kept together, and a second sales effort is made at a later time. Once the promotion is successful, the program can move forward.

Finally, and most important: the geologist who examines and likes the potential of a mineral prospect or concept has an *obligation* to approach his superiors in the organization and to promote or sell them the idea. If he cannot or will not do so, then he is doing a disservice to his employer. For example, let's take a hypothetical gold prospect in Nevada. Two geologists from different companies, at different times, go to look at it. Both like the prospect and believe that substantial production could be obtained from the property.

Geologist A is conservative; he prefers not to oversell or excite management unduly. If a deal were to be made and the company were to proceed with costly exploration on the property which then produced negative results, Geologist A fears that he might be considered too optimistic or that his judgment might be regarded as questionable. He considers it a safer decision to be somewhat negative. Consequently, he reports to the company in cautiously optimistic, but guarded, phrases. He places extra emphasis on possible pitfalls, and downgrades potential reserves. When company management reviews the report they cannot become enthusiastic because the report is not encouraging. It seems that better gold opportunities should be sought elsewhere. No deal is made.

Geologist B, on the other hand, is an enthusiastic person who believes in his own abilities to recognize a good prospect. When a good prospect is found, he is determined that the company should explore it if the proper deal can be made. To accomplish this, he must promote or sell company management on the idea. He prepares good maps and an accurate, enthusiastic report. He welcomes a discussion with management about the geology of the prospect, and answers their questions readily. As a result, a deal is made and the prospect is explored.

This story would have a happy ending if Geologist B's company made a profitable discovery. However, the point here is that even if no discovery results, Geologist B has performed his job in a superior manner. The management of good exploration companies wants and expects its geologists to become enthusiastic when they like a prospect, and executives encourage and appreciate good promotional efforts directed at them. Unfortunately, many geologists or explorationists do not have the imagination or the ability to originate or otherwise recognize good prospects, and too often they have little or no promotional ability.

Administration

As with any profession in which the allocating of funds and the directing of people's activities is concerned, mineral exploration necessitates a certain amount of administrative work. Many explorationists rightfully resent the administrative duties, which remove them from the physical, more exciting aspects of exploration. Nevertheless, administration is an inherent and essential part of most exploration programs.

We have all heard of the Peter Principle, which states that in a given organization an individual will rise to his level of incompetence. The principle applies to mineral explorationists. Those of us who have been involved with exploration for several years have known successful geologists who have been highly proficient in finding ore bodies, and some may even have succeeded in carrying through the entire chain of events which we described earlier. Frequently, these very people, once they become successful, find themselves "advanced" to a position where they become so overwhelmed with administrative work that they lose not only their motivation for finding new deposits, but also their skill at originating new ideas and concepts which might lead to a discovery. Certain good explorationists may rise through the ranks until they become district geologists or exploration managers. Here, their administrative responsibilities frequently loom so large and may occupy so much of their time that their capabilities as explorationists fall by the wayside. This syndrome seems particularly prevalent in large companies, where massive volumes of paperwork are generated, and the shuffling of personnel goes on incessantly.

To allow enough time for conducting meaningful exploration, the explorationist must keep his administrative duties and responsibilities to an absolute minimum. Companies should recognize that if an employee has significant talent as an explorationist, he should not be moved into a position where he is overburdened with time-consuming administrative responsibilities. The individual may either lose his motivation or be forced to resign and move to a new position where he can perform exploration work unhindered by the duties of administration.

The organizational structure of many companies is at fault here, in that often the only way an explorationist can advance through the corporate hierarchy is by assuming ever greater administrative responsibilities. Wise company management will not tolerate this situation, since imaginative work may become stifled, and talented employees may be lost through resignation.

Idea Generation

As we discussed above in relation to the chain-of-events diagram, we believe that idea generation or concept development (the first link) is of extreme importance; without this link, there can be no chain. The explorationist who contributes significantly to his company will occasionally originate exploration ideas, one or more of which may lead to a successful discovery— although this is by no means the sole requirement for a successful explorationist. Someone must carry out the basic research, mapping, and so forth; but the concept geologist deserves a special mention here, because this role is so critical.

It has been our experience that persons with "number-oriented" training, such as mining engineers, and even some geologists with specialized graduate work, generally are not successful in generating ideas for new exploration projects. These new exploration concepts materialize when geological observations are placed into context vis-à-vis a working theory. Many of the factors important to a new exploration concept defy mathematical analysis or formulation.

The person who originates exploration ideas utilizes wide-ranging knowledge of ore deposits and general geology of prospective areas. Engineers and development geologists are usually more interested in details of specific occurrences than they are in broad relationships. The exploration concept geologist is interested in details of specific occurrences only to the extent that they can be useful in developing a generalized perspective.

A successful concept geologist must be adept at fitting observations into theory and modifying theory to conform with new data. Theory should never be permitted to filter observations. The concept geologist should be optimistic and imaginative and, at the same time, scientific and objective.

Thoughts for the Beginner

For the beginning explorationist who may be aghast at the apparently complex tangle of geology and laws facing him, we offer encouragement. If he wonders whether there is a place, somewhere in the morass, for a person with talent, energy, and ambition, the answer is emphatically yes. He can become successful in this business by exerting effort and focusing his skills on a specific goal.

For the explorationist looking for a place in industry, there are two important factors working in his favor to create opportunities and to make niches presently filled available once again. First, some small companies actively engaged in putting together deals and promoting them become successful. They thereby attain their own substantial cash flow from production, and subsequently invest their own money for exploration, rather than using seed money to put together a deal and carrying out the exploration with funding from another party. When companies are obliged to spend their own money, they become much more conservative and begin to guide their investments with care; this effectively removes them from the area of strong competition for new wildcat plays. Second, today's leading explorationists and strong competitors likely will become successful after surmounting such obstacles as grueling negotiations to sell a tough deal. They may then decide to retire to a South Pacific island, or otherwise take it easy, while the royalty payments accrue.

These factors effectively make room for the eager, energetic new explorationist. Those of us already in the arena welcome any newcomers who care to join us. As long as the free enterprise system prevails, there will be abundant opportunities for those with sufficient intelligence and motivation to pursue them.

In addition, the beginning explorationist or student may wonder whether there are still opportunities for those with intelligence and ambition to explore and make discoveries in known districts. Although a review of the literature might imply that most areas have been so thoroughly examined and drilled that any deposits which exist have already been found, this is not so. We firmly believe that an ambitious person with good education and experience, who is also an intelligent observer, could begin geologic investigation in and around most of the known districts and, with adequate time (say, two to four years), find some good prospects and perhaps an ore body. Even though some explorationists are currently doing this, there are relatively few who meet the described qualifications. As a result, there are many available opportunities.

Some explorationists stress the need for luck in the search for mineral deposits. We do not believe in luck as it is defined in the

dictionary: "the *force* that seems to operate for good or ill in a person's life, as in shaping circumstances, events, or opportunities" (italics added). Good results may occur when preparation meets opportunity. The reality of the matter is that there is no magic force in mineral exploration; we are really historians trying to figure out what has happened geologically in a particular area over millions or hundreds of millions of years. Most mineral deposits are at least 25 million years old, and it is not luck when we discover one of these deposits. Perhaps we prospected or drilled in the right place because our interpretation of the geology was correct, or as the saying goes, we may have drilled in the right place for the wrong reason. In any event, discovery of a mineral deposit is not due to any mystical force, as the word "luck" connotes.

The beginner must also become familiar with the hard political and economic facts of exploration. Walthier (1976) included some excellent comments on this subject:

Exploration is a complex mixture of science, economics, art, business acumen, experience, and calculated risk-taking. Established mining companies have spent a lot of time and money to assemble and develop the special staff, corporate structure, and the management philosophy needed to explore successfully.

Exploration is more than geology, geophysical devices, and drill rigs. Geology is a basic input, but it is a specialized field of geology. Even a well-educated geologist requires considerable additional training before he becomes an exploration geologist; the same is true for geophysicists and geochemists. Many good earth scientists simply never make it in exploration. Their approach and personality are wrong. For this reason, I object to the Congress-mandated once-over mineral potential evaluation by the US Geological Survey and the Bureau of Mines of proposed Wilderness Withdrawal Areas. The evaluation *cannot be done properly*, because even if the evaluation teams were composed of well-trained and experienced explorationists (they usually aren't), adequate mineral evaluation requires a long sequence of qualified specialists. Each person looks at the area from a slightly different perspective, uses the data collected by prior workers, adds some of his own, and synthesizes it in his own way. This is the slow process of how ore targets are first conceptualized, and ore bodies then found.

The next critical ingredient in exploration is management. It requires a breed of people who are production oriented, who recognize the high risk involved in the business, and who have the courage and ability to keep the money coming to the specialists even when there is a string of failures. Lastly, it takes a special kind of investor who will provide the funds on a continuing basis—and it takes continuity to be successful. Very few people can adjust to this type of investment, or really understand it.

Successful exploration is the result of the synergistic interplay of unusual people, specially trained, oriented, directed, and financed.

This combination, in my experience, is extremely difficult to find and to put together. It is all the more difficult in countries lacking a private sector.

I have worked for several government agencies, and have worked closely with many government employees, from clerks to ministers. In my opinion, the organization, the attitude, training, job motivation, the decision-making process and the reward system seemingly inherent in government are antithetical to good exploration and mine development requirements.

CHARACTERISTICS OF A GOOD EXPLORATIONIST

Some Industry Views

First, let us point out that there is a difference between a successful explorationist (one who can find ore), and the explorationist who is capable only of routine, uncreative work. As we stated earlier, the concept geologist must develop ideas and concepts, either for his company's follow-up or, if he is independent, for his own property acquisitions. Valuable ideas usually result from familiarity with the geology of known mineral occurrences, knowledge of the literature, and ability to recognize new opportunities when they appear (fig. 1-3). For example, in the course of field work, recognition of an altered outcrop might generate an idea for a play *if* it is recognized as such.

How many *good* exploration geologists are there in the industry? Responses to our question during interviews suggest that there are not many. One exploration manager we spoke to has been engaged in uranium exploration and development for about 30 years, and in that time he has worked closely with 50 to 60 geologists. Of this number, he believes perhaps 5 could be considered good geologists, the rest mediocre to poor. A district geologist for a major oil and mineral company commented that, in his opinion, 75 to 80 percent of the geologists with whom he has worked are technicians only, incapable of imaginative work. Other interviews suggest similar conclusions, as does our own experience.

However, geologists in general are probably no less competent, nor more competent than any other professional group. A close look at any other profession—medicine, for example—could lead to the

Figure 1-3. Co-author Milton O. Childers excavates a test pit to expose the source of a radioactive anomaly detected by low-level aerial radioactivity survey conducted from a helicopter (in background). Helicopter pilot looks on. Note ever-present scintillation counter to right of pit.

conclusion that incompetence runs rampant in that field. It is not our intention to offend or criticize any particular profession. In mineral exploration, however, more people are needed with the characteristics and the will to become good exploration geologists. In addition, the industry needs good managers who can properly utilize talent available to them. Another necessary ingredient is adequate budgets to carry the work forward.

If you are a student or an inexperienced geologist, you will be extremely fortunate to have an opportunity to work closely with a truly competent geologist. If you do, you should profit from the experience and learn all you can as an apprentice. If you do not have this opportunity, you can still become a competent geologist as long as you have the basic abilities and *use* them. Further,

you will have an opportunity to work with many different people as your career progresses. In each situation, you should improve on poor techniques or procedures and adopt good ones; invent some of your own; and learn to welcome constructive criticism. If incompetence runs rampant, so do good opportunities.

The explorationist who is alert, aggressive, optimistic, and not afraid to take chances is the one who can generate ideas. Combine this with a strong desire to win, and the competition will recognize this person as a serious competitor. Earth scientists may be at work in the lab, and a knowledgeable management may be in support, but it is the individual explorationist who is out and about—looking, listening, reading, thinking—who will recommend that good new plays be made, that exciting new areas be staked or leased, or that worthy properties be acquired from other parties.

In many professions, a positive mental attitude is often stressed as an important characteristic of a successful person; mineral exploration is no exception. What is this attitude, though, and how is it developed? In our view, the attitude results from four factors: (1) having an ambitious but realistic goal within a realistic time frame, (2) having the determination to reach the goal, (3) being prepared for the task, and (4) being able to see a just reward upon reaching that goal.

Of these four factors, being prepared is the most important. In mineral exploration, this requires a solid knowledge of known deposits of a particular type, as well as theoretical aspects, before undertaking exploration for that mineral. An explorationist should strive to attain a firm, realistic understanding of known deposits before undertaking exploration for a given mineral or suite of minerals. This usually requires in-depth literature studies and visits to existing mining operations for observation of ore bodies. Mining companies' cooperation in permitting mine visits is extremely valuable in this connection. Literature studies and mine visits are both time-consuming activities and consequently expensive. Company management should have the foresight to recognize that such expense is well justified.

With a keen understanding of known deposits and an ability to grasp the geological history of a given area, the explorationist is prepared to explore. The positive mental attitude will follow when a realistic goal, determination, and promise of a just reward are added.

The just reward deserves special consideration, because the reward (if any) is often determined by individuals, such as supervisors, or groups, such as company management, who are outside the direct

influence of the explorationist. The explorationist in the corporate structure is usually provided an opportunity to make mineral discoveries for the company by means of the company furnishing funds for the exploration. What happens, however, when the explorationist makes a significant discovery, or takes part in one? What is the just reward? A gold watch? A small salary increase? Miller (1976) addressed this problem, and while we do not agree with some of the points in his paper, certain comments strike the heart of the issue. Miller (p. 836) stated:

> The successful explorationist, as a rule, receives little credit for an ore discovery, due to dilution within the corporate management system and, therefore, he changes corporations. Since the value and life of a major discovery is so great, corporations have not suffered financially on a short-term basis. However, the graveyard of mining corporations is growing, and this is due to the loss of their unique explorationists.
>
> The mining corporations must realize that to maintain continued growth, successful explorationists must be financially rewarded. The rewards should be significant stock awards, high salaries, and increased freedom to explore. With the proper credit and sufficient reward to the explorationist, a multinational corporation will have continued growth, limited only by the excellence of competitive corporations.

Many corporations have lost excellent explorationists because of lack of a just reward after an important discovery has been made and proven to be profitable. Such explorationists sometimes resign to become independent, begin a consulting practice, form a small company, or join another corporation in a similar or higher capacity.

Miller also suggested recognition by the Society of Economic Geologists, which is a commendable suggestion, but one which has no direct effect on the pocketbook. Somehow corporations must reward successful explorationists monetarily through a bonus system. Some of the smaller companies have such systems, but they are nonexistent among the major companies.

Yet some of these same large companies (and many others, also) have cash bonus systems for employees who turn in ideas which save money for the company. Eastman Kodak, among others, encourages employees to submit ideas and pays them thousands of dollars annually for those which are accepted. A company panel reviews the ideas and determines the rewards, which are deemed generous by the employees. We see no reason why this approach would not function well in exploration, thereby fulfilling the additional needs of a just reward.

We suggest that mineral discovery bonuses should be calculated as a percentage of the discovery determined by a company group within a certain time after data have been assembled on the size, grade, and apparent economics of the discovery. Perhaps a sliding scale could be employed so that smaller discoveries would result in a higher percentage. Payment to the discoverer would be made over a period of a few years, or, at the option of the discoverer, would be made in corporate stock.

A small company involved in assembling and selling deals could pay bonuses based on the income from deals put together and sold, as well as a small retained royalty on production from the prospect. Such arrangements now exist in some of the small exploration organizations in the United States, and many of them function very well. In particular, we know of several petroleum geologists who played a key role in the discovery of new petroleum reserves and who are now wealthy because of such bonus and/or royalty plans.

Some companies will immediately argue that if they pay substantial bonuses to employees who make discoveries, the employees will leave the company anyway because they will be financially independent. We would respond that if an important discovery is made, the discoverer *deserves* financial independence.

Those who employ mineral explorationists are well advised to review their internal reward system to determine if it is adequate. Salaries, insurance, retirement plans, and other fringe benefits should be considered, but these programs are usually available to all upper-level salaried employees and do not constitute a bonus for the outstanding explorationist who makes a significant ore discovery. The argument could be raised that geologists are on the payroll to explore, and if they make a discovery, they are simply performing the job they were hired to do. That is the equivalent of telling architect Frank Lloyd Wright (were he an employee) that just because he designed some outstanding buildings he deserves no special reward, because he was *hired* to design buildings. An outstanding performer must be justly rewarded.

Education

We have already indicated that an explorationist need not have a degree in order to explore for minerals. However, those with a more advanced education in geology should have an infinitely better chance of discovering ore because of this specialized training. Thus, in discussing education, we shall concern ourselves with the B.S. and more advanced degrees. How much education does a good explorationist require? We have posed this question to exploration managers of several companies and, combining their responses with our own experience, have reached the following conclusions. (1) If a geologist is to be involved in research, rather than the day-to-day problems of a practicing explorationist, then a master's degree or doctorate is desirable. If he is not to be engaged in research, then a bachelor's degree is adequate, provided that the other desirable attributes described earlier in this chapter are present. (2) Grades are quite important, but not the overall average; it is progression which matters most. A student may be required to make adjustments or cope with other problems during the freshman and sophomore years. However, in the junior, senior, and postgraduate years, good grades should appear, especially in geological subjects.

Of far greater importance than the degree or grades are the attitude and motivation of the individual geologist. If he is a hard worker, has a desire to continue learning, is willing to discuss uncertainties, is a good observer, is willing to accept responsibility, and wants to do first-class work, then he has the potential to be a good explorationist. These qualities, combined with the optimism and enthusiasm of a true prospector, usually produce a good exploration geologist.

Today, classified advertisements for geologists or other scientists often read, "Must have master's degree or higher education." We would agree with this stipulation if the company is seeking someone primarily for research work. However, the geologist or explorationist who works mostly outdoors with drilling rigs, staking holes, logging samples, and interpreting logs is not likely to have a Ph.D. Most frequently, we find that the geologist who is working in the field on a daily basis and is engaged in the myriad tasks essential to an efficient exploration program, has only a bachelor's degree. He may be a dedicated individual, though, devoted to the physical, practical aspects of geology.

Compounding the educational problem is the "inexperienced staff syndrome" which is often found in geology departments of many colleges and universities. It is generally acknowledged that an instructor who has had experience in industry should be better qualified to teach students who hope to enter industry than an instructor with no such experience. Yet colleges seem to have difficulty attracting experienced instructors, either through insufficient funds or lack of interest. Some colleges shun industry experience in favor of inexperienced Ph.D. graduates from the alma mater of the current staff, or in favor of their own inexperienced graduates. In our opinion, this practice results in students receiving degrees when they are inadequately prepared to become good explorationists.

We have seen numerous advertisements in *Geotimes* placed by colleges and universities seeking professors to teach various geology courses. Invariably a Ph.D. degree is specified as an essential qualification. When was the last time you saw such an ad that included "candidates with lesser degrees who have demonstrated outstanding mineral exploration ability will also be considered"? We have never seen an ad of this type. Our schools apparently attach such significance to advanced degrees that a person's track record of successful exploration has no bearing in an application. They will hire a Ph.D. with no experience and will not even consider a B.S. or M.S. Is it any wonder that some of the schools seem out of touch with industry?

We have also observed that many graduate geologists are often unwilling to engage in practical, physical work such as using a pick and shovel to uncover critical outcrops (some exploration managers blame the educational process for this factor); typically, as the level of education advances toward the doctorate level, the inclination to perform physical work decreases.

Thus, we find very few geologists with doctorate degrees who are willing to do field work around the drilling rigs. Yet this is one of the most important aspects of exploring for mineral deposits. If an explorationist lacks the competence or the desire to interpret samples and logs, or to plan an efficient drilling program, then the entire project may be lost. The geologist working outdoors with the rigs (the familiar term in the industry is "chasing rigs") is performing an extremely valuable service. Well-site work or chasing rigs requires good observations, interpretations, and decisions which are vital to the success of the program, and the importance of this aspect of exploration should never be underestimated.

It is disappointing to hear a geologist with two or three years' experience, who has been moved from field work into the office, commenting, "I hope I never have to chase drilling rigs again. I've advanced from that position!" We find a similar philosophy prevailing in the oil industry, where it is often considered demeaning by more experienced company geologists to be doing well-site work. The "old hands" like to delegate well-site work to the "new hands" or to consultants. Yet, as in the uranium industry, field work is an extremely crucial function.

Field supervision of drilling is fundamental to an exploration program, and, in many cases, exploration geologists get their first experience in the field with the drilling rigs. This is the most efficient way to begin learning about mineral deposits. A drilling operation is simply an extension of geologic field mapping. It is also true that drilling operations consume most of the exploration expenditures, so a young geologist may learn much about this aspect of exploration economics while working with the drilling rigs.

As is the case with any field work, a competent scientist with any amount of experience may always learn more about the geology of ore deposits by working with a drilling operation. We never fail to take advantage of an opportunity to work with the rigs and observe the results of a drilling operation. Competent, experienced geologists should spend some time on a continuing basis with field operations to maintain familiarity with the basic aspects of exploration. At the same time, it is important that competent and experienced geologists be available to help the beginners to learn the correct methods.

If a geologist "advances" to jobs with responsibilities such as doing regional studies, designing drilling programs, or even generating new prospects or promotional efforts without thoroughly learning the fundamental aspects of exploration, he will be lacking in the background necessary to do a good job. This is analogous to working in a field like nuclear physics without first getting a good background in mathematics.

How do you tell from an interview whether an applicant with a bachelor's, master's, or doctorate degree will become a competent explorationist? This is an important question, one which employers have been wrestling with for many years. There are no pat answers, but we believe it to be a serious mistake to rule out arbitrarily those applicants without advanced degrees; employers should be careful not to pass up a potentially successful explorationist on the basis of lack of advanced degree alone. Some of the most successful explorationists we know in the Rocky Mountain region, both in hard minerals and in petroleum, do not have advanced degrees in geology.

In hiring recent graduates, it must be recognized that schools tend to homogenize students because the students are all doing the same work and learning essentially the same things. After graduation, individual traits, talents, and interest usually emerge. Many explorationists, meeting a former classmate eight or ten years later, are surprised to observe how he has developed professionally during that time.

Experience

Experience is one of the most important and crucial assets for an explorationist. In effect, it is a continuation of the education process, and the explorationist who has good experience has continued to learn in each new situation. Formal education provides the foundation to which additional learning can be added, and if an explorationist does not add significantly and constantly to the growing base of knowledge, he is unlikely to be successful.

Many employers are finding a genuine scarcity of successful explorationists. Those who are available, sometimes hired from other companies, rightly command high salaries. Ideally, companies ought to select some of their most promising junior exploration personnel and continue educating them through an exploration training program, including worldwide experience in various types of exploration programs; unfortunately, this seldom happens. Geologists usually are assigned to various projects, as needed, in the best interests of the company; varied experience thus derives from company necessity rather than from a benevolent plan for the employee's (and thereby the employer's) benefit.

When hiring an explorationist with broad experience, an employer usually seeks the ability to make accurate judgments and decisions concerning property submittals. An employer also expects the explorationist to be able to evaluate the potential of new ideas.

If an explorationist intends to work with only one commodity in one specific area, such as porphyry copper in Arizona, then it is vital that he have experience with the geography and geology of that area.

Knowledge of other parts of the country is less important, although it is useful to have seen other mineral occurrences. Concentrating on one particular area, an explorationist becomes well acquainted with it and can perform proficiently there as a specialist. However, if he has knowledge of only one area, he may tend to make decisions based on inadequate experience if given an assignment in another area. In such event, he should qualify any opinions rendered or decisions made. Better still, he should postpone important decisions until he has become more familiar with the new area and his feet are firmly on the ground. If necessary, a consulting geologist with significant experience in that area should be contacted.

If an explorationist is familiar with roll-front uranium deposits, such as those found in Wyoming or Texas, the deposits are sufficiently similar to one another that experience in one area can easily be applied to another area. An exception, however, is the Grants mineral belt, where deposits have different characteristics. A geologist who is familiar only with roll-front geology (a roll-front specialist) should familiarize himself with characteristics of the New Mexican deposits before proceeding with exploration for uranium deposits in the Grants area.

The explorationist with a broad base of experience theoretically can make the best judgments in a given mineral exploration situation. Further, a competent geologist who gains broad experience should advance steadily within a corporate situation, eventually supervising the activities of others with lesser experience.

Individual Capabilities and Motivation

In mineral exploration, as in any other scientific discipline, the capability and motivation of an individual determine how successfully he will function. Measuring these characteristics is a difficult and complex task. Generally speaking, capability is finite and cannot be significantly improved, since it is directly related to basic intelligence in an individual. Motivation, in many instances, can be increased.

How, then, does a person who is intelligent (and therefore assumed to be capable), and who has a basic interest in mineral exploration, become motivated strongly enough to become a good mineral explorationist? What stimulating force helps an individual become a valuable employee, one who can become successful in finding mineral deposits?

Many papers and voluminous texts have been written about motivation: what it is, and how it can be acquired and developed. It is not within the scope of this book to dwell on this matter. However, we would be remiss not to discuss the apparent motivation of some explorationists. Often a successful explorationist has had a basic scientific curiosity from childhood. Throughout his education, he has had a desire to make an extra effort to succeed. Having developed at an early age a basic drive and positive attitude, he frequently retains these qualities throughout his professional life. Thus, he may be more likely to be strongly motivated and, consequently, more successful. This theory applies not only to exploration geology but to other fields of endeavor. Successful explorationists like to *win*—and thus are poor losers. Perhaps an old adage could be appropriately reworded: "Show me an explorationist who does not mind losing an ore discovery to the competition, and I'll show you a poor explorationist."

Once he is involved in a working environment, the fast pace of hard-mineral—especially uranium—exploration can be very motivating to a geologist. For example, in a uranium exploration program, it is possible to generate an idea or make an interpretation one day and drill a hole or holes to test the concept the following day. This is in contrast to the oil industry, where a geologist can work up a play and not see it drilled for eight or ten years—if it is drilled at all. Another motivating factor of uranium exploration is the "shows" encountered in the course of drilling; these can provide an incentive to discovering whether they are a lead to an ore body, and if so, why.

Although it is a vast generalization, we might characterize the successful explorationist as one who, first, has the basic intelligence and capability to perform a complex scientific task well. Second, the individual has a strong desire, competitive spirit, and the will to succeed in whatever he undertakes. We believe that a person with these qualities, combined with the proper motivation, is likely to be successful in mineral exploration. We should point out, however, that exploration has become so complex that one individual can no longer handle the entire spectrum of activities. Interests, abilities, and motivation of the individual will determine which aspect of exploration is to become his specialty.

KEY PEOPLE IN AN EXPLORATION PROGRAM

The organization chart (fig. 1-4) shows a typical personnel arrangement within an exploration company, from corporate management to supporting personnel, such as landmen. Typical duties of those persons most directly connected with the exploration program (manager or chief geologist, project geologist or field supervisor, field geologist, and landman) are described on the following pages. Before proceeding with these descriptions, we shall comment on physical fitness.

An exploration program is only as good as the persons in charge of the program. For a program to be operated effectively,

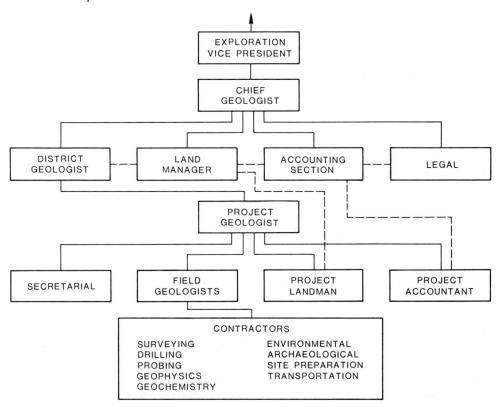

Figure 1-4. Organization chart.

the personnel must be in fine physical condition. It has been correctly stated that "the payoff for physical fitness is increased productivity." Many geologists, by their nature, enjoy being outdoors. They tend to become involved in pastimes which require physical exertion, such as backpacking, climbing, hiking, and fishing. Younger geologists, in particular, tend to be in good physical condition. As they grow older, however, they are often obliged to spend more time sitting at desks; staying in good shape requires a determination and a willingness to work at it. In our opinion, those who impose bad habits and conditions on their bodies (such as smoking or being overweight) may find their working efficiency faltering at a time when, as a result of their experience, they should be most

effective and generate the most income.

How many company mangers consider a job applicant's physical condition (exclusive of truly disabled persons) when conducting an employment interview? Many of us do; we believe, for instance, that a slovenly, overweight, or out-of-shape person reflects undesirable personal traits which forecast poor or inadequate work, not to mention lack of ambition.

Exploration Manager or Chief Geologist

The exploration manager or chief geologist, as mentioned earlier, must have an in-depth knowledge of mineral exploration and development on a wide, perhaps even worldwide, basis. Among other responsibilities, this individual establishes guidelines for the project geologist, in coordination with company management. The chief geologist usually establishes budgets, again in cooperation with company management, and generally reports to the partners, if any, and to company management on progress being made on a particular project. He is constantly in touch with projects, and he reviews the work carried out by employees under his supervision. The chief geologist is responsible for bringing to the attention of company management significant new ideas of the field crews and for attempting to get approval of them.

The chief geologist is usually in charge of more than one project at any given time. If the company is very active, he may be responsible for overseeing several projects which are being conducted concurrently. Major decisions must be made concerning the progress of each project, from exploration to development (or abandonment), and the chief geologist is very much involved at each step of the program, as shown on the chain-of-events diagram.

The chief geologist must be adept at report writing and conveying information to others. He must also be able to promote ideas, including those of his subordinates, to company management. He should be able to sell joint venture proposals to the management of other companies, and he must understand the economic considerations of the minerals or metals sought, as well as their transportation and milling and the costs thereof.

The chief geologist is usually based in the company's headquarters, or perhaps in a regional office. Despite his usually substantial administrative duties, he must remain a concept geologist to a certain extent, capable of generating new ideas for possible exploration ventures for his company. If he lacks this quality, then concept geologists are an important part of his staff.

If a company lacks adequate concept personnel, it will be unable to generate original exploration programs. Consequently, such a company might have to rely on reviewing submittals or properties which have been worked by others and then abandoned, or it might resort to surveillance of the activities of others and thus attempt to participate in some plays. This approach is far from satisfactory if the company wishes to be involved in aggressive mineral exploration programs.

Project Geologist or Field Supervisor

This individual must also have a broad background in mineral exploration and development, and familiarity with the geologic characteristics of known commercial deposits of the mineral being sought. Part of his role is to establish priorities and to plan programs. Naturally, fluent communication must be maintained with supervisors, such as the managers, as we discussed above. He must coordinate the activities of the geologic staff, including geologists supervised, and other individuals (such as landmen) who might be working on a specific project.

The project geologist must direct land acquisition and relinquishment. He must troubleshoot miscellaneous problems which may arise on a field project, particularly during periods of intense drilling. Drilling contracts must be negotiated with contractors, within guidelines established by the company. Drilling results must be planned and monitored on a continual basis, which usually requires a day-by-day check.

The project geologist may arrange the purchase or lease of vehicles, if the company has no purchasing agent. He supervises or participates in the preparation of reserve estimates, and may perform some preliminary mining studies. Like the manager, the project geologist must be capable of writing accurate, informative reports. A certain amount of promotional capability is required, so that ideas may be successfully conveyed to persons of higher authority.

The project geologist is usually based at one of the company's field offices; we have found, in fact, that it is usually detrimental to the efficient functioning of an exploration program for this key person to be situated in the company headquarters where he is not in daily contact with the activities of the field office. The project geologist also carries a certain amount of administrative responsibility, although to a lesser extent than the manager.

We believe the project geologist and others in the company responsible for dealing with contractors should be aggressive persons who are not easily intimidated. They should be tough when the situation calls for it, with no hesitancy about firing or otherwise taking corrective action with poor or careless contractors or inept employees.

Field Geologist

This individual may have extensive experience or essentially none. The amount of responsibility assigned to the field geologist on a given project *should* be directly related to his degree of experience and ability. Yet inexperienced or otherwise incapable persons are often assigned tasks they cannot perform well, either because the supervisor is thoughtless or because no one else is available. On a drilling project, the field geologist works with drilling rigs in the field, and must be able to deal firmly with drillers and other contractors. He logs samples and does log calculations, interpretations, and correlations.

A solid education in mineralogy, stratigraphy, and structure (as the need may be) is of value; and accurate map reading is essential. In addition, the field geologist must have the ability to design and con-

struct work maps of various types, including structural, isopach, and contour maps. He must be able to stake drill-hole locations based on his own or others' interpretations, utilizing a minimum number of holes to adequately evaluate mineralization or ore which may be present. At times he will need the ingenuity to extricate himself from difficult situations in the field (fig. 1-5).

The field geologist must be proficient at report writing, although not necessarily to the extent required of the project geologist or manager. Promotional ability is a distinct advantage, better enabling the sale of ideas to supervisors.

Regardless of what level they may have attained on the organizational chart, all exploration geologists should be receptive to constructive suggestions and comments of others, especially of those working in a lesser capacity. If a project geologist or field geologist has some worthy ideas which are transmitted to company management, he will soon become discouraged if management constantly turns a deaf ear. We have seen frequent instances where company geologists had no access to executive level management and no way to effectively communicate their ideas through the organizational chain. Some companies have lost many good geologists to other organizations as a result.

Similarly, company management should be aware of managers or project geologists who are inadequate or ineffective in their jobs and, consequently, may filter out valuable suggestions originating from persons under their supervision. If management is ignorant of such practices and the situations remain uncorrected, the company again runs the risk of losing good people. Any manager who filters out worthwhile suggestions from his subordinates, or who fails to communicate new ideas to his superiors, should be promptly relieved of his

Figure 1-5. What appeared to be a firm crossing because of ice cover turned out instead to be a soft quagmire. The geologist must get the vehicle out quickly and be on his way.

responsibilities. We know of many companies where able, talented employees have resigned, leaving behind the so-called dead wood. Does management of these companies then understand why the exploration program falters? It is doubtful that many do, and certain that some do not.

Landman

Generally, the landman deals with any land-related problems which may arise in the course of an exploration program. He must have a keen understanding of laws pertaining to surface and mineral rights and of the various procedures involved in obtaining and relinquishing leases, staking claims, paying rentals, obtaining mineral prospecting permits, and so forth. The landman deals extensively with people, from geologists and engineers within his own company to surface lessees and landowners (either fee mineral owners or surface owners). He should fully comprehend

various types of mineral lease and permit forms, and he should be able to utilize these forms and explain them effectively in the field.

The landman must also be familiar with correct recording procedures, so that instruments may be recorded promptly and accurately in the appropriate recording place, such as the county courthouse. He must be able to deal effectively with federal and state agencies, such as the Bureau of Land Management, the Forest Service, the Geological Survey, and the Environmental Protection Agency. The well-experienced landman should have the ability to properly negotiate farm-outs, farm-ins, sales, joint ventures, and other types of mineral ventures.

The landman should also act as a scout for his organization, constantly gathering information about the activities of others: who is leasing and in which area, who is staking claims, what deals are being made, who is buying, and who is selling. This sort of information is of important practical value to an aggressive exploration group, and the company should insist on *written* scouting reports (as discussed in chapter 5).

The landman employed by a company usually supervises the activities of other landmen who may be working on a consulting basis (frequently referred to as "day work" or custom landmen), insuring that their work is done efficiently and at the lowest possible cost.

Certain mining companies have not yet recognized the value of a land specialist. As we point out in chapter 5, landmen should do land work, accountants should do accounting, lawyers should do legal work, and *explorationists should explore for minerals.* This may sound simple, yet it is astounding to see how much time is spent on other pursuits by those who are supposed to be explorationists. Often, corporate procedures require explorationists to spend time and effort outside their specialty field—in land work, for example, if the company has no landmen. We also know of independent geologists who have become so engrossed in accounting procedures that they spent months learning mineral accounting (unnecessarily, in our opinion). In the meantime, several accountants could have been hired with the money that would have been earned had the explorationists' time been spent in geology.

The Field Office

To carry out a field operation, which includes mapping, drilling, and evaluation of prospects and deposits, it is necessary to have a field office nearby (that is, not more than one hour's travel—flight or drive—from the project). Repeatedly we have seen attempts at directing day-by-day exploration from a remote office. While sometimes necessary in the short term, this will not work in the long term; too much time is wasted traveling to and from the office, and the geologists are unable to keep in close touch with happenings on the project. Whenever possible, an office should be located near the project so that geologists may spend sufficient time in data gathering, making studies, mapping, and working with other materials which are gathered in the course of an exploration program. The office should be where the action is, particularly if the project is important.

A married employee usually is more content on a project and works more effectively if his spouse and children are living nearby. If a project approaches the stage where a viable ore deposit is found, and adequate housing within reasonable proximity to the project is unavailable, some companies may construct housing and schools for employees and their dependents, if the size of the planned project warrants it.

2
Types of Uranium Deposits

GENERAL DISCUSSION

Knowledge about known uranium deposits is fundamental to any exploration program, and therefore any reasonably complete treatment of the exploration process must include a basic description of known types of deposits. In this book, we are not attempting to make inclusive descriptions of known uranium deposits, nor are we endeavoring to discuss all of the plausible theories bearing on the origin of uranium deposits. However, exploration programs must always be designed to look for known types of uranium deposits using theoretical and empirical models.

Most of our knowledge and most of the past exploration and development work have been directed at uranium deposits which occur in sedimentary rocks, and these have been the most important economically. However, large low-grade deposits occur in granitic rocks, and in recent years, very large high-grade deposits have been discovered in metamorphic rocks

in northern Australia and Saskatchewan. Many uranium exploration organizations are currently developing programs designed to evaluate the potential of granite and metamorphic rocks. Concepts of origin and ore controls for these occurrences are still vague, but as they are developed and refined, it is likely that they will turn out to be substantially different in some aspects from those now understood for uranium deposits in sedimentary rocks.

Classification of Uranium Deposits

Uranium deposits are so varied in detail that they defy any simple inclusive classification. Indeed, it is tempting to join Joralemon (1975) in his tongue-in-cheek classification in which all ores are divided into categories according to who is best qualified to find them.

It is interesting to note that Ruzicka (1975, p. 14) has proposed a classification for uranium deposits based on genetic types. His scheme classifies all deposits as either syngenetic or epigenetic, depending on whether they are interpreted to have formed together with their host or after the host rock was formed. From an exploration standpoint, the Ruzicka classification is unacceptable. First, we feel that his grouping can be counterproductive to a better understanding of uranium deposits, because it does not systematically group them according to their most important characteristics. For example, he groups together deposits in sedimentary rocks and igneous rocks if they are interpreted to be syngenetic, but a similar sedimentary deposit thought to be epigenetic might be classified as a close kin to a vein deposit. Also, the genetic classification is based on interpretations which have been, and continue to be, controversial. For example, as discussed in this chapter, there is a growing mass of data which indicates that many of the deposits in sandstones of the western United States are early diagenetic in origin. Under these circumstances, the designation of syngenetic versus epigenetic becomes arbitrary and of little importance.

A classification which emphasizes observable, important differences and similarities (with genetic implications) is needed, because misleading and contradictory terms have informally come into use. Indeed, two of these terms have been formally adopted by the International Atomic Energy Agency (1974). In this case, the confusion results from contrasting "sedimentary basins and sandstone-type deposits" with "uranium in quartz-pebble conglomerates." Many of the occurrences which are included in the former "sandstone-type deposits" are more conglomeratic than those which are referred to as the "quartz-pebble conglomerates." Both types occur in sandstones and conglomerates and are strata controlled.

The main difference between these two types is as follows. (1) The "sandstone-type deposits" consist of uranium minerals which occupy interstitial spaces and appear to have been formed by precipitation or fixation from solution or colloidal fluids which percolated through the pore spaces of the hosts, whereas (2) the "quartz-pebble conglomerates" contain uraninite grains together with other heavy minerals in placer concentrations. Some uranium has moved very locally by diffusion or possibly in hydrous solution, but the "quartz-pebble conglomerate" deposits are distinguished by their detrital pyrite and uraninite grains. This is why we have classified them as heavy-mineral deposits.

Our classification (table 2-1) is based on three main types of uranium deposits depending on the principal control of mineralization; i.e., strata controlled, structure controlled, and intrusive controlled. Most of the known deposits occur in sedimentary hosts with strata control, and this book is primarily concerned with those deposits. As more is learned about deposits occurring in metamorphic and igneous hosts, the classification should be amplified and modified to accommodate new data resulting from the study of such deposits.

STRATA–CONTROLLED DEPOSITS

The strata-controlled deposits with the greatest known reserves of uranium occur in formations which include both oxidized and reduced facies. Such formations are often called "varicolored" because of red, green, and gray components. The most prominent of these formations in the western United States are the Triassic Chinle Formation of northwestern Arizona, eastern Utah, and western Colorado; the Jurassic Morrison Formation of northwestern New Mexico, eastern Utah, and western Colorado; the Eocene Wind River, Wasatch, and Battle Spring Formations of Wyoming; and the Miocene and Pliocene Catahoula Tuff, Oakville Sandstone, Fleming Formation, and Goliad Sand of Texas (table 2-2).

TABLE 2-1. CLASSIFICATION OF URANIUM DEPOSITS

I. Strata controlled
  A. Sandstone-conglomerate hosts
    1. *Trend deposits:* ore bodies distributed along mineralized belts or trends; ore bodies generally tabular in habit subparallel to gross stratification.
      a. Mineralized trends parallel to trends of hosts (paleodrainage control). Examples: Westwater Canyon deposits in Ambrosia Lake trend, New Mexico; Chinle deposits in Lisbon Valley, White Canyon, and Monument Valley trends, Utah and Arizona.
      b. Mineralized trends generally crossing paleodrainage patterns. Examples: Salt Wash deposits in Uravan mineral belt, Colorado and Utah.
    2. *Roll-front deposits:* ore bodies distributed along lateral and distal margins of altered complexes in sandstones and conglomerates; ore bodies occur in permeable hosts generally discordant to stratification.
      a. *Bifacies* roll fronts occur in formations with both oxidized and reduced facies. Examples: uranium occurrences in Wind River, Wasatch, and Battle Spring Formations of Wyoming; Goliad, Oakville, and Catahoula deposits in south Texas.
      b. *Monofacies* roll fronts occur in formations which are uniformly reducing. Examples: Inyan Kara, Fox Hills, Lance, and Fort Union deposits of Wyoming and nearby areas; Jacksonville deposits of south Texas.
    3. *Stack deposits:* irregular shapes associated with trend ore frequently controlled in part by structures. Example: Westwater Canyon "redistributed ore" occurrences in Grants uranium region, New Mexico.
    4. *Precambrian heavy-mineral deposits:* uraninite and other heavy minerals concentrated along stratification in sandstones and pebbly conglomerates. (Appear to represent placer concentrations in ancient reducing atmosphere.) Examples: Witwatersrand, South Africa; Elliot Lake, Canada.
  B. Carbonate hosts
    1. *Reef-trend deposits:* ore bodies occur along reef fronts. Example: Todilto deposits in Grants uranium region, New Mexico.
    2. *Calcrete deposits:* irregular distribution of carnotite in calcareous material (caliche or calcrete) distributed in roughly horizontal tabular bodies along major drainageways in arid to semiarid regions. Example: Yeelirrie, Western Australia.
  C. Lignite, black shale, and phosphatic rock hosts
    1. *Uraniferous lignites:* irregular dissemination in lignites and associated carbonaceous shales and siltstones.
      a. Low-grade wide distribution in lignites marginal to varicolored fluvial environments. Example: Wasatch–Green River deposits of the Great Divide Basin, Wyoming.
      b. Deposits (including high-grade) in lignites associated with permeable sandstones below a regional unconformity. Example: Fort Union–Hell Creek deposits of the southwest Williston Basin in Montana, South Dakota, and North Dakota.
    2. *Uraniferous black shales:* low-grade but widely disseminated deposits in reducing environments marginal to varicolored or oxidizing environments. Examples:

TABLE 2-1 (*continued*).

Chattanooga Shale in central Tennessee, Kentucky, and Alabama; Phosphoria Formation in western Utah and Idaho; lacustrine beds of Eocene age in Wyoming and Utah; Lodeve area, southern France; black shales of Sweden.

II. Structure or fracture controlled (vein-type and similar deposits)

    A. Fractured and brecciated host rocks with pitchblende and other uranium minerals filling voids, coating fractures, and partially replacing host rock. Examples: Rabbit Lake, Key Lake, and Cluff Lake in Saskatchewan, Canada; Rum Jungle–Alligator Rivers province, Northern Territory, Australia; Front Range district, Colorado; Midnite Mine, Washington.

III. Intrusive controlled

    A. Igneous disseminated: finely disseminated uranium minerals in alaskite, nepheline syenite, and other felsic igneous rocks. Examples: Ilimaussaq, Greenland, where uranium occurs in nepheline syenite; Rössing, South West Africa, where uraninite is disseminated in syntectic alaskite; Colorado Front Range, where uranium is disseminated in dikes of bostonite.

TABLE 2-2. STRATIGRAPHIC SUMMARY OF URANIUM-BEARING FORMATIONS IN WYOMING, COLORADO PLATEAU, AND TEXAS

Wyoming

Early Eocene: Battle Spring, Wind River, and Wasatch Formations (age equivalents). Major uranium deposits occur in arkosic sandstones and conglomerates in *reduced facies*. Formations also include *oxidized facies*.

—Unconformity—

Paleocene: Fort Union Formation. Uranium deposits occur locally in sandstones and lignites which have been in contact with the overlying unconformity. Formation is entirely reduced unless altered or weathered.

Cretaceous: Lance, Hell Creek, Fox Hills, Fall River, and Lakota Formations. All reduced. Uranium deposits occur locally in sandstones exposed to the unconformity.

Colorado Plateau (New Mexico, Colorado, Utah, Arizona)

Jurassic: Morrison Formation (includes Brushy Basin, Westwater Canyon, Salt Wash, and Recapture Members). Major uranium deposits occur in reduced sandstones, but formation includes both *oxidized* and *reduced facies*.
Todilto Limestone. Reduced facies (lacustrine or marine) marginal to dominantly oxidized facies. Uranium deposits occur locally in reef fronts.

Triassic: Chinle Formation (includes Mossback and Shinarump Members). Major uranium deposits occur in reduced sandstones and conglomerates. Formation includes both *oxidized* and *reduced facies*.

TABLE 2-2 *(continued)*.

Texas

Pliocene: Goliad Sand. Uranium deposits occur in reduced sandstone. Formation includes oxidized facies.

Miocene: Fleming Formation, Oakville Sandstone, Catahoula Tuff. Uranium deposits occur in reduced sandstones and conglomerates. Formations include both *oxidized* and *reduced facies.*

—Unconformity—

Eocene: Whitsett Formation (of Jackson Group). Entirely *reduced.* Uranium deposits occur in sandstones which have been exposed to the unconformity.

Note: Formations which do not host significant uranium deposits have been omitted.

The oxidized facies of these formations consists mainly of red and gray banded mudstone, red siltstone, and red hematite-stained sandstone. The reduced facies consists mainly of gray and green mudstone, gray siltstone, and gray sandstone. Carbonaceous material is common in the reduced facies. Vertical changes from dominantly reduced to dominantly oxidized facies are believed to reflect major variations in precipitation, and these formations record geologic intervals when weathering, erosion, and deposition were taking place under the influence of a savannah climate. The term "savannah climate" is used in sedimentology to refer to a hot or warm climate with pronounced fluctuation in precipitation. In their discussion of red beds, Dunbar and Rodgers (1957, p. 210) suggested that the savannah climate with its "strongly seasonal rains and widely fluctuating water table" is ideal for the development of red soils. The geologic record also suggests that the seasonal cycles of rainfall are superimposed upon more pronounced cycles of much greater duration (fig. 2-1). In this context, our use of the term savannah climate is an extension of previous usage. For example, we are suggesting that a savannah climate persisted in Wyoming throughout Eocene and early Oligocene time. Major dry or wet phases within this time frame, if isolated and classified alone, would be respectively hot arid or humid tropical climates.

Evidence for a savannah climate includes the alternating thin layers of red and pale gray mudstones in the oxidized facies (Van Houten, 1948). The hot, arid phases resulted in extremely high oxidation potentials on the flood plains of the intermittent small tributary streams, where hematite was developed in the muds and silts. Water tables dropped during arid phases in the upland areas of tributary streams. These conditions developed ground waters with sufficiently high Eh values to produce hematite in subsurface sands.

On the other hand, the warm, humid phases promoted intensified chemical weathering in areas where erosion was occurring. Plant life flourished in most environments during the humid phases.

In theory, a fluvial environment with small, intermittent tributary streams draining areas of uranium bearing source rocks and large, permanent trunk streams with stable water tables and permanent vegetation under the influence of a savannah climate is ideal for concentration of uranium. Under the fluvial-savannah model, uranium is dissolved and easily removed

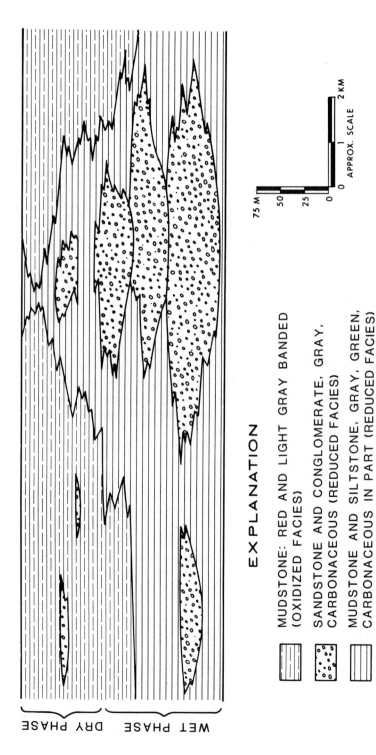

Figure 2-1. Diagrammatic cross section showing the sedimentary facies which characterize deposits accumulated during a major cycle of a savannah climate. Annual cycles are recorded by red and gray bands in oxidized facies.

from the oxidized environments and subsequently concentrated in the humate-rich,
reduced sands beneath the trunk streams.
The alternating humid phases and dry
phases, with fluctuating water tables, promote deep chemical weathering. This theoretical model explains why most of the
strata-controlled uranium deposits occur in
some association with varicolored formations characterized by both oxidized and
reduced facies. In the following discussion,
we often refer to these varicolored formations simply as "bifacies" formations.

A uniformly humid climate results in
ubiquitous vegetation and humus-rich soil
which fixes uranium in any part of the environment. Chemical weathering is enhanced, but uranium is not systematically
leached out of any large part of the terrane
because of its strong affinity with the ubiquitous humates.

It is noteworthy that Krasnikov and
Sharkov (1962) concluded that large exogenetic uranium deposits have formed only
under arid conditions. We concur with
their general observations, but, in our opinion, they do not adequately consider the
role of humid phases in weathering.

Trend Deposits

Most of the significant uranium deposits
which have been discovered and described
in the Westwater Canyon Member of the
Morrison Formation in the Grants uranium
region of northwestern New Mexico are
generally called trend deposits. The term
"trend deposits" has been used extensively
in an informal way, and we are formaliz-
ing it and expanding its application to
cover those uranium deposits occurring in
sandstones and conglomerates and having
the following characteristics:

1. Ore bodies are distributed along mineralized belts or trends which may be
   several kilometres wide and tens of
   kilometres in length.

2. Ore bodies are generally tabular in
   habit and oriented subparallel to the
   gross stratification.
3. Commonly, individual ore bodies are
   elongated parallel to the long dimensions of the enclosing sandstone or
   conglomerate bodies, and this paleodrainage pattern controls the trend
   of mineralization. However, in a
   complex of interconnecting channel
   sandstone bodies, the trend of mineralization might cross the paleodrainage, suggesting some other dominant control. When the trend of
   mineralization crosses the paleodrainage, the individual ore bodies are typically amoebic in plan with very little
   preferred orientation.
4. Uranium mineralization occurs in
   gray to black reduced rock which is
   usually carbonaceous. Most of the
   carbonaceous material is amorphous
   and finely disseminated, and it coats
   grains and partially fills interstices
   between sand grains.
5. Uranium-ore bodies commonly form
   the nuclei of larger volumes of gray
   reduced rock; this contrasts sharply
   with the more extensively developed
   red hematite-stained or white
   bleached sandstone or conglomerate
   of the Morrison or Chinle. These reduced rocks which enclose ore bodies
   are interpreted by many geologists
   to be altered because they are closely associated with ore and because
   they are different in appearance from
   the typical oxidized unmineralized
   rocks farther removed from mineralization.

Grants Uranium Region

The largest known concentration of high-
grade uranium in trend deposits occurs in
the Grants uranium region of northwestern
New Mexico (fig. 2-2), where the Westwater

32

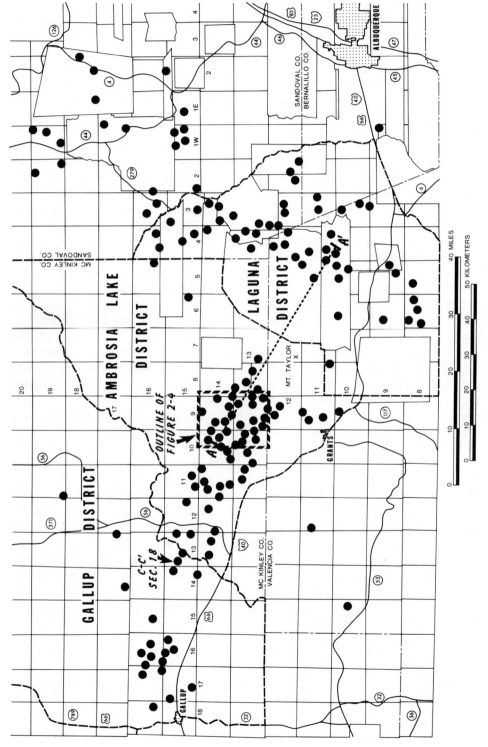

Figure 2-2. Map of Grants uranium region, northwestern New Mexico. Dots indicate uranium occurrences. Cross section A-A' is figure 2-3.

Canyon Member of the Morrison Formation hosts most of the ore bodies (fig. 2-3).
One belt or trend in which the ore deposits
are concentrated is shown in figure 2-4 and
is called the Ambrosia Lake trend. This
mineralized trend is as much as 2.5 km
wide and exceeds 6 km in length. The
trends in the Grants uranium region generally follow old paleodrainage systems in
which the streams flowed easterly. The
Westwater Canyon Member is about 60 m
thick along this trend.

The trend deposits in the Westwater
Canyon Member generally consist of
tabular ore bodies which are frequently
elongate parallel to the paleodrainage
direction, and this is also the direction in
which the aggregate of ore bodies is concentrated (fig. 2-5). The trend deposits of
the Grants region range from thin, small
bodies to large ore bodies 9 m thick, 240 m
wide, and 2 km long, which include some
of the largest high-grade uranium-ore
bodies in sedimentary rocks in the world.

A hydrous uranium silicate, coffinite, is
often reported to be an important ore mineral in the trend deposits of the Grants
region (Corbett, 1963; Hilpert, 1969;
Granger, 1963). The coffinite is extremely
fine and can best be identified by x-ray
diffraction. Uraninite has only rarely been
identified in the trend deposits of the Westwater Canyon Member, but this might be
due to the fine character of the ore. In
the case of the Jackpile and Paguate mines,
Valencia County, New Mexico, Kittel
(1963, p. 170) states that although coffinite is often reported as the most important
mineral, "metallurgical testing indicates
that mill ore from those deposits normally
contains about two percent coffinite, 80
percent unidentifiable oxidized uranium
complexes, 15 percent uraninite, and three
percent organo-uranium complexes." It is
doubtful that 80 percent of the uranium is
complexed in the hexavalent state. If this

is based on solubility tests, the fineness
must be considered. Uraninite globules
have been measured with the electron
microscope, with diameters ranging from
less than 1 micron to about 4 microns (see
Langen and Kidwell, 1974).

In many places, black to dark gray
masses of nonuraniferous mineralization occur marginal to the trend ore deposits.
These have been found to contain an amorphous molybdenum sulphide, believed to be
jordisite, together with vanadium and
manganese minerals (Granger, 1963). These
uranium-deficient black masses are usually
low in carbonaceous material also. Selenium is sometimes concentrated with the
jordisite.

The ore boundaries in the trend deposits
of the Grants region are often sharp. Bedding surfaces frequently form the lower
boundaries of ore bodies (Santos, 1963),
but the upper boundaries are more frequently gradational. Other stratigraphic
controls include scour contacts or disconformities, claystone beds, and clay-pebble
conglomerates. There is a tendency for ore
to concentrate parallel to intraformational
disconformities and along mudstone-sand-
stone contacts (fig. 2-6). In some cases, the
ore ends against sharply defined curved surfaces called "rolls." John Motica (personal
communication) has observed numerous
occurrences of relatively high-grade ore in a
lower-grade ore body concentrated along
cross-stratification and truncated by scour
surfaces (fig. 2-7).

There is no obvious close relationship
between oxidized sandstone and the trend
ore deposits in the Grants region. Some
studies have indicated that the Westwater
Canyon sands were originally deposited in
an oxidized state with ubiquitous limonite
and hematite staining, and that "favorable"
gray sand with pyrite was a product of
diagenetic alteration (Young and Ealy,
1956). Granger and others (1961) and

34

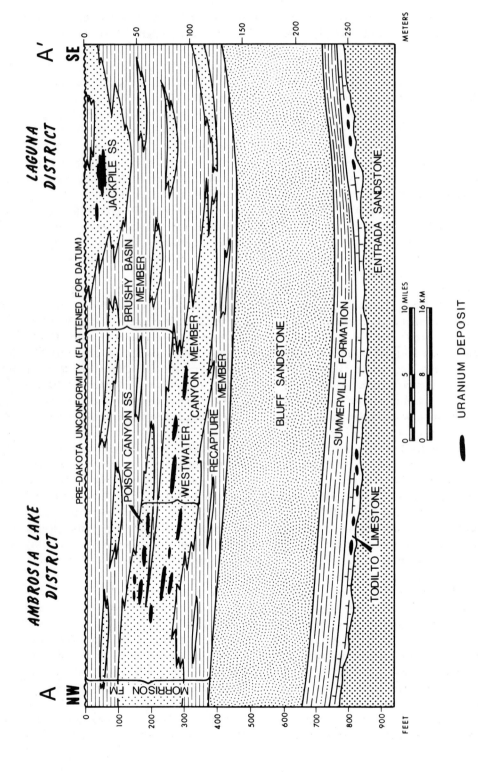

Figure 2-3. Diagrammatic cross section showing generalized relationships between major uranium-bearing formations and interbedded or equivalent formations in Grants uranium region, New Mexico. Location in figure 2-2. Modified in part after Hilpert (1963).

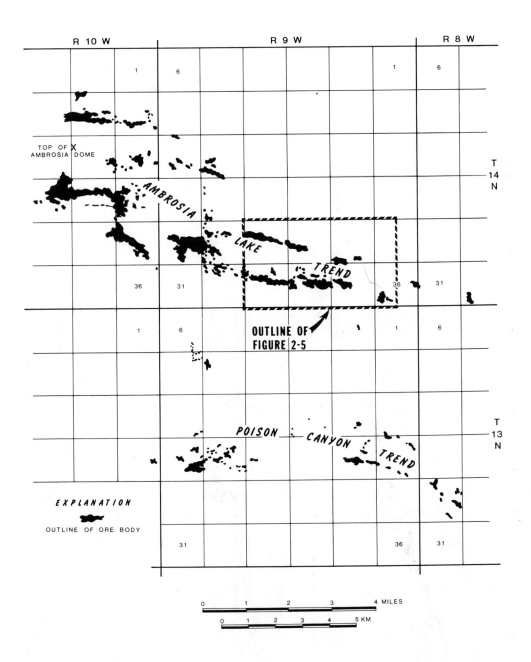

Figure 2-4. Map of ore deposits in the Morrison Formation, Ambrosia Lake area, New Mexico. Modified after Santos (1963).

36

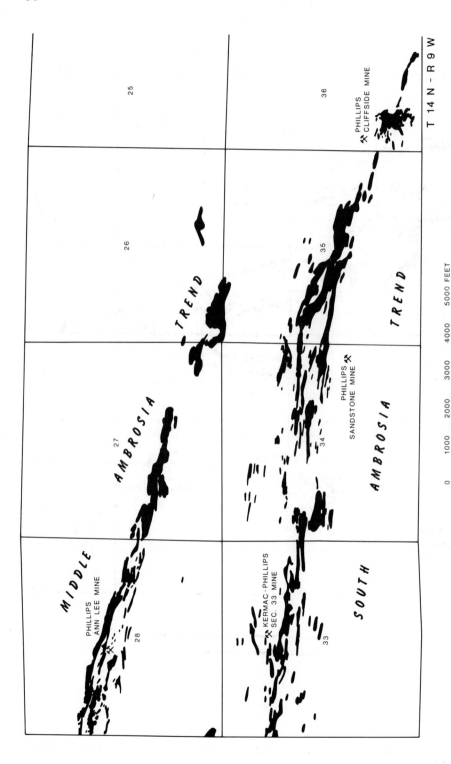

Figure 2-5. Map showing ore trends of the southeastern part of Ambrosia Lake district, New Mexico. Note elongate ore bodies which parallel the general ore trends. Ore trends generally follow paleodrainage (streams flowed east-southeast). Modified after Hazlett and Kreek (1963).

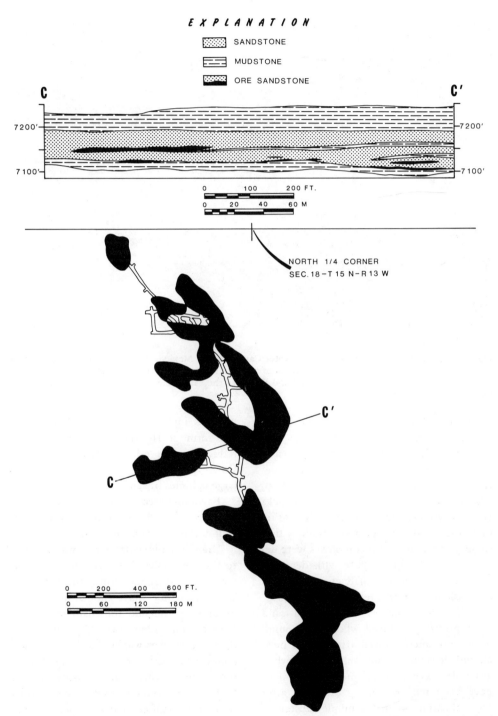

SANDSTONE

MUDSTONE

ORE SANDSTONE

Figure 2-6. Map and cross section of the Black Jack No. 2 mine, Smith Lake area (Gallup district) of northwestern New Mexico, illustrating the tabular, peneconcordant habit of trend ore deposits in the Westwater Canyon Member of the Morrison Formation. Modified after Hoskins (1963).

Figure 2-7. Mineralized sandstone in Westwater Canyon, New Mexico, showing scour surface or intra-formational disconformity cutting off a streak of high-grade mineralization within an ore body. Such relationships have been commonly observed and illustrate the early diagenetic origin of the uranium-ore trend. Photo courtesy of John Motica.

Clark and Havenstrite (1963) described hematite staining in close association with trend ore (and redistributed stack ore) as a later epigenetic alteration of gray sandstone. However, as discussed by Hilpert (1969, p. 69), the color relations of these host rocks are incompletely understood, especially as regards the trend ore deposits. It is likely that some parts of the sand complexes were deposited in a reduced state with decaying organic material in sufficient supply to establish and preserve reducing conditions.

Additional volumes of oxidized sand were probably reduced when the groundwater movement slowed after deeper burial, and decaying organic material absorbed available oxygen. Even more sand might have been reduced during the years after mineralization when the humic substances lost their volatile components and became more like kerogen. These volatile components include hydrogen and sulphur in the form of H_2 and H_2S. Such strongly reducing gases might spread through the slowly percolating ground water to reduce large volumes of sand.

Favorable areas for uranium mineralization are limited to localities which have well-developed reduced sandstones.

Uranium mineralization is not confined to the ore bodies themselves in the Grants area. Certain zones within the Westwater Canyon Member have anomalous amounts of uranium outside the main ore bodies. Laverty and others (1963, p. 203) state that "the Westwater Canyon averages about 100 parts per million in uranium in the Grants district but outside the ore bodies." Laverty (personal communication) has qualified that statement to refer only to several zones within the Westwater Canyon Member.

Uravan Mineral Belt

Another major development of trend deposits is the region which has come to be known as the Uravan mineral belt. This belt of uranium mineralization is about 115 km long and as much as 13 km wide, extending in an arcuate pattern from Polar Mesa in southeastern Utah to just beyond the Slick Rock district in southwestern Colorado (fig. 2-8). The boundaries of the belt are indistinct and are subject to some interpretation (Motica, 1968, p. 806). Most of the uranium in the Uravan mineral belt occurs in the Salt Wash Member of the Morrison Formation (table 2-3). The main ore-bearing sandstone of the upper Salt Wash has been interpreted to have been deposited as a system of aggrading braided streams flowing in an easterly direction generally normal to the mineral belt. Individual sandstones in this system are as much as 1.6 km in width, 30 m in thickness, and several kilometres in length. The color of the sandstones in the Salt Wash varies from red to brown and light gray. The mudstones vary from red-brown to gray-green. Ore-bearing sandstones are usually dark gray to black and carbonaceous and are surrounded by light gray sandstones. The associated mudstones and clays are some shade of gray near the ore bodies and in the favorable areas in the vicinity of the ore bodies. In unfavorable areas more removed from ore deposits, the sandstones are typically red, and the associated mudstones are principally red as well.

The most favorable sandstone hosts for ore bodies range from 15 to 27 m in thickness, but ore does occur in thinner sandstones. Most of the ore deposits are nearly concordant to the bedding of the host rock, and most of them are relatively small (averaging about 3,000 tons), tabular or podlike bodies. Ore bodies are often elongate with the long dimension parallel to the paleodrainage direction.

The uranium minerals which have been identified in the unweathered ores are uraninite and coffinite. Chlorite and hydromica are the main vanadium-bearing minerals in both the weathered and the unweathered ores. The uranium and vanadium minerals almost always occur intimately mixed together and generally the vanadium exceeds the uranium in ratios ranging from 3:1 to 10:1. Almost all of the ore deposits are associated directly or indirectly with carbonaceous material. However, some uranium-vanadium deposits appear to be deficient in carbonaceous material, while some carbonaceous zones are uranium deficient.

The Uravan mineral belt had produced 12.2 million short tons of ore, which averaged 1.33 percent V_2O_5 and 0.25 percent U_3O_8, by 1974. Ore bodies average less than 1.5 m in thickness but range up to a maximum thickness of 7.5 m. Low-grade mineralization is associated with some of the ore deposits, but generally the ore boundaries are sharp and well defined.

Other Areas with Morrison Deposits

The Salt Wash Member of the Morrison Formation also hosts uranium and vanadium ore deposits in the Henry Mountains district and in the Green River Desert area. The deposits have most of the characteristics which typify the Uravan mineral belt just described, but on the average, they are probably smaller. The deposits of the Henry Mountains area appear to occur lower in the Salt Wash than do the deposits in the Uravan area, and they appear to be aligned parallel to a north-trending paleodrainage system.

Chinle Deposits

The Chinle Formation of Triassic age contains ore deposits in Arizona and Utah which are best classified as trend deposits.

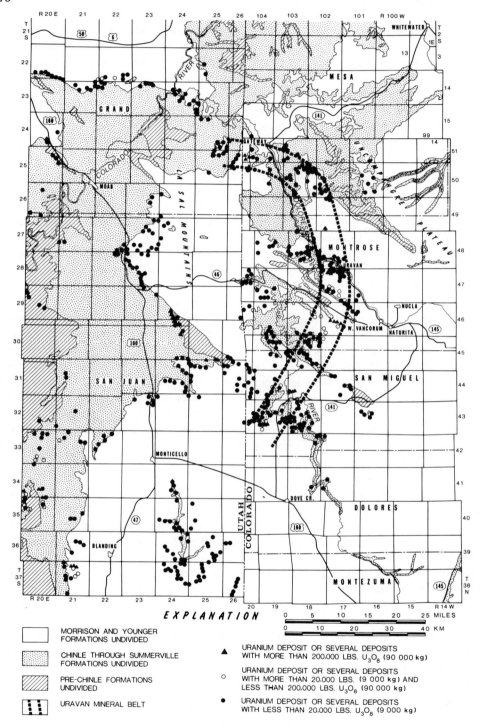

Figure 2-8. Map of southwestern Colorado and southeastern Utah showing generalized geology and uranium occurrences. Triassic uranium deposits occur along base of Chinle through Summerville Formations undivided, and Morrison (Salt Wash) deposits occur along the base of Morrison and younger formations undivided. Note outline of Uravan mineral belt.

TABLE 2-3. GENERALIZED STRATIGRAPHIC CROSS SECTION OF
INTERVAL THAT HOSTS MAJOR URANIUM DEPOSITS OF
COLORADO PLATEAU EXCLUSIVE OF NEW MEXICO

Jurassic

Morrison Formation
  Brushy Basin Member: variegated mudstone with lenses of siltstone, sandstone, and con-
    glomerate. Mudstone increases upward. Oxidized and reduced facies (mainly oxi-
    dized). Thickness: 120–150 m.
  Salt Wash Member:* sandstone and mudstone with pronounced conglomerate units in
    southwestern Utah. Sandstones are red-brown to light gray; mudstones red-brown to
    gray-green. A sandstone unit near the top hosts large and small uranium deposits in
    Uravan mineral belt. Other sandstones host uranium deposits in Utah and northeast
    Arizona. Oxidized and reduced facies. Thickness: 80 m.
  Recapture Member (not recognized in the Uravan mineral belt or most of Utah): alter-
    nating beds of gray sandstone and grayish-red siltstone or mudstone.
Summerville Formation
  Reddish shale, mudstone, and thin-bedded sandstone. Red-brown and pale-brown
    colors predominate. Uniform bedding suggests tidal-flat deposits. Oxidized.
    Thickness: 3-30 m.
Entrada Sandstone
  Reddish sandstone, cross-bedded. Thickness: 60–100 m.
Carmel Formation
  Marine shale and siltstone. Thickness: 10 m.
Navajo Sandstone
  Massive, cross-bedded sandstone. Thickness: 50–100 m.

Triassic

Kayenta Formation
  Interbedded red sandstone, red-brown siltstone, and red-brown mudstone. Thickness:
    40–80 m.
Wingate Sandstone
  Massive red-brown to orange sandstone. Thickness: 80–100 m.
Chinle Formation
  Upper members: reddish-brown, thin-bedded, calcareous mudstones, siltstones, and sand-
    stones overlying gray to green, bentonitic clays and purple to chocolate brown mud-
    stones with numerous sandstone lenses. *Oxidized* and *reduced facies* (mainly oxi-
    dized). Thickness: 100–280 m.
  Mossback Member:* gray sandstone and conglomerate with lenses of siltstone and mud-
    stone. Quartzite, chert, and limestone pebbles in conglomerates. Forms the base of
    the Chinle in Lisbon Valley area of Utah. *Oxidized* and *reduced facies.* Thickness:
    0–45 m.
  Monitor Butte Member: variegated mudstone with sandstone lenses. *Oxidized* and *re-
    duced facies.* Thickness: 0–80 m.

TABLE 2-3 (*continued*).

Shinarump Member:* conglomerate and sandstone with carbonaceous shale or siltstone. Pebbles of chert, quartzite, quartz, limestone, mudstone. Forms the base of the formation in southern Utah and northeastern Arizona, where it unconformably overlies the Moenkopi Formation.

*Major uranium host.

Most of the deposits occur in an arcuate belt of favorable sandstone 5 to 20 km wide and about 210 km long. This favorable belt extends from Monument Valley on the south through White Canyon and Elk Ridge to Indian Creek on the north (fig. 2-9). The Chinle contains similar ore deposits in other areas as well, notably on the flanks of the San Rafael Swell west of the Green River Desert in Utah. Malan (1968) and Chenowith and Malan (1973) published reports on the Chinle and its uranium deposits in northern Arizona and southern Utah and described paleodrainage systems which generally control the best uranium mineralization. As shown on the map (fig. 2-9), the paleodrainage systems of the Chinle, according to Malan, flowed northwestward, southwestward, and westward in southern Utah and northeastern Arizona. Individual channel sandstone bodies average between .8 and 1.6 km in width and can be traced for several miles along their trends. The channel deposits consist of quartz-pebble conglomerates and sandstones, and the individual sandstone beds range up to 12 m in thickness. Units dominated by sandstones and conglomerates have been named the Shinarump Member and the Mossback Member of the Chinle, and these members are as much as 75 m thick. Carbonaceous material is common in these units. Most of the associated mudstones are gray-green, but some thin mudstone beds are reddish to red-brown. The thick mudstones in the upper Chinle above the Mossback Member are dominantly red-brown.

Most of the uranium deposits occur in carbonaceous sandstone and conglomerate beds. Ore bodies are generally closely spaced and tabular, and are usually concordant with the bedding. Single ore deposits range from a few metres to a few hundred metres in length and up to 3.5 m in thickness. The average length ranges from about 5 to 10 times the average width. Mineable ore is continuous in one mine for 2,100 m (Malan, 1968, p. 800). Uraninite and coffinite have been identified in the unoxidized ore bodies together with vanadium minerals such as montroseite, corvusite, doloresite, and vanadium hydromica. Copper sulphide minerals are also present, and the copper increases westward at the expense of vanadium.

Another arcuate belt of uranium deposits occurs in the Lisbon Valley area, where more than 23 million kg (50 million lb) of U_3O_8 have been mined and milled. The Lisbon Valley uranium belt is 25 km long in a northwest trend and 0.8 km wide (Wood, 1968). Most of the uranium deposits occur in the Mossback Member of the Chinle Formation just above the Triassic/Permian unconformity. The Mossback Member consists primarily of fluvial crossbedded calcareous arkosic sandstones with interbedded mudstone and conglomerate lenses. Coalified plant material is common in association with the uranium deposits. The Mossback was deposited on the post-Cutler erosion surface by streams flowing northwesterly to westerly. The Mossback

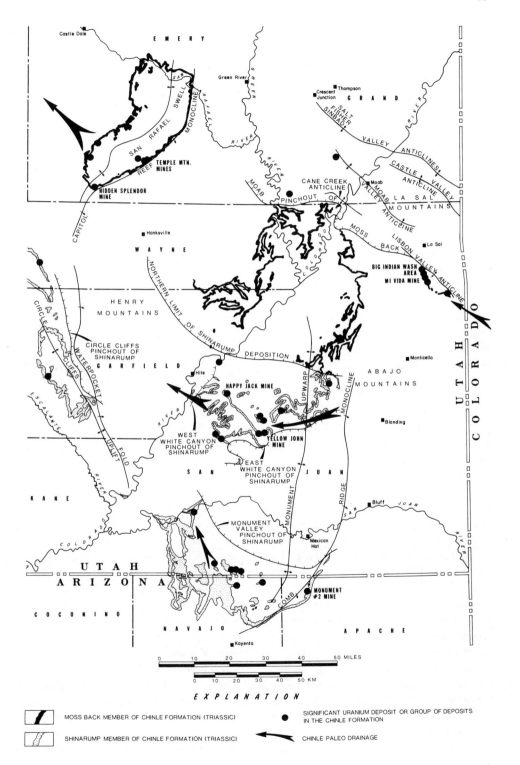

Figure 2-9. Map of southeastern Utah and northeastern Arizona showing Chinle outcrop pattern and uranium occurrences.

is generally gray-green to dark brown or a greenish-gray and light gray, in contrast with the dominantly red and varicolored overlying Chinle mudstone and thin sandstones.

In general, the ore deposits measure up to 14 m in thickness and average about 2 m in thickness. The ore averages 0.39 percent U_3O_8 and generally contains calcium carbonate and minor amounts of carbonaceous material. In the south part of the Lisbon Valley area, uranium is associated with vanadium, and the ratios of vanadium to uranium range from less than 1:1 to 3:1. In the north part of the Lisbon Valley area, vanadium is insignificant.

The uranium deposits are generally irregular in plan but concordant to the bedding. Wood (1968, p. 786) described persistent bleaching in the sandstones of the Mossback and underlying Cutler in the vicinity of uranium ore and suggested that it is probably significant relative to the origin of the deposits.

Other deposits in the Chinle Formation occur on the flanks of the San Rafael Swell in Utah. These deposits are generally similar to the other Chinle occurrences just described.

Origin of Trend Deposits

Geologists have proposed many theories to explain the origin of the trend deposits of the Colorado Plateau area. These include the mixing of hydrothermal solutions with ground water in lateral transport through the sandstone host, leaching from granitic or volcanic source materials, syngenetic origin, epigenetic origin, and combinations of the above.

The following characteristics of trend deposits are significant when considering the origin of the deposits:

1. All trend deposits occur in sandstones and conglomerates deposited by streams.

2. The formations in which the deposits occur are bifacies in character with both oxidized and reduced aspects (red beds and gray carbonaceous beds).

3. Although the formations are mainly oxidized on a regional scale, the ore bodies always occur in a reduced host. In the Salt Wash Member of the Morrison Formation, this reduced zone is frequently very local in character.

4. All deposits are strata controlled and generally concordant with the strata.

5. Most individual deposits are elongate and tabular, commonly with irregular shapes in plan. Many show elongations aligned with paleodrainage patterns.

6. The deposits tend to be concentrated in specific stratigraphic and lithologic units within the formation.

7. Most deposits are aligned in belts or trends. Some follow paleodrainage systems and some have less obvious control.

8. Most deposits have some association with carbonaceous material or humates.

There is mounting evidence that hexavalent uranium was transported by oxidizing ground water and concentrated in some places where the host sand was reducing. Evidence is also mounting that the mineralization took place soon after the host sands were deposited, often as an early diagenetic process (Hilpert, 1969; Malan, 1968; Wood, 1968; Squires, 1972; Miller and Kulp, 1963; Lee, 1975). In most cases, the uranium-bearing ground water probably moved under the influence of the parent streams which deposited the host sands and gravels. The climate during Chinle and Morrison times was consistently very warm, similar to the climates in the tropics of today; but the precipitation varied in a cyclical fashion similar to the precipitation in the savannahs of Africa and parts of Brazil today. Annual variations in precipitation were significantly magnified by long-range cyclic variations in the amount of precipitation. This is reflected by the banded mudstones of the Morrison and Chinle, which record short-period cycles; the long-range cycles are indicated by the major changes in zones within these formations, varying from green and gray to red. The Shinarump and Mossback Members of the Chinle Formation represent relatively wet phases, and the upper member of the Chinle generally represents a more arid phase of the climate.

The Chinle climate was dominantly arid, and many of the tributaries to the Shinarump and Mossback paleodrainage system were probably intermittent in nature, with substantial fluctuations in their water tables. It is reasonable to deduce that only the lowlands along the trunk streams were able to maintain a constant growth of healthy vegetation, whereas vegetation along the tributaries was repeatedly wiped out by drought. The decaying organic matter would be washed into the main stream-channel system. A natural concentration system for uranium existed because all of the intermittent tributaries supplied aerated alkaline waters with uranium and humic substances to the carbonaceous sands and gravels of the main trunk-stream alluvium. The repeated wetting and drying of the sediments in the upland and in the drainages of the intermittent tributary streams resulted in thorough leaching of the uranium which was present in the volcanic and/or granitic source rocks. Most of the uranium deposits in the Chinle occur in the lower portions of the deepest scours at the base of the formation. It appears that the oxidizing waters bypassed the tabular deposits of uranium in the basal parts of the channel system as the hydraulic conditions changed and sedimentation continued

to take place. The net result was a series of tabular ore bodies distributed along the trunk of the old paleodrainage system.

In most places, the sands in the Salt Wash Member of the Morrison are oxidized with ferric iron-oxide staining. Reduced mudstones and sandstones in the Salt Wash are developed locally, and in some cases, these reduced sandstones host uranium deposits. Because the uranium deposits almost always occur in reduced sandstones, geologists have interpreted the reduced rock surrounding the ore bodies as alteration. Fisher and others (1970) suggested that the altered rock usually includes the addition of finely disseminated pyrite and a change in color of the sandstone from pale reddish to pale gray or white and the mudstone from red to green or gray. Hostetler and Garrels (1962) suggested that the water carrying the uranium was weakly alkaline and moderately reducing and that precipitation took place in more strongly reducing environments with abundant carbon or H_2S or even H_2. Fisher and others (1970, p. 782) suggested that the waters were (1) neutral in Eh in unaltered rock where solutions apparently did not react with the rock, (2) mildly reducing in large masses of altered rocks where carbonaceous wood was present and where red oxide was reduced to pyrite, and (3) strongly reduced where ore formed.

However, the almost ubiquitous occurrence of hematite in both the sandstones and enclosing mudstones of the Salt Wash suggests to us that most of the ground water moving through these sands was characterized by high oxidation potentials and probably was alkaline in character. The surface water and ground water were capable of dissolving organic complexes produced from decaying vegetation and transporting them in solution as humic acids; such water was also capable of transporting uranium in solution.

In many cases, finely disseminated carbon is present with the uranium deposits of the Uravan area, but it has not been noticed by miners or geologists. The reports which describe the Salt Wash uranium deposits as often lacking carbonaceous material are valid to the extent that coarse carbonized wood and coaly material are often absent, but finely disseminated carbon is probably almost always present in the unweathered black ores. There is also evidence that some of the carbon is oxidized to form CO_2 and $CaCO_3$. Recent work by geochemists of the Uranium and Thorium Branch of the U.S. Geological Survey suggests that some $CaCO_3$ in ore-bearing sandstones of Ambrosia Lake area derived its carbon from organic carbonaceous material (Joel Leventhal, personal communication).

When a living plant reduces and combines carbon dioxide and water with the assistance of solar energy the chemical reaction is expressed by the following equation:

$$6CO_2 + 5H_2O + 678,000 \text{ cal} = C_6H_{10}O_5 + 6O_2.$$

There are other elements involved in terrestrial plant chemistry, such as sulphur, but this generalized equation gives a realistic picture. Plants convert solar energy into chemical energy stored in organic compounds.

After the plants die, bacterial action begins the reversal of the process. Indeed, in the case of humates in uranium deposits of the Colorado Plateau, the disseminated organic material once was soluble or at least capable of colloidal suspension and aqueous transport. The material is insoluble now, however, having lost almost all of the hydrogen.

We believe this is how the rock was reduced in the altered zones which surround black ore in the Colorado Plateau.

The volatile components of aging organic material, as it changes from humate to carbon (and eventually even $CaCO_3$?), diffuse outward into the rock pores, thereby reducing iron minerals.

Adler (1974, p. 146) proposed a twofold subdivision of tabular or peneconcordant (trend) deposits: (1) Salt Wash or Chinle deposits in which uranium was reduced and concentrated by indigenous plant material, and (2) Westwater Canyon deposits in which the reductants were "water-transported humic residues believed to have been derived from boggy vegetated terrain at one time directly overlying the outcrops of the uranium host rocks." Adler admitted a preference for relatively late origin of these uranium deposits when he espoused the tuff-leach concept which has been propounded by some geologists. This probably added to his preference for the Dakota source hypothesis to explain the large masses of carbonaceous material in the Westwater Canyon deposits. Granger and others (1961, p. 1197) outlined three hypothetical origins for the carbonaceous masses in the Westwater Canyon: the Dakota source, the syngenetic source, and the internal source. They did not cite evidence which would support any one of the hypotheses to the exclusion of the others, but Adler (1974, 1975) apparently adopted the Dakota souce. Indeed, he went further to conclude that the Westwater Canyon deposits probably represent a very unusual combination of circumstances in nature, making it unlikely that similar deposits will be found elsewhere.

There are several lines of evidence which make the Dakota source (or any post-diagenetic source) unlikely:

1. In the Ambrosia Lake trend, most of the trend ore bodies (with coextensive carbonaceous material) are elongate and oriented with their long dimensions parallel to the directions

of depositional trends (Granger and others, 1961, p. 1191). This suggests a ground-water flow or underflow when the carbonaceous material and uranium were deposited in the same direction as the flow of the parent streams.

2. "Intraformational disconformities and scour surfaces are the dominant local control of the shapes and positions of primary ore bodies" (Granger and others, 1961, p. 1191).

3. If the carbonaceous material had been introduced as humic substances by influent ground water from swamps overlying the eroded outcrop of Westwater Canyon sands during Dakota time, we should see greater concentrations of the carbonaceous material southward toward the outcrops of the Westwater Canyon. The Dakota source hypothesis seems to be incongruous with the present distribution of the carbonaceous masses.

4. The trend deposits are probably older than the pre-Dakota erosion surface. According to Hilpert (1969, p. 140), "none of the primary deposits are known to transect the Morrison-Dakota contact, and some data suggest that such deposits locally are beveled under the pre-Dakota erosion surface." In their extensive study of the isotopic evidence, Miller and Kulp (1963) concluded that the uranium deposits are probably as old as the enclosing formations, but that some were affected by chemical alteration less than 20 million years ago. Moench (1963) also cited evidence for the pre-Dakota age of trend deposits in New Mexico.

5. The carbonaceous strata in the Dakota Formation are not extensively mineralized, yet the "intimate and total association of uranium and black carbonaceous matter in the Cliffside deposit (Ambrosia Lake area) suggests that these materials were deposited by the same solution" (Clark and Havenstrite, 1963, p. 116).

Most of the known environments of northwestern New Mexico during the time of Westwater Canyon deposition were oxidizing, as evidenced by the widespread occurrences of hematite-stained sandstones. In most of the area, wide fluctuation of the ground-water table and high oxidation potentials resulted in ideal conditions for complete dissolution and removal of decaying organic matter and uranium. The complexes of humic substances and uranium which were leached from the tributary drainages concentrated in the sands beneath the channels of the main trunk streams. The actual process of concentration might have been a gradual thickening of colloidal or gelatinous masses, as Squires proposed (1972). Again, a savannah climate characterized by repeated cycles with wet and dry phases would be ideal for generating and concentrating the large masses of carbonaceous material found in the Westwater Canyon Member.

This interpretation suggests that the Westwater Canyon trend deposits of New Mexico had an origin very similar to the other trend deposits of the Colorado Plateau. It also suggests that the New Mexico deposits are not as unique in nature, as Adler proposed. Finally, it indicates that, in these large fluvial systems which are dominantly oxidized, trunk streams are favorable hunting grounds for reducing environments which may contain uranium deposits.

Roll-Front Deposits

General Description

Uranium-vanadium miners in the Colorado

48

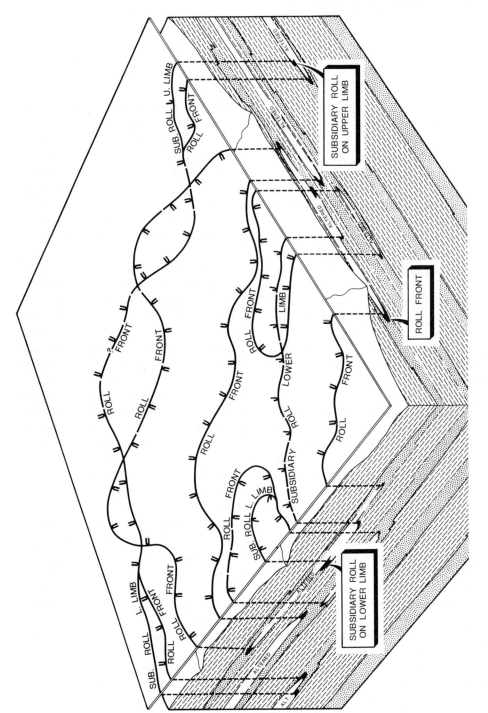

Figure 2-10. Idealized block diagram with superimposed map projection illustrating altered sandstone complex with roll fronts and subsidiary rolls.

Plateau area first used the term "roll" probably to describe the habit of some ore which "rolled" up or down as they tried to follow it underground. The term was subsequently adopted by Fischer (1942) to describe vanadium-uranium ore bodies with curved configurations which occur in the Colorado Plateau. The roll-type deposits which occur in the Colorado Plateau are secondary in importance to the trend deposits already described, but their relatives in the Wyoming basins and in south Texas are the most important ore deposits of those regions.

In the Wyoming area, several terms have been used interchangeably with the term "roll front" and with very little formal definition. These include "geochemical interface," "roll," "subsidiary roll," "roll-

type uranium deposits," "geochemical cell," and "solution front."

We believe the term roll front should be formalized to refer to the outermost margin of alteration within a sand (fig. 2-10). The outermost margin of an alteration complex within a complex of sands is a roll-front complex. The term geochemical interface can refer to the upper or lower edge of alteration within a sand or to any contact between altered and unaltered sand (fig. 2-11). The term roll can refer to any crescent-shaped or curved interface between altered and unaltered sand, whether it is near the outer lateral margin of the alteration or not (figs. 2-12, 2-13, 2-14). Subsidiary roll should refer to a roll feature associated with either the upper or lower limb of an altered complex,

Figure 2-11. Geochemical interface between altered sandstone below and unaltered sandstone above, with black high-grade uranium ore just above the contact. A pencil has been placed across the contact to show the scale. This photograph was taken in a mine in Shirley Basin, Wyoming. The mineralization freely crosses the bedding in the coarse-to-pebble, very permeable sandstone. The darkest spots in the mineralized sandstone are rich concentrations of uranium minerals around organic matter. The dark spots in the altered sandstone in the lower half of the photo are mineralized sandstone which has fallen down from above during mining operations.

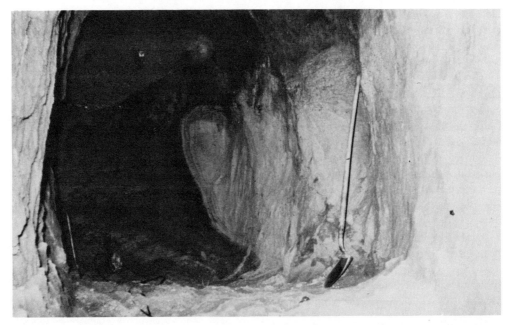

Figure 2-12. A uranium roll front exposed in an adit being driven into a wall of one of the open pits (Lucky Mc) in the Gas Hills, Wyoming. Altered sandstone is on the right near the shovel. Solution banding is evident near the dark ore, and additional solution phenomena can be seen in the ore itself. Such individual roll fronts have been followed underground for hundreds of metres.

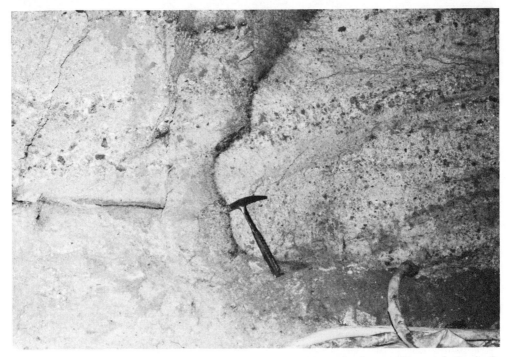

Figure 2-13. A roll front in conglomerate in the Battle Spring Formation in an underground mine in the Crook's Gap area of Wyoming. Altered sandstone is on the right side of the photograph.

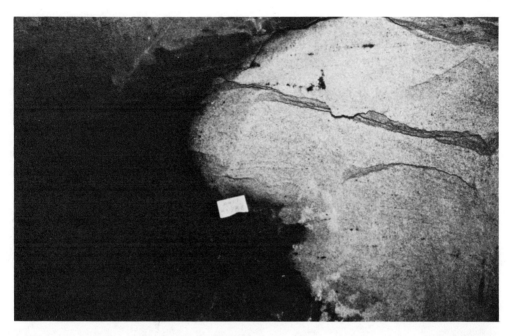

Figure 2-14. Extremely high grade ore characterizes this roll front mined underground in the Wind River Formation (Eocene) in Shirley Basin, Wyoming. The grade of some of the ore near the 10 x 15–cm card exceeds 10 percent U_3O_8. Altered sandstone is on the right.

Figure 2-15. Altered sandstone below mineralized sandstone in an underground uranium mine at Crook's Gap, Wyoming. Much of the black material occurring with and near ore in these mines has been described by many authors as "humate."

Figure 2-16. View of a wall of an open pit in Shirley Basin, Wyoming, where a roll front is exposed in the left center of the photograph. Diesel smoke from the tractor-scraper obscures the upper limb of mineralization, and the lower limb has not yet been exposed.

but not the outermost margin. The term geochemical cell (Rackley and others, 1968), was introduced with genetic implications; it will be discussed in more detail below.

Roll fronts and roll-front uranium deposits occur marginal to alteration within sands, and the alteration is inferred to have been formed by mineralizing solutions (figs. 2-15, 2-16, 2-17, 2-18). This basic relationship was first recognized and described by Vickers (1957), and its true significance was later proven in detail by Bailey (1965). Below the zone of recent weathering, unaltered sands are typically reduced in the favorable facies. Unaltered sands commonly are gray and contain disseminated carbonaceous material and pyrite. In contrast, the altered sand is usually oxidized to the extent that most, if not all, of the carbonaceous material has been removed. Usually the iron has been oxidized to some degree to form limonite or hematite and to produce colors which contrast sharply with the surrounding unaltered sandstone. Many alteration complexes in Wyoming show evidence of much stronger oxidation potentials existing during the time of mineralization than exist today. Another feature distinctive of altered sands is solution banding (fig. 2-19). Solution banding usually consists of concentric color bands smoothly cross-cutting a sand body in arcuate patterns. These bands consist of various tints of red, yellow, rusty brown, etc. Solution bands probably reflect slight changes in the chemical composition of the ground water during mineralization. Solution bands near a roll front or other geochemical interface

Figure 2-17. A roll front in a wall of an open-pit mine in Gas Hills, Wyoming. Dark gray sandstone to the right of the geologist's hand is ore containing about 0.20 percent U_3O_8.

frequently parallel the interface itself.

Color zonation in alteration complexes is common. Red hematite staining may characterize parts of an alteration complex, whereas yellow-brown or rusty limonite staining may dominate other parts, and white bleached sandstone may characterize other parts in an alteration complex.

Alteration complexes or tongues of alteration within a sandstone body vary considerably in size. A major alteration complex within a large fluvial sand system might require years of drilling, correlation, and mapping before it is fully delimited. Some individual tongues of alteration within sandstones in such a complex might have dimensions of several kilometres in length and several thousand metres in width. The complex itself might be tens of kilometres in length and several kilometres in width. Harshman (1972) published a de-

tailed description with maps of the Shirley Basin district that is the most complete description of an alteration complex published to date. Less detailed and partial descriptions of the Gas Hills alteration complex have been published by Anderson (1969) and Armstrong (1970).

Observations made in the course of our exploration and development programs strongly suggest that roll-front deposits should be subdivided into two major types with very important genetic significance: (1) *bifacies roll fronts,* which occur in formations that have both oxidized and reduced facies; and (2) *monofacies roll fronts,* which occur beneath unconformities in uniformly reduced formations. We interpret the bifacies roll-front deposits to be early diagenetic in origin, whereas monofacies roll fronts formed beneath unconformities in uniformly reduced formations and are almost certainly epigenetic and related to the unconformities.

The location and configuration of most alteration tongues and alteration complexes within bifacies formations are controlled mainly by permeable sandstones resulting from paleodrainage and paleodepositional systems. Where the relationships have been worked out, the mineralizing solutions which formed the alteration complexes appear to have followed the main stream course and to have moved in the same general direction as the streams which deposited the host sands. This results from (1) greatest gross permeability in the main part of the stream system, and (2) mineralization shortly after sediment deposition. In contrast, the monfacies roll fronts and their associated alteration complexes are controlled both by channels in the unconformity surface and permeability in the underlying strata.

Roll fronts and subsidiary rolls show all degrees of sinuosity in map plan. Some are simple and linear, and the associated ore bodies are also simple and linear. However, others are very complex with extreme sinuosity, and the associated ore bodies are erratic and usually difficult to delimit with drilling. Because roll fronts are generally the primary ore controls, exploration drilling is generally designed to delimit and thoroughly define their configuration. However, in certain cases, subsidiary rolls become the dominant ore controls along parts of an alteration complex (fig. 2-20). Factors which seem to control the location of ore along a roll front or subsidiary roll include facies changes within the sand, permeability changes, carbon trash build-ups, and transported humate concentrations. Some geologists have suggested that irregularities in plan along a roll front may tend to concentrate ore also. Our experience has not confirmed this.

Ore bodies associated with roll fronts

Figure 2-18. In this pit wall in the Gas Hills, it was obvious that a mudstone bed (just above geologist) had played an important role in influencing the movement of mineralizing solutions and the resulting roll front.

and subsidiary rolls are generally crescent-shaped in vertical cross section with the sharp concave margin facing the alteration (figs. 2-15, 2-16, 2-17, 2-18). The highest grade ore generally occurs very near the contact with the altered sandstone, and the grade tapers off outward away from the alteration. Low-grade mineralization may extend for hundreds of feet away from the roll front itself. The ore bodies are generally elongate with the long dimension paralleling the roll front. The larger ore bodies may be as much as 30 m wide and 2 km long (along the roll front) and may have thicknesses as much as 10 m. However, many mineable ore deposits along roll fronts and subsidiary rolls have dimensions of less than 5 m in width, 3 m in thickness, and several hundred metres in length.

All of the large bifacies roll front uranium deposits in Wyoming occur in forma-

Figure 2-19. Solution banding in a well-sorted fine sandstone in Wyoming. The banding is not as pronounced here as it is in certain other illustrations in this text, but it nonetheless is important that the explorationist recognize this phenomenon when it is present. This sandstone is entirely altered.

tions of early Eocene age with prominent oxidized and reduced aspects (fig. 2-21). The largest known deposits occur in the Wind River Formation, the Wasatch Formation, and the Battle Spring Formation (table 2-4).

In recent years, significant reserves of uranium have been discovered in monofacies roll-front deposits of the Fort Union Formation of Paleocene age. We believe that with new concepts and continuing exploration, substantial reserves will be developed in formations of Late Cretaceous age as well. Monofacies roll-front deposits have been mined for years in Lower Cretaceous sandstones on the flanks of the Black Hills (fig. 2-21).

Gas Hills District

The uranium deposits of the Gas Hills district occur in the upper part of the Wind River Formation (fig. 2-22). Armstrong (1970, p. 44) estimated that about 68 million kg (150 million lb) of U_3O_8 ultimately will be recovered from the Gas Hills area. The ore deposits occur marginal to a large complex of altered sandstone and conglomerate in the large-stream facies that Soister (1966) named the Puddle Springs Member of the Wind River Formation (fig. 2-23). The alteration in the Gas Hills is typically pale yellowish-gray with a slight white cast (bleached appearance), whereas the unaltered sand is dark gray and carbonaceous (Anderson, 1969). Drilling has documented an extension of the altered arkose complex southward from the main ore occurrences for a distance of more than 10 km. Northward projections have been removed by erosion.

Evidence is quite conclusive that the streams which deposited the host sands

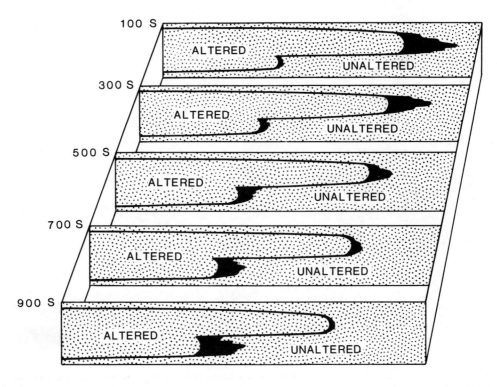

Figure 2-20. Series of cross sections on 200-m spacing illustrating how ore developed along a roll front may fade out as a subsidiary roll begins to control mineralization.

flowed northward from the Granite Mountains highland which was 18 km south of Gas Hills. The configuration of the roll fronts indicates that the mineralizing solutions also moved northward though the Gas Hills. The fluvial sands of the Puddle Springs grade southward into piedmont fans on the flanks of Granite Mountains. The geology of the Gas Hills district has been described by Soister (1967), Anderson (1969), Armstrong (1970), and Childers (1974).

Shirley Basin District

The Shirley Basin contains 38 million kg (85 million lb) of U_3O_8, according to estimates of government sources, and it has not been completely evaluated. A well-developed large-stream facies of the upper Wind River Formation contains two thick,

very coarse grained arkosic sandstone complexes which host major uranium deposits (Harshman, 1961, 1962, 1966, 1972; Bailey, 1965). Thin lignites and carbonaceous siltstone and mudstone are associated with the massive sandstones.

The streams which deposited the arkosic sands that host the large, high-grade uranium deposits of the Shirley Basin headed in the ancient Granite Mountains highland, located about 10 to 50 km west of the mining area. These streams flowed eastward from the mountains and, in Shirley Basin, were diverted northward by topographic influences.

The altered complex in the Shirley Basin has been described in detail by Harshman (1972) (fig. 2-24). The mineralizing solutions moved northward (fig. 2-24), but present ground water is moving

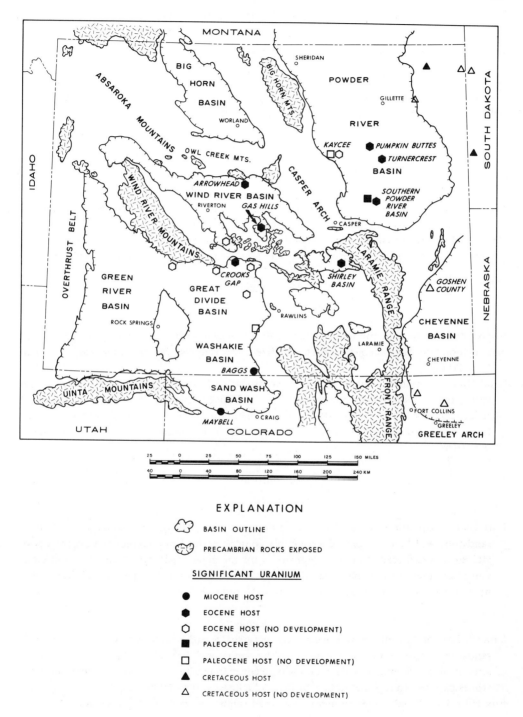

Figure 2-21. Map of Wyoming uranium province showing uranium deposits which occur in Cretaceous and Tertiary strata.

58

TABLE 2-4. SEQUENCE OF STRATIGRAPHIC UNITS CONTAINING URANIUM DEPOSITS IN THE WYOMING URANIUM REGION

Miocene

Brown's Park and Arikaree Formations*: sandstone, conglomerate, and mudstone, gray to yellow-brown, tuffaceous in part. Thickness: 0–100 m.

—Unconformity—

Oligocene

White River Group
Brule Formation: mudstone, argillaceous limestone, siltstone, and lenticular conglomerate and sandstone, pale pinkish-gray. Tuffaceous in part. Thickness: 0–120 m.
Chadron Formation: mudstone, conglomerate, and sandstone, varicolored red-brown and green-gray, arkosic, *oxidized* and *reduced facies* (mainly oxidized). Thickness: 0–50 m.

—Unconformity—

Eocene

Wagon Bed Formation: mudstone, green and red-brown with some arkosic sandstone and conglomerate. Tuffaceous in part. Oxidized and reduced facies. Thickness: 0–200 m (avg. 50 m).
Wind River*, Wasatch*, Battle Spring Formations*: arkosic sandstone, conglomerate, and variegated mudstone with some lignite and carbonaceous shale. Prominent *oxidized* and *reduced facies.* Mainly fluvial deposits with piedmont-fan facies near Granite Mountains. Large-stream deposits in reduced facies host major uranium deposits. Thickness: 0–500+ m.

—Unconformity—

Paleocene

Fort Union Formation* (has group status north of Black Hills): gray mudstone, siltstone, sandstone and lignite. Some chert-pebble conglomerate. Entirely reduced except where altered or weathered. Uranium deposits occur marginal to alteration compleses in sandstones exposed to regional unconformity. Uranium-bearing lignites occur directly beneath the unconformity. Thickness: 300 to 1000+ m.

Cretaceous

Lance*, Laramie*, Hell Creek Formations: gray shale, carbonaceous shale, siltstone, coal, sandstone, with some local conglomeratic sandstone. Entirely reduced except where altered or weathered. Uranium deposits occur marginal to alteration complexes in sandstones exposed to regional unconformity. Thickness: 50 to 1000+ m.
Fox Hills Sandstone*: fine- to medium-grained sandstone, gray, carbonaceous in part with some carbonaceous shale and siltstone. Uranium deposits occur marginal to altered tongues which extend downdip from the regional unconformity. Thickness: 30–80 m.

TABLE 2-4 *(continued).*

(Several marine and nonmarine formations are omitted here. They are not known to host significant uranium deposits.)

Inyan Kara Group

Fall River Formation*: fine-grained, thin-bedded, gray sandstone and gray to black shale and siltstone; carbonaceous in part. Entirely reduced. Uranium occurs marginal to alteration complexes which extend downdip from outcrops. Thickness: 45 m.

Lakota Formation*: fine- to coarse-grained, gray to white sandstone (some conglomeratic) with some interbedded varicolored mudstone. Chert, quartzite, and quartz pebbles. Uranium mineralization occurs marginal to alteration complexes which extend downdip from outcrops. Thickness: 80–150 m.

*Significant uranium host.

southward through the mining district. Altered sand has a slightly bleached greenish-yellow appearance, in contrast to unaltered equivalents which are medium gray and commonly carbonaceous. In some areas, altered sands are rusty to orange. Hematite-stained red sand is rare.

Crooks Gap–Great Divide Basin

The large mineable uranium deposits of Crooks Gap and the Great Divide Basin area occur in fluvial sandstone facies of the Battle Spring Formation, which is a thick complex of piedmont fan and fluvial deposits with prominent oxidized and reduced facies. The geology of this area has been described by Masursky (1962), Pipiringos (1961), Bailey (1969), Love (1970), Groth (1970), Pipiringos and Denson (1970), and Childers (1974). As shown in figure 2-25, fluvial facies representing several paleodrainages in which large streams flowed southward from the ancient Granite Mountains highland occur in the north part of the Great Divide Basin. All of these large-stream facies are mineralized with uranium marginal to alteration complexes. The mineralizing solutions appear to have moved southward. In Crooks Gap and to the southwest on Cyclone Rim, the interior parts of the altered complexes are mainly reddish, with abundant hematite staining; but near the roll fronts, altered rock is more often pale yellowish-gray or greenish-gray, with a slight bleached appearance (Bailey, 1969).

Powder River Basin

The uranium deposits of the Powder River Basin (fig. 2-26) occur in the Wasatch Formation, which has prominent oxidized and reduced facies, and in the underlying Fort Union Formation, which is uniformly reduced. The geology of this area has been described by Sharp and Gibbons (1964), Mrak (1968), Davis (1969), Childers (1970), and Langen and Kidwell (1974). A large alteration complex occurs within an extensive fluvial system of early Eocene age in the southern and middle parts of the basin, near the axis of the basin. Regional correlations indicate that these altered and mineralized sands are older than the uranium hosts in Gas Hills and Shirley Basin. The streams of the paleodrainage flowed northward, and they included the downstream portions of the rivers which flowed across the Wind River Basin and Shirley Basin before sediments began accumulating in Gas Hills and Shirley Basin areas. Marginal to the large-stream facies, the Wasatch also contains an oxidized mudstone facies in the southern half of the Powder River Basin. In the northern half

60

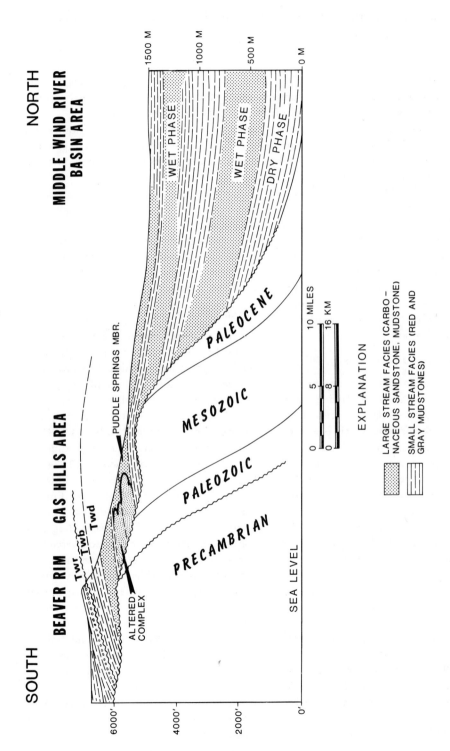

Figure 2-22. Generalized cross section through the Gas Hills uranium district, central Wyoming. Twd: Wind River Formation (Eocene); Twb: Wagon Bed Formation (Eocene); Twr: White River Formation (Oligocene). Note pronounced onlap relationships of Wind River Formation over older strata.

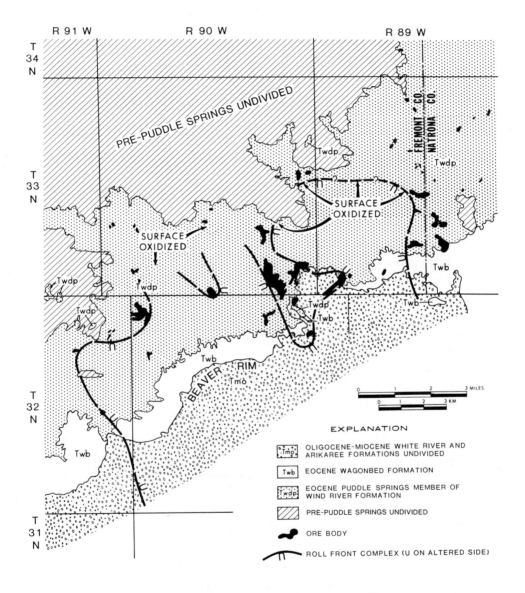

Figure 2-23. Generalized map, Gas Hills district, Wyoming. Modified in part after Anderson (1969).

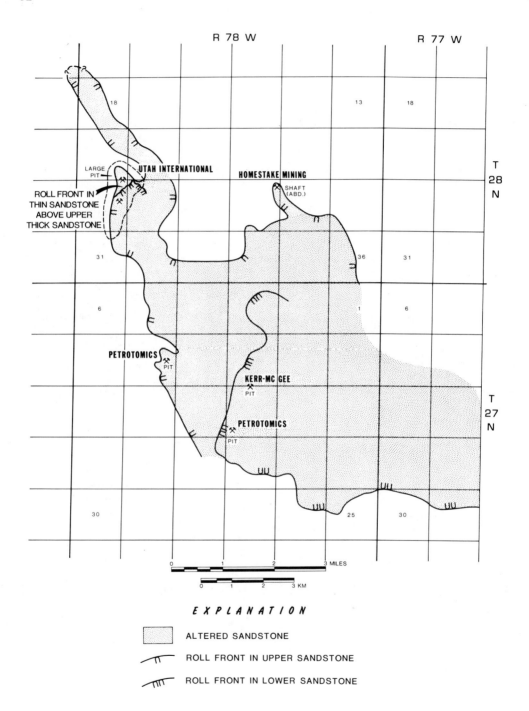

Figure 2-24. Map of Shirley Basin uranium district, Wyoming, showing generalized outline of roll fronts as mapped by Harshman (1972) with slight modifications.

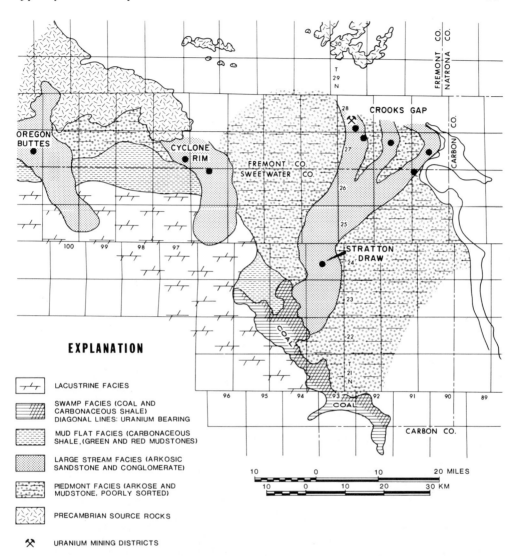

Figure 2-25. Map of northern part of Great Divide Basin, Wyoming, showing uranium occurrences in fluvial and swamp facies.

of the basin, the entire Wasatch Formation is coaly and carbonaceous, representing broad swamp and low flood-plain environments where vegetation was able to thrive even during the major dry phases of the Eocene climate. No uranium mineralization is known to occur within the Wasatch north of the mapped alteration in the north part of the basin.

Along the southwestern and western margins of the basin, sandstones of the Fort Union are altered and mineralized with uranium at several localities. The mineralization in the Fort Union is restricted to permeable sandstones which were exposed to the post-Paleocene unconformity.

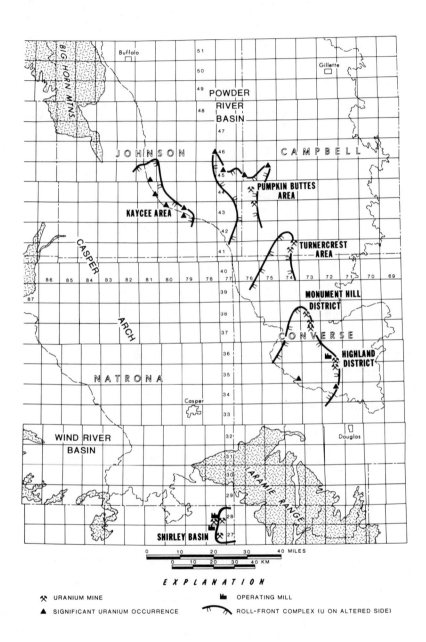

Figure 2-26. Map of southern part of Powder River Basin, Wyoming. Roll-front complexes are broadly generalized and interpretive.

Lobes of alteration project downdip from the unconformity or from the outcrop to depths in excess of 450 m in some places.

The alteration in the Fort Union and Wasatch ranges in color from heavy red hematite staining to pale yellowish tints with a white bleached aspect. A color called "old rose" has been found to be very common in both the Fort Union and Wasatch; it is pale purple-red. In many Fort Union sands, the alteration extends well down into ground water which is now stagnant.

Many roll fronts in the Powder River Basin remain unevaluated by drilling, and many years will be required to fully evaluate the uranium mineralization. When this is completed, we expect the ultimate mineable reserves to exceed 45 million kg (100 million lb) of U_3O_8.

Black Hills Area

Uranium deposits occur in sandstones and conglomerates of the Lower Cretaceous Inyan Kara Group around the flanks of the Black Hills uplift. The best known mineralization occurs on the northern, western, and southern flanks of the uplift (fig. 2-27). Vanadium occurs with uranium in most of the deposits, which occur as monofacies roll fronts marginal to tongues of alteration which project from the outcrops of sandstones in the Inyan Kara down to depths in excess of 300 m in some areas. Altered sandstone is typically red to pink as a result of hematite staining (Vickers, 1957; Hart, 1968; Gott and others, 1974).

The Inyan Kara is subdivided into two formations: the Lakota and the Fall River. The Lakota, from 60 to 90 m thick, is a fluvial complex with medium- to coarse-grained carbonaceous sandstones and associated carbonaceous shales and siltstones. It conformably overlies the Morrison Formation. Usually some varicolored mudstone occurs in the upper part of the Lako-

ta. The overlying Fall River is a nearshore complex of carbonaceous marine sandstones and siltstones which were deposited along the shoreline as the Skull Creek sea expanded southeastward over the area. The Fall River is 30 to 45 m thick. Uranium mineralization and alteration occur in both the Fall River and Lakota, but the best deposits are found in the Lakota.

Cheyenne and Powder River Basins (Lance and Fox Hills)

Uranium mineralization has been partially delineated by drilling along monofacies roll fronts in sandstones of the Lance Formation and Fox Hills Sandstone in the eastern flank of the Powder River Basin and in the Cheyenne Basin in southeastern Wyoming and north central Colorado (fig. 2-21). No significant data have been released on the occurrences in the Powder River Basin, but the drill patterns indicate that alteration tongues extend downdip from the outcrop similar to those described around the Black Hills.

In the Cheyenne Basin, the mineralized sandstones of the Fox Hills and Lance are truncated by a regional angular unconformity, over which the Chadron Formation was deposited. Arkosic conglomerates in the basal part of the Chadron fill channels cut into the underlying Cretaceous beds. These Chadron channel deposits are frequently altered, but the Chadron generally lacks any significant reducing facies in which uranium could be precipitated. The altered sandstones in the Cretaceous strata are commonly stained red with hematite, but some are yellowish and even bleached white.

In Goshen County, Wyoming, the Lance Formation and Fox Hills Sandstone dip gently toward the basin axis under the Chadron, which is essentially horizontal (except where influenced by normal faulting). The Chadron in turn is overlain by the

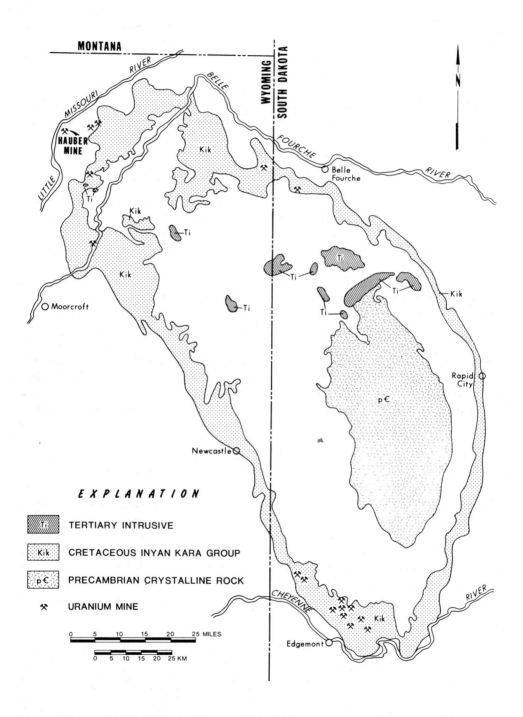

Figure 2-27. Map of Black Hills uplift showing principal uranium occurrences and outcrop of Inyan Kara Group.

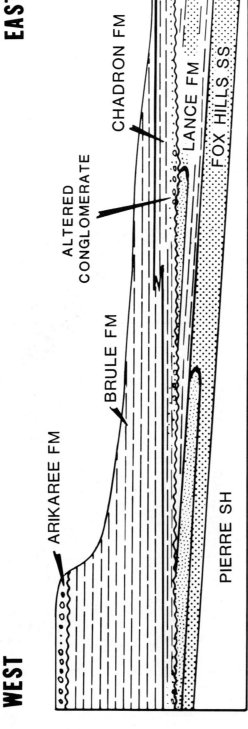
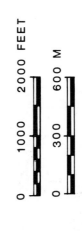

WEST

EAST

ARIKAREE FM

ALTERED
CONGLOMERATE

CHADRON FM

BRULE FM

LANCE FM

FOX HILLS SS

PIERRE SH

NO VERTICAL EXAGGERATION

| 0 | 1000 | 2000 FEET |

| 0 | 300 | 600 M |

Figure 2-28. Cross section in Goshen Hole area, southeastern Wyoming, showing generalized roll-front development in Fox Hills Sandstone and Lance Formation (Cretaceous) beneath pre-Oligocene unconformity. Conglomerates in basal Chadron are altered, but upper Chadron and Brule mudstones appear to preclude post-Chadron alteration and mineralization.

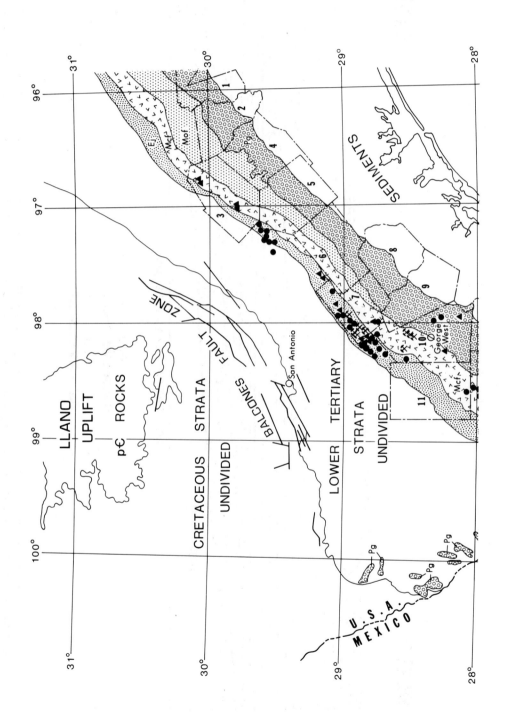

69

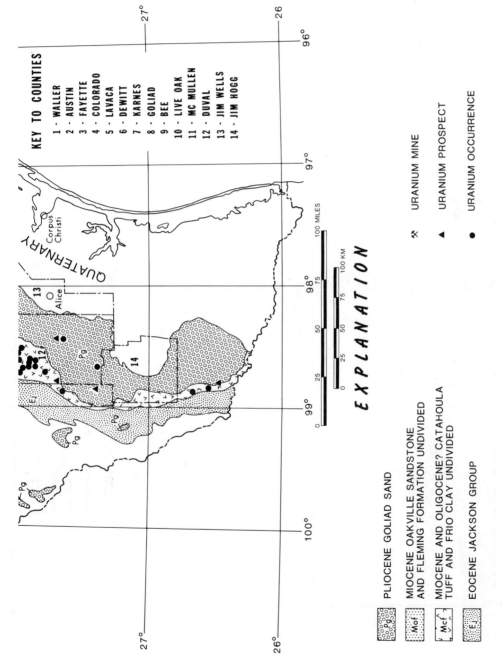

Figure 2-29. Map of south Texas uranium province showing uranium occurrences and outcrops of principal uranium-bearing formations.

EXPLANATION

PLIOCENE GOLIAD SAND

MIOCENE OAKVILLE SANDSTONE AND FLEMING FORMATION UNDIVIDED

MIOCENE AND OLIGOCENE? CATAHOULA TUFF AND FRIO CLAY UNDIVIDED

EOCENE JACKSON GROUP

✗ URANIUM MINE

▲ URANIUM PROSPECT

● URANIUM OCCURRENCE

KEY TO COUNTIES

1 - WALLER
2 - AUSTIN
3 - FAYETTE
4 - COLORADO
5 - LAVACA
6 - DEWITT
7 - KARNES
8 - GOLIAD
9 - BEE
10 - LIVE OAK
11 - MC MULLEN
12 - DUVAL
13 - JIM WELLS
14 - JIM HOGG

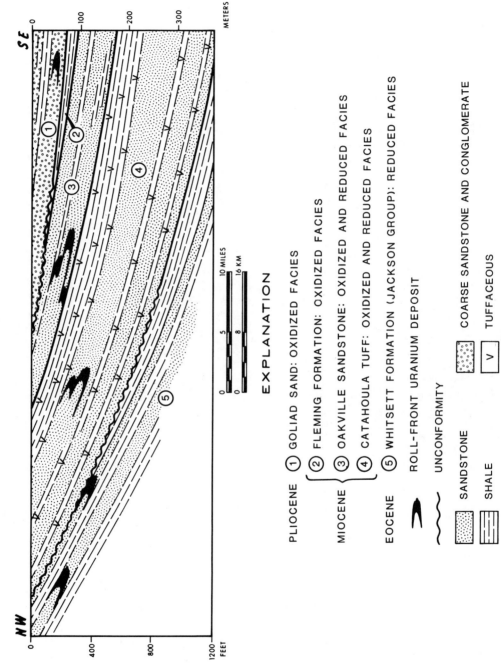

EXPLANATION

PLIOCENE ① GOLIAD SAND: OXIDIZED FACIES

MIOCENE ② FLEMING FORMATION: OXIDIZED FACIES

③ OAKVILLE SANDSTONE: OXIDIZED AND REDUCED FACIES

④ CATAHOULA TUFF: OXIDIZED AND REDUCED FACIES

EOCENE ⑤ WHITSETT FORMATION (JACKSON GROUP): REDUCED FACIES

⬥ ROLL-FRONT URANIUM DEPOSIT

〰 UNCONFORMITY

SANDSTONE COARSE SANDSTONE AND CONGLOMERATE

SHALE V TUFFACEOUS

Figure 2-30. Diagrammatic cross section showing uranium hosts in south Texas.

Brule Formation, which consists of pink-ish-gray calcareous and tuffaceous siltstone and mudstone (fig. 2-28). The upper Chadron and the Brule are essentially impervious and seem to have effectively sealed, in many places, the altered channel deposits of the basal Chadron and the altered sandstones of the underlying Cretaceous formations from influent ground water. Present depths of alteration in the Cretaceous sandstones where the Brule has not been removed by erosion sometimes exceed 400 m.

Texas

Uranium deposits occur marginal to alteration complexes in late Eocene, Oligocene, Miocene, and Pliocene sandstones of south Texas (fig. 2-29) (Eargle and others, 1975; Klohn and Pickens, 1970). The Eocene Whitsett Formation (upper unit of the Jackson Group) contains several sandstone members which host monofacies roll-front uranium deposits. The Whitsett is uniformly reduced and contains both marine and continental deposits. The Catahoula Tuff (Miocene) overlaps the Whitsett unconformably northwestward (fig. 2-30). Miocene and Pliocene strata, which contain bifacies roll-front deposits, are mainly of fluvial origin and were deposited by southeast-flowing streams. The Catahoula contains some pink, calcareous claystone in the upper part, and the overlying formations have increasing amounts of oxidized, red, hematite-stained claystones and sandstones. These Miocene and Pliocene strata have both oxidized and reduced facies and probably record deposition under the influence of a savannah climate.

The uranium ore bodies occur along roll fronts developed in permeable sandstones where ground-water movement was apparently active. Some of the host sandstones are reported to contain very little organic carbon (Eargle and others, 1975,

p. 778). Some geologists have suggested that the reductant is hydrogen sulphide that seeps up from deeper petroleum deposits (Eargle and Weeks, 1961; Klohn and Pickens, 1970). Many of the known uranium deposits occur close to oil fields and faulted structures.

Origin of Roll-Front Deposits

Most geologists agree that the uranium deposits associated with roll fronts were formed by oxidizing ground water carrying hexavalent uranyl ions in solution. There are several widely held hypotheses, however, concerning the sources of the uranium, the ages of mineralization, the paleohydrodynamics, and in some cases, the modes and chemical nature of precipitation.

Two hypothetical sources of uranium in the Wyoming area have survived the years of discussion and remain popular. One group of geologists believes the uranium was leached from overlying tuffaceous rocks of Oligocene and Miocene age; this concept requires that the uranium was entirely exogenous and was emplaced a considerable length of time after the host sands were deposited (Love, 1970; Davis, 1969; Masursky, 1962).

Another group believes that the uranium was leached during weathering and erosion from the granitic rocks which were also the source rocks from which the host sands were derived. Even the granite-leach advocates for the most part seem to favor a late development of the alteration complexes and associated uranium deposits. They conceive uplift, tilting of the basin margins, and erosion of the uplifted edges of the host formations to be the events which initiated influent movement of uraniferous ground water. Commonly these events are loosely tied to the Pliocene, because that is interpreted to be the time when normal faulting in Wyoming was

Figure 2-31. A roll front in the wall of an open-pit uranium mine in Shirley Basin, Wyoming. The authors believe that if geologists and engineers working with the Wyoming roll-front deposits will be observant, it is possible that exposures will be found that exhibit a situation where the uranium mineralization has been scoured out by slightly later stream action. Such an observation would provide nearly conclusive evidence concerning the time of mineralization. A bench has been cut just above this face exposed in 1964, but a suggestion exists here that erosion and deposition may have closely followed mineralization in the fluvial environment in which the sands themselves were deposited.

most active (Shockey and others, 1967; Rackley and others, 1968).

There is some mounting evidence, however, that the ore deposits and alteration complexes of the Wyoming area were formed during Eocene and early Oligocene (pre-Brule) time, when the area was under the influence of a savannah climate (Childers, 1974). All of the bifacies roll-front uranium deposits in the Wind River, Battle Spring, and Wasatch Formations appear to have been formed by ground water moving as underflows beneath the parent streams which deposited the host sands. The uranium deposits were formed soon after the sediments were deposited (fig. 2-31).

A fluvial environment, under the control of a savannah climate, with pronounced cycles of precipitation and consistent warmth, persisted in Wyoming and northern Colorado through early Eocene and, with moderating temperatures, into early Oligocene time. Small and intermediate-sized streams tributary to the large streams were repeatedly reduced to subsurface ground-water movement during dry phases of the savannah cycles. As the water table dropped in the intermittent-stream flood plains, plant life was destroyed, and the hot sun baked the muds sufficiently to raise the oxidation potential and produce the red, hematite-stained bands which characterize the oxidized facies.

The large *permanent* streams generally maintained near-surface ground-water tables and supported abundant vegetation, and so the sediments of their flood plains were consistently rich in humus. The large streams carried abundant sand and gravel derived from granitic source areas, depositing thick, permeable, and extensive bodies of sand which were capable of transmitting large volumes of ground water.

Uranium was soluble and readily transported by both surface and ground water in the oxidized environments of intermittent tributary streams. We believe that such processes took place during Eocene and early Oligocene time in the Rocky Mountain area. Available data indicate that many of the granitic source rocks in Wyoming and northern Colorado had anomalously high uranium contents. Conditions were favorable in many places for the development of roll fronts or typical Wyoming uranium-ore deposits in the reducing facies or large-stream sands of the Wind River, Wasatch, and Battle Spring Formations before they were deeply buried and *during the period when the parent streams were still flowing under the influence of the savannah climate.* Uranium, in the form of the uranyl ion, moved downstream through the oxidized environments; this strongly oxidized water continually invaded the reducing environment of the alluvial sands in the large-stream facies. Conditions were more favorable for uranium transport in massive ground-water movement (and over a longer time period with higher oxidation potentials) during the savannah climate of Eocene and early Oligocene time than during any subsequent epoch. Most of the fluvial sands of Eocene age in Wyoming probably carried more ground water during the life of the parent stream systems than in all subsequent time. The ideal time for optimum expansion of the alteration (development of a roll front) was during the transition from a wet to a dry phase. Well-developed sands in the reducing large-stream

facies had been deposited, and as the climate became more arid, the oxidizing ground water of the tributary streams invaded the reducing environment.

Harshman (1972, p. 67-68) studied the effects of alkaline water on samples of granite from the Laramie and Shirley Mountains (source rocks for the Shirley Basin) and concluded that the available uranium was probably leached from the source rock during weathering and erosion and carried in solution by the waters of the parent streams. Harshman did not speculate as to the importance of this source of uranium. Houston (1969) pointed out that higher levels of granite which may have been exposed to erosion during Eocene time could have contained higher percentages of uranium than the granite presently exposed. Rankama (1963, p. 36-37) reported that in many igneous bodies studied by means of autoradioactivity, uranium increases toward the border zones. A study of silicic and subsilicic granodiorites showed that 95 percent of radioactivity is in the accessory minerals, most of which are relatively easily broken down by weathering (Rankama, 1963).

All of the uranium did not concentrate along roll fronts during Eocene time; indeed, syngenetic uranium deposits formed in strongly reducing environments such as swamps, where uranium-bearing surface water lost uranium to organic material. These deposits are generally of low grade but are widespread. Examples include the uranium-bearing lignites of the Great Divide Basin and the uraniferous phosphatic lake beds described by Love (1964, 1970).

The altered sandstones and associated monofacies roll fronts of pre-Eocene age in the Powder River Basin and the Cheyenne Basin were probably mineralized by Eocene streams flowing over the erosional surface and/or by the early Oligocene

Chadron streams and their associated ground water. It appears unlikely that sufficient concentrations of influent ground water gained access to some of the altered sands in post-Chadron time, because the Brule Formation had effectively blanketed and sealed off many of the mineralized sands.

Evidence that favors an Eocene to early Oligocene (pre-Brule) origin of the uranium deposits in the Great Divide, Wind River, Shirley, southern Powder River, and Cheyenne Basins includes the following:

1. Syngenetic uranium concentrations are found in numerous carbonaceous shales, coals, and phosphatic rocks of Eocene age and in various rocks of Oligocene, Miocene, and Pliocene ages; however, no syngenetic uranium concentrations have been found in Late Cretaceous and Paleocene strata. This suggests that the local oxidizing environments that prevailed during Eocene time and the ubiquitous oxidizing conditions during post-Eocene times favored solution and transport of uranium—a requisite for uranium concentration. On the other hand, ubiquitous humates during the humid and tropical climate of Late Cretaceous and Paleocene time precluded extensive leaching, transport, and concentration of uranium.

2. Alteration complexes with associated uranium deposits are found in sandstones of Late Cretaceous, Paleocene, and Eocene ages; sandstones of the early Oligocene (Chadron) are locally altered. However, such complexes are not found in post-Chadron strata. The uranium deposits in the Miocene sandstones of the Sand Wash Basin, Moffat County, Colorado, are not associated with alteration complexes. Even though low-grade syngenetic uranium concentrations are common in post-Chadron strata, and anomalous amounts of uranium are now in solution in post-Chadron aquifers (as evidenced by numerous analy-

ses of ground-water samples), we have found that no alteration complexes are developed in these strata.

3. Evidence from exploration drilling has shown that bifacies roll fronts in the Eocene sandstones of the Cyclone Rim area (northern Great Divide Basin) were developed before the normal faulting. In addition, thickening of the units in the White River Formation suggests that the normal faulting began during Oligocene time (Childers, 1974).

4. Many of the altered sands are characterized by red to purple and old rose hematite staining. This indicates Eh values higher than those now present in Wyoming ground water. The hot, dry phases of the Eocene and early Oligocene savannah climate, which produced the red-banded deposits, developed highly oxidized ground water capable of producing the alteration observed.

5. Hematite-stained altered sandstones have been shown by exploration drilling to extend to depths of more than 450 m in some areas where the movement of post-Eocene ground water has been inhibited. In some instances, these altered sands extend into ground water which is presently stagnant and reducing.

6. The extensive altered complex in the Puddle Springs Member of the Wind River Formation in the Gas Hills and in the large area south of the Gas Hills is difficult to explain in terms of post-Eocene development. Much significance has been attached to two local channels exposed on Beaver Rim south of the Gas Hills, where basal White River channel conglomerates are in contact with the upper part of the Wind River Formation (Love, 1970, p. C132). The conglomerates in the White River are not altered and mineralized themselves, and they do not provide an adequate conduit between the Pliocene source proposed by Love and the altered complex. Both

upper Wind River mudstones and White River mudstones appear to reduce the effectiveness of the basal White River channel deposits as conduits. Love (1970, p. C123) proposed a Pleistocene age for the Gas Hills uranium deposits, but such a relatively recent age is unlikely for the entire altered complex with which the Gas Hills deposits are associated. In our opinion, the main development of the Gas Hills alteration complex probably ceased before the major faulting that produced the reversal of dip from north to south.

7. Several occurrences of altered and mineralized Cretaceous sands in the Cheyenne Basin are unconformably overlain by altered Chadron sandstones. However, the unaltered, relatively impermeable Brule Formation blankets the Chadron and inhibits the downward movement of influent ground water to most of the altered complexes.

8. Most of the sandstone uranium deposits of the world are associated with formations which include some oxidized red strata, although the hosts themselves are usually reducing. The best ore deposits of Wyoming are associated with formations of Eocene age. These formations include both reduced and oxidized facies.

The roll-front deposits in uniformly reduced formations such as the Fort Union, Lance, Laramie, and Fox Hills are exogenetic, and we have proposed the term monofacies roll front for this type of deposit. In all cases, it is apparent to us that mineralizing solutions moved downward through an erosion surface into these reduced sandstones of pre-Eocene age in northern Colorado, eastern Wyoming, and nearby states.

The tuff-leach proponents, who espouse Miocene or later mineralization, envisage deposition of the Oligocene and Miocene strata over a regional unconformity and later leaching of uranium from these for-

mations by downward-percolating ground water. They suggest that this water, charged with uranium, continued downward through the regional unconformity into reducing permeable strata below (Denson and Gill, 1965). They cite anomalous uranium contents in ground water within the Miocene and Oligocene strata as their strongest support for this concept.

We first began doubting the inclusive and unqualified validity of the tuff-leach hypothesis when we encountered altered complexes and uranium-mineralized roll fronts in the Upper Cretaceous Fox Hills and Lance of southeast Wyoming. It is very unlikely that these occurrences could have been developed after Chadron and Brule were deposited, because thick mudstones of these formations effectively seal off the underlying altered sandstones from influent ground water. Also, in most instances, conditions for post-Miocene recharge into some of the altered sandstones appear to be inadequate to explain the extensive development of alteration. In our opinion, the most likely time of mineralization was the Eocene and early Oligocene (Chadron) when streams with uranium in solution were flowing across the region and eroding the Paleocene and Cretaceous terrane. This was a time when climatic conditions favored uranium dissolution and transport in surface and ground water. It is also the time most favorable for extensive influent ground-water movement into the Cretaceous and Paleocene sandstones (with development of monofacies roll fronts).

There is some difference of opinion as to the origin of the roll fronts in the Inyan Kara on the flanks of the Black Hills uplift. Hart (1968, p. 837) suggested that the uranium was leached from overlying tuffaceous strata of Oligocene and Miocene age. Hart postulated that the three major drainage basins controlled movement of ground water as the Oligocene and Miocene strata were eroded from the flanks of the Black Hills (see fig. 2-27).

Renfro (1969, p. 91), citing a differential in uranium content between the unaltered sand of the Inyan Kara (average 14 ppm U_3O_8) and its altered equivalents (average 5 ppm U_3O_8), postulated that oxidizing ground water leached the indigenous disseminated uranium from updip parts of the hosts and concentrated it along the redox boundaries (roll fronts). If, in fact, the typical unaltered sands beyond influence of existing roll fronts have 14 ppm U_3O_8, and a true average for the altered sand is only 5 ppm, Renfro has a very strong case. He gives no details, however, about the loci of samples, types of samples, and number of samples used to determine these important averages. Renfro cites an apparent lack of significant uranium in permeable, pre-Tertiary sandstones below and above the Inyan Kara as an argument against the exogenetic origin of the deposits. This argument has been refuted by discoveries in the Fox Hills, Hell Creek, and Fort Union since 1969.

We have found that uranium tends to be irregularly distributed in sandstones, whether altered or unaltered, to the extent that it would require an extensive and carefully planned sampling program to establish average values for reliable comparison of altered and unaltered sandstones. For any particular area and environment, this would be a costly and difficult undertaking indeed. Davis (1969, p. 135) reported that analyses of cores from Eocene sandstones from the Powder River Basin showed that altered sands contained more uranium (6 ppm) than the unaltered sands ahead of the mineralized area (2 ppm). Davis noted that his data were insufficient, however, to preclude large areas of altered sands deficient in uranium.

Harshman (1972, p. 67) tested samples of granite from likely source areas for host

sands in the Shirley Basin and found that uranium is readily leached by water similar to that found in the ground water of the Shirley Basin today. He concluded that during weathering, erosion, transport, and deposition the uranium of the granitic source rocks would for the most part be taken into solution, and that diagenetic or epigenetic alteration would have no detrital uranium to redistribute and concentrate.

Harshman (1972) also reported that the selenium content may be higher in altered sandstone than in equivalent unaltered sandstone. These limited samplings do not support the concept of roll-front development in which uranium is leached from weakly mineralized hosts. Indeed, the term "barren interior" proposed by King and Austin (1966) is misleading. Altered sandstone is often enriched in certain elements.

This brings us to the hypothesis first proposed by Shockey and others (1967) and later reiterated by Rackley and others (1968) in which the geochemical cell concept is advanced. Shockey and others proposed the term "geochemical cell" to refer to any alteration tongue or alteration complex with uranium-bearing roll fronts. The geochemical cell concept holds that uranium was anomalously high but disseminated through the arkosic sands when they were deposited and that later basin-flank tilting or regional uplift triggered ground-water movement, which resulted in leaching of the disseminated uranium and concentration along the margins of the leached mass. The term "geochemical cell" implies that the alteration complexes grew from within themselves. In general, we feel that the evidence does not support this concept, and we reject the term.

The roll-front deposits of Texas are developed in Miocene and Pliocene strata that have both oxidized and reduced facies (bifacies), and they are also developed in underlying Eocene strata which are uniformly reducing (monofacies). Most of the geologists who have published opinions seem to favor the Miocene Catahoula Tuff as the main source of uranium for deposits which occur in strata both younger and older than the Catahoula. Evidence favors a savannah climate in south Texas during Miocene and into Pliocene time, and conditions were favorable for leaching uranium from the tuffaceous terranes and transporting it in surface and ground water to reducing environments.

Some of the deposits occur in clean marine sands which generally lack carbonaceous material similar to that often seen as the reductant in roll-front deposits in other areas. Because the Texas deposits are proximal to oil and gas fields and associated faults, it has been postulated that fluids leaking upward from the hydrocarbon accumulations served as the reductants.

Stack Deposits

The term "stack deposits" has been used extensively to refer to uranium-ore deposits associated with trend deposits in the Westwater Canyon of the Grants region, New Mexico. These stack deposits are also sometimes called "redistributed" or "post-fault" ore, because they appear to represent uranium which was redistributed subsequent to the formation of the trend deposits (fig. 2-32). The geometry of stack deposits frequently is controlled by faults or fractures which postdate the formation of trend ore. Stack deposits generally have greater thicknesses than the associated trend ore, but the dimensions in plan are erratic.

Red sandstone with hematite staining is commonly in close association with stack deposits, and this has been interpreted to mean that oxidizing ground water invaded the environment and redistributed the uranium. In this respect, stack deposits are similar to roll-front deposits.

Precambrian Heavy-Mineral Deposits

These deposits have been called "conglomerate-type," "Precambrian conglomerate deposits," or "quartz-pebble conglomerate deposits" by various authors. We prefer to use the more specific designation "Precambrian heavy-mineral deposits" because there are numerous uranium deposits in conglomerates and even in Precambrian conglomerates that are not similar to these which occur in heavy mineral assemblages. Robertson (1974) briefly described the most significant known deposits of this type, and suggested that they formed "after the development of an extensive acid crust . . . but before the development of an oxidizing atmosphere." The suggested time interval in absolute age before present is between 2,800 and 2,200 m.y.

Robertson included areas in Canada, Brazil, Western Australia, and South Africa, where he and others have studied similar uranium deposits. According to Robertson (1974, p. 708), Precambrian heavy-mineral uranium ores now make up about 40 percent of the world's low-cost uranium reserves. The most important deposits to date are in the Elliot Lake area, Ontario, Canada, and the Witwatersrand in South Africa.

Elliot Lake Area

Elliot Lake reserves have been estimated at between 300,000 and 400,000 tons of contained U_3O_8 (including some inferred reserves). The Elliot Lake deposits occur in sheets of quartz-pebble conglomerate which contain abraded grains of uraninite and pyrite together with other heavy minerals, including monazite and brannerite. These uraniferous conglomerates are in the lower part of the Huronian Supergroup (Frarey and Roscoe, 1970).

The deposits are deficient in magnetite

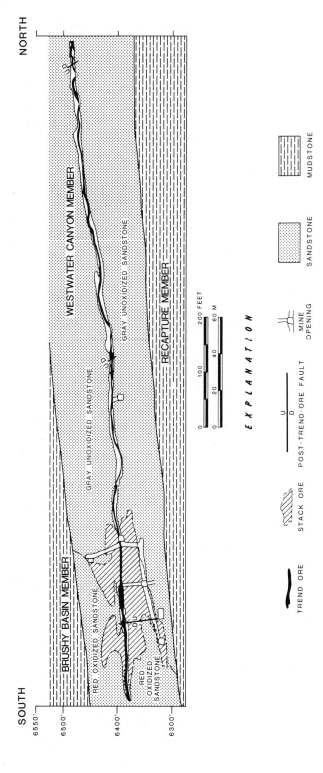

Figure 2-32. Cross section of trend and stack ore in Ambrosia Lake area, New Mexico. Note similarity of stack ore to typical roll front. Red oxidized sandstone resembles altered sandstone of roll-front occurrences. Modified from Hilpert (1969, p. 66) and Gould and others (1963).

and other iron oxides, and some of the pyrite (which is abundant) is anomalously magnetic, suggesting sulphidization of detrital magnetite. The uraninite grains are extremely small, ranging from 100 to 200 microns. Carbon is present in some of the heavy-mineral concentrations, and carbon sometimes partially replaces uraninite and fills microfractures in uraninite grains. Some uranium appears to have locally dissolved and reprecipitated as pitchblende with carbonaceous material called thucolite.

Witwatersrand Area

The Witwatersrand Supergroup of South Africa includes a sequence of quartzites, conglomerates, and siltstones, with interbedded volcanics, that is about 7,500 m thick (Robertson, 1974; Brock and Pretorius, 1964; Minter, 1976). Uranium and gold occur together with pyrite, zircon, chromite, leucoxene, and other heavy minerals in pebbly quartz-arenites. The Witwatersrand strata are about 2,500 m.y. old. The uraninite and other heavy minerals occur in concentrations on cross-bedded foresets and scour surfaces; they tend to be more concentrated in pebbly layers. The uranium and gold are also entrapped in networks of hydrocarbon filaments in carbon seams.

The uranium minerals appear to have been transported and concentrated as very fine detrital grains together with gold, pyrite, and many other heavy minerals by shallow braided streams on sandy and pebbly paleoslopes during early Proterozoic time (2,800 to 2,200 m.y. ago) when the atmosphere was not yet oxidizing. These deposits represent placer concentrations of uranium.

Reef-Trend Deposits

Limestone of the Todilto Formation (Jurassic) contains uranium-ore deposits in the Grants district of New Mexico. The most important primary mineral is pitchblende, but weathering has produced secondary minerals of significance to miners. In the Grants mineralized area, the Todilto is about 7.5 m thick, with three units: (1) a lower "platy zone" of thin-bedded fine-crystalline limestone; (2) a middle "crinkly zone" of thin-bedded, finely crenulated clay-laminated limestone; and (3) an upper "massive zone" of coarse-crystalline limestone. According to Perry (1963), the upper massive zone almost occupies the entire formation in some places, but in other places, the massive and crinkly zones are absent. Perry presented evidence supporting reef development in the Todilto. Indeed, he suggested that the reefs and associated breccias along the reef edges served as favorable loci for primary ore development.

The structural controls of uranium mineralization described by Hilpert and Moench (1960), Moench (1963), and McLaughlin (1963) appear to be in large part at least related to reef development as described by Perry (1963).

It appears that the pitchblende mineralization, which fills voids and partially replaces coarse-crystalline limestone in the massive zone, is early diagenetic. Perry (1963, p. 152) noted that "many completely isolated breccia boulders are highly mineralized; yet, the surrounding shales contain no uranium minerals. However, when the breccia fragments are enclosed by reef detritus, it is common for both the detrital material and the breccia to be mineralized." Also, Moench (1963, p. 163–164) showed that uranium mineralization accompanied the development of folds where unlithified lime flowed down the flanks of the growing structures.

No hypothesis has been advanced to explain the selective concentration in the reefs at such an early time. The mineralized breccia fragments described by Perry suggest syngenetic mineralization of the reefs —before complete burial.

Calcrete Deposits

Major deposits containing more than 46,000 tons of U_3O_8 occur in calcrete (caliche) at Yeelirrie in Western Australia (fig. 2-35). Calcrete is limestone formed by shallow ground water in major drainageways (Langford, 1974). Usually calcrete is well indurated, with some laminated layers separated by coarse layers with lenticular cavities. The limestone is very fine grained, white, pale brown, or gray, and porcellaneous. Sometimes it resembles breccia. At depth, pea-sized nodular limestone is interlayered with laminated limestone. Opaline silica layers appear to record varying positions of the water table.

Calcretes are porous and very permeable aquifers. Calcrete may rest on scour surfaces cut in bedrock or on gravel alluvium with calcite cement. At Yeelirrie, carnotite fills fractures and forms coatings in cavities and on surfaces of the calcrete.

The carnotite-bearing calcretes occur in stream channels where the streams have been choked with alluvium to the extent that subsurface drainage and evaporation prevails. There are high-level laterites and underlying kaolinite in the area which record a long period of deep chemical weathering under warm, humid conditions (fig. 2-33). This laterite development took place in early Tertiary time. In Miocene or Pliocene time, uplift resulted in erosional downcutting of drainageways through the laterites and into bedrock. The laterites contain nodular hydrous oxides of iron and aluminum with 1,120 ppm vanadium (Langford, 1974, p. 519). Langford suggested that the laterites are the source of vanadium for the carnotite ores.

During Miocene and Pliocene time, the climate became progressively more arid, with alternating wet and dry seasons (a savannah climate). During dry periods, evaporation resulted in salt concentrations, but the more soluble salts were flushed away during the wet seasons. Calcrete was the end result.

The deeply weathered granites of the area are a good source for the uranium and potassium which are required for carnotite precipitation. Langford reported that some granites in the Yeelirrie area are abnormally radioactive.

In summary, the Yeelirrie carnotite-calcrete deposits appear to have formed as a result of deep chemical weathering of granite terrane followed by a change in climate from hot-humid to hot-arid (with an intermediate savannah). The initial concentration of vanadium in the laterite might have been essential to the chemical conditions which resulted in carnotite precipitation in the calcrete.

Uraniferous Lignites

There are two important types of uraniferous lignites: (1) lignites and associated carbonaceous shales with widely distributed low-grade uranium mineralization marginal to varicolored fluvial environments, and (2) lignites and associated sandstones, dark shales, and carbonaceous siltstones with high- and low-grade uranium mineralization below a regional unconformity.

Great Divide Basin

Low-grade uraniferous lignites are extensively developed in the Red Desert of Wyoming's Great Divide Basin marginal to the Battle Spring Formation of early Eocene age (fig. 2-25) (Wyant and others, 1956; Pipiringos, 1961; Masursky, 1962; Childers, 1974). Most of these uranium-bearing lignites and carbonaceous shales appear to represent accumulations in swamps and mud flats marginal to Lake Gosiute, which occupied most of the

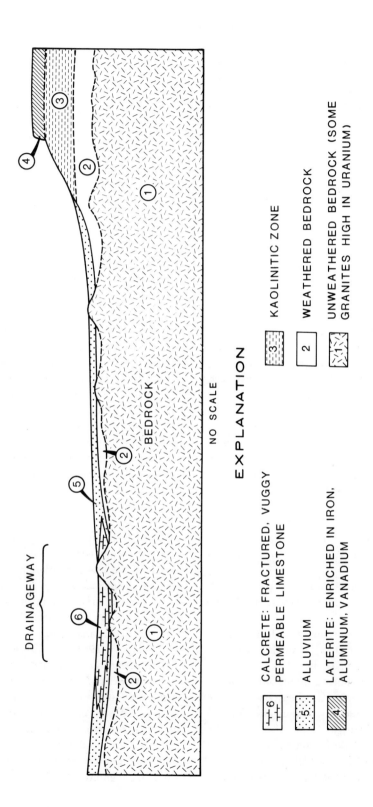

NO SCALE

EXPLANATION

⬜6	CALCRETE: FRACTURED, VUGGY PERMEABLE LIMESTONE
⬜5	ALLUVIUM
⬜4	LATERITE: ENRICHED IN IRON, ALUMINUM, VANADIUM
⬜3	KAOLINITIC ZONE
⬜2	WEATHERED BEDROCK
⬜1	UNWEATHERED BEDROCK (SOME GRANITES HIGH IN URANIUM)

Figure 2-33. Idealized cross section of calcrete uranium-vanadium deposit typical of Yeelirrie in Western Australia. Deep weathering with laterite and kaolinite development records warm humid climate in Late Cretaceous–early Tertiary. Note that calcrete deposits occur along choked drainageways. Modified from Langford (1974).

southwest part of Wyoming in Eocene time. Streams which headed in the Granite Mountains highlands to the north and northeast were depositing coarse arkosic sands north of the swamps and along channelways built out through the swamps and mud flats.

The uranium mineralization in the Red Desert lignites is low grade (ranging up to 0.05 percent U_3O_8), with the best grades occurring near permeable sandstones (along channelways where streams flowed through the swamps). Some anomalously high uranium grades occur in lignites exposed to recent weathering and to ancient weathering related to the post-Eocene regional unconformity. Wyant and others (1956) first proposed the theory that the uranium was concentrated in the Red Desert lignites during deposition (syngenetic). However, subsequent authors have, for the most part, indicated a preference for the hypothesis that the uranium in these lignites and shales was leached from tuffaceous Miocene beds and transported into the subsurface by alkaline influent waters. Masursky (1962) listed several observations in support of this hypothesis:

1. The highest concentration of uranium is in the highest carbonaceous bed directly below the unconformity on which Miocene strata were deposited.
2. The uranium content of coal is greater adjacent to permeable sandstone beds.
3. The uranium content is higher in impure coals than in thick, pure, and impermeable coals.
4. The uranium content of coal is higher to the northeast, where the coals grade into sandstones of the Battle Spring.
5. Organic shale in contact with permeable sandstone is high in uranium.
6. Red Desert coal is effective in ex-

tracting uranium from solution.
7. Artesian water from aquifers interbedded with coal carries as much as 47 ppb uranium at the present time. The Red Desert coal will extract uranium from this solution.

None of Masursky's observations are in conflict with a syngenetic origin with the possible exception of the last listed, and data which have been obtained by drilling subsequent to Masursky's study show that this observation is not necessarily inconsistent with a syngenetic origin. It is to be expected that uranium would be concentrated and enriched in coals directly beneath the pre-Miocene unconformity. As Vine and others (1966, p. 189) have proven, weathering locally causes supergene enrichment a short distance in from the outcrop of a uranium-bearing lignite. If surface waters of the parent streams depositing the Battle Spring sands carried uranium in dilute solution into the peat bogs and stagnant swamps (which resulted in the Red Desert coals), this uranium would be concentrated by the peat with increase in concentration toward the source. Szalay (1966, p. 184) showed experimentally that peat absorbs uranium almost perfectly even from the very dilute solutions in nature. Although, as Masursky states, uranium tends to be more concentrated in the less pure margins of coals, it remains a fact, as seen in his logs and as seen in other drill hole information, that uranium also is concentrated in *very impermeable* coals and shales.

Pipiringos (1961, p. 72) pointed out that impermeable shale beds preclude vertical movement of influent water into some of the uranium-bearing coals. An exogenetic, post-Eocene origin requires in some instances several kilometres of solution transport in ground water traveling through sandstone aquifers which are now carbonaceous throughout and strongly reducing, with no evidence of alteration. Indeed, several wildcat wells drilled for oil and gas have logged anomalous radioactivity in Eocene coals at depths of as much as 900 m in the Great Divide Basin. Ground water in the deep sandstones associated with these coals probably became stagnant before the end of Eocene time.

The uranium-bearing artesian water which Masursky noted (observation 7 above) would be difficult to explain if the aquifer were typical of the sands associated with the uranium-bearing lignites. Uranium should not be circulating in solution through sands with ubiquitous fixed humates and carbonaceous matter. The phenomenon can probably be explained, however, because exploration drilling in recent years has discovered the presence of several altered sands with roll-front development in the area where Masursky encountered artesian water. These are not typical of the sands generally interbedded with the lignites of the Great Divide Basin.

Carbonaceous shales which represent mud flats and shallow, nearshore lacustrine muds of Tipton or Luman age in the area southeast of Oregon Buttes (northwest Great Divide Basin) contain anomalous amounts of uranium disseminated over large areas. The field relationships in that area strongly favor a syngenetic origin for the uranium. Data from exploration drilling has shown that most of the uranium here is totally isolated in impermeable strata.

Williston Basin

Lignite, carbonaceous shale, and carbonaceous siltstone containing as much as 5 percent U_3O_8 occur in the Hell Creek, Ludlow, Tongue River, and Sentinel Butte Formations (fig. 2-34) of southwestern Williston Basin (Denson and Gill, 1965; Denson and others, 1959). The best known uranium mineralization of this area is in thin lignites

or carbonaceous siltstones that are in contact with permeable sandstones and less than 60 m below the regional pre-Oligocene unconformity. The formations in which the uraniferous lignites occur are uniformly reduced with no primary oxidized facies.

Several generalizations bearing on the genesis of the uranium deposits can be made. (1) Most lignites which are interbedded with shales and impervious siltstones are not uraniferous. (2) Lignites with the strongest uranium mineralization are associated with permeable sandstones. (3) Uraniferous lignites are not restricted to any particular stratigraphic zones. They occur in the Hell Creek, Ludlow, Tongue River, and Sentinel Butte Formations. (4) Some uraniferous lignites in exposures very near the old pre-Oligocene unconformity are not in obvious contact with permeable sandstones. (5) Where good exposures of the formations directly below the pre-Oligocene unconformity occur, it is common to observe deep weathering characterized by intense oxidation.

The evidence suggests to us that the uranium was transported by oxidizing, influent ground water which moved downward through the pre-Oligocene unconformity.

Uraniferous Phosphatic Rocks and Black Shales

Low-grade occurrences of uranium are sometimes widely disseminated in dark phosphatic and other black shales rich in organic material which have oxidized facies equivalents. Uranium-bearing phosphatic rocks are widespread in the Phosphoria Formation (Permian) in Utah, Idaho, and western Wyoming (McKelvey and Carswell, 1956). These marine phosphatic rocks grade eastward into nonmarine beds, and this results in a major bifacies formation composed of strongly oxidized beds to the east and strongly reduced beds to the west, where beds 1.5 to 3 m thick with grades

84

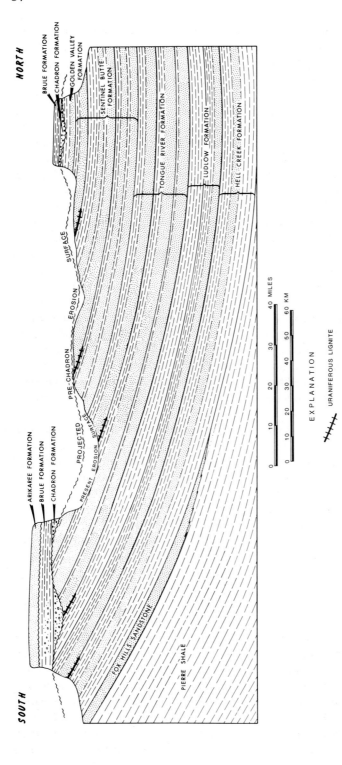

Figure 2-34. Diagrammatic cross section, southwestern Williston Basin, North Dakota and South Dakota, illustrating the relationship of uranium-bearing lignites to regional unconformity.

ranging from 0.007 to 0.07 percent U_3O_8 underlie hundreds of square kilometres.

Phosphatic siltstones and sandstones in the shallow margins of the lacustrine Green River Formation in southwestern Wyoming contain disseminated uranium in certain areas (Love, 1964). Love describes other occurrences of uranium-bearing phosphatic rocks of Eocene age in the Rocky Mountain area as well. All of these occurrences are low grade, widely disseminated, and located in reduced facies marginal to oxidized facies equivalents.

The Bone Valley Formation (Pliocene) phosphate rock of central Florida, which accounts for 80 percent of domestic phosphate production, contains 0.01 to 0.02 percent uranium (Altschuler and others, 1956).

The Chattanooga Shale (Devonian and Mississippian) in central Tennessee, Kentucky, and Alabama contains widespread, low-grade, disseminated uranium (Swanson, 1961). A bed 4 to 6 m thick underlies 10,000 km^2 and averages 0.007 percent U_3O_8. The Chattanooga Shale has red-bed equivalents to the north.

In south central Sweden, black shales of Cambrian and Ordovician age contain a horizontally bedded uraniferous zone 2.5 to 4 m thick that covers about 500 km^2 (Svenke, 1956). This bed averages about 0.03 percent U_3O_8. It also contains about 20 percent organic carbonaceous material. The area is estimated to contain about 1 million tons U_3O_8, and a demonstration mine and mill for processing 800,000 tons of shale per year has been built. Plans call for extending the operation to mine and process 6 million tons per year. Recovery is estimated to be 235 ppm, or 1,275 tons of U_3O_8 per year.

STRUCTURE- OR FRACTURE-CONTROLLED DEPOSITS (VEIN-TYPE AND SIMILAR DEPOSITS)

Some of the largest high-grade uranium deposits which have been discovered in recent years consist of stockwork fracture fillings, breccia fillings, and partial replacement in metamorphic and well-lithified sedimentary rocks. Many of these occurrences appear to be controlled in part by lithology, although the primary control is structural. The dominant lithologic control appears to be rocks with low oxidation potential, similar to strata-controlled deposits described above. Carbonaceous black shales, slates, phyllites, and schists are common host rocks. Carbonate rocks with clay minerals and micas are also favorable host rocks. In addition to carbon, other reductants are pyrite, marcasite, and other sulphides.

Most of the largest and highest-grade deposits occur in lower Proterozoic sedimentary or metasedimentary rocks, which were intensely deformed during mineralization or before mineralization occurred. The mineralized fracture complexes are generally elongate and steeply dipping. The fracturing or brecciation and associated mineralization generally do not extend to depths greater than 300 m; however, there are significant exceptions to this, such as Schwartzwalder in Colorado, Beaverlodge in Saskatechewan, and Jabiluka in Australia.

In most fracture-controlled deposits, pitchblende is the only important mineral below the zone of weathering (Rich and others, 1975, p. 16). The pitchblende occurs generally as sooty or massive black fillings or botryoidal coatings, but crystalline uraninite has also been noted. Usually some quartz or calcite occurs as gangue in the veins, together with some minor sulphide such as pyrite. Hematite is very common in altered wall rock; Rich and others (1975) emphasized the importance of hematite in theoretical origin of the uranium deposits.

The largest and most significant deposits occur in northern Australia and northern

Saskatchewan, Canada. These large, high-grade uranium deposits occur in lower Proterozoic sedimentary and metamorphic rocks below regional unconformities. There is also evidence that those ancient erosional surfaces were exposed to long periods of weathering under influence of alternating hot-arid and hot-humid (savannah) climate. The savannah climate may have played a critical role in the formation of these deposits comparable to the role it appears to have had in the formation of many important strata-controlled deposits.

Rum Jungle–Alligator Rivers Province, Northern Territory, Australia

Six uranium districts with pitchblende in fractures and breccia zones have been discovered and partly defined in Northern Territory, Australia (Dodson and others, 1974; Langford, 1974; Rich and others, 1975; Dodson, 1972). All of the significant ore bodies occur in lower Proterozoic chloritic and carbonaceous schists or carbonaceous shales and siltstones. Pitchblende fills fractures and voids along faults, shears, and breccias.

The Rum Jungle–Alligator Rivers province (figs. 2-35, 2-36) includes five significant deposits with an aggregate reserve of about 450 million kg (1 billion lb) U_3O_8. Pitchblende fills fractures along north-trending faults in tightly folded carbonaceous shales and chloritic slates of the lower Proterozoic Golden Dyke Formation (fig. 2-37). The lower Proterozoic formations unconformably overlie the Rum Jungle complex of Archean crystalline basement rocks. These basement rocks average 2 to 30 ppm uranium.

Nabarlek consists of a high-grade ore body containing an estimated reserve of 10 million kg (22 million lb) U_3O_8. Average grade is greater than 2 percent U_3O_8. The host rock is quartz-chlorite-muscovite schist of the Koolpin Formation equivalent

(lower Proterozoic). The ore body is as much as 20 m wide and 230 m long and extends to a depth of about 70 m.

Jabiluka 1 and Jabiluka 2, located about 500 m apart (east-west), include an estimated reserve of 205 million kg (450 million lb) U_3O_8. The deposits have not been fully defined by drilling at the time of this writing. The mineralized fracture zone trends east-northeast and dips south at about 45 degrees. The ore zone is as much as 49 m thick and extends to depths of 190 m. Gold occurs with the uranium in parts of the trend. Host rock is quartz-chlorite carbonaceous schist of the Koolpin Formation equivalent.

Ranger 1 includes six deposits grouped in a north-trending arcuate belt 6.5 km long and 1 km wide. The deposits are not fully evaluated at this time, but estimated reserves confirmed by drilling so far are 108 million kg (228 million lb) U_3O_8. The pitchblende occurs in chloritized biotite-feldspar-quartz schist which is locally carbonaceous. The host rock has been tentatively correlated to be part of the Koolpin Formation. The ore bodies are in the form of stockworks of fine uraniferous chloritic veinlets.

The Koongarra deposits consist of a series of *en echelon* zones of disseminated uranium enclosing cores of higher-grade ore. Host rock is quartz-chlorite-muscovite schist of Koolpin Formation equivalent. Mineralization consists of pitchblende in shear planes paralleling a reverse fault which dips about 60 degrees to the southeast. The fault is the footwall of the mineralized zone, and the hanging wall is a carbonaceous zone about 70 m above the fault. The foliation of the schist almost parallels the fault. Reserves have been reported to be about 40 million kg (88 million lb) U_3O_8 (L. R. Reimer, unpub. data).

Dodson and others (1974) suggested that these Northern Territory deposits

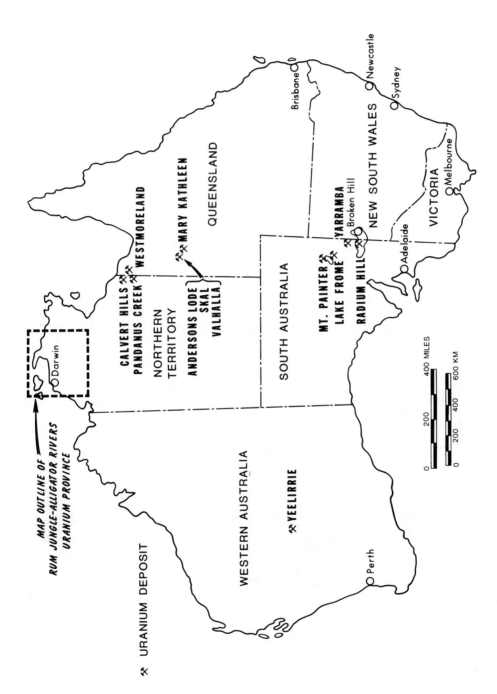

Figure 2-35. Map of Australia with major uranium districts.

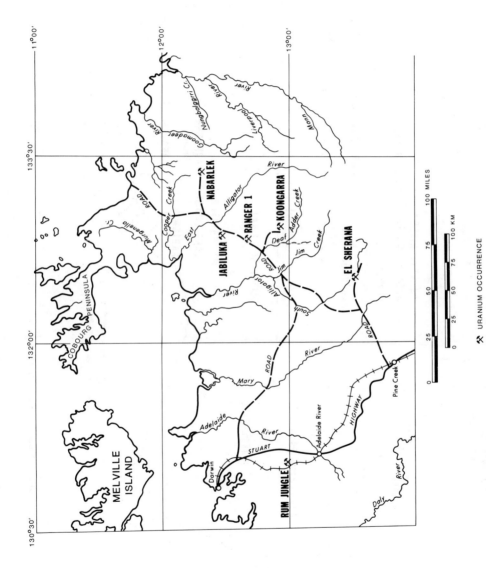

Figure 2-36. Map of Rum Jungle–Alligator Rivers uranium province, Northern Territory, Australia.

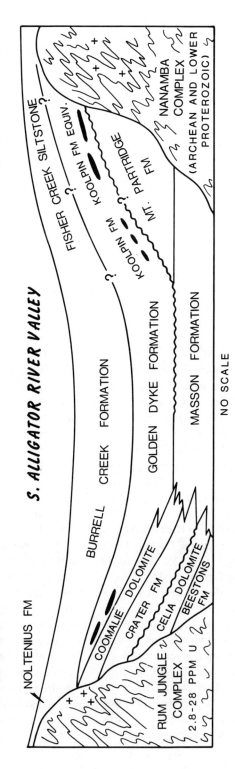

Figure 2-37. Correlation diagram showing interpreted relationships between uranium-bearing Proterozoic rocks in Rum Jungle–Alligator Rivers uranium province, Australia. Modified after Dodson and others, 1974.

were originally low-grade syngenetic con-centrations in carbonaceous sedimentary strata of early Proterozoic age. They fur-ther postulated that a major igneous phase (1,700 to 1,800 m.y. ago) accompanied by tectonic activity resulted in remobilization of uranium and concentration in low-stress loci (breccia and open-fracture zones). They finally suggested that locally some supergene enrichment might have occurred. Langford (1974) postulated a much more important role for supergene mineraliza-tion. It seems that the term "supergene" is appropriate for the concept proposed by Langford in which uranium was carried downward into the fracture systems by in-fluent oxidizing meteoric waters. The ura-nium is postulated to have been leached by weathering and erosion from the uranium-rich Archean granitic terranes.

Northern Saskatchewan Province

Large high-grade uranium deposits occur as fracture fillings and stockworks, pitch-blende-mineralized breccias, and red vein-type pitchblende ore bodies in lower Proterozoic and Archean rocks of northern Saskatchewan. The following four districts contain most of the uranium reserves: Beaverlodge (Uranium City), Cluff Lake, Rabbit Lake, and Key Lake (fig. 2-38). Descriptions of the geology of these dis-tricts are summarized from Rich and others (1975), Lintott and others (1976), Little (1970), Knipping (1974), Tremblay (1972), and von Backstrom (1974).

The Beaverlodge district contains minor occurrences of uranium and thorium in pegmatites and granites and larger mine-able deposits of pitchblende in fracture fillings, stockworks, and breccia zones de-veloped in altered rocks of Archean age (the Tazin Group). The Tazin Group is a thick complex of regionally metamor-phosed graywackes, sandstones, mafic tuffs, and carbonates. The metamorphic

rocks of the sequence include gneisses, schists, argillites, quartzites, and red gran-ites, which are tightly folded, faulted, and brecciated. Tazin rocks are typically high in uranium content throughout the area. The Martin Formation of early Proterozoic age overlies the Tazin unconformably, and these younger rocks are only moderately folded. The Martin includes some red beds. Hema-tite is prominent in the altered host rocks, and chloritization is also common. The age of uranium mineralization appears to be about 1,780 m.y., which is the age of the overlying Martin Formation.

The Cluff Lake district includes three known ore bodies, all of which are located on the south flank of the Carswell dome structure in the Athabasca Basin (fig. 2-39). The inner core of the Carswell structure is 18 km in diameter and consists of Archean and lower Proterozoic leucocratic meta-morphic rocks and minor amphibolites. The periphery of the Carswell structure has dropped along a ring fault about 40 km in diameter. A ring of crumpled carbonates, siltstones, and sandstones of the Carswell and Douglas Formations, which are young-er than the Athabasca Formation, has been preserved. K/Ar measurements on Cluff breccias (which include fragments of all rock types in the area) indicate that the structure was formed during Ordovician time (478 m.y. ago).

Uranium mineralization at Cluff Lake is mainly pitchblende and uraninite, which oc-cur together with gold tellurides, native gold, cobalt, nickel, and several sulphides. Some mineralization occurs in both base-ment metamorphic rocks and younger Atha-basca pelites and arenites; however, most of the ore is in stockworks along shear zones in the leucocratic metamorphic rocks of Ar-chean and early Proterozoic age near the pre-Athabasca unconformity. The mineral-ized shear zones are frequently graphitic and chloritized. The age of uranium mineraliza-

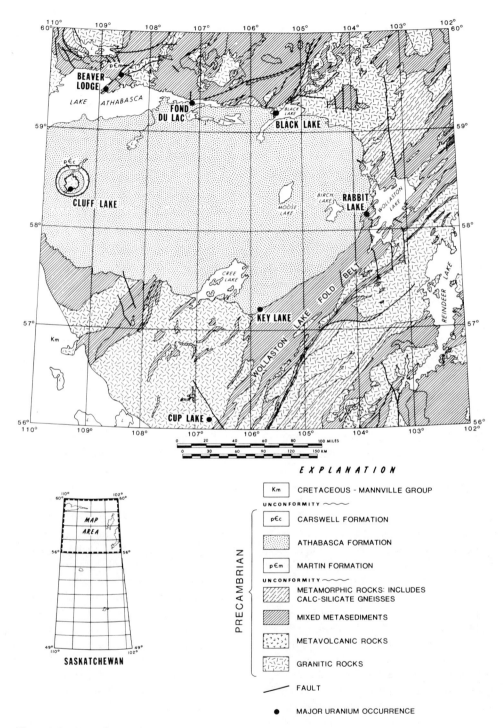

Figure 2-38. Map of northern Saskatchewan uranium province.

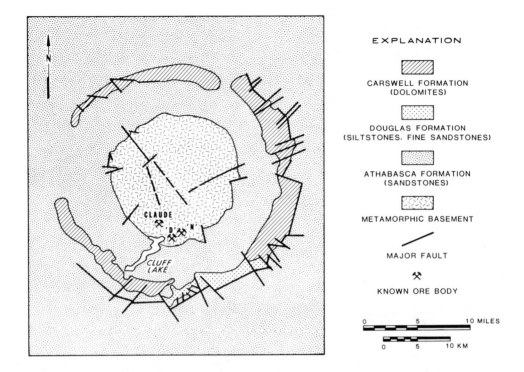

EXPLANATION

CARSWELL FORMATION
(DOLOMITES)

DOUGLAS FORMATION
(SILTSTONES, FINE SANDSTONES)

ATHABASCA FORMATION
(SANDSTONES)

METAMORPHIC BASEMENT

MAJOR FAULT

KNOWN ORE BODY

Figure 2-39. Map of Cluff Lake uranium district, Saskatchewan. Note uranium occurrences clustered on south flank of Carswell dome.

tion has been dated at 1,100 to 1,400 m.y., about contemporaneous with deposition of the Athabasca Formation. Lintott and others (1976) postulated that prolonged weathering of the pre-Athabasca (Judsonian) surface resulted in solution and transport of uranium to the fractured hosts where it was reduced and precipitated. Estimated reserves at Cluff Lake are reported to be 18 million kg (40 million lb) U_3O_8.

The Rabbit Lake district contains high-grade reserves reported to exceed 19 million kg (42 million lb) U_3O_8. The deposit is located on the Wollaston Lake fold belt, one of several northeast-trending belts of lower Proterozoic rocks. Meta-arkose, biotite paragneiss, and calc-silicate rocks characterize the lower Proterozoic rocks of the Rabbit Lake area. These rocks were tightly folded and faulted with extensive breccia develop-

ment during the Hudsonian orogeny (1,800 m.y. ago). Concurrently, metamorphism of the lower Proterozoic rocks took place, and the structure which controls uranium mineralization was developed. Uranium mineralization is localized in a fractured and brecciated zone in the axial region of an overturned synform. The host rock is dolomite, which is strongly altered. Alteration minerals include kaolinite, vermiculite, chlorite, dickite, hematite, calcite, and quartz. Alteration is most intense in the upper 90 m from the surface and decreases downward. No alteration has been observed below 135 m depth, and the uranium mineralization bottoms at 105 m. The pre-Athabasca unconformity is within 15 m above the top of the ore body. The synform which hosts the ore body was faulted up over the Athabasca by a southeast-dipping reverse fault, and no mineralization has been found within the footwall block of the fault. The fault itself is unmineralized, but some alteration extends through the fault zone into the footwall. Knipping (1974) proposed a supergene origin for the uranium deposits.

The Key Lake deposit is the last major uranium discovery to be reported in Saskatchewan. It is located about 150 km southwest of Rabbit Lake in the Wollaston Lake fold belt (fig. 2-38). Very little information is available, but it is reported that the mineralization is similar to that at Rabbit Lake with pitchblende mineralization in calc-silicate rocks. Published drill information reports 5 m of massive pitchblende within 33 m of mineralized section. High-grade mineralization has been reported, but average grades and tonnages have not been released as yet.

There are conflicting data relating to the origin of the uranium deposits in northern Saskatchewan, but most of the filed evidence supports a supergene origin for the large high-grade deposits. The host rocks in the Saskatchewan area are not as carbonaceous (reducing) as are the hosts in northern Australia, but other characteristics are strikingly similar in the two areas. Knipping (1974, p. 543) reported that the pre-Athabasca unconformity is marked by variable thicknesses (as much as 90 m) of weathered regolith within the Athabasca Basin. He interpreted the data to indicate a "hot humid climate interspersed with arid hot conditions" during peneplanation after the Judsonian orogeny. Once again we have a savannah climate in evidence during the time when uranium was being dissolved and transported.

There are unanswered questions bearing on the controls of fluid movement downward and along fracture systems in these metamorphic and generally impervious rocks. In the case of Saskatchewan, it is difficult to explain the reduction of large volumes of uraniferous oxidizing water. Other questions remain unanswered, but as the deposits are discovered, drilled, mined, and studied, we should accumulate valuable data to increase our efficiency in locating and defining additional ore bodies of this type.

Most of the Saskatchewan deposits were discovered by radiometric surveys detecting anomalies in glacial deposits over the bedrock occurrences. Field mapping and drilling was necessary to confirm the presence of ore.

INTRUSIVE–CONTROLLED DEPOSITS

Most intrusive-controlled deposits occur in felsic igneous rocks such as alaskite and nepheline syenite. Most of the known uranium occurrences of this type are either small deposits in pegmatites or large low-grade deposits; mine and mill development is in progress on one large low-grade occurrence. Intrusive-controlled deposits may be important producers in the future.

Rössing Uranium Deposit

The most important intrusive-controlled uranium-ore body now defined is Rössing in South West Africa (fig. 2-40). This deposit has been described by von Backstrom (1974), Berning and others (1976), and Ruzicka (1975). It is reported that the Rössing deposit contains about 140 million kg (300 million lb) of U_3O_8 in low-grade ore which averages 0.035 percent U_3O_8.

The Rössing deposit is located about 65 km northeast of Swakopmund in the Namib Desert. The host rocks are intensely folded and faulted and lie in the upper Precambrian Damaran orogenic belt, which trends northeasterly across the northern part of South West Africa. The Rössing uranium deposit is situated on the south flank of a large domal structure.

The Rössing deposit is a concentration of uraninite, betafite, and associated

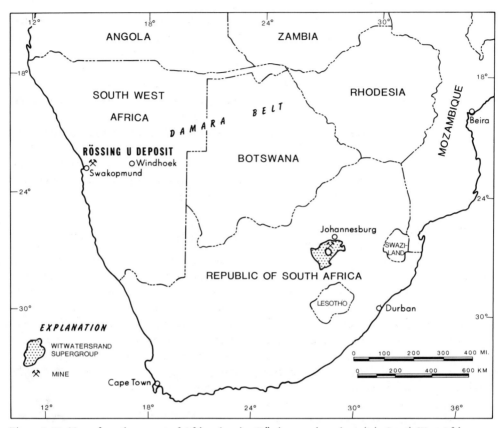

Figure 2-40. Map of southern part of Africa showing Rössing uranium deposit in South West Africa.

secondary minerals disseminated in intrusive bodies of syntectic alaskite. The alaskite is part of a migmatite zone in which the intrusive is aplitic, granitic, and pegmatitic in texture and evidences concordant, discordant, and replacement relationships to the heavily folded gneisses, schists, marbles, and limestones of the upper Precambrian formations.

The primary uranium minerals are found only in the alaskite intrusives, and this is the only known ore control. Uraninite is the dominant uranium mineral, and it occurs mainly as grains 0.05 to 0.1 mm across associated with biotite and zircon. It also occurs within quartz and feldspar and interstitially. Monazite is also associated with the uranium minerals. Ratios of

uranium to thorium average about 10:1.

Berning and others (1976) postulated a syntectic origin of the alaskite and its included accessory minerals, including uraninite. They suggested that "the localization of mineralization in certain sectors only of the alaskite spread is probably a simple reflection of the uraniferous content of the basement rocks before melting resulted in their upward emplacement into cover rocks."

Ilimaussaq

Another interesting but uneconomic occurrence is the uraniferous nepheline syenites of the Ilimaussaq intrusion in south Greenland (Sørensen, 1974). Small zones have as much as 0.3 percent U_3O_8, but most of the host has less than 400 ppm. The uranium occurs in refractory minerals and is difficult to recover.

SUMMARY AND CONCLUSIONS

Types of Deposits: Recent Exploration

As outlined in table 2-1, uranium deposits can be classified according to their dominant controls, their host rocks, and other characteristics. Strata-controlled deposits in sandstones and conglomerates have produced more than 90 percent of the uranium to date; also, more than 90 percent of exploratory drilling has been directed at targets in sandstones and conglomerates.

Exploration for strata-controlled deposits has been successful in defining substantial new reserves along mineralized belts in the Grants region of New Mexico; also, a new trend, located north of the Ambrosia Lake trend, has recently been discovered. This new trend, called the Nose Rock occurrence, was reportedly first discovered by deep drilling in T. 19 N., R. 11 W. (fig. 2-2). Many explorationists believe the Nose Rock trend has the potential for reserves comparable to those which have already been developed along the Ambrosia Lake trend. Definition of the new trend will be slow due to its depth (about 900 m).

In Wyoming, exploration drilling since 1968 has been successful in defining substantial new reserves along bifacies roll fronts in the Powder River and Great Divide Basins (fig. 2-15). Also, in Wyoming and nearby areas, monofacies roll fronts that have substantial uranium potential have been discovered and partially defined in Paleocene and Cretaceous strata. The process of defining reserves along these monofacies roll fronts is slow, because dimensions and grades of individual ore bodies are typically smaller than in bifacies deposits.

The most exciting recent developments in uranium exploration have been the discoveries of large high-grade structure- or fracture-controlled deposits in northern Australia and northern Saskatchewan. These discoveries have proven that structure-controlled deposits may compare favorably in size and grade with the largest and highest-grade strata-controlled deposits known. If the political obstacles are not too severe, it is probable that structure-controlled deposits will soon make a substantial contribution to the world's energy requirements.

The large structure-controlled deposits should be carefully studied; theories of origin should be developed. A sound foundation must be constructed on which to build viable exploration programs aimed at finding more structure-controlled deposits.

Origin of Strata- and Structure-controlled Deposits

We first began studying uranium deposits 22 years ago, when most of the influential economic geologists espoused theories based on hydrothermal origins for almost all metallic deposits, including uranium. A small minority still favors some hydrothermal influence on strata-controlled deposits, but most geologists who are familiar with the pertinent data recognize the overwhelming evidence which favors the ground-water origins discussed earlier in this chapter.

We are convinced that the role of savannah climates in the origin of strata-controlled deposits (excluding, of course, the Precambrian heavy-mineral deposits) is extremely important. High temperatures and high precipitation and humidity are requisites for deep chemical weathering; then the transition to a hot arid phase (which may continue over thousands of years) is important for leaching humates and uranium (as well as other metals) from the upland areas.

There is some evidence that savannah climates may have been important in the development of some large structure-controlled deposits as well. The known deposits in the Wollaston Lake fold belt occur in fracture systems which were exposed to the pre-Athabasca erosion surface, and the regolith on this surface records a savannah climate. Age determinations based on lead: lead and lead:uranium ratios measured on pitchblende samples from the Rabbit Lake deposit vary from 190 to 1,320 m.y. (Knipping, 1974, p. 541). The older determinations suggest that the uranium deposits may have been forming about the same time that the Athabasca deposits were accumulating on the unconformity. (The wide range of lead and uranium isotope ages is typical of geochronology based on uranium samples in any type of deposit.) The geochronology is not sufficiently accurate to preclude the development of the Rabbit Lake uranium deposits under the influence of the pre-Athabasca savannah climate.

Isotopic Ages of Uranium Deposits

Isotopic age determinations have become widely accepted by the geologic profession during the past 25 years, and many applications of isotopic dating seem to be reasonably accurate and consistent with other evidence. However, we have been misled by some isotopic ages of uranium deposits. Even though the ages calculated for a deposit may vary widely from sample to sample, it has been common practice to select one determination as a factual age, after an equivocal accounting for discordances.

An example of this is discussed by Miller and Kulp (1963) in their excellent paper on the origin of the Colorado Plateau uranium deposits. They stated that, although isotopic ages calculated during the 1950s for some Colorado Plateau deposits were widely discordant, the published age was about 65 m.y. (Late Cretaceous or younger). After publication of this age, numerous geologists accepted it as a fact and used it in their interpretations of the origin of Colorado Plateau deposits.

Miller and Kulp found the earlier hypotheses for explaining the discordant ages to be invalid; further, they concluded that the time of uranium mineralization was probably during or soon after deposition of the enclosing strata in Lisbon Valley, Utah, and Cameron, Arizona—that is, about 210 m.y. ago. They found that the ages were in part complicated by alteration of the pitchblende by loss of radiogenic lead less than 20 m.y. ago.

Harshman (1972, p. 73) showed total uranium-lead ages for 15 high-grade uranium samples from Shirley Basin, Wyoming, which range from about 11 m.y. to 40 m.y. In a later study, Dooley and others (1974) analyzed a high-grade uraninite sample from the Gas Hills and one from Shirley Basin. On the basis of $^{206}Pb/^{238}U$ and $^{207}Pb/^{235}U$ dating, they calculated an age of 22 m.y. for both samples. In addition, they analyzed three more samples from Shirley Basin which indicated ages ranging from 20 to 29 m.y. We are convinced that additional analyses of uranium samples from these districts will result in wider-ranging ages. The uranium was originally concentrated through a process of repeated partial solution, transport, and reprecipitation. Uranium minerals in these deposits are locally transient in the perspective of geologic time. Dissolution and local transport is all that is needed to reset the time clock; these time clocks are so easily reset that they are poor indicators of the age of mineralization.

3
Developing an Idea and Preliminary Investigation: Uranium

FORMULATING IDEAS: THEORY

Developing ideas for new uranium projects is the work of a concept geologist, who should have a good general background in uranium geology as well as the overall geology of the region under consideration. It is essential that the concept geologist be familiar with the most important controls and general aspects of important known uranium deposits which have been defined and studied in environments similar to the one under consideration. In addition to looking for "more of the same thing," the successful explorationist is obliged to hunt for new varieties with some characteristics of known deposits. To do this effectively, the explorationist must know what the objects of his search look like, what their geologic setting is like, and what aspects of that setting are really important to the deposits.

The explorationist's theory of the origin

of uranium deposits plays an important role in formulating ideas for exploration projects. A simplified example of this might be illustrated by two hypothetical competing exploration projects in Wyoming. In the first instance, the geologist favors the theory that the uranium in Wyoming roll-front deposits was derived from overlying tuffaceous beds in the White River and Arikaree Formations. This geologist would concentrate his exploration in areas where White River and Arikaree had once been deposited or where these formations are still present. Further, he would look for places where potential hosts had been in contact with the pre-Oligocene erosion surface and where these hosts had not been extensively eroded in post–White River time.

In the second instance, the explorationist favors the theory that the uranium in the Wyoming roll-front deposits was derived from ancient Precambrian terranes of the Granite Mountains which supplied the debris for the Eocene host rocks. This geologist would be more concerned with Eocene paleodrainage systems than with the erosion surface beneath the Oligocene rocks.

Deep drilling in the Powder River Basin was probably delayed for several years because of the general acceptance of the tuff-leach theory. Some geologists, including the authors, find it difficult to conceive how uranium-bearing solutions could move downward in post-Oligocene time to some of the deeper host sandstones in the axial parts of the basin, but the deposits now have been found regardless of which theory, if any, the discoverers used.

Several exploration companies bypassed the Powder River Basin altogether in the middle 1960s because their ruling theory was that all of the uranium in the basin had been tied up with vanadium in small near-surface deposits. This theory was completely in error, and some large but deep uranium deposits were discovered by other companies. If those who stayed out of the Powder River Basin had instead used the approach of a working hypothesis, they might have taken the position that even though the uranium had probably been tied up with vanadium, drilling in favorable trends should be done to see if the hypothesis was correct.

Theories always play a key role in exploration, and the experience we have gained from years of exploration in various areas convinces us that a theory should be used to strengthen and supplement an exploration program; however, a theory should never be allowed to limit an exploration program. An explorationist should always be on guard against the ruling theory.

LITERATURE RESEARCH

When a concept for a new play begins to take form, a review of the literature pertinent to the concept should be made. Descriptive material on known uranium occurrences which are similar to the anticipated targets may be helpful. Details of the stratigraphy, structure, and even occurrences of known mineralization in the area of interest might be found in the literature. We have found that descriptions of isolated occurrences of mineralization, which would not appear to be significant to the casual reader, sometimes take on significance when considered in the context of a developing exploration concept. The importance of this fact cannot be overstated.

Prior to reviewing the literature, it is a good idea to acquire base maps on which to post data. These maps can also be helpful for general orientation during the literature study. U.S. Geological Survey topographic maps, state or county highway maps, or even aerial photographs are usually available for these purposes.

Sources of information include good geological libraries such as those maintained by a few public libraries and by geology departments at colleges and universities. The U.S. Geological Survey often has a library at various field offices, and the Energy Research and Development Administration (ERDA) has a geology library at Grand Junction, Colorado. The ERDA library has all of the old Atomic Energy Commission short-form prospect reports (described in more detail near the end of this chapter). Listings of USGS and U.S. Bureau of Mines publications should be referred to for data concerning the area under consideration. State geological surveys usually maintain libraries and can prove very helpful in answering questions about information available for areas within their state. Various geological organizations prepare and publish guidebooks which often contain valuable papers and maps concerning geology in their state or region—for example, the New Mexico Geological Society (in Albuquerque), the Rocky Mountain Association of Geologists (in Denver), the Wyoming Geological Association (in Casper), and the Intermountain Association of Geologists (in Salt Lake City).

REGIONAL SUBSURFACE STUDIES

If the area of interest is located in a sedimentary basin where petroleum exploration has been conducted, a regional subsurface study might be considered. In many instances, gamma-ray and electric logs of oil-exploration drill holes can be used to establish generalized correlations and to map regional geology. Structure-contour maps can sometimes be prepared to show the broad tectonic setting of the prospect area. Various companies also maintain and sell maps showing the locations of oil wells and dry holes, and some of them also prepare and maintain structure-contour maps which may be of value to the explorationist searching for mineral deposits in sedimentary rocks. If certain oil tests appear to have been drilled where some data of value might be obtained from the logs, the logs can be ordered from one of the firms which store and reproduce these logs, such as Rocky Mountain Well Log Service or Rocky Mountain Geological Library.

In most instances, however, the logs of oil-exploration drill holes do not include the shallow strata of interest to the uranium geologist. There are two reasons for this: (1) surface casing is set on every test, and this often extends down through the zone of interest; and (2) the oil explorationists are seldom interested in zones within 500 m of the surface. Further, oil-exploration drill holes are usually not closely spaced. For these reasons, oil-well data will not usually be adequate to put together detailed maps and cross sections

useful in uranium exploration. However, the subsurface information gained from oil wells might be adequate to put together a regional picture and thereby formulate a setting for the uranium play.

On one occasion, we were reviewing logs of oil tests in a certain part of a basin to determine which logs included a gamma curve; we wanted to examine all those with gamma curves. We became quite excited when we found that a log of a test which had been completed as a gas well had a gamma curve to within 30 m of the surface, and showed very strong gamma through the casing from 330 to 450 m (1,100 to 1,500 ft). We attempted to make some calculations for grade but found this to be virtually impossible with a Schlumberger log.

The geology of the area seemed favorable, and so we decided to offset the well with a uranium-exploration hole to evaluate the radioactive zone. We offset 30 m and drilled a hole to 450 m. We noted very little sand while the hole was being drilled and, after the log was available, found no radioactivity above background in our hole. We could only conclude that some component of the cement that had been placed behind the casing in the gas well was radioactive. After that experience, we looked askance at every radioactive anomaly in a cased well.

Unfortunately for the mining industry, and the United States as well, there is no good way at the present time to get drilling information from other companies. Mining companies have always been very secretive about information they may have concerning a prospect or area, and they remain secretive today. Oil companies release logs for use by the public, but mining companies as a whole do not, and, in fact, object to any suggestion that a cooperative effort be made to create a depository and reproduction facility, so that drill hole logs and sample descriptions could be available to anyone who is willing to pay for the reproduction. We fear that unless something is done by industry promptly, legislation will force such requirements upon the industry either at the state or national level. Much as we abhor any additional regulations, it would be preferable to have the data made available by means of legislation than to have extremely valuable data gathering dust in someone's files. The petroleum business has shown that the concept will work; now the mining business must break through the secrecy barrier. The argument of companies "trading logs" is made only by those who do not realize how concepts develop, how regional or district studies progress, or how the independent explorationist functions. What would the condition of the petroleum exploration industry be if the major companies exchanged logs only with each other?

AERIAL PHOTOGRAPHY

Black and White or Natural Color

Aerial photographs can be useful in mapping surface geology, as well as evaluating and locating (1) exploration activities of other operators; (2) cultural developments, including agriculture; (3) forested areas; and (4) access routes to points of interest. The explorationist may purchase aerial photographs from the USGS, the U.S. Department of Agriculture, or from several service companies. If the quality of photography available from these sources is not adequate, it may be reasonable to have the aerial photography done by one of the service companies.

Most of the older government photographs are available only in black and white at scales of 1:24,000 or smaller (before enlargement). However, since 1966 many areas have been covered by natural color photography at 1:24,000. Natural color

photography is useful in mapping alteration, oxidized and reduced facies, and other aspects of geology.

In several basins of Wyoming, natural color photography flown at an elevation of 3,600 m (12,000 ft) above the ground surface has been used successfully to map altered sandstone outcrops where they are characterized by red hematite coloration. This mapping requires good photographs with proper color balance. It has been our experience that color photography flown at elevations higher than 3,600 m above the ground surface lacks the sharpness of detail and color quality necessary for mapping alteration and lithologic units. If economic considerations permit, the photography should be flown at lower elevations (larger scale).

The general geology and gross oxidized and reduced facies of uranium-bearing formations such as the Chinle, Morrison, Wind River, Wasatch, or Battle Spring can easily be mapped on natural color photographs. We prefer photography at a scale of 1:24,000 or larger, but higher-altitude photography can be used for this mapping if necessary.

At one prospect on the western side of Wyoming's Powder River Basin, we participated in initial evaluation which included the use of natural color photography. Altered sandstones had been found in the Fort Union and Wasatch Formations, dipping into the basin at angles of 14° to 18° (Fort Union) to 5° to 7° (Wasatch). For the most part, the altered sands were well exposed and colored in tints of red, orange, pink, lavender, and old rose. The area was flown at low altitude (3,600 m above ground surface), and excellent color balance was obtained on the prints. These photographs proved to be extremely valuable for mapping the sandstones, and several mineralized areas were found on the ground where the photographs showed a color change from reds to buff.

Black-and-white or color photography may also be used to prepare maps which show drill patterns and other exploration activities (figs. 3-1a, 3-1b). These maps are frequently helpful in projecting mineral trends, evaluating land acquisition programs, or in other practical aspects of exploration.

Color Infrared (False Color)

The U.S. Forest Service has used infrared false color, because it exaggerates the distinction between healthy and unhealthy vegetation (forests). Explorationists have tried to use this type of false color, because reflected energy in the near-infrared bands of the spectrum penetrates haze more effectively and results in clearer photographs at high altitudes than total-visible-spectrum photography. A disadvantage of the false color infrared is that the true rock colors are difficult to interpret.

LANDSAT ORBITAL SYSTEMS: MULTISPECTRAL DATA

Remote-sensing studies of areas that are known to contain uranium and that have extensive exposures of altered sandstones were reviewed by Offield (1976). The Landsat orbital systems measure energy reflected from the surface of the earth with multispectral scanners. The data are transmitted from the satellites to ground stations where they are recorded on magnetic tapes and framed into photographic images. The data can be computer processed to enhance spectral contrasts which are known from independent measurements to be typical of altered rocks. Unfortunately, these contrasts are subtle and not unique to alteration.

Landsat data generally do not have sufficiently fine resolution capability to permit mapping of individual altered sandstone outcrops; however, the data should be utilized in regional reconnaissance studies and, under certain circumstances, may be useful to map major structural features and to detect favorable formations and relatively extensive exposures of altered rocks. The data are available for many areas and should prove to be of special value for early stages of exploration in remote areas.

AIRBORNE GEOLOGICAL AND STRATEGIC RECONNAISSANCE

Early in the program of investigation, airborne reconnaissance should be considered. This is an efficient way to get a quick overview of the terrain, evidence of activity by competitors, and general geologic aspects of the area. We have used small aircraft such as the Cessna 182 or Supercub for early reconnaissance work. If a large area must be examined and time is limited, then a twin-engine aircraft such as the Cessna Skymaster may be used. All of the aircraft mentioned are high-winged, and we strongly recommend against using low-winged craft because visibility is severely restricted.

The Supercub is a small aircraft with a powerful engine, has good maneuverability, and can land and take off on short runways. Radiometric surveys are often carried out with the help of a Supercub, and the geologist must sit in a small compartment directly behind the pilot. The craft has a rather slow airspeed, which is good for observations, and for the first hour or two the geologist is usually pleased that such an aircraft is available for reconnaissance. After three hours he may begin to wonder if circulation will ever return to his posterior. After four hours he is convinced he will never walk again, and when five hours have been spent in a Supercub he will probably have to be pried out of the seat and will be unable to straighten up for at least an hour. Hearing may be impaired for an equal amount of time. It is difficult

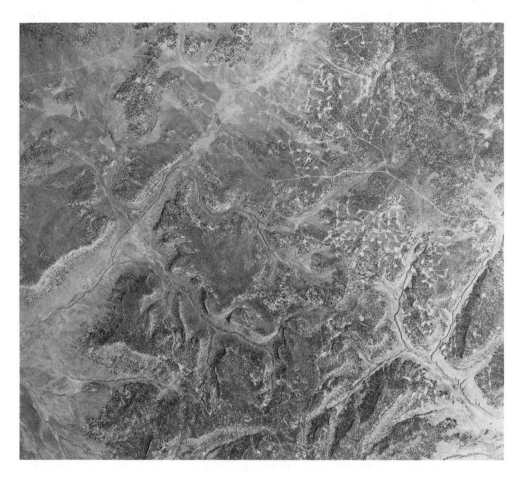

Figure 3-1a. Black-and-white aerial photograph. Courtesy of IntraSearch, Denver, Colo.

to concentrate on geology under such conditions.

The geologist should have maps of varying scales for navigating and plotting information. Generally a small-scale map with topography, roads, and some buildings is best for navigation. We frequently utilize Army Map Service (AMS) maps (scale 1:250,000) in wild, undeveloped areas with little or no culture. State or county highway maps (at a scale of 1:126,720) may be used for navigating and plotting reconnaissance information in areas where it is possible to navigate on the basis of drainage patterns, roads, and other culture

shown on the maps. In some areas, it is advisable to use several types of maps together. Larger-scale maps (for example, 1:24,000) are generally not suitable for reconnaissance flying because of the rapid and repeated movement necessary from one such map to another.

The geologist should carry a field notebook on the airborne reconnaissance. Points where observations require a note can be numbered on the map and keyed to the field notes (figs. 3-2, 3-3). A camera might also be useful to record observations in more detail; photographs should also be keyed to the field notes and maps.

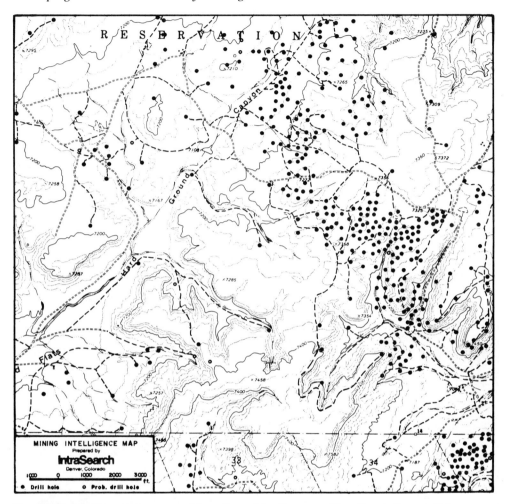

Figure 3-1b. Intelligence map prepared from figure 3-1a. Courtesy of IntraSearch, Denver, Colo.

The primary purpose of airborne reconnaissance is usually to evaluate the exposed geology from a broad perspective. It is often possible to view large areas, noting outcrop loci and briefly describing the geologic relationships in a short span of time. We have covered as much as 1,000 km^2 in three hours' flying time. Another advantage of airborne reconnaissance is that facies relationships and other major stratigraphic features are frequently more apparent when viewed from the air than they are from ground observations. Important outcrops which are hidden from roads

and difficult of access can be spotted from the air, located on the maps, and later examined on the ground.

General descriptions of the exposed geology should be prepared during an air-reconnaissance survey. An effort may be made to map predominant lithologies and interpret facies relationships. Favorable facies and potential host rocks should be carefully noted on maps by symbols, together with notations of access routes to the critical exposures. Evidence for alteration is almost always at the top of the explorationist's list of priorities and should

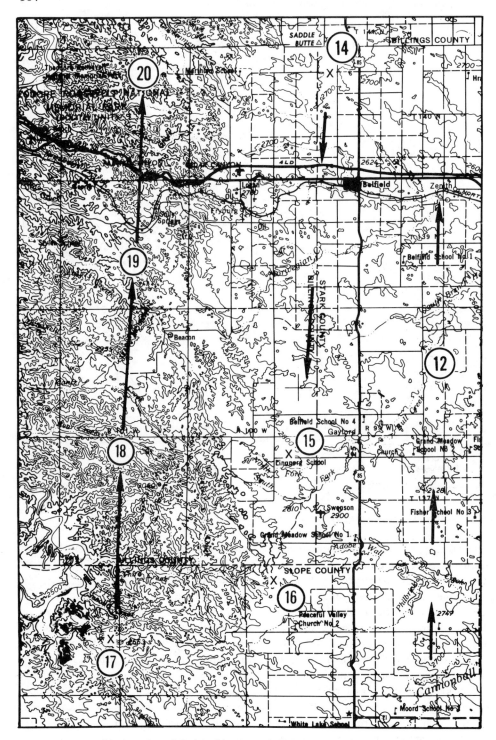

Figure 3-2. Map with data plotted during airborne geologic reconnaissance. Army Map Service base.

be carefully located and described.

As the geology is being observed, it is also important to look for evidence of competitors' activities and to note problems of access and other information which may be pertinent to ground operations.

Reconnaissance flying is conducted too high above the ground surface to detect any but very strong gamma-ray anomalies, but a scintillation counter (frequently called a scintillometer) should be carried in the aircraft if it is convenient. When interesting exposures with potential for mineralization are observed, it is frequently a good practice to circle around and bring the aircraft down to a lower altitude to get a better view of the geology. At such times, the scintillation counter may be used. If the project continues, more extensive aeroradiometry should be carried out, perhaps by helicopter, sometime after the early airborne reconnaissance is completed.

Observations from the air reconnaissance should be plotted on the maps that will be used for field work, and the notes should be summarized into an integrated report after the flying is completed. These data are usually invaluable during later phases of the exploration program.

Many exploration groups at one time or another consider purchasing an aircraft for business use. Our experience has led us to several firm conclusions in this regard. (1) Twin-engine aircraft are much safer than single-engine planes, and whenever possible corporate personnel should be transported in twin-engine craft. Prospect examination where slower speeds are desired is an exception. (2) Owning and maintaining an aircraft is very expensive, especially a twin-engine craft, and it must be used extensively to justify the cost. This is particularly true if a pilot must be retained on the staff (even though the pilot may draft or do other work when not piloting). (3) Most small to medium-sized exploration groups

Figure 3-3. Field notes describing observations spotted on map (fig. 3-2). Notes are usually brief, because geologist is also required to navigate.

should use commercial airlines and charter aircraft when needed rather than own an aircraft.

These views would not apply to operations in Alaska or Australia, for example, where vast distances must be traveled just to get from prospect to prospect; but they are valid in the United States, where one may fly on scheduled airlines to any section of the country.

AERORADIOMETRY

The airborne gamma-ray survey has been the most important preliminary investigation technique in many successful uranium exploration programs. Several of the large, high-grade, fracture-controlled deposits in Saskatchewan were discovered by drilling

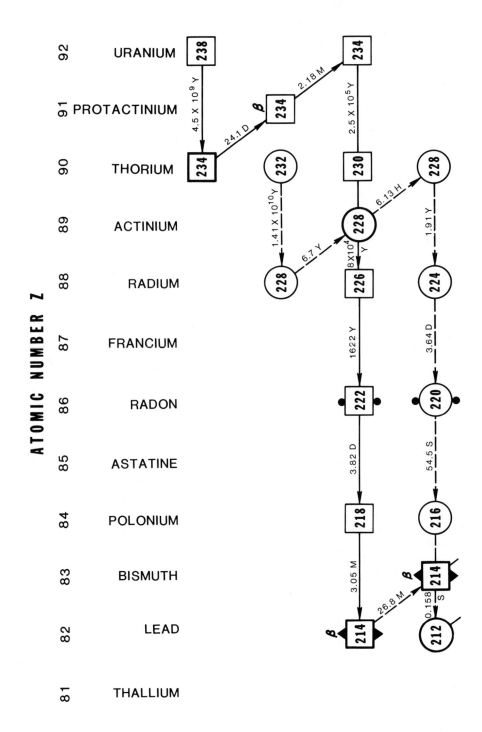

A

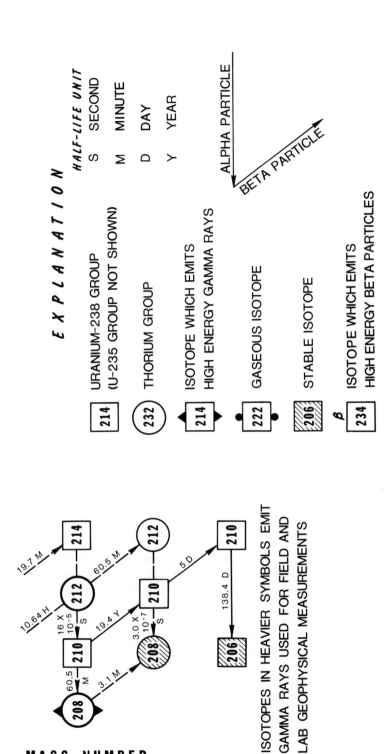

Figure 3-4. Chart of uranium-238 and thorium-232 decay series.

Figure 3-5. Illustration of one type of scintillation counter used in the industry for total gamma-radiation detection. The 2.54 x 3.8–cm (1 x 1.5–in.) thallium-activated sodium iodide crystal commonly used in an instrument of this type is shown beside the scintillation counter. The crystal is pale yellow and clear unless it has been damaged.

in areas where airborne gamma-ray surveys detected uranium mineralization in glacial drift. Many deposits in various parts of the world have been pinpointed by aeroradiometry.

Darnley (1972) reviewed the airborne gamma-ray survey techniques in some detail, and there are numerous other published reports on the subject (Gregory, 1960; Darnley, 1975). Equipment ranges in sophistication from the relatively simple total-gamma scintillation counter to complex multichannel gamma-ray spectrometers with electronic data processing.

Gamma Radiation and Instrumentation

The gamma radiation which is measured in airborne surveys includes background radiation and radiation originating within a thin layer immediately below the ground

surface. Soil, rock, and even water attenuate gamma radiation very effectively: approximately 90 percent of the measured radiation, which originates beneath the ground surface, comes from the upper 0.3 m of soil or 0.7 m of water.

Gamma radiation with energies sufficient to penetrate 100 to 200 m of air and be detected by a NaI (Tl) crystal is given off by the decay of several isotopes of the uranium and thorium decay series (fig. 3-4), as well as potassium (^{40}K). The important gamma emitter for the uranium series is bismuth-214 (^{214}Bi), and thallium-208 (^{208}Tl) is the important contributor from the thorium series. Gamma rays are high-energy photons with numerous photo peaks ranging from several KeV (thousand electron volts) to 2.8 MeV (million electron volts) (Adams and Gasparini, 1970).

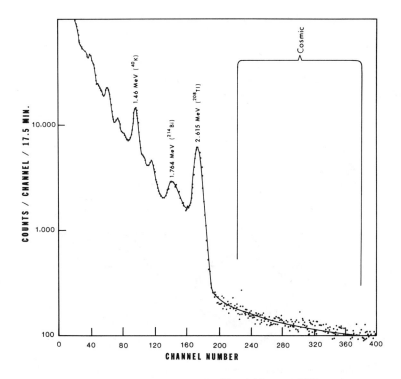

Figure 3-6. Gamma-ray spectrum showing ^{40}K, ^{214}Bi, and ^{208}Tl energies.

The energy levels of the three important photo peaks are 1.46 MeV (^{40}K), 1.76 MeV (^{214}Bi), and 2.62 MeV (^{208}Tl). If instrumentation is good, total gamma-radiation surveys detect and measure indiscriminately all of these photons, and others as well, with energies above an energy threshold which is determined by the instrument setting (fig. 3-5). As illustrated in figure 3-6, the number of photons increases rapidly with decreasing energy at the lower end of the spectrum; thus, the energy threshold setting must be stable.

Some scintillation counters have three settings for the energy threshold: (1) 0.5 MeV to count ^{40}K + ^{214}Bi + ^{208}Tl; (2) 1.6 MeV to count only ^{214}Bi + ^{208}Tl; or (3) 2.4 MeV to count only ^{208}Tl. These counters require only a single, medium-sized crystal and are small and simple in design (fig. 3-7).

"Window spectrometers" are used in most airborne spectrometric surveys. These instruments usually are designed with four windows to count photons with energies only at the photo peaks of interest. More sophisticated (and costly) is the multichannel spectrometer which defines the full spectrum (fig. 3-8). The data are recorded on magnetic tape.

Total Gamma-Radiation Surveys

Total gamma surveys may be flown in straight, parallel flight lines where counts are mechanically recorded together with elevations, or they can be flown along geologic outcrops or drainage patterns. We have found that in sedimentary basins, where geology is adequately exposed, the method of following geologic outcrops sometimes works well. A good pilot should handle the aircraft, and the geologist-

Figure 3-7. Discriminating scintillation counter of a type used in many aerial and ground radioactivity surveys. A large NaI (Tl) crystal and photomultiplier tube is contained in the piece of equipment in the left half of the photo. This scintillation counter is suitable for use in a small aircraft. Photo by Roberts Commercial Photo.

navigator can handle the maps and scintillation counter and give directions. The geologist usually has little difficulty pinpointing his location if he has a geologic map with topography (scale 1:24,000). Radiometric anomalies can be plotted directly on the geologic map. In areas where favorable zones or formations are known from published work or from earlier reconnaissance, the geologist-navigator may fly a concentrated flight pattern over and *along* outcrops of such stratigraphic units. The same advantages of this method may also apply in metamorphic fold belts when looking for structure-controlled deposits. Topography often is largely controlled by lithologic variations and structure; thus, flying along strike valleys or fault-controlled valleys (or ridges) simplifies the task of maintaining a relatively uniform elevation above the ground surface (fig. 3-9).

In some areas, thick recent deposits, such as pediment gravels or regolith, cover interstream uplands uniformly and effectively. When this is the case, field geologists follow drainages in search of exposed bedrock, and this is also a good method of airborne gamma-radiation surveying. Sometimes the opposite is the case: valley alluvium covers bedrock, but ridges are exposed.

Advantages of the total gamma survey are (1) low relative cost (compared to gamma-ray spectrometry); (2) flexibility to concentrate on favorable areas; (3) ability to fly low and directly over specific outcrops; (4) ability to circle and check out apparent anomalies; and (5) ability to modify flight lines to conform to exposures, geologic structure, or outcrop patterns.

111

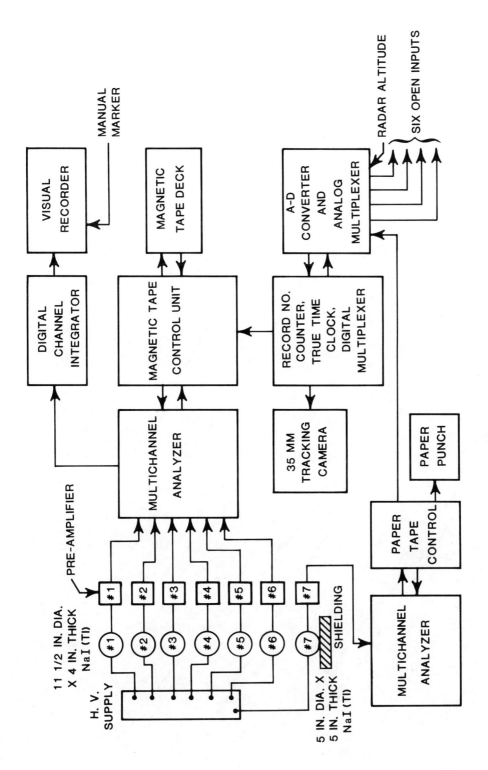

Figure 3-8. Diagram of multichannel gamma-ray spectrometer.

112

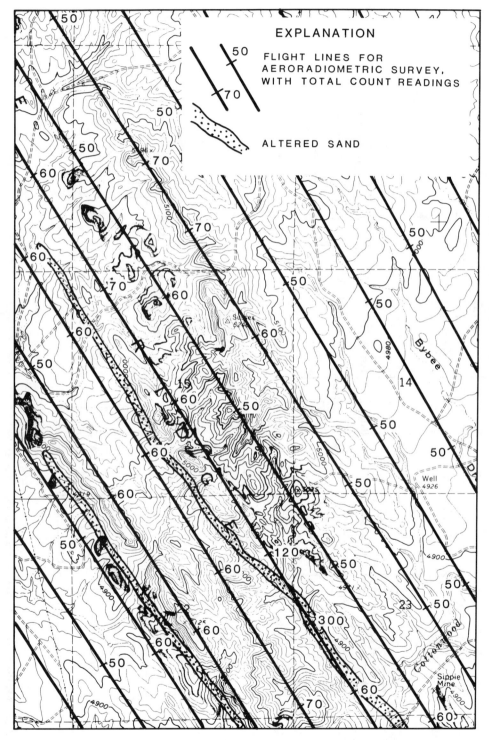

Figure 3-9. Airborne radiometric data plotted on topographic map. Flight lines parallel topographic features which also parallel geologic outcrop patterns.

Gamma-Ray Spectrometry

The recent increase in exploration for igneous-disseminated and structure-controlled uranium deposits has resulted in great demand for airborne gamma-ray spectrometric surveys and surveying equipment. The spectrometric survey has several advantages in igneous and metamorphic terranes that justify its high cost relative to total gamma. First, it is obviously important to differentiate between uranium and thorium when both may be present in significant amounts. Also, many of these areas are wild and remote, making surface work difficult and expensive; thus, the explorationist is obliged to get all information possible from airborne surveys. In jungle, desert, or other areas where it is difficult to obtain an accurate location from visual observation, the well-equipped aerial survey will enable the preparation of maps or charts showing reasonably accurate locations of anomalies.

Gamma-ray spectrometry also has the advantage that isotope-ratio maps can be prepared from the data. Ground-surface configuration and other factors can produce weak anomalies in ^{214}Bi or ^{208}Tl, but the ratio of ^{214}Bi/^{208}Tl is only affected by real differences in one isotope relative to the other (when elevation variations are corrected).

Computer processing of spectrometric data results in more accuracy and flexibility in presentation. However, due to the cost, some density of coverage is usually compromised.

FIELD WORK

Geologic Reconnaissance and Mapping

After completion of the airborne reconnaissance, but early in the investigation of an area, preliminary field work should be carried out to determine the general potential for good mineable deposits. If the area is remote, wooded, and not easily accessible, it might be necessary to contract for a helicopter for the preliminary field work. This is often a good investment even in areas with moderately well developed access roads, because it saves time and is usually conducive to a more thorough examination. We have used helicopters in areas of the western United States where field work would have been possible with four-wheel-drive vehicles, and in each instance we were satisfied that the time savings and improved perspectives were well worth the costs (fig. 3-10).

The field geologist should have the best available maps and aerial photographs for reconnaissance field work. Data from the airborne reconnaissance, together with other pertinent data from the published literature, should be plotted on these maps and photos. The field geologist should also be equipped with a Brunton compass and surveyor's chain to measure attitudes of bedding, foliation, joints, and faults. In addition, his equipment should include tools for sampling and sample bags. A good scintillometer is always extremely important to any uranium exploration field work. In remote areas, two scintillometers and extra batteries for each should be carried in the event that one breaks down. Unfortunately, many uranium explorationists who use these instruments on a frequent basis do not realize the sensitivity to shock of the thallium-activated sodium iodide crystal, which is the heart of the instrument. These crystals are sensitive to thermal as well as mechanical shock, and scintillation counters therefore should never be taken directly from a very warm area, such as a heated automobile interior, to a very cold area, such as an outcrop in subfreezing weather. Likewise, when they are used in cold outdoor weather, they should not be heated up quickly. We have observed scintil-

Figure 3-10. Co-author R. V. Bailey prepares to field-check a radioactive anomaly detected by flying drainage systems and favorable geology. Pick and shovel, scintillation counter, camera, and sample bags (in backpack) are standard field equipment. Pilot is a professional employed by the contract firm owning the helicopter.

lation counters being shipped in boxes as baggage on airlines, and we definitely recommend against such practice, because damage to the crystal is almost certain to result from the rough handling. Instead, the scintillometer should be carried aboard the aircraft either by a carrying handle or by means of a shoulder strap.

The first objective of the preliminary field work is usually to determine if surface evidence is favorable or unfavorable for the primary controls hypothesized for the occurrence of mineralization. For example, if it is a prospect where strata-controlled deposits are expected, the geologist must first determine if a favorable reducing facies is present and if this facies contains good potential host rocks (fig. 3-11). Similarly, if the prospect is for structure-controlled deposits, the first task is to look for evidence

of fracturing and to decipher the structure of the area. When the explorationist is looking for intrusive-controlled deposits, he is obliged to determine if a favorable intrusive is present and if it in fact contains uranium values. The objectives and techniques of field work vary with the geologic environment and type of uranium deposits anticipated.

Alteration Complexes and Roll Fronts

Specific objectives of the geologic reconnaissance in search of evidence for roll-front deposits include stratigraphy, facies, unconformities, potential host sandstones, alteration, mineralization, and structure. A reduced facies with thick, laterally extensive sandstones (potential hosts) is essential to a roll-front prospect. Mudstones and shales in the reduced facies are typically

Figure 3-11. Potential host rocks, such as this soft friable sandstone exposed in a wind-eroded depression in Wyoming's Powder River Basin, are sought by explorationists in their efforts to determine if an area appears favorable for the occurrence of uranium deposits. Were it not for the wind erosion, this sandstone would be invisible from the surface because of sand, grass, and sod cover as seen near the vehicle. Humate, coalified wood, and sulphide minerals are among the constituents looked for in addition to radioactivity. It is usually very helpful to recognize alteration.

gray or greenish with carbonaceous components. Good potential host sandstones should reach thicknesses of at least 10 m, but thickness is often difficult to determine by surface work because of the weathering characteristics of favorable sandstones. We have observed that the best potential hosts are friable sandstones which, due to a lack of cement or binder, do not stand up well in outcrops. Favorable sandstones frequently become disaggregated in the weathered zone and support vegetation well (figs. 3-11, 3-12). Clay-bound, or calcite-cemented, cliff-forming sandstones generally are not favorable hosts. The practical effect of these criteria for roll-front prospects is that the good or favorable sandstones are hidden

while the poor or unfavorable ones cap ridges or are otherwise exposed for examination.

The geologist should attempt to determine how porous and permeable the sandstones are during the preliminary field examination. This can best be done by putting a few drops of water on a specimen of the sandstone. A permeable sandstone will absorb the water quickly through capillary action, but a "tight" sandstone will not.

While looking at a prospect in South Dakota, we noted a white sandstone in the bank of a pond which had been dug for watering livestock. Examination of the sandstone with a hand lens revealed a fine, well-sorted, mostly quartz sand with apparently excellent permeability. It

Figure 3-12. A small near-surface uranium pit was excavated at this location in Wyoming's Powder River Basin. The occurrence was first detected as a surface anomaly and then was drilled prior to mining. The sandstone is poorly resistant and unidentifiable from the surface before mining. The faintly bleached sandstone in the left center part of the photograph is altered, and the best ore was taken from the area in the middle of the photograph.

appeared to be an excellent host rock for uranium deposits. We were astounded then to see that water from a recent rain was still standing in the erosional cavity. At first we assumed that a clay, such as bentonite, had washed in with the water to seal off the permeability. We tested the idea by digging our own hole in the sand and pouring water from the pond in it. The sand would not absorb the water. Apparently interstitial clay existed in the sand in sufficient quantity to swell when water contacted it, thereby causing nearly total loss of permeability.

In examining sandstones, cross-stratification, grain size, conglomerate composition, quartz:feldspar ratios, accessories such as micas and chert, clay galls, calcite concretions, carbonaceous zones, siliceous zones, channel scours, color staining, and other features should be described.

Because the best hosts are often friable and poorly exposed, we always include picks and shovels in our field equipment (fig. 3-13). As a matter of fact, we consider this excellent advice to all explorationists looking for sandstone-type mineral deposits. In order to examine critical contacts, check for radioactivity, look at lithologies and so forth, you are well advised to carry, and *use* a pick and shovel as important components of field equipment. Too many explorationists fail to make the effort to expose important features in this manner. We believe that critical but observable features often exist near the surface, and when they do, we want to see them, even if it means digging through a metre or

Figure 3-13. Co-author M. O. Childers wields a pick in a strong wind to expose a source of radioactivity detected during a car-borne scintillation survey. A landman stands by to assist. Note the ever-present scintillation counter.

more of soil or overburden to see them. Please refill the holes after making your examination.

It is difficult or impossible to determine the thickness of sandstones in many cases, particularly those which are nonresistant (hence favorable), and it is even more difficult to determine the lateral extent of these bodies from surface work. The strata can sometimes be projected from one outcrop to another with some degree of reliability. Sometimes a hard, ridge-forming unit will provide a reference which can be mapped and used to correlate isolated outcrops of the potential hosts. We have learned from experience that if *any* of the nonresistant sandstone units can be observed in an area of poor outcrops, then there are probably *many* sandstones of that type in the area, particularly if they are altered. A good rule of thumb in the

Rocky Mountain states is that in an area of poor exposures you will usually find evidence for less than 10 percent of the sandstones actually present. For example, in such an area in Wyoming, if you found an outcrop of one altered sandstone 3 m thick, it is likely that sandstones totaling 30 m or more in thickness would exist in the area, whether you could observe them directly or not.

In many good prospective areas, no sandstones at all will be seen at the surface, and the geologist must obtain evidence from indirect sources, such as drill holes, to assess the potential.

As the potential hosts are being mapped and described, it is important that evidence of alteration be sought. Alteration related to roll fronts is frequently highlighted by bright colors which include tints of red, orange, purple, rusty yellow-brown, and

Figure 3-14. Solution banding, important evidence for the passage of reactive solutions through the sandstone, is evident at this exposure in Wyoming. Bedding is almost horizontal at this location, with abundant cross-bedding and evidence of slumping, and the dip is essentially zero. Solution bands are dark streaks which cross the exposure from upper right to lower left. The bands freely cross the bedding. All of the sandstone in this photograph is altered.

our old standby, old rose. These colors are related to hematite and hydrous iron oxides which coat grains and permeate clay minerals in the interstices of the sandstones. Altered sandstones are also sometimes white, with a bleached appearance. In some cases, alteration is difficult to distinguish from ordinary weathering of sandstone in outcrop. The alteration is characterized by extensive development of red hematite staining in some areas, whereas in other areas, such intense oxidation is very local or not at all in evidence. The degree of oxidation and apparent intensity of alteration does not seem to bear any consistent relationship to the size and grade of uranium deposits on the associated roll fronts. Indeed, there is no consistent relationship between the size of the alteration

complex and the size and grade of associated uranium deposits.

Unless certain observable phenomena are present in outcrops, we are quite unsure about whether a sandstone is altered, and it is also true that finding an altered sandstone and an interface between such sandstone and an unaltered sandstone does *not* necessarily mean ore will be found. Indeed, we have drilled a number of interfaces which, by all the usual criteria should have had ore on them—but they did not, and we are still speculating why. As mentioned above, finding a large altered complex does not guarantee that ore will be located. We have drilled some large ones without finding ore and have found ore on some smaller systems. More study is needed in this area.

Figure 3-15. A dramatic example of solution banding from Wyoming's Powder River Basin. This sandstone is in the Wasatch Formation (Eocene) and contains uranium ore less than 200 m from this location. The overall sand color at this site is mostly red to pink, while the solution bands are maroon, purple, old rose, pale red, and rusty yellow-brown. All of the sandstone in this photograph is altered; the sinuosity of the solution bands indicates areas within the sand where reactions took place, probably as a result of porosity and permeability changes. The sharp curve in the lower right part of the photograph is shown in closer detail in figure 3-16.

"Solution banding" is common in altered complexes and serves as an excellent criterion for alteration (figs. 3-14, 3-15, 3-16). Solution banding generally consists of concentric and arcuate color bands or stains in the sandstone. Such banding is graphic proof that solutions capable of reacting with the iron minerals of the sandstones once percolated through the permeable beds. Solution banding is sometimes very weakly expressed, especially where the sandstone is white or only slightly stained with limonite. In other instances, it consists of bright bands in various tints of red, purple, orange, and rusty colors.

Oxides of iron are very resistant to dissolution in oxidized, neutral, or slightly alkaline waters. It is likely, therefore, that solution banding was developed by ground-water action in the subsurface where ferrous-iron compounds were dissolved, locally leached and concentrated, and finally oxidized and fixed. Solution banding is not direct evidence of alteration associated with uranium roll fronts; it merely records active ground-water movement and solution effects. Together with other evidence, however, solution banding serves as a very favorable indicator of alteration related to roll fronts.

When a sandstone outcrop is suspected of being altered, the contacts with shales or mudstones should be carefullly checked for radioactivity. Carbonaceous zones or

Figure 3-16. Close-up part of solution banding shown in figure 3-15. Sandstone is coarse to pebble, poorly sorted, and very permeable. Solution banding freely crosses the bedding. This is not a roll front, but it is a result of chemical reactions within the altered sandstone.

beds which are in contact with the "altered" sand are most likely to be mineralized and radioactive. We have often seen inexperienced geologists carefully examining thick, porous, and obviously altered sandstone outcrops with the scintillation counter. This is a waste of time. If mineralization is present, it will be at the margins of alteration (geochemical interfaces), such as contacts with overlying or underlying shales, or in relatively impermeable clayey or carbonaceous zones within the altered sandstone.

A contact between an altered sandstone and an overlying or underlying shale is not always radioactive. Typically, in fact, there is little or no radioactivity at the contact, and geologists who have limited experience with roll-front deposits often have difficulty with this. Even in the subsurface, urani-

um mineralization in significant amounts very often does not extend far back from the roll front on limbs above and below the alteration. In drill holes, a very sensitive gamma detector is required to pick up the weak radioactivity. Little wonder that not much is found under the severe conditions of weathering in the outcrop. A repeatable reading of 10 percent or more above background at a sand-shale contact is significant.

The manner in which a scintillation counter is used in the field is quite important, especially in the early stages of investigation. As we approach a prospective area, we turn the instrument on to get a background determination, and then we leave it on as we travel around the area both in and out of vehicles. We turn it off only when we are not traveling over new ground or checking outcrops.

To facilitate extended surveys, we have found that the type of instrument with an alarm which may be set to alert the operator at various readings is preferable over other types. Some instruments emit a continuous noise with variable intensities, others sound off when a full-scale reading is reached, and others have no audio alarm.

When a high-background broad anomaly is detected, and point sources are being sought, it is important to have the scintillometer close to the ground (within about 0.3 m). This procedure allows close definition of the point source of radioactivity. A carrying strap may be used to suspend the scintillometer near the ground (fig. 3-17). When walking cross-country, the instrument may be carried by means of the shoulder strap.

It is always important to establish the background count when surveying an area for radioactive anomalies. In the same area, the background will change from time to time due to variable atmospheric, cosmic, and ground conditions. These temporal variations commonly exceed 50 percent of the total background count, and such variation can take place during a rain storm or overnight when atmospheric changes occur. We record our background readings on a map and key them to field notes, in order to facilitate determination of background on repeated surveys or extensions of a survey.

It is frequently important to determine that an area has a high radioactive background relative to surrounding areas, even without specific point-source anomalies. If the area of high background coincides with favorable facies determined from geologic mapping, it might be very significant to the prospect.

Uranium mineralization in contact with a sandstone which appears to be altered is the very best evidence that roll-front de-

Figure 3-17. In seeking point sources of radioactivity, it is advisable for the geologist to have the scintillation counter near the ground. The carrying strap can be used conveniently for this purpose.

posits will be found by subsequent drilling. Frequently, when a sandstone is friable, the contacts are not well exposed, and the source of a gamma anomaly will not be known. When this is the case, and the relationships cannot be induced with confidence, we recommend that the explorationist dig into the most anomalous spot and locate the source (fig. 3-18).

When uranium mineralization is indicated by gamma anomalies in a sandstone or siltstone, such mineralization can be confirmed and details of its distribution determined by the potassium ferrocyanide test. We carry two one-litre plastic bottles with trigger-type spray attachments for this test: first we spray the apparently mineralized rock with dilute nitric acid (5 to 10 percent); then we spray the same area with

Figure 3-18. Explorationists use pick and shovel to expose an anomalous interval at the base of an altered sandstone. Such surface exposures may provide information which will allow the explorationist to formulate thoughts about the magnitude of exploration targets in the area.

potassium ferrocyanide solution (10 percent). The reaction is immediate: a red-brown uranyl ferrocyanide precipitate will show in detail where uranium occurs in the rock. If soluble iron is present, it might interfere by yielding a bright blue precipitate. The presence of copper will result in a brown ferrocyanide as uranium does, but copper is usually obvious in weathered rock because of its brightly colored oxides. Blue iron precipitate will be minimal if dilute acid is used sparingly and the potassium ferrocyanide solution is sprayed on immediately. Dilute hydrochloric and sulphuric acids may also be used, and in fact are preferable where nitric acid oxidizes too many interfering ions. This test is valuable in mining operations where background radioactivity may be so high that the geologist has difficulty determining where urani-

um mineralization exists in a certain face.

In addition to providing data concerning favorable hosts and mineralization, the geologic reconnaissance for roll-front prospects should lay the foundation for placing broad limits on the leasing play by mapping unfavorable facies, unconformity controls, and possible structural controls. If it is a bifacies formation, an effort should be made to map the surface occurrences of both oxidized and reduced facies. Frequently a favorable reduced facies with thick, friable, and permeable sandstones will be poorly exposed and impossible to confirm by surface work. The oxidized facies are typically more resistant to weathering and do not develop good soil profiles that support vegetation to the extent that friable sandstones do. The oxidized facies can be mapped, and the size of the reduced facies can be inferred by the process of elimination.

Unconformities are important controls on the distribution of monofacies roll fronts and associated alteration complexes, and the explorationist should attempt to recognize and map these unconformities during the reconnaissance field work. As discussed in chapter 2, the known occurrences of monofacies roll-front deposits appear to be exogenous. Uranium-bearing solutions moved downward through erosion surfaces or unconformities into permeable, reducing host rocks to form these deposits. It is further postulated that in many cases the paleodrainages which were developed on the old erosion surfaces played dual roles by scouring down into potential hosts and by controlling the movement of surface and ground water. Unconformities might also truncate the potential hosts and thus determine the limit of a prospect in one direction.

During the early reconnaissance, bedding attitudes, faulting, and stratigraphic changes such as pinch-outs or facies changes should be mapped, not necessarily in great detail, but at least such features should be noted. In connection with observations of bedding attitudes, we should note that it is never too early to begin making estimates of expected depths to mineralization. This is important to geochemical and geophysical work, as well as to any drilling program which may follow.

Trend Deposits

All known trend deposits occur in bifacies formations in which the oxidized facies are the more extensively developed. The reduced facies or even locally reduced sandstone masses are favorable for uranium deposits. Consequently, it is important to locate and map reduced rocks. The reduced sandstones are gray and frequently carbonaceous, but they may be yellowish-brown in weathered outcrops. The unfavorable sandstones are typically reddish due to hematite staining. Mudstones and shales in the reduced facies are typically greenish-gray or gray and they may be locally carbonaceous.

In the areas where trend deposits have been defined, the sandstones are generally moderately well cemented and well exposed (in sharp contrast to the soft sandstones of most roll-front environments). One of the common controls in the development of trend deposits appears to be sandstone thickness. Many Chinle Formation deposits occur along the depo-axes of fluvial complexes where the deepest channel scouring occurred. Mapping of the paleodrainages seems to be a good method of defining favorable trends for uranium deposits. This might also serve as a basis for projecting known mineralization into the subsurface.

Reduced sandstones should be carefully examined with a scintillation counter, because mineralization at the surface is the best known indicator of a favorable area for drilling.

The structure should be mapped so that depth projections can be made for drilling.

Structure-controlled Deposits

Conditions vary substantially in areas which seem to have potential for structure-controlled deposits, and the objectives of reconnaissance field work are, in large part, controlled by these terrain conditions. For example, the Canadian Shield areas, where substantial uranium deposits have been found, are largely glaciated, with numerous lakes, gravel and boulder drift, and extensive forest cover. Other areas, such as northern Australia, are subtropical forests; still others might be arid deserts.

In cold forested regions, some areas of good exposure are usually found where glacial drift was not deposited or has been removed by erosion. Weathering is superficial, and bedrock exposures reveal identifiable minerals and structures. It is usually possible to map lithologic units and interpret the structure where scattered exposures exist in temperate or cold climates. The geologist should attempt to map the geology and structure, looking for fracture or breccia zones which might control mineralization.

On-the-ground field geology in areas such as the Canadian Shield or northern Australia usually follows extensive airborne reconnaissance and some airborne geophysics including gamma-ray spectrometry, magnetics, and possibly some electrical methods. Aeroradiometric anomalies should be located and evaluated by a ground radiometric survey, and the geology of the significant areas where anomalies were located should be mapped. In the case of mineralized gravel float, the sources of the float should be sought.

In temperate climates or desert areas where ground access is not difficult and exposures are good, field work might reveal structural features which justify additional study, even though radiometric anomalies are not detected. In such studies, the geologist is likely to rely on a working hypothesis to guide much of his effort.

Geochemistry

Regional or reconnaissance geochemical surveys are used extensively in both Precambrian shield areas and sedimentary basins. The type of geochemical survey and method of carrying it out depends on the terrain conditions and the geology of the area. For example, in Canadian Shield areas, numerous lakes are accessible to pontoon-equipped aircraft; some streams are also accessible; but the forested land areas are relatively inaccessible. It follows that for this area, sampling surface waters and bottom sediment is easier than sampling soils or soil gas. Indeed, in most of the Canadian Shield areas, surface-water and bottom-sediment sampling are most commonly used in reconnaissance geochemical surveys. For similar reasons, this is probably true for tropical forest areas as well.

In sedimentary basins, ground water may often be sampled at springs and water wells, and it is frequently possible to determine the formation from which the ground water came. This can be very useful information in a reconnaissance program.

There are other reasons why surface- and ground-water sampling programs have been used extensively in early reconnaissance uranium exploration. Perhaps the most important of these is the solubility and mobility of uranium in surface water and shallow ground water. Uranium is likely to be widely dispersed by moving water which has come into contact with a soluble source of the element. This characteristic makes it possible to detect uranium anomalies with widely spaced water samples.

The sampling of soil gas is not usually considered in reconnaissance work, because helium and radon anomalies in soil are typically confined to areas not much larger than the sources. Radon does not move far from its immediate parent (due to its short half-life) unless it is dispersed in fast-moving water or air. Helium moves vertically through the soil to disperse into the atmosphere, but it often is concentrated in ground water within a confined aquifer to form widespread anomalies.

Our knowledge of the dispersal of uranium in soils is limited, because very few surveys of this type have been conducted and described in the literature. In most cases, we believe that soil-sampling surveys are not ideally suited for reconnaissance work, because the anomalies are probably not much larger than the bedrock source. Depending on the topography, erosional characteristics, and humate content of the soil, we would expect some variation in the size and intensity of uranium anomalies in soil. Pathfinder elements in soils might hold some potential for reconnaissance surveys, but more needs to be learned before such surveys can be recommended. Soil sampling and soil-gas sampling are more completely considered in chapter 6, where detailed surveys are discussed.

Water Samples

Uranium is quite soluble in much surface water and shallow ground water, and this solubility is very much enhanced by CO_3 or SO_4, which may also be in solution (Garrels and Christ, 1959, p. 86). Many aquifers in sedimentary basins of the western United States have active shallow ground water containing carbonate or sulphate ions capable of combining with uranium to form soluble complexes which remain in solution under wide-ranging conditions. These reactions may be generalized as follows:

$$UO_2^{++} + 3CO_3^{-2} \longrightarrow [UO_2(CO_3)_3]^{-4}$$

$$UO_2^{++} + 2CO_3^{-2} + 2H_2O$$
$$\longrightarrow [UO_2(CO_3)_2(H_2O)_2]^{-2}$$

$$UO_2^{++} + SO_4^{-2} \longrightarrow UO_2SO_4.$$

Regional ground-water and surface-water geochemical surveys we have been involved with, followed by drilling programs and subsurface geologic studies, indicate that uranium is commonly transported for tens of kilometres in carbonate or sulphate complexes. Dyck (1972, 1975) showed that uranium transport is substantial even on the Canadian Shield, where carbonate bedrock increases the carbonate ion in surface water. However, ubiquitous decaying organic material on the Canadian Shield generally reduces transport of

uranium in that water, relative to water in sedimentary basins of the western United States and other semiarid temperate areas.

The mobility of uranium in surface water and shallow ground water (in the absence of extensive decaying organic matter) makes it a potentially useful pathfinder element for regional hydrogeochemical surveys in many environments. However, it must be used with caution.

Consider for a moment the process by which roll-front uranium deposits are formed. Hexavalent uranyl ions, often complexed with carbonate ions, in oxidizing ground water move through aquifers until they are fixed by reduction-precipitation or by complexing with humic substances. Once precipitated, the ions will not remobilize unless geochemical conditions change. Under our working hypothesis, many of the roll fronts formed in Eocene and Oligocene time have been stable ever since; that is, they have not moved significantly in post-Oligocene time, and they are not moving now, because they are in an environment of reduced ground water. Can one expect to detect such deposits by sampling "downstream" in the same aquifer? No, probably not. Only if some of the uranium is remobilized and moving will an anomaly be detected. In areas where the aquifers contain humic substances, or are reduced due to H_2S or other strong reductants, the uranium content of the water is consistently low, that is, less than 5 parts per billion (ppb).

High uranium contents of ground water frequently correlate simply with oxidized (sometimes altered) sandstones; conversely, low uranium contents *may* indicate that the aquifer is unaltered and reduced. Proximity to uranium ore is more speculative. A water sample taken from an unaltered sandstone aquifer near a uranium ore body may contain only 2 ppb uranium, whereas a sample taken from the same aquifer far removed from the ore but in an altered or oxidized area might contain 200 ppb uranium.

Another problem which the explorationist faces in interpreting uranium anomalies in water is to determine the source of the element. Uranium concentrations commonly build up in ground water which probably has no contact with any ore bodies. Indeed, most of the anomalous uranium-bearing ground water in the sedimentary basins of the western United States probably acquired its uranium from disseminated sources of no economic value. We have conducted regional hydrogeochemical surveys in which uranium anomalies ranging up to 500 ppb were encountered in eastern Wyoming. (We should mention in this connection that some of the samples with very high uranium values were taken from water taps in the kitchens of ranches or farms which got their water from nearby wells. While our interest was focused primarily on the search for mineable uranium deposits, we could not help wondering about the long-range effects that drinking and otherwise consuming this water may have on those persons using it. In most instances, they drink it, cook with it, water their vegetable garden with it, water cattle with it [they may later eat the meat], and water their milk cow with it. We did not test the radon and radium content of these waters, and these isotopes may pose a greater hazard than the uranium itself. We were hesitant to drink any of the water, even from those wells which contained "only" 100 ppb. Because of our concern, we advised the State Health Department, even to the extent of sending a well-water anomaly map, but so far as we know, no action has been taken.)

Denson and Gill (1965, p. 46) report uranium contents up to 598 ppb in ground water of the Williston Basin. Samples collected from water wells scattered over large areas in eastern Wyoming and the southwestern part of the Williston Basin commonly contain 20 to 60 ppb uranium, which in many areas would be good to strong anomalies. However, regional geologic studies and drilling indicate that most of the uranium in solution is derived from disseminated sources of no economic value.

Some high uranium concentrations in ground water *are* derived from ore bodies. Very high concentrations occur in spring water coming from the Gas Hills area of Wyoming, for example, because near-surface oxidation is currently attacking some of the previously reduced uranium concentrations.

Similarly, high uranium anomalies have been reported in ground water associated closely with ore bodies in the southern Powder River Basin, Wyoming. Several samples from these localities have been found to contain more than 1,000 ppb uranium. Perhaps these unusually high values can be assumed to be indicative of an ore body nearby.

In general the geochemical survey utilizing uranium in water must be evaluated in conjunction with a knowledge of the geology. If wells in a basin are sampled, an attempt should be made to determine from what depth the well produces. Later, this information should be used to identify the aquifers and relate samples to specific aquifers. It may be possible to roughly outline altered and unaltered areas with a regional hydrogeochemical survey utilizing uranium analyses (fig. 3-19).

In our uranium-hydrogeochemical surveys, we collect two samples (each about 250 to 500 ml) at each site. If the samples are collected from wells which have not been flowing, it is best to pump the well for a few minutes to draw fresh water from the formation. Many domestic wells have small surge or storage tanks and it is best to

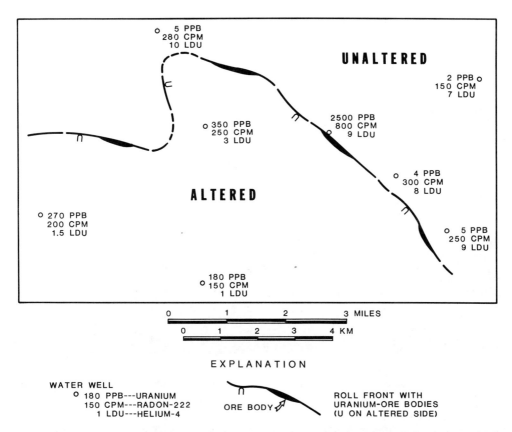

Figure 3-19. Idealized map showing ground-water analyses. Note that moderately high uranium anomalies occur in oxidized or altered areas. Uranium is consistently low in unoxidized (unaltered) areas, even near uranium mineralization. Helium values are high near uranium and carry for substantial distances in the direction of ground-water movement. Radon values are high near uranium occurrences and for moderate distances in the direction of ground-water movement.

get a fresh water sample which has not spent time in the tank. Clean glass sample containers should be rinsed in the water sampled. The sample should be marked with an identifying number which is also plotted on the map. Field notes should describe each location with appearance of the water, odor (for example, hydrogen sulphide), capacity of well, flow of spring or stream, depth of well, depth of screen, and probable formation from which the water came. The owner of the well may be able to answer many of these questions and in addition provide other information about the water, such as color changes, quality at

various depths, corrosiveness, and so forth. A further source of water-well information might be the state engineer's office or other depository of such records. Many states require that well drillers file records showing rock formations encountered in the course of drilling water wells.

Surface water should not be sampled during high water, spring thaw, or other unusual circumstances. We have found that acidifying the samples is not necessary if the samples are properly handled after collection. Indeed, we have observed that iron will often precipitate (and presumably some uranium with it) from an alkaline

Figure 3-20. G. M. Reimer, geochemist, agitating a water sample to remove gas from solution. This is a typical stock well with a windmill. Water flows uniformly from pipe in lower part of photo into stock-watering tank.

water sample if it is acidified slightly. In such event, the sample must be discarded or the iron taken back into solution through pH adjustments. Samples should be analyzed as soon as possible after collection; samples should not be stored in warm, light places, because this may result in bacterial or algal growth which may take uranium out of solution. If several months' storage is necessary, samples should be placed in boxes in a cool place. We usually analyze one sample and hold the other for additional analysis if an anomaly is indicated. Sulphate and carbonate may be included in the second analysis.

Some explorationists run Eh and pH determinations on water samples as soon as they are collected in the field. We have considered doing this and have thus far concluded that the information obtained

would not be of sufficient value to justify the time and expense involved. We have by no means ruled it out, however, because there may be situations where such determinations would be important in defining favorable areas.

Radon-222 has a half-life of 3.8 days and does not appear to migrate far from its parent (radium-226) in *subsurface* environments where fluid movements are slow. It is also a heavy gas (the heaviest known) which does not appear to diffuse rapidly in standing water, water-saturated rocks, or even air. According to Dyck (1975, p. 34), radon will move, at most, only 6 m in soils and even less in still water. Of course, in moving, agitated surface water, radon is lost to the atmosphere; in most cases, radon will escape from a stream within 100 m of its radium source. Radon is closely

Figure 3-21. After water sample has been agitated and allowed to sit for one minute (gas bubbles will rise to the surface) sample for helium analysis is drawn off. Radon may be drawn from sample into zinc-sulphide detector after helium has been sampled.

tied to radium-226, mainly because of the short half-life of radon-222, but radium-226 has a half life of 1,622 years. When sulphate and carbonate ions are present (as in most ground water), radium forms stable, insoluble compounds:

$$Ra^{++} + CO_3^{-2} \longrightarrow RaCO_3 \downarrow$$

$$Ra^{++} + SO_4^{-2} \longrightarrow RaSO_4 \downarrow$$

Although the carbonate and sulphate ions increase the mobility of uranium, these same ions combine with radium to firmly fix it and render it immobile in most environments.

Radon might be used as a qualitative supplement to uranium in a regional hydrogeochemical survey, to give a better focus on the sources of anomalies. Immeasurable amounts of radon are lost from water when sampling from wells, and similarly, immeasurable amounts are lost from natural springs before the water is sampled. Because of these losses, radon measurements on water samples are always qualitative.

Dyck (1972, p. 218) removed radon from the water sample by passing air through it, and then he measured the radon in the radon-air mixture. Dyck (1975) stated that silver-activated zinc sulphide cells are more accurate in radon analysis than is the scintillation counter.

G. M. Reimer (personal communication) has had good results in measuring helium contents of ground-water samples by simply agitating the sample in a sealed container containing four times as much water as air, and then drawing off a sample of the gas from the top of the container (figs. 3-20, 3-21). This, in effect, is the same procedure

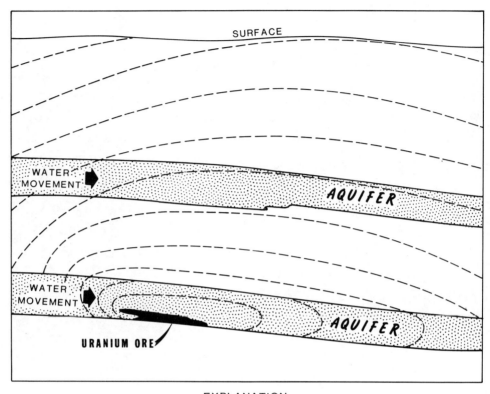

EXPLANATION

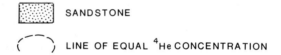   SANDSTONE

(— —) LINE OF EQUAL ^{4}He CONCENTRATION

Figure 3-22. Idealized cross section illustrating hypothetical migration of helium from a uranium deposit in sedimentary strata.

described by Dyck.

Helium-4 is produced as a by-product of uranium and thorium radioactive decay. Alpha particles, which are given off by the decay of several of the radioactive isotopes (fig. 3-4), pick up electrons from the environment and form stable ^{4}He. An excellent summary of the theory behind helium as a pathfinder element for uranium and techniques for its measurement was given by G. M. Reimer (1976a).

Some helium produced in the earth es-

capes to the atmosphere, but much is trapped in rock pores and crystal lattices within the earth. Impermeable rocks trap helium in its upward migration almost as effectively as natural gas and other fluids are trapped to form oil and gas fields. This process, of course, has led to accumulation of helium gas in structural and stratigraphic traps, some of which are commercially exploited.

Even though helium is very light and therefore diffuses rapidly, it tends to form measurable concentration gradients or haloes around uranium deposits beneath the ground surface (fig. 3-22). G. M. Reimer (personal communication) has measured very distinct high helium concentrations in samples of ground water taken from water wells producing from sandstones near known uranium deposits. He has also detected distinct helium anomalies in soil gas over known altered and mineralized sandstones which occur at depths of 60 to 80 m.

G. M. Reimer (1976b) and co-workers with the U.S. Geological Survey have designed and built a mobile helium detector with a sensitivity better than 50 parts of helium per 10^9 parts of gas. The detector utilizes a mass spectrometer, tuned for helium-4.

The new technology for helium sampling and analysis has not been available for extensive use by industry; indeed, the U.S. Geological Survey has completed only preliminary field tests at this time. We are encouraged by these preliminary tests, and consider helium to have excellent potential for regional hydrogeochemical surveys (ground water) and detailed soil-gas sampling surveys. We want to express caution here, however, that some recently reported field studies for helium anomalies near known uranium occurrences strongly suggest either faulty instrumentation or faulty sampling procedures in those studies. Well-designed instrumentation and extreme care in sampling are essential to a successful helium survey. Soil-gas sampling for helium and radon is discussed in chapter 6, because this is a method suitable for detailed study.

Bottom-Sediment Sampling

The sediments in lake and stream bottoms have been sampled and analyzed for uranium, radium, and other elements with variable results (Dyck, 1974; Wenrich-Verbeek, 1976). There are more uncertainties involved in sediment sampling than there are in water sampling. Uranium and several other metals tend to concentrate in decaying organic matter and, to a lesser extent, in adsorptive clays. Radium, on the other hand, is rendered insoluble by combining with CO_3 or SO_4, and therefore does not move freely in most environments.

In clear-water areas with ubiquitous bottom organic material, such as the Canadian Shield, uranium in sediments tends to be concentrated within 2 to 4 km from its bedrock sources (Dyck, 1974, p. 17). However, in warmer semiarid regions, uranium may be carried in solution (usually complexed with CO_3) for tens of kilometres, leaving small concentrations dispersed in adsorptive clays and organic material erratically distributed within the bottom sediments.

If the sediments are poorly sorted and show evidence of rapid transport during floods with partial sorting during "normal" periods, the distribution of uranium and other soluble metals is even more difficult to interpret. The extreme case is the intermittent stream with a high sand:clay ratio. A *total* uranium analysis might detect some uranium which is locked up in zircon, biotite, and other resistant minerals, but most of the readily soluble uranium will be leached out of these sediments.

In most areas, sediment sampling has too many uncertainties to be a good method for exploration. Sediment samples may be collected, however, in remote areas (such as parts of the Canadian Shield) as a supplement to water sampling. Sediment samples can be preserved for longer periods than water samples, and they might be useful in future studies.

PROSPECT EVALUATION

Review and Decision

After regional reconnaissance geology,

geophysics, and geochemical work have been completed, the results must be reviewed before deciding whether a land acquisition program should be launched. If the decision is made to move forward with the project, an area of acquisition is determined and, in some large prospects, priorities are established. This is Link 2 of the chain-of-events diagram discussed in chapter 1. In many cases, the priorities established at the end of Link 2 are modified substantially during or after the next exploration phase, which is described in chapter 6.

The review usually is based on maps and cross sections which integrate the data from the regional studies. This is the first important time of decision after the exploration activity has been put in motion and after substantial expenditures have been made.

Figure 3-23 is the first half of the exploration flow sheet (fig. 7-1 is the second half). The concept for these diagrams came from Bailly (1976). It should be noted here that a decision either to move forward or to reject the project might be founded on poor judgment or inadequate observation. Also, one exploration group might drop a project, and a second group with more (or less) data or better (or worse) insight might pick it up and carry it forward.

Maps and Cross Sections

As the regional reconnaissance work is carried out, various types of maps are developed. In most cases, several scales will be used, and it is often difficult to relate the information shown on one map to that shown on another. We have found that composite maps are sometimes useful, and, in other cases, transparent overlays are better.

When a composite map is prepared, showing several types of geochemical, geophysical, and geological data, it is usually illustrated in several colors to facilitate visual analysis. For example, the surface geology might be shown in pale colors, with brightly colored contours showing radiometric anomalies or geochemical anomalies superimposed. Mineralization is always very important and should be highlighted with bright colors (we often use fluorescent orange or red).

Geologic cross sections should be prepared from the data obtained during the reconnaissance field mapping and any other supportive information available. Through the implementation of a working hypothesis, an attempt should be made to illustrate the concept of mineralization which is expected at depth. If possible, this should be accomplished on the maps and geologic cross sections (fig. 3-24). This is where the geologist is obliged to assess all the available information and then put forth his best conceptual model.

Geophysical and geochemical anomalies may frequently be highlighted with profiles (fig. 3-24). As in the case of cross sections, these profiles are sometimes drafted alongside the map or may be prepared separately. These profiles are often most effective when inserted with the written report.

Written Report

When the reconnaissance work is completed in an area, and the maps and cross sections are prepared, a report should be written. The effort put into the report usually reflects the interest the investigator has in the prospect—that is, not much interest will result in a short report. Such written reports should include a description of the area which includes access; proximity to towns, services, and special characteristics of terrain. If mineralization or mining activities occur in or near the area, they should be described.

The general stratigraphic and tectonic

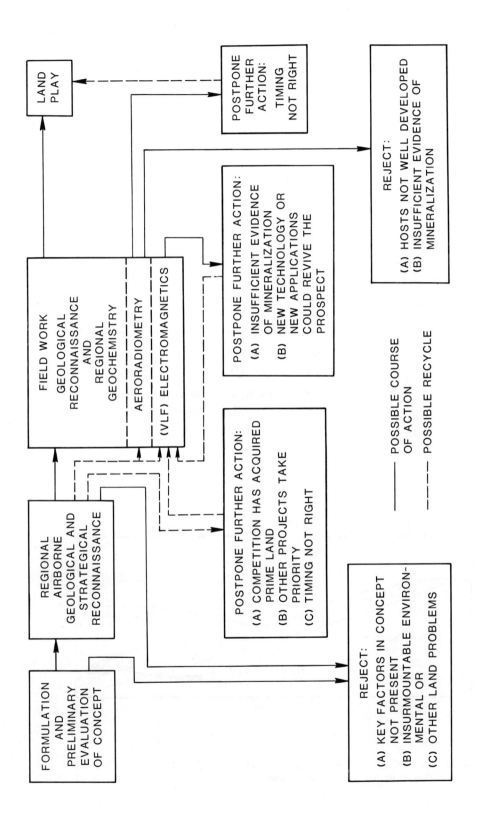

Figure 3-23. First half of exploration flow sheet (fig. 7-1 is second half).

134

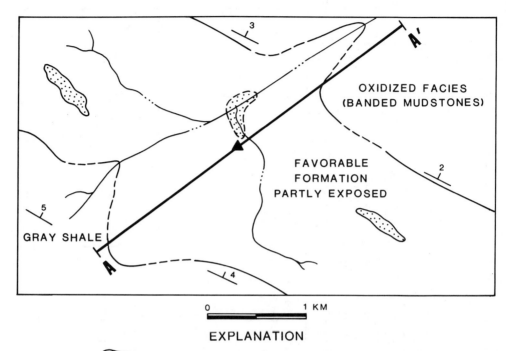

0   1 KM

EXPLANATION

SANDSTONE, COARSE GRAINED SOLUTION
BANDING, LIMONITE AND HEMATITE

▲   SURFACE RADIOMETRIC ANOMALY

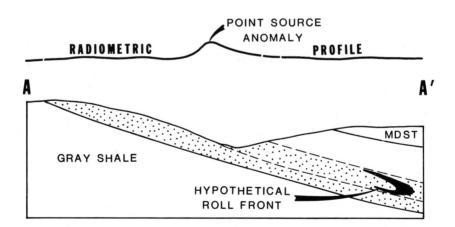

Figure 3-24. Idealized map and cross section showing surface geology and hypothetical roll front in sub-surface. Profile highlights surface radiometric anomaly.

framework should be described, and the geology of the area itself should be detailed. Special attention should be paid to mineralization, possible controls on mineralization, and the results of geochemical and geophysical surveys. Photographs are important, and they should definitely be included.

The report should appraise the odds for the prospect having deposits of economic value, and the appraisal should be substantiated by a general interpretation of all the accumulated data. The probable size of the potential resources should be estimated, and the basis for the estimate should be detailed. It is important to make interpretations and describe the results of prospect examinations even if the area appears to be of no further interest. The information could prevent future duplication of the work if the project remains unworthy. On the other side of the coin, if new economics, new technology, or new scientific knowledge (fig. 3-23) brings the area back as a favorable prospect, the written report and supportive maps might serve as a nucleus for the renewed work.

Whether or not plans call for continuation of the project, the report should be carefully prepared, with no details ignored. Unfortunately, our industry usually drops a rejected prospect like a hot potato, with the philosophy that good money should not be expended on a bad investment. The small investment required to systematically record and interpret the results of an exploration project should always be made. Indeed, it would be a major step forward for the industry as a whole, and for the nation, if exploration organizations would cooperate enough to publish reports on rejected prospects.

If the project is still considered viable at the end of the reconnaissance phase, it is important to prepare a plan for land acquisition and continued exploration and development. The report should include recommended targets for detailed geochemical, geophysical, and geological work. The specific types of studies and surveys planned, suggested, or proposed for detailed exploration should be outlined in the report.

A budget for the next phase of exploration should be prepared, and the time required for its completion should be estimated.

The subject of written prospect examination reports reminds us of two happenings concerning such reports. The first involves a substantial mining company which got into the uranium business in the early 1950s and developed several mines. The company always had a good-sized geological staff, and at an early date the exploration manager established a procedure for the geologists to prepare written reports on all prospects they examined. This procedure was followed to the extent that 15 years later a large set of files had been built up consisting of prospect examination reports, maps, photographs, assays, and so forth. The files represented tens of thousands of man-hours and hundreds of thousands of dollars in expenditures. Some of the data were irreplaceable; there were included examinations of many small operating mines which later were mined out or abandoned and would never again be accessible.

Then a tragedy occurred. A new chief geologist was appointed, and he quickly determined that the "old files" were worthless and were taking up a lot of valuable space. He ordered that they be hauled to the garbage dump and burned, and this was done. There is little doubt in our minds that those records, in the hands of competent explorationists, would now be worth tens of millions of dollars, because they would lead to additional, or new, uranium ore discoveries.

We must remember that there was a

very meager understanding of uranium geology back in the 1950s and early 1960s, but now good geological observations from a prospect or mine operated in the past can be coupled with new theories to lead to ore discovery. In addition, simple changes in economics, for instance, the change in the value of concentrate from $17.60 per kilogram ($8 per pound) then to $88 per kilogram ($40 per pound) in 1976, may cause a prospect to become a mineable deposit.

The second incident involves the records of ERDA in Grand Junction, Colorado. In the 1950s, the Atomic Energy Commission (AEC; now ERDA) had a short-form report policy, and required that its geologists and engineers prepare a report for every uranium prospect examined. The reports were grouped by states and by counties within those states. Until the middle 1960s, these short-form reports had been kept confidential by the AEC. When they were finally released for public inspection, the reports proved to be very interesting to read. Not only were uranium prospects examined then which may be of economic interest now, but the reports enabled a comparison of the state of knowledge then with what we now believe. Prospects where thick altered sandstones and even roll fronts themselves, crop out at the surface often were described simply as "uranium mineralization in drab sandstone." The geologists didn't see the alteration and its relationship to the mineralization because they weren't looking for it, and even if they saw sandstone of a different color, it had no significance to them.

Tuning In

Even today, as we examine prospects or carry out reconnaissance in various areas, we find it safe to say that if you aren't tuned in to the significant geologic phenomena in that area with respect to mineral deposits, then you aren't going to see many of the phenomena. There are so many geologic features to observe in most areas that it is impossible in a short period of time to assimilate them all into a meaningful picture. The explorationist must *focus* on the important factors, whatever they may be. In examining roll-front uranium deposits, the factors are (1) character of host rock (porosity, permeability, thickness, composition, lateral continuity); (2) evidence of alteration, including solution banding; and (3) evidence of mineralization. If a geologist is looking only at factor 3 and part of factor 1, he cannot make a full appraisal of the prospect.

When we begin working in a new area, we usually have much to learn about phenomena related to the mineralization. As we work with the prospect or deposit (provided we are good observers), we learn more about the significance of certain features of the geology. In many cases, it is possible to suddenly realize later the importance of faulting, stratigraphic changes, color differences, and so forth, even though we might have observed the features earlier without realizing their significance.

There are many classic examples of this in the oil business. Typically, one company will drill an oil test and will get a show of oil in a good sand but a "dry hole," and the prospect will be abandoned. Geologists for another company will examine the data gathered by the first company and conclude that a test should be drilled updip, and the test will be drilled and result in an oil discovery. Geologists for the second company recognized the significance of the information and the company acted upon their recommendation.

PHILOSOPHICAL COMMENTS

Perhaps the most important problem facing

the explorationist as he examines a prospect or area is the problem of properly assessing the reality of the situation. There is one very strong factor on his side if it is properly used, and that is the field evidence. Each aspect of field evidence—the rocks, structure, mineralization, and so forth—has an unbiased, truthful story to tell, but the explorationist must be observant enough to note and assemble the silent clues into a true picture of the geological history of the area, including the mineralization process. When you think about it, if geologists had all the clues (field evidence) assembled and properly interpreted, including time considerations, there would be no more dry holes drilled for oil tests, every exploration hole would be an ore hole, mining operators would know precisely how much ore remained in place, and metal and mineral supplies could be forecast with confidence.

The reality of the situation is, however, that geologists seldom if ever have all the clues. Except for a wide-open type of deposit such as salt in the Great Salt Lake, many facets of a prospect, or even a mine, are never seen by anyone, mostly because such clues may be related to the broad geological picture and only indirectly to the deposit in question. They may be deep-seated or for other reasons never are investigated. If we acknowledge that we will never have all the clues, then we will realize that we must work with those we do have to assemble a reasonably accurate picture.

Another point to remember is that a prospect will not become a good prospect just because someone, such as the prospector, wishes it to be so. Many a prospector has "broken his pick" on a worthless prospect because he refused to accept, or would not face, the reality that the prospect was of no value for economically recoverable minerals. Despite overwhelming evidence to the contrary, they have

visions of making a "big strike" and may decide that a particular location is the place where the ore will be found, and focus all of their attention on that one prospect. In other words, it is possible to sell oneself on one's own deal or prospect. We have known a number of explorationists, geologists included, who have become so convinced that their own prospect was a winner that good sense fell by the wayside. They began seeing field evidence the way they wished it to be rather than the way it was. Favorable evidence was quickly grasped, but unfavorable evidence was ignored. Their wish, which became their belief, was that an ore deposit or petroleum accumulation was present on the property; the reality was that it was just another prospect with very high odds against a commercially valuable discovery.

This may seem to be a pessimistic approach, and mineral exploration can indeed have its adverse times—but explorationists should have an attitude of *optimism balanced with realism*. Pessimists have no place in mineral exploration, but neither do blind optimists. There are exciting things to do in exploration, and there are exciting and valuable mineral discoveries to be made, but most prospects will not prove to be mineable deposits, and many hypotheses will be proven wrong. The opportunities are there, however, and deposits remain to be discovered by those who can gather and interpret the clues.

We have considered from time to time setting up a grading system for uranium prospects whereby the prospects would be given a grade ranging from A+ (the best) down to F (no potential). Had we stayed in consulting, we likely would have carried this through for the convenience of clients. Now, however, we do not spend any time or effort on prospects which are not at least an A, and so the need for the remaining grades no longer exists for us.

We encourage the explorationist to avoid spending time and effort on factors which are not related to the discovery of ore. Persons in the industry, including prospectors and engineers, can be observed doing all kinds of things around a property, but very often little of their attention is directed at what should be the primary objective of proving or disproving the existence of ore.

In following chapters, we discuss land and land problems in considerable detail, but a preliminary investigation stage is not too early to give some thought to possible problems relating to the land picture. Indeed, land problems may immediately loom so large as to discourage any geological work whatsoever. Not long ago, we developed an idea for a uranium prospect in part of a sedimentary basin where, before a field examination, we assumed that our only land problems would be either that the play had been made already or that landowners in the area might not be receptive to leasing. When we checked the area from the air, we found that the geology looked good, but a major part had been subdivided into small lots, and mobile homes dotted the landscape. Another area had been effectively withdrawn from exploration.

Walthier (1976) summarized many of the land problems facing explorationists, and we quote selectively from his excellent article:

Exploration is the initial stage in developing raw materials to meet future world demand. Government actions that shrink profits, or nullify existing agreements, or expropriate mines, curtail exploration (and *future* mines) as effectively as direct prohibitions on exploration. In the US, vast areas are banned to exploration and mining, while increased taxation, safety, and environ-

mental strictures add further disincentives. Abroad, high taxes, plus laws limiting mine ownership vie with expropriation in scaring off exploration funds. There simply are relatively fewer and fewer places where sizable private mining investments (and consequently exploration) can be considered prudent.

Affecting all land in the US, regardless of ownership, is the massive concern that has developed recently over "the environment." Whether this is good or bad is irrelevant for our purposes here. The effect, however, is to remove large areas of the country from places where we are *willing* to look for the geologic rarities we call ore bodies. I am not referring to specific prohibitions (which I'll discuss later), but simply to the antimining attitude that crowds us out of many areas.

I admit a reluctance to authorize exploration near parks, wilderness areas, scenic and roadless areas, wildlife sanctuaries, and the like; other company managers share my hesitancy. In general, if the area of proposed exploration can be easily seen from the withdrawn areas, or from the main access routes into them, I require an especially attractive geologic opportunity.

But the "indirect" problem is more complex. Large scenic and wild areas far removed from "withdrawn areas" simply engender such emotional reaction to preclude any industrial development, particularly by the extractive industries. It can be foolish to explore in these places.

Before an ore can be mined and processed, it must be found. The search is exceedingly difficult and expensive, and becoming more so all the time. Each discovery reduces by one the finite number remaining unfound, and in general the more easily found ones have been discovered. Private mining companies spend several hundred million dollars annually just looking for metallic ores around the world. Literally thousands of prospects must be examined, and regional surveys carried out, before a success is registered.

During exploration, especially at the early stages, very large areas must be searched. Clues to where ore may lie hidden are vague and elusive. Hence, access to large regions at low cost is essential for reconnaissance, geologic, geophysical and related studies. Now, to me, it is self-evident that the terms and conditions for developing and operating a discovery must be known before one undertakes all the expense and bother of trying to find an ore body. And,

further, one must have confidence that these conditions will be respected during the life of the operating mine.

In sponsoring mineral exploration today, we must have a locale that possesses a geology favorable for discovery, that has social and political institutions that permit proper technical operating procedures, that has governmental stability and integrity, plus a positive attitude toward private (foreign) resource investment. Moreover, the government and people must engender confidence that reasonable conditions will prevail for 20 to 30 years so that we will risk millions upon millions of dollars and years of our time.

Where are such places? Do they exist?

4

A Review of Mining Law as It Pertains to Mineral Development in the United States

INTRODUCTION

In determining the content and organization of this chapter, we concluded that simply reciting some mining law was not enough, and that the explorationist could benefit from an understanding of how mining law came into being in the first place. Consequently, we discuss briefly the origin of some legal concepts from which mining law has derived. We could have done so without including Latin terms, but lawyers like to use such terms and we have found that they are often convenient; therefore, we have included a few of them.

We should point out that this discussion is not intended to be comprehensive. Nor does it offer a detailed analysis of the law of a particular jurisdiction. The purpose is rather to provide an introduction to the concept of mineral ownership, and to highlight the major legal problems one is likely to encounter in the acquisition of mineral rights. We refer the reader to the publication by

the Rocky Mountain Mineral Law Foundation consisting of papers presented at its Uranium Exploration and Development Institute in Denver in November 1976 (see Bloomenthal, 1976).

In recent years, the mining industry, like most businesses, has faced an increasing array of state and federal regulations. All such regulations are, in a broad sense, "mining law." The scope of this chapter, however, is much narrower. The major concern here will be with that aspect of mining law that first concerns the entrepreneur—the acquisition and retention of mineral rights or ownership.

OWNERSHIP OF MINERALS

Whenever we say that a person "owns property," whether the property be an automobile, a piece of land, or minerals in place, we define both a relationship and an expectation. The relationship describes a quantum of "property" rights that the "owner" is entitled to assert against the

141

world at large. The expectation is that the state (that is, the government, whether federal, state, or local) will enforce these rights if they are violated by third persons.

The particular rights associated with property ownership can, and do, vary with time and place. At an earlier time, these rights were primarily established by custom. Today they are often said to be defined by law—that is, the power to define ownership is seen as residing in the government. For example, in the United States today, this power is described as resting with the state and federal governments, subject only to limitations imposed by the state and federal constitutions. Ultimately, however, the nature of rights to be associated with property ownership is a political question; at any given time, governments can be expected to add to or subtract from these rights to promote the prevailing political philosophy. Thus, a change from an elected government to a dictatorship may drastically alter mineral leasing or permit arrangements, including outright expropriation.

Subsurface minerals have been extracted from the earth for literally thousands of years. The various approaches that have been taken toward the problem of defining mineral ownership reflect the widely varying conditions in which mining has taken place.

So-called civil law countries are countries whose legal systems have been derived, in whole or in part, from the Roman law. It is perhaps more correct to refer to this system as "Romano-Germanic." Included in this system are most of the countries of continental Europe and South America. In the civil law countries, three general systems of mineral ownership have been recognized at one time or another. First (here comes a Latin term), the *res nullius* approach, rarely applied today, regarded minerals as being unowned until

they were either discovered or reduced to actual possession. Thus, mere ownership of land overlying the minerals did not necessarily guarantee ownership. Second, the *regalian* theory, still applied by Spanish law and in most Latin American countries, regards minerals as being, in one form or another, the property of the state. Thus, surface ownership does not vest mineral ownership. Third, under English common law, which was the direct predecessor of most American common law, mineral ownership was normally determined according to the *ad caelum* theory, by which mineral wealth, unless it had been separately conveyed, belonged to the owner of the surface estate. It was not, however, applied without exception. For instance, gold and silver mines were said to be the property of the Crown—and many districts recognized forms of the early Germanic custom of "free mining," by which miners could enter unappropriated waste land, even if actually "owned" by another, and remove minerals for their own benefit, subject only to payment of a small royalty and rent.

OWNERSHIP OF MINERALS IN THE UNITED STATES

As noted above, the United States absorbed most of its early common law from England—and, consistent with the *ad caelum* theory of ownership, minerals in the United States, once lands have passed into private ownership, are normally deemed to be included with the surface estate. Thus, if the owner is a private individual, he can acquire mineral rights by license, lease, or purchase. If the owner is a governmental unit, then it must comply with whatever rules that body has established for transfer of title or right to use. Subsequent portions of this chapter will describe these rules.

It must be emphasized, however, that this general rule of ownership is useful only as a starting point. Ownership in land (defined here to include subsurface mineral rights) can be fragmented in many ways; in particular, mineral rights can be separated from the surface estate. For instance, a private owner can sell the mineral rights under his land, yet retain the surface for farming or some other use. Conversely, the federal government often retained subsurface mineral rights when it disposed of public land. The explorationist or mining developer, when acquiring land, must therefore make a thorough search of records to insure that he has identified everyone with a legally protectable interest in that land. Such searches, and land acquisition as well, are described in detail in chapter 5.

FEDERAL MINING LAW

By far the greatest mineral wealth in the United States today is found in the western states. At one time, this area was occupied by Native Americans (Indians), who did not recognize land ownership as such; the surface was available for use by any who occupied it. Later, the federal government, as western lands were acquired by purchase, treaty, and otherwise, claimed ownership to virtually all of this land. It still owns a substantial portion. For that reason, it is important to be familiar with the federal general mining law, which governs the acquisition of mineral interests on the federal public domain (that is, federal lands acquired initially by purchase, cession, or treaty, or by other methods where the lands have specifically been declared by Congress to be public lands. Indian lands established as reservations are not covered by this law.)

Development of Federal Mining Law

When the United States was first formed, there were no public lands. The first such lands, those east of the Mississippi, were acquired by cession from certain of the original 13 states. Public lands west of the Mississippi were acquired later by purchase, conquest, and treaty.

For many years, little was done by the federal government to control mineral disposition on these lands. The land was viewed primarily as a source of revenue, the idea being to simply sell it. Certain lands were sold specifically as mineral properties, and for a short time, a mineral leasing program was instituted. In general, however, there was no coherent program of disposition.

The success of the various sales acts during this early period was very limited; only minimal revenues were produced. By 1830, therefore, it was decided that the emphasis should shift from a policy of raising money to one of encouraging western settlement. The result of this shift in policy was the enactment of the general preemption act of 1841. Under this act, public land continued to be sold, but at a fairly low price and only to actual settlers. Subsequent legislation continued this policy. Nonetheless, the scheme was not successful in encouraging western migration.

Finally, in 1862, the policy of requiring sales was ended with passage of the first homestead law. This law provided for outright transfer of "nonmining" land to settlers as long as they complied with certain conditions of residence, cultivation, and use.

During this extended period of inactivity with regard to mineral disposition, mineral prospectors did not simply bide their time, waiting for establishment of a federal minerals program. The public domain was a vast area, and effective federal control was virtually lacking. Miners therefore prospected wherever they saw fit, and, as a practical matter, conflicts between miners and other settlers were few. Technically, of course, they were trespassers, because the land belonged to the government. However, the miners generally benefitted from a policy of benign neglect.

This sort of arrangement was viable, of course, only as long as the public domain was sparsely populated and interest in particular mining areas was minimal. These conditions ended with the discovery of gold in California in 1847. By 1848, the California Gold Rush had begun. Perhaps as many as 10,000 people were in the region of the discovery by 1848. By 1849, there were probably 40,000.

Due to the absence of federal statutes governing acquisition of mining rights, the miners of the Gold Rush resorted to self-help as a method of maintaining order. They organized mining districts, which might include one or several mining camps. These districts then drafted by-laws to govern the conduct of mining within their territorial jurisdictions. Rules governing the maximum number and size of claims, methods of location, abandonment and forfeiture, use of water, and methods of conflict resolution could usually be found. Most districts also set up camp recorders to keep track of the boundaries of the various claims and their ownership.

It is, of course, impossible to trace the geneaology of mining district rules with any certainty. It can be assumed that there was significant influence from all the major legal systems—miners came to California from all over the world. It can also be assumed, however, that the rules of these systems, whenever adopted, were modified to suit the exigencies of these camps.

By 1866, there were more than 1,000 mining districts in the Gold Rush area. There was naturally considerable variation in rules among these districts. Nonetheless, at least three important principles seem to have emerged by this time: the foundation of "title" was discovery of a mineral, fol-

lowed by appropriation; priority in time created priority in right; and development of a claim was a condition to its retention.

Notwithstanding the formation of these mining districts, the miners in California were still technically trespassers. There persisted, nonetheless, the feeling that they were entitled to quiet possession of their claims and the minerals removed from them. As a result, the federal policy of noninterference continued. In addition, the miners were protected by California state law, at least to the extent that the state exerted effective control over public land within its boundaries. However, despite such benevolent policies, the uncertainty of title that resulted from the situation was clearly unsatisfactory. Perhaps the greatest problem was that, due to the absence of legal titles, miners found it extremely difficult to raise sufficient capital to carry on large-scale operations.

In reaction to this situation, the first federal mining law, that of 1866, was finally passed. Passage followed a long debate between western and eastern interests as to the method by which mining rights should be granted. The West favored legislation that would confirm "title" to existing locations, and that would promote free mining on the public domain. The East, on the other hand, generally favored sales or leases, in order to provide revenue for defraying the cost of the Civil War. In the end, the West prevailed. Although the bill was crudely drafted, and in many respects deficient, it confirmed existing claims, and clearly established a federal policy of open exploration and occupation of the public lands. The bill also contained provisions for locating lode claims and acquiring title from the government—although by implication, removal of minerals prior to patent was to be allowed. Customs of local mining districts were to be retained insofar as they did not conflict with the law. This law was supplemented in 1870 by the Placer Act, which reaffirmed the policy of open mining and which generally extended the provisions of the 1866 act to the location and patenting of placer claims.

Important as these acts were, they need not be considered in detail, for they were superseded by the General Mining Law of 1872. This law expanded and refined the earlier laws, although it reaffirmed the underlying policies—and with surprisingly few changes, it continues to this day to govern the means by which hard-rock mineral interests are acquired on the public domain.

A Summary of Federal Mining Law

Act of May 10, 1872

Be it enacted by the Senate and House of Representatives of the United States of America in Congress assembled, That all valuable mineral deposits in lands belonging to the United States, both surveyed and unsurveyed, are hereby declared to be free and open to exploration and purchase, and the lands in which they are found to occupation and purchase, by citizens of the United States and those who have declared their intention to become such, under regulations prescribed by law, and according to the local customs or rules of miners, in the several mining-districts, so far as the same are applicable and not inconsistent with the laws of the United States.

Sec. 2. That mining-claims upon veins or lodes of quartz or other rock in place bearing gold, silver, cinnabar, lead, tin, copper, or other valuable mineral deposits heretofore located, shall be governed as to length along the vein or lode by the customs, regulations, and laws in force at the date of their location.

A mining-claim located after the passage of this act, whether located by one or more persons, may equal, but shall not exceed, one thousand five hundred feet in length along the vein or lode; but no location of a mining-claim shall be made until the discovery of the vein or lode within the limits of the claim located. No claim shall extend more than three hundred feet on each side of the middle of the vein at the surface, nor shall any claim be limited by any mining regulation to less than twenty-five feet on each side of the middle of the vein at the surface, except where adverse rights existing at the passage of this act shall render such limitation necessary. The end-lines of each claim shall be parallel to each other.

Sec. 3. That the locators of all mining locations . . . which shall hereafter be made, on any mineral vein, lode, or ledge, situated in the public domain, . . . so long as they comply with the laws of the United States, and with State, territorial, and local regulations . . . shall have the exclusive right of possession and enjoyment of all the surface included within the line of their locations, and of all veins, lodes, and ledges throughout their entire depth, the top or apex of which lies inside of such surface-lines extended downward vertically, although such veins, lodes, or ledges may so far depart from a perpendicular in their course downward as to extend outside the vertical side-lines of said surface locations. . . .

Sec. 5. That the miners of each mining district may make rules and regulations not in conflict with the laws of the United States . . . governing the location, manner of recording, amount of work necessary to hold possession of a mining-claim, subject to the following requirements: The location must be distinctly marked on the ground so that its boundaries can be readily traced. All records of mining-claims hereafter made shall contain the name or names of the locators, the date of the location, and such a description of the claim or claims located by reference to some natural object or permanent monument as will identify the claim. On each claim located after the passage of this act, and until a patent shall have been issued therefor, not less than one hundred dollars' worth of labor shall be performed or improvements made during each year. . . . (U)pon a failure to comply with these conditions, the claim or mine upon which such failure occurred shall be open to relocation in the same manner as if no location of the same had ever been made: *Provided,* That the original locators . . . have not resumed work upon the claim after such failure and before such location. . . .

Sec. 6. That a patent for any land claimed and located for valuable deposits may be obtained. . . .

Sec. 10. That [the Placer Act of 1870] . . . shall remain in full force, except as to the proceedings to obtain a patent, which shall be similar to the proceedings prescribed by (this Act) . . . but where said placer-claims shall be upon surveyed lands, and conform to legal subdivisions, no further survey or plat shall be required, and all placer mining-claims hereafter located shall conform as near as practicable with the United States system of public land surveys . . . and no such location shall include more than twenty acres for each individual claimant.

Lands Subject to Mineral Entry

Possibly the most important task of a mining entrepreneur, prior to incurring substantial development expense, is to in-sure that the land he seeks to work is actually open to mineral entry. If it is not, then all else is for naught.

The 1872 mining law opened all "lands belonging to the United States" to mineral exploration and purchase. As noted earlier, these "public lands" are, in general, those initially acquired by the United States via purchase, cession, and treaty (although certain tracts, known as "acquired lands," are subject only to special land laws). At one time, these lands made up practically the entire country west of the Mississippi. However, due to the large number of land dispositions over the years, the pattern of mineral ownership is a complicated one. The public domain still remains a primary source of mineral rights. However, as a result of various additional exclusions and withdrawals, not all such land is now open to mineral entry. As is to be expected, a small amount of this land is subject to presently valid mining claims (even though, since they are unpatented, paramount title remains with the government). In addition, large areas are withdrawn from entry because they have been set aside for uses that are incompatible with mining operations. For example, such withdrawals are often associated with national parks or monuments, reclamation districts, water power sites, and a number of other uses.

In many cases, lands have been disposed of by the government for nonmineral uses, but the mineral estate has been retained by the government (the Powder River basin in Wyoming contains many such lands). These lands are open to mineral entry, although adequate steps must be taken to protect the interests of the owner of the surface estate, and this as a practical matter often requires an agreement with the surface owner.

Due to the great variety of dispositions and uses affecting public lands, it has become exceedingly difficult to determine whether any particular parcel is open to mineral entry. There are few valid rules of thumb. After deciding upon an area where one would like to locate claims, and obtaining a workable description of the land, the best course is to check the public land records. As described in chapter 5, the needed information can be obtained from the state office of the Bureau of Land Management (which is part of the U.S. Department of the Interior) and from the clerk and recorder's office of the county in which the land is located. Records at the BLM should indicate whether the land is still owned by the federal government and whether it is subject to mineral entry. If so, it is then necessary to check the county records (in most states) to see if it is covered by an existing claim. The BLM records will also indicate whether, notwithstanding issuance of a surface patent, mineral rights were reserved by the government. Mineral title plats, available from the appropriate office of the Bureau of Land Management, show the general status of lands as being public domain, patented with a reservation of minerals to the United States, and patented without reservation of minerals. Unless one is experienced in such record searching, it is best to farm the job out to a competent landman, abstractor, or attorney.

It will occasionally happen that conflicts will arise between mineral and non-mineral users of the public domain. For the mineral entryman to maintain his claim in such a case, he must establish that the land is "mineral" in character—which may be difficult if he has not yet made a full-fledged discovery. To establish that the land is "mineral," he must show that the known conditions are such as reasonably to engender the belief that the lands contain minerals of such quality and in such quantity as will render extraction profitable and justify expenditures to that end. If the land

is located within a national forest, the Forest Service may challenge the validity of the claim, and in such a case the standard for establishing a discovery of valuable minerals is probably higher than with respect to public domain lands generally.

Since 1934, the U.S. Geological Survey has engaged in a program of classifying unappropriated portions of the public domain.

Minerals Subject to Location

The 1872 act provided for the location of mining claims for "gold, silver, cinnabar, lead, tin, copper, or other valuable deposits." The term "other valuable deposits" has been construed to refer to both metalliferous and nonmetalliferous deposits, although the outcomes of the legal tests have been extremely vague; the decisions used such language as "whatever is recognized as mineral by the standard authorities"; "whatever is classified as such in trade or commerce"; and "such substances as possess economic value for use in trade, manufacture, the sciences, or mechanical or ornamental arts."

The major group of minerals that are excluded from location under the 1872 act are those now covered by the Mineral Leasing Act of 1920. These include oil, oil shale, coal, phosphate, sodium, potassium, natural gas, potash, and sulphur (in Louisiana and New Mexico). Rights to these minerals can now be obtained only by complying with the requirements of the 1920 act. However, some of these minerals were locatable before the passage of that law, and location rights existing as of that date were expressly saved.

The other major group of minerals excluded from location under the 1872 act are the so-called common varieties, that is, sand, stone, gravel, pumice, pumicite, cinders, or petrified wood. Clay is also viewed as a common variety. This exclusion was applied administratively by the Department of Interior almost from the very passage of the act. It was made explicit by the Surface Resources Act of July 23, 1955. A material will not be viewed as a common variety, even if it falls within one of the generic classes listed above, if it possesses "distinct and separate properties" which make it valuable for use in manufacturing, industrial, or processing operations. These properties, however, must make the material valuable for entirely different classes of *use* than the true common variety—that is, it is not enough to show merely that the claim contains a superior grade of material. It must also be shown that the material will actually be marketed for the nontypical use. In addition, the person who claims a "special and distinct" class of common variety must also make a more thorough and convincing showing that the claim is "valuable"; that is, that the material is truly "marketable." This problem is discussed later in this chapter.

Types of Claims: Lodes and Placers

Mining locations under the 1872 act must be made either as lodes or placers. Claims upon "veins or lodes of quartz or other rock in place" must be located as lodes. Literally all others must be located as placers, although as noted below, "lode" has been interpreted to include many mineral occurrences which would not be regarded as such by a geologist. Choice of the wrong method of location can invalidate a claim; unfortunately, these terms are not clearly defined in the act. In borderline cases, it can be difficult to know what method of location will be required, and in such cases, the explorationist might be well advised to stake the ground with *both* lode and placer claims. Such dual staking has been done in recent years for zeolites because the law is vague in regard to such minerals.

Early judicial decisions guided the course of subsequent interpretation by giving a practical interpretation to these terms, in which the interests of miners overrode chemical and geological reasoning. Under these decisions, lodes were viewed as "formations by which the miner could be led or guided," and "continuous bod[ies] of mineralized rock lying within well-defined boundaries, apparently from the same geological source." These decisions have generally been followed to the present, although not with complete uniformity. It is best, however, to follow historical practices. Uranium in sandstone, for instance, is now almost always located as a lode claim, although in the past, arguments were made that placer location might be more appropriate.

According to the law, lode claims must be located along the axis of the vein or fissure of mineralization. However, we must remember that the law was written at a time when prospectors were able to make a discovery by exposing mineralized rock at or near the surface. Today's techniques of drilling for ore bodies which lie 900 m below the surface were not imagined.

The locator has a right to 1,500 ft along the vein, and 300 ft along each side (although this can be reduced to as little as 25 ft by local regulation). To protect the possibility of acquiring extra-lateral rights, the locator should mark the claim along the apex of the lode (if one is identifiable), and insure that the endlines of the claim are parallel with one another. This is known as the apex concept, and while it is an extremely rare situation in which this concept comes into play, the explorationist should be aware that such a legal concept exists. If you are wondering what the apex is, the following is a definition as handed down by a court: "If the vein is viewed as a plane of mineralization running at an angle to the plane of the surface, then the apex can be viewed as the line defining the intersection of these two planes; or, if there is no intersection, then the line of mineralization closest to the surface." Now you are probably *still* wondering what the apex is.

Placer claims are limited to 20 acres in size per locator; group claims of as much as 160 acres (8 locators) are also allowed. As opposed to lodes, whose boundaries must follow the strike of the vein, the boundaries of a placer claim must follow as nearly as practicable the lines of the public land survey. In addition, to prevent excessive cutting up of the public lands, the placer claim must "fit" within a contiguous 40-acre tract; that is, it must be compact. It should be noted that there is no limitation on the number of either lode or placer claims which can be staked by any one company or individual, provided all legal requirements are met.

Valuable Mineral Discovery

A prerequisite to valid location under the federal mining laws is discovery of a valuable mineral deposit. One of the greatest problems under the 1872 act has been to define adequately what constitutes a "valuable mineral discovery."

The classic test for discovery has been what is known as the "prudent man" test:

> Where minerals have been found and the evidence is of such a character that a person of ordinary prudence would be justified in the further expenditure of his labor and means, with a reasonable prospect of success, in developing a valuable mine, the requirements of the statute have been met.

As a matter of proof, however, in the context of an application for patent or a contest with the government, the problem is at least threefold. First, the claimant must demonstrate that a locatable mineral has been found on the claim. Second, he must establish that a market for this mineral presently exists. Third, he must demonstrate that the mineral is present on the claim in sufficient quantities to make operation of a mine profitable. (Note: this section discusses discovery only in terms of proof to establish a valid location.)

As to the first problem—that of proving that mineralization exists—it is clear that there must be some physical exposure of the mineral. This requirement can be inconvenient in light of modern exploration and development techniques, which rely heavily on geological and geophysical inference—especially in view of the requirement that independent proof of mineralization be presented for each individual claim in a group. However, core-sample "exposures" are sufficient to meet this requirement. In the case of uranium, readings from a Geiger counter or scintillometer are not adequate for discovery purposes unless it can be conclusively shown that the radiometric readings reflect the presence of uranium and that disequilibrium (discussed more fully in chapter 11) is not a significant factor. Obviously, the possibility of doing this is significantly greater for uranium occurrences which are well cemented, below the water table, or otherwise protected from conditions conducive to the development of a disequilibrium condition. In such an event, the authors believe that drill holes probed with a calibrated instrument can establish the presence of uranium if the proof is properly presented. However, it would ordinarily be desirable to corroborate such information with adequate core data, especially chemical analyses, establishing the disequilibrium factor. For these reasons, as well as good exploration technique, the explorationist is cautioned to adequately and accurately identify all samples and to maintain good records of assay results.

The second problem—proof of present marketability—is often the most difficult to deal with. First, it should be emphasized that this requirement extends to *all* mineral discoveries. However, the existence of a market is generally assumed for the so-called intrinsically valuable minerals, such as gold and silver. The real problems of proof extend to two other classes: minerals with cyclical market, including those resulting from government purchase programs; and the more common minerals of widespread occurrence. Uranium is a good example of the former class. For many years, even though it was clear that uranium had great potential value, the federal government was the only customer and, when the government wasn't buying, no present market could be proved. The latter problem is usually encountered with allegedly distinct and special classes of common varieties. When a claim is made to these minerals, affirmative proof of a present market must be made. Factors to consider in this regard are the size of the demand, the present supply from other sources, and the proximity of the claim to both existing users and existing suppliers. Under current interpretation, however, it is not necessary to prove an actual commitment to buy the minerals in question (that is, from the claim). It is enough to rely on general, but local, market conditions.

The actual burden of proof that must be met to establish present marketability can vary, depending on the status of the land in question. For example, a stronger-than-normal showing of marketability is required if the claim is located on land administered by the U.S. Forest Service.

The third problem—proof that the mineral exists in sufficient quantities to make the claim profitable—tends in practice to merge with the first and second. However, the problem here is distinct—to demonstrate, in light of proven market conditions, that the deposits are sufficiently extensive and of a sufficiently high grade. In this regard, it is generally permissible to rely fairly heavily on geological and geophysical inference. However, it is still wise to confirm these inferences by assays of core samples.

Proof of profitability must meet objective criteria; that is, a locator cannot lay claim to an otherwise inadequate discovery by asserting that he is willing to operate at a lower rate of return than is reasonable under accepted commercial standards.

Location Procedures

The 1872 act requires only that the locator of a lode or placer claim mark the boundaries of the claim "distinctly" on the ground so that they can be readily traced. It also requires that all records of claims set forth the names of the locators, the date of location, and a description of the claim by reference to natural objects or permanent monuments—although the act, by its own terms, does not require that such records be kept. In addition, however, the act allows "local mining districts" to prescribe more detailed procedures for location. This task has everywhere been taken over by the states, and it is to state regulations that the locator must turn in order to learn the exact requirements.

In general, these state laws require a locator to post a location notice, perform certain discovery work, mark the boundaries of the claim, and file a certificate of location. However, the requirements vary considerably from state to state, so local laws *must* be consulted. Only a general discussion can be undertaken here.

Most states require the mining locator to post a so-called location notice on the ground at or near the point of discovery. Such requirements, which derive directly from the early mining district regulations, visualize the prospector as making a discovery (of an outcrop or vein) and then claiming this discovery by posting notice. In connection with lode claims there are two general types of notice required, depending upon the state: first, a notice which merely sets forth the name of the claim, the names of the locators, and the date of discovery; and second, the more detailed form which requires information of the kind normally found in location certificates—such as the general course of the vein, length and width of the claim, and a description. Some states require both, in a two-stage posting procedure. A similar division is found with respect to placer location notices: some states require the notice to contain only the name of the claim, the names of the locators, the date of the location, and the number of acres claimed; others require a more formal description of the claim. The purpose of the location notice is to put other potential claimants on notice of the discovery. For this reason, substantial and good-faith compliance is normally deemed sufficient to satisfy the requirement—particularly as against rival locators with actual knowledge of the prior location. However, to back-date a notice (post a date which is earlier than the day the notice is placed) constitutes fraud.

Most states also require some sort of discovery work to be performed, at least as to lode claims, within a specified time after the date of discovery. The details vary widely but generally consist of a requirement that drilling or some sort of shaft, open cut, or tunnel be cut on or near the point of discovery. In addition, many states require that this work further expose the valuable mineral deposit. The purpose of this requirement seems primarily to require the locator to demonstrate his good faith; however, in many instances such discovery work (including drilling) does not expose a valuable mineral, and the token performance of the work raises serious questions

as to the validity of the claim. In light of modern techniques, the discovery-work requirement is truly anachronistic. In addition, it has in some areas led to an extensive and unnecessary destruction of surface resources. Following the uranium rush of the 1950s, for example, statutes in Wyoming were changed to permit drilling validation holes rather than requiring a pit to be dug, many of which had scarred the landscape (fig. 4-1).

The federal statute requires the boundaries of the claims to be distinctly marked on the ground. The purpose of this requirement is to "fix" the claim—to prevent the locator from "floating" or "swinging" the claim around the original point of discovery. The boundaries also serve as notice of the claim to other prospectors. In furtherance of these functions, some states require that the boundaries be marked within a relatively short time after discovery. Others set no specific time limits; but in any event, the boundaries must be fixed by the time the location certificate is filed. Many states also specify the type, number, and location of boundary monuments. Once these monuments are set, there is no general requirement that they be maintained. However, extant monuments will prevail over erroneous descriptions of the claim. Thus, maintenance of the monuments will protect the locator from inadvertent errors in his description.

As noted earlier, the 1872 act does not require that any filing be made with a federal agency; however, virtually every state requires that location certificates be filed within a specified time after the date of discovery. The Bureau of Land Management Organic Act, if adopted, will require that a filing be made with the BLM and will prescribe a period of time within which filings must be made for claims located prior to the adoption of the act.

This completes the traditional mining

Figure 4-1. Federal and state mining laws typically require an excavation of some type in which valuable minerals are exposed before a mining claim can be located on public domain. In attempting to comply with the law, many explorationists dug pits either with a bulldozer or backhoe. This pit was dug in about 1960 and is still open for inspection in 1976. The fencing is to prevent livestock from falling into the pit.

district scenario: discovery, posting of notice, digging the discovery shaft, fixing and marking of boundaries, and filing. The location certificates must include, in addition to the information in the location notice, a description of the claim. This requirement is often a pitfall for the unwary. As noted above, the federal law requires that the description be tied to some natural or permanent monument. This does not—despite the language of 30 U.S.C. (United States Code) Section 34—require that the description be tied in with the public land surveys. However, it must be sufficiently detailed to identify the claim adequately. Early cases were often very lenient in allowing marginal descriptions. However, it is unlikely that a modern court would be so tolerant. Current practice in the industry is to tie the claim to a public land survey, even if doing so requires surveying a tie line for several miles.

The purpose of the filing requirement is to give constructive notice of the claim. Because of this, in some states there exists a curious tension between the constructive notice of the location certificate recorded in the county records and actual notice resulting from location notices and boundary markers placed on the ground. The filing "requirement" in some states may be deemed to be merely permissive as to junior (secondary) locators if these locators have actual notice of the claim. As stated earlier, the Bureau of Land Management Organic Act, if adopted by Congress, will require a filing with the BLM.

Rights of the Locator

Upon completion of a valid location, the mining claimant obtains the rights to all locatable minerals on and under the surface of the claim, any extralateral rights that might exist (if all the requirements have been complied with), and the right to use surface resources to the extent necessary to carry on mining operations except as limited by the Surface Resources Act. This act prohibits the removal of "common varieties," noted above, which can be obtained only by purchase pursuant to the Materials Disposal Act. The act also limits the use of timber resources. Prior to patent, the locator cannot use the surface for purposes not related to mining operations. However, the owner of an unpatented claim has the right to remove and sell locatable minerals.

In theory, discovery should precede location—and, as noted, the 1872 act treats discovery as the initial act of location. In practice, however, it is common for mining locations to be made in advance of discovery (note that the recitals of discovery in location notices and certificates are totally self-serving and need not be accompanied by any proof). The courts, recognizing this fact, have held that this reversal of the statutory order is unobjectionable as long as the rights of third parties have not intervened. If a discovery is subsequently made, the location becomes effective from the date of discovery.

In addition to the above, the courts have developed a doctrine known as (more Latin) *pedis possessio*, which protects the premature locator in his physical possession of the claim prior to discovery. Under this doctrine, an explorer in actual occupation (who is diligently searching for mineralization) is treated much like a licensee or tenant at will, and no rights can be acquired against him through forcible, fraudulent, or clandestine intrusion upon his possession. He acquires, of course, no rights against the government by such possession unless a valid discovery is subsequently made.

Once a valid location is made, rights in the claim are essentially fixed. The original locator can amend, or relocate, to correct errors in his original location, but only if the rights of third parties have not intervened. Third parties can relocate only if the claim has been abandoned.

Assessment Work

After location, but prior to patent, $100 of assessment work must be performed annually on every mining claim to protect it from relocation by third parties. The assessment year now runs from September 1 through August 31 of the following year. Where contiguous claims are held in common, the work may be done on a single claim, as long as it tends to develop the block of claims as a whole. Failure to perform this work leaves the claim subject to relocation. Usually, this means that a failure to perform assessment work operates only in favor of subsequent locators, and cannot affect a claimant's rights against the government. However, where claims are located on land that is subsequently withdrawn from mineral entry, the government may assert the right to reclaim unworked land. This happened, for example, with many oil-shale claims following passage of the Mineral Leasing Act of 1920.

The 1872 mining act does not define what type of labor can be applied toward this annual requirement. A large number of judicial decisions on that question, however, have indicated that almost any expenditure can be applied as long as it tends to develop and benefit the claim. Allowable activities include excavation, construction, purchase of equipment, hiring of personnel, and many more. If the activity was performed on the claim, then proof of compliance is complete. If the activity was conducted *off* the claim, then the presumption is that the work did not benefit the claim. However, this can be rebutted by evidence to the contrary.

A major pitfall, prior to 1958, was the rule that geological, geophysical, and geochemical work could not count as assessment labor. In that year, however, a federal statute became effective which allows geological, geochemical, and geophysical surveys to be included as assessment labor as long as they are conducted by qualified experts, verified by the filing of a detailed report, and are not applied toward the assessment work requirement for more than two consecutive years, or for five years total.

Legal fees, travel expenses, and expenses relating to mills cannot be applied toward the requirement.

Some states require that annual affidavits be filed, describing the nature and value of assessment work performed during the preceding year, but other states do not. Details vary considerably and local requirements must be met.

Under certain circumstances, the performance of assessment work may be deferred or waived. Waiver has occurred, for instance, during war years, and persons in military service can generally be relieved of the requirement by complying with appropriate procedures. Performance may also be deferred if, for good reason, the locator is unable to enter the claim. Such deferrals are, however, relatively rare.

If a co-owner fails to contribute his proportionate share of the cost of annual assessment work, his partners may, after demand and appropriate publication of notice, take over his interest.

A locator who has become delinquent in his assessment work may redeem himself, if no third-party rights have intervened, by

resuming assessment work. In the event of such resumption, it is necessary only to do the work for the current year; there is no necessity of making up the work for prior years. However, the assessment work must be resumed before rights of third parties have intervened, and the work must be resumed in good faith. Failure to do such work for several years may be evidence of abandonment. If a claim has been abandoned, it cannot be revived by resumption of assessment work; a relocation would be necessary in that event.

Taking the Claim to Patent

As long as a locator maintains his claim, there is no requirement that he ever take his claim to patent (to patent, of course, means to acquire the legal fee simple title to public land through a conveyance from the United States). However, if development of his operation will require the investment of substantial capital, it is likely that investors will require that their money be protected by the legal title conveyed by patent. Further, a patent will protect the claim against a subsequent withdrawal of the lands by the United States. The patent conveys title to all minerals on the claim except leasing-act minerals "known" to exist, or covered by permit or lease. In the case of placer deposits, the patent will not include lodes that were "known" at the time of application and not specifically applied for. In addition, the patent carries title to all surface resources (as long as they have not previously been disposed of pursuant to a nonmineral entry), and relieves the locator of performing annual assessment work.

Application for a patent is a long and complex process, and the claimant will usually require the assistance of a competent attorney. The greatest danger that an applicant faces in this process is a finding that there has been no adequate discovery. If this happens, of course, the claimant loses his right to remain on the land by virtue of a mining claim.

Environmental Constraints

Recognition of the rights of mineral locators and the issuance of patents (if all requirements have been met) are nondiscretionary with the government. For this reason, the provisions of the National Environmental Policy Act do not apply, and no environmental impact statement need be prepared in conjunction with these actions. However, other environmental constraints do apply. No discharges into "navigable waters" (as broadly defined) may be made without obtaining an NPDES (National Pollution Discharge Elimination System) permit pursuant to the requirements of the Federal Water Pollution Control Act Amendments of 1972. No underground "injections" may be made in violation of the Safe Drinking Water Act. Airborne emissions must comply with constraints imposed by the Clean Air Act.

It is highly unlikely that the states actually have the power, absent the consent of the federal government, to impose their own environmental laws on users of public lands. To date, however, the government has consented to the exercise of jurisdiction as a matter of mutual courtesy. For this reason, the mining operator must also comply with parallel state regulations on air and water emissions. In addition, many states now have detailed siting and reclamation acts that must be followed. These acts often require the filing and approval of mining and reclamation plans prior to the undertaking of surface-disturbing work, along with the posting of substantial performance bonds.

The level of federal and state environmental control has now reached the point that mining operators can no longer consider such regulations only as an after-

thought. Unless the requirements of these statutes are kept in mind from the beginning of the planning process, the explorationist is very likely to run into substantial delays in the acquisition of all necessary permits and approvals. Experience has shown, however, that even with good planning by the explorationist, substantial delays are likely to result in *any* program where prior approval by federal, state, or local bureaucrats is required. More will be said about this in subsequent chapters.

ACQUIRING MINERAL RIGHTS ON STATE LANDS

As each western state was admitted to the Union, selected tracts of land were granted to the state. Revenues from this land were generally earmarked for support of schools.

Where state lands are undeveloped, it is often possible to obtain exploration and mining rights (described further in chapter 5). The requirements for obtaining these rights vary substantially, because they are managed by state agencies. However, many states conduct a leasing program patterned after the federal Mineral Leasing Act of 1920. Under these programs, mineral rights are acquired for a certain number of years. This term may usually be extended, but title to the land never vests in the locator—and when the lease expires, all rights revert to the state.

Several of the western states have published summaries of mining law in general and mining law in their state in particular. One of the best we have seen is *Laws and Regulations Governing Mineral Rights in New Mexico,* by Victor H. Verity and Robert J. Young, published as Bulletin 104 in 1973 by the New Mexico Bureau of Mines and Mineral Resources, Socorro, NM 87801.

Very often, operating and reclamation requirements are incorporated into state leases. A reclamation bond might also have to be posted. Annual rentals and royalties are usually payable to the state.

Unlike federal lands, state lands are not generally open to free prospecting. In states which conduct a leasing program, prospecting rights are generally obtained by permit.

ACQUIRING MINERAL RIGHTS ON PRIVATE LANDS

As indicated earlier, mining rights on private land can be acquired by license, lease, or purchase. The terms of such acquisition are matters for negotiation between the parties. Structurally, they are often patterned after standardized oil and gas leases—that is, they will provide for an initial term, renewal rights, delay rentals, a royalty, and covenants to assure reasonable development. To insure the "legality" of the transaction, it is necessary only to comply with the real property–conveyancing laws of the state in which the land is located. Lease acquisition is discussed in detail in chapter 5.

The explorationist should bear in mind that when acquiring private mineral rights, it is most important to consider the problem of surface damage. To avoid future problems, it is preferable to contractually define the rights and liabilities of all affected parties *before* any surface-disturbing activity is undertaken.

Lastly, it should be remembered that mining operations, even when carried out on private land, are still subject to most state and federal laws governing siting, operations, and air and water emission standards.

5
Land Ownership and Leasing

INTRODUCTION

There is one basic, cardinal rule in mineral exploration and development: if you do not have a valid mineral lease or claim to the land where exploration is planned and where a mineral deposit may be located, then you cannot control the mineral deposit. Consequently, it is crucial that the land and mineral ownership be carefully examined before any exploration program is embarked upon. In fact, it is advisable to consider the mineral ownership of an area before an idea gets past the initial concept stage, so that no time is wasted planning an exploration program in an area where mineral exploration and exploitation would be prohibited or stifled. Large-scale mineral exploration by foreign persons or companies is prohibited, or severely restricted, in most countries, as summarized by Walthier (1976); this was discussed in chapter 1.

LAND SITUATION IN THE UNITED STATES

Let's examine the situation in the United States. The political stability of this country, combined with substantial reserves and good additional potential, make it most attractive to the explorationist. Investment in exploration and development in the United States has been welcomed in the past, and this general policy does not seem likely to change in the foreseeable future, even though far too much land once classified as public domain has been withdrawn from exploration. Consequently, exploration groups representing several foreign countries are working in the United States, in both joint ventures and individual efforts.

Exploration is prohibited in wilderness areas, primitive areas, wild river areas, "withdrawn" areas, and in most wildlife refuges. Alaska, with its vast acreage, has been dissected so extensively by such restrictions, as well as by native claims withdrawals, that exploration has been stifled in that state. Exploration may take place (with acceptable exploration procedures) in certain other areas, provided that permission of the appropriate party in control of the surface and/or minerals is granted.

The following are some land-ownership classifications and definitions:

Fee simple estate, or simply fee land. This is land where the surface and minerals are owned by an individual or individuals, with title usually originating from the federal government.

State lands. These are lands owned by various states, particularly the western states. They can usually be leased for mineral exploration and development, provided the state owns the minerals. This will be discussed later in more detail.

Federal lands (public domain). These are lands owned by the federal government. They are mostly located in the western United States and are lands upon which mining claims must be staked.

Federal acquired lands. These are lands which were once deeded from the federal government to an individual or individuals; the government later reacquired the lands, usually because of default by the purchaser or homesteader. Many lands in the western United States reverted to the government during the 1930s (the depression years). Mining claims cannot be located on these lands. The explorationist must apply for a prospecting permit; if the permit is issued, and if a discovery is made, he must then apply for a "preference right lease."

Complicating Factors

Complications do exist in terms of land-ownership classifications. For example, a landowner might decide to sell the surface of his land but retain the mineral owner-ship. The law allows such a division or severance; thus arose the concept of having the surface estate *severed* from the mineral estate. We shall not discuss in detail the historical background of this concept; however, the explorationist should know that this situation exists, and that it can complicate matters considerably.

Not only did individual landowners sever surface and minerals when they sold or deeded land; the states, the federal government, and land-grant railroads did likewise in many areas. The net result is that it is impossible today to look at a mineral ownership map and tell who owns the surface; similarly, examining a surface ownership map reveals little about the mineral ownership. In some areas, it is possible to assume who the various owners are, but the county or other appropriate records must still be checked for confirmation prior to entering into a lease.

For example, let's assume that a rancher homesteaded in McKinley County, New Mexico, in 1894 and received a deed to 1,280 acres from the federal government. In 1939, he decided to sell the surface and retain the minerals. In so doing, he retained rights of ingress and egress for mineral development. He died in 1950 and willed the mineral rights equally to each of his eight surviving children. Each child then owned one-eighth of the mineral estate. If one child were to die, he might will his one-eighth share equally to his four surviving children; each would therefore have one thirty-second.

Such multiple mineral ownership situations are not uncommon; consequently, some residual mineral interests are so small (such as 1/164) that it is hardly worth the effort to contact all the mineral owners for a lease (especially for small tracts) unless a discovery has been made on the property. In such multiple ownership cases, heirs to small fractional mineral estates are well advised to consolidate their interests, buy the other fractional interests, or place as many interests as possible in a trust to facilitate leasing and rental payments by explorationists.

As it remains true that most exploration for minerals does not result in a commercial discovery, if inexpensive drilling is possible, it is usually more economical to drill one or two test holes than to spend hours or days (and large amounts of money) attempting to run down minority interest holders. For example, you may hold a uranium lease on five-eighths of the mineral interest on a 640-acre tract of land. The other three-eighths is divided among 25 individuals, many of whom cannot be readily located. As long as you have *some* of the mineral interest under lease, you are legally entitled to conduct exploration on the property. If it is a wildcat prospect, you may wish to enter the property and drill one or two holes to determine whether it is worth the effort to locate the remaining mineral interest owners. If the results of the drill test are negative, there is no point in pursuing the remaining interests. If competitors are actively leasing and drilling in the area, you may either wish, as a precautionary measure, to delay the drilling until all the interest is leased, or else you might conduct drilling while the landmen are running down the remaining interests. Naturally, the landmen should be halted if the drilling results are negative.

In another situation, you may have five-eighths interest leased and discover that the owners of the remaining three-eighths are "holdouts." They may have various reasons for not wishing to lease: the lease form may be unacceptable; perhaps the royalty arrangement is not satisfactory; they may not agree to the bonus or rentals; possibly their lawyers have not examined the lease form, or, if they have, they may not agree to some aspect therein. If you are faced

with holdouts, you must decide whether to meet the demand, negotiate a compromise, or let the holdouts sit until you have a better idea of what the property is worth. This latter course involves some risk: if you should drill and make a discovery, it may be more difficult, or even impossible, to lease the interest; perhaps worse, a competitor may meanwhile meet the original demand of the holdout, thus becoming your co-lessor. More often, however, when mineral interest owners refuse to lease for the current or "going" rates in a wildcat or remote exploration area, they forfeit any potential bonuses or rentals, and their land remains unleased as drilling activity fades away in the area and the explorationists go elsewhere.

SOURCES OF INFORMATION

Generally speaking, when a landman is dealing in an unfamiliar area, he is advised to check multiple sources for land ownership. First, he should check the courthouse records and get a feeling for how their titles are set up. He should then go to an abstract office if he is unable to discover the precise information he is seeking at the courthouse. In some states, such as Wyoming and Colorado, where there are many different types of federal lands, it is often necessary to check Bureau of Land Management records in addition, and perhaps also the land records at the state capitol.

Determining ownership of a given land area can sometimes be complex and time consuming. County records and abstract companies are usually the best sources of land ownership information, and these should be consulted first. This task usually falls to the landman, whether company-employed or operating as an independent consultant.

County Records

In many instances, county records are adequate for checking title, and the services of an abstract office are not required. Many counties, particularly in Wyoming and Colorado, maintain tract books which list all land according to section, township, range, grantor, grantee, date of transaction, and the volume and page number of the instruments recorded, with a complete description of the particular tract. If the records revealed 10-year-term leases, taken just a year or two prior to that time, it is a fair assumption that the leases are still valid, as long as the rentals have been paid.

The validity of the leases, however, cannot be determined from county records. The landowner or the lessor must be contacted for such information, because, in many instances, companies or individuals may cease to pay rentals without filing a release of record to indicate that they have relinquished the lease. There is law in the state of Colorado which requires that a party must place on record the release of a mineral lease; unfortunately, this is not strictly enforced. As a courtesy, companies and individuals should always file of record a relinquishment in order to clear the title for the landowner. This practice also insures a continuing good relationship between the lessee and the landowner.

Some states, and some counties within states, have far better record-keeping systems than others. Counties in Wyoming and Colorado have superior record-keeping systems for land ownership, whereas counties in Utah, North Dakota, and South Dakota are considered by most landmen to have inadequate systems. The most useable systems are those where each township is listed in one book, and all transactions relating to a particular section of land are listed according to the acreage involved.

For example, the southeast quarter of a given section may show a mineral deed on that quarter; it should include the date of the instrument, the date the instrument was filed of record in the courthouse, and the names of the grantor and grantee. It should be possible to establish the "chain of title," as it is called, and clearly follow it from its origin to its present ownership. If the specific instruments are not designated as mineral deeds, they may easily be checked by noting their numbers to determine whether there were any mineral reservations.

In addition, the listing of each land description should show the warranty deed and should then give the book and page number of the county record filing system for that entry. Each listing should also be cross-referenced to the filing system of the abstract office. The county records are sometimes available on microfilm, and the book and page number can easily be located for a description of the instrument. However, in each of the better systems, the key is a tract book for each township and range, listing all 36 sections individually.

We found an example of one particularly ineffective system in a county in North Dakota, where they list the instruments according to sections but do not designate the *nature* of each instrument (that is, whether an instrument is a deed, lease, assignment, or other); they merely refer to a volume and page number where the instrument was recorded. The landman is then obliged to go to the proper records, locate each individual instrument which is on file there, and peruse the document to discover the nature of the transaction. This method consumes an enormous amount of time.

We encountered another very poor system in some counties in Utah, where records are filed according to grantor/grantee. It is therefore necessary to know, for example, the name of a specific mining claim in order to discover its location from the records, or the name of the person involved in a transaction in order to locate a particular instrument. In some cases, this requires going back through records covering five or six years before information pertaining to existence (or lack thereof) of mining claims in a given area can be ascertained.

We have concluded, then, that some counties have far more efficient, workable record-keeping systems than others. Occasionally when a new county clerk is elected to office, he may modify or completely change the system which he inherited—and generally with some improvement. Other times, the same inefficient systems are retained and perpetuated.

Another source for checking land ownership is the tax assessor's office. If the name of the landowner is known, it is simply a matter of checking to see what lands that owner has listed on the county records for tax assessment purposes. Although this method can give some insight into a particular piece of land, information derived from files in the tax assessor's office in most cases only identifies surface ownership. It is usually not possible to acquire mineral leases on the basis of this information. Also, the information is often incomplete.

For example, a landowner may have sold a piece of property under a contract according to an escrow agreement with a local bank; such a situation will not appear on the tax assessment records. It is then necessary to find the original owner to determine the precise terms of the transaction in which he relinquished title to land; frequently, a copy of the agreement must be obtained from the bank in order to facilitate discussing a lease with the present owner. In cases where land is in the process of changing hands, but the new owner has not completed payment, the lease must be executed by the new owner and ratified by the original owner. In this way, should the new owner default on payment and the land revert to the original

owner, the original owner's ratification of the lease remains in effect; thus the lessee is protected.

Abstract Companies

Abstract offices are privately owned operations, frequently run by retired judges or attorneys. They may be found in the county seat of each county, and they maintain records for the entire land area of that particular county. Abstracters maintain current ownership information by making daily trips to the county office where such information is filed and placed on record. They extract all pertinent information and transfer it to their own records. A landman has simply to know the section, township, and range of a particular tract in order to readily find information such as who has made leases, who has sold the land, and whether any mineral deeds were made.

Abstracters' records are usually available to landmen free of charge, but occasionally an hourly fee is charged for the use of the records. The minimal fee often can be negotiated (but usually is less than $5 per hour), and much time can be saved in seeking ownership of title and also, sometimes, mineral ownership. Many abstract offices do not rely upon this title service for a substantial part of their income; instead they are often actively engaged in real estate transactions, or they may sell title insurance. However, not all abstracters make their records available—even for a fee. They sometimes maintain a policy that no one may enter their offices and have access to their records; instead, they research the required information, utilizing their own personnel, and then they charge accordingly. Naturally, this would be at a rate higher than the hourly rate of $5 for records alone. Sometimes it is possible to make a deal with abstracters. For example, if a "hot play" is rumored in a particular area, the landmen who arrive on the scene first have been known to make a deal with the abstract office whereby they tied up all records for three or four weeks, insuring that no other party would have access to them during that time to check titles. Exclusive entree to an abstract office's records can usually be had for a certain price.

This practice definitely provides a competitive edge when it comes to leasing. If we were the victims of such circumstances —that is to say, if our landmen arrived in an area where a hot play was under way and discovered that the abstract office's records were tied up—we very well might advise our landmen to go directly to the land in question and talk to the landowners, asking questions about mineral and surface ownership, leases already taken, and so forth. We would probably take their word as to which mineral rights they owned but were unleased, and we would perhaps sign a lease based on such information (if terms could be agreed upon). Such action is certainly justifiable in cases where a nominal bonus and rental (such as $1 per acre) is required to obtain the lease. In cases involving a hot play lease, where the bonus and rental might be $25, $50, or even $100 per acre, it is somewhat risky to execute a lease without first checking titles. However, if the necessary records are inaccessible, we suggest that the landmen write a 60-day sight draft containing a lesser interest clause which states that the lease will be final subject to approval of title. After checking the titles within the 60-day period, there is no obligation to pay the sight draft (and the lease is void) if the titles were incorrectly given.

For example, if the landman is informed (verbally) that a particular rancher owns 1,280 acres of minerals, the landman may issue a sight draft as a bonus and first year's rental for a lease covering all 1,280 acres. Later, upon checking the records, it may be discovered that that rancher owns

only 640 acres. If this situation should occur, the original sight draft should be returned and a new one issued covering the precise acreage owned.

Sight Draft

A sight draft is handled by banks in a manner similar to a check except that a sight draft is a *collection item* while a check is a *cash item.* For example, you may issue a 30-day sight draft to a landowner (30 days is an arbitrary time designation). He then takes it to his bank for collection. The landowner's bank then sends the draft to your bank, and your bank, in turn, notifies you of its arrival. You then have 30 days (after *sight* by the bank, hence the term) in which to honor the payment. The landowner's bank, anticipating the credit to be returned there, will notify your bank promptly if the payment is not received by the thirtieth day.

The files of any company or individual engaged in buying leases must contain ownership information specifying title and gross and net acres for each specific tract to be leased. Accordingly, when a broker or landman is assigned the task of buying a lease, he should make certain notations on the draft. First he should note the designation of the lessor—whether he is a surface owner, owner of surface and minerals, or strictly the minerals owner. Second, he should indicate the gross acreage and the net acreage. For example, for a lease covering 160 acres in a quarter section of land where the lessor owns the surface and an undivided half of the minerals, the landman should indicate a gross of 160 acres and a net of 80 acres. The information concerning the net 80 acres is vital, because this is the basis upon which the rentals are paid.

If there are different tracts of lands in different sections, the lessor may own a one-third interest in one section, perhaps a one-fourth interest in another, and a one-eighth interest in a third parcel. These should be set out in the draft so that the lessor's undivided interests in these sections are clearly stated; the landman can then immediately ascertain that this particular lease does not cover all of the property, and that there may be other parties from whom he should seek leases. Such information placed on a draft for quick reference is extremely valuable for accounting purposes, or for a quick check on the title without having to resort to voluminous title notes taken from the abstract office or the register-of-deeds office.

The draft should be prepared with an original and three carbon copies. The lessee should receive the original at the time he honors the draft at the bank; the copies should be placed in the lease file, together with the lease and other related information. The draft plays an important part in title work as well as in accounting, and is usually required by auditors when performing an audit, to check that the draft has been paid. Every lease should be numbered; any cost or capitalized item can be filed accordingly. The lease numbers should be cross-referenced in a lease schedule, wherein any lease may be researched and the appropriate number found.

Situations arise where an explorationist may, in good faith, issue a sight draft, perhaps in the amount of $50 per acre, anticipating a hot play on a given area. During the 30 days, he may be unsuccessful in making a deal to sell the play (including the unpaid-for acreage), and thus he would be unable to pay the draft upon the due date. If he is sincerely attempting to acquire the necessary funding to put together a deal, the landowner's bank (with permission of the landowner) may allow him an extension beyond the original 30-day period.

There are certain "hot-draft artists," as they are known in the industry, who frequently have no intentions of honoring sight drafts unless they can quickly sell the acreage at a profit. They may claim that they require additional time for checking title; or perhaps they may declare that they are dissatisfied with their findings from a title check. If they can't make a sale, however, they generally refuse to honor the sight draft; in this case, of course, the lease they took is not valid.

In Texas, it is common practice for the final lease to be executed at the same time as the sight draft, and *both* documents are sent together to the bank. Upon paying off the draft, the explorationist receives his copy of the executed lease. In some cases, the explorationist's landman and the landowner might meet at the bank; the landowner would endorse the sight draft, sign the lease and have it notarized, and the total package would be sent to the explorationist's bank for collection. This enables the landman to see the lease and insure that it is properly executed and notarized when the landowner gives it to the collecting bank; however, the lease would not be in the landman's possession until after payment of the draft had been honored. Such practice evolved to protect the landowner from the potentially fraudulent lease speculator or explorationist who might take the executed lease and attempt to negotiate or sell it without first having paid the fee.

Aside from discovery of defective title, or an unanswered question relating to title, there are no other grounds for defaulting on payment of a sight draft. The landman or explorationist who may decide within the 30-day period that he has changed his mind about making a play on that particular piece of land is leaving himself open to litigation, because the sight draft has bound him to honor the executed lease. In addition, landmen and explorationists have to consider their reputations. If they were to default on a sight draft, knowing that the title was satisfactory, they would jeopardize their good standing with financial institutions and landowners alike in that community, and undoubtedly would find it extremely difficult to lease in that area again. Few, if any, honest explorationists are willing to forfeit the privilege of doing business in the future.

As stated earlier, the 30-day period is simply an arbitrary figure. Sight drafts may be issued for any period of time from one day to six months; 30 days, however, is usually considered a standard period in which to permit the explorationist an opportunity to verify title. Figure 5-1 shows a sample of a typical sight draft form. As you can see, the duration of the draft is left blank, to be filled in by the respective parties when an agreement has been made.

Various types of deals have been made concerning leases and sight drafts in some inactive or "cold" areas. For instance, if the landman or explorationist has been able to gain the confidence of a landowner, he might agree to the use of a slight draft as a type of option on the property, and provide for a lengthy duration prior to payment becoming due on the lease. It has also been known for a landman to offer a landowner a certain percentage of his deal, in addition to the lease fee, although such an arrangement is quite rare and would probably only arise in a relatively inactive area.

Generally speaking, once a lease has been executed, there are legal responsibilities and obligations on the part of both the landowner and the explorationist. The explorationist who delays payment of a draft, for whatever reasons beyond title failure, is usually employing a deceptive technique, attempting to get a free option, in effect, by tying up the landowner's land when he is aware that his play may possibly not materialize.

State Leases

In most cases, state land records are available to anyone who wishes to check them to determine where state lands are located and whether the lands are under lease. If they are leased, the name of the lessee, date of the lease, and amount of acreage can be obtained. If the lands are not leased, an application for lease may be filed. It should be noted that different states have different classifications of state lands—for example, public school lands, capitol buildings, military homes, and so forth. Many states insist on separate leases for each category, so that the funds may be appropriated accordingly. Some states have awkward or complicated leasing systems. Wyoming, for example, while otherwise operating efficiently, will lease only two sections on one lease. In addition, the board of land commissioners stipulates that the two sections be within a six-mile radius of each other. Colorado will issue one lease covering a large number of acres and will indicate several categories of state lands on one lease, identifying the particular fund beside each tract. However, if one portion is dropped by the lessee, an entirely new lease for all retained tracts must be issued.

Other states have different stipulations for issuing leases on state lands. Many states will not issue one lease covering a large acreage, such as 20,000 acres; they insist that the acreage be broken down into small areas, with a different lease to cover each area. Their rationale for this stipulation is that if the lessee wishes at a later date to drop one portion, the state is saved considerable paperwork in issuing amended leases. On the other hand, regulations pertaining to the filing of federal prospecting permit applications for acquired lands specify a maximum acreage per permit of 2,560 acres, or four sections. The sections do not have to be contiguous; however,

165

CUSTOMER'S DRAFT
With privilege of Re-Draft

COLLECT DIRECTLY THROUGH Farmers National Bank

N? **18793**

Casper, Wyoming September 10, 1976

Thirty (30)_____ Days After Sight and Subject to Approval of Title

Pay to the
Order of _____John Q. Rancher_____ $1,110.03

---One thousand one hundred ten and 03/100--- DOLLARS

WITH EXCHANGE

Consideration for Execution of mining lease, of even date, covering lands in Morgan
County, Colorado, described below

Robert J. Miner

TO: Able Exploration Corp.
1000 S. Broadway
Denver, Co. 80000

County Morgan
State   Colorado
Gross Acres   1902.11
Net Acres   1110.03

DO NOT USE THIS SPACE

- - - - - - DETACH BEFORE PRESENTING FOR PAYMENT - - - - - -

Township 8 South, Range 67 East, 6th P.M.
Section 14: SE 1/4

Township 9 South, Range 67 East, 6th P.M.
Section 10: W 1/2, SE 1/4
Section 14: All

Township 10 South, Range 67 East, 6th P.M.
Section : 30: Lots 1, 2, 3, 4, E 1/2
           W 1/2, E 1/2

Figure 5-1. Sight draft.

they must be within a six-mile radius of one another.

BLM Data: MT, OG, and MIN Plats

Maps provide an essential first step in checking title to lands. The preparation of accurate mineral ownership and surface ownership maps of prospective exploration areas is a time-consuming chore, and it was therefore with interest that we learned that the Bureau of Land Management (BLM) has recently undertaken the mapping of certain key areas of the United States to show federal surface and mineral ownership (*only* federal ownership is shown). The maps designate the following categories: all minerals; coal only; oil and gas only; oil, gas, and coal only; and other. It should be noted that although the maps show minerals *ownership*, this does not mean that minerals do exist. It means that the owners possess the right to exercise mineral exploration or to lease the land to others for exploration purposes.

Also available from the BLM are mineral title plats (commonly refered to as MT plats); these show areas where the federal government owns the minerals (and state minerals deeded from the government) in individual townships (36 square miles). Once the federal and state lands have been located, the balance of the acreage can be identified as either patented lands or fee lands. MT plats are obtainable from the BLM branch office which has supervision of the area in question. They may be requested by township and range, and the present cost is $2 per map. The scale is 1 in. = 2,000 ft.

Two other types of maps are available from the BLM office: hard-mineral (MIN) plats and oil and gas (OG) plats. MIN plats show the status of hard-mineral titles on federal land within the township being checked. For example, if uranium prospecting permit applications had been filed

on federal acquired land within the area, the applications would be shown. Or, if permits had been issued, these would be shown. OG plats show mineral ownership of oil, gas, and associated hydrocarbons and the leases covering same.

BLM maps sometimes reveal that a particular area of interest is part of an Indian reservation, and other good ownership maps of Indian lands are usually available. Each reservation has a business office, usually similar to a Forest Service office, located on the reservation.

Indian Lands

Because explorationists must deal through an Indian agency to acquire a lease on Indian reservation lands, Indian lands pose a different kind of problem. The Bureau of Indian Affairs (BIA) has personnel whose duties include dealing with mineral leasing. Such officials are capable of giving information concerning specific ownership as well as some comments concerning terms.

While many Indian lands have been held under lease and explored in the past, indications are that Indians are becoming dissatisfied with past practices, even to the point of considering attempts to nullify existing leases. Many existing leases have been very beneficial to Indian tribes through production royalties. Examples of this include oil and gas production in Oklahoma and coal production in New Mexico.

Commercially Available Maps

In most medium-sized to large cities in the United States where mineral exploration is active, there are firms which specialize in drafting and map-preparation services. Certain of these firms prepare maps of active mining areas showing claims, leases, mines, mills, and so forth, and such maps are often available to the public for a reasonable charge. Those exploring for uranium in sedimentary rocks may also find some of the commercially prepared petroleum ownership maps to be of significant value in the determination of mineral owners, although such maps must be used with caution because oil and gas may be severed from other minerals within certain tracts.

Firms which sell U.S. Geological Survey and other government publications are good places to ask about commercial or other maps showing ownership, geology, etc. For example, in Albuquerque, New Mexico, one might determine where topographic maps are sold and then inquire at that firm about other available maps or publications related to exploration for a certain mineral in New Mexico. However, once such mining claim information is posted on a map, it is generally not removed, nor is it updated. Even though there may be systematic checking of county records, in most states there is no way to insure that the claims posted on the maps are still in effect, because most states do not require that assessment work be recorded. Thus it is often necessary to carry out a search *on the land* to see if work was done, and even this procedure is not reliable. Nevertheless, information posted on commercial maps may be of value in terms of revealing whether claims had been staked in a given area or whether land had been leased. However, such information is not to be relied upon for staking or leasing purposes.

Certain remote areas have very poor map coverage. In such areas, we suggest obtaining county maps published by the state highway department, if such maps are available. Although the scale may be somewhat small, these maps usually provide adequate information for locating landowners in the county and, if necessary, such maps may be enlarged. Highways and county roads are usually clearly marked and categorized (e.g., better-class roads, private roads, and unpaved trails). In addition,

these maps frequently show ranch buildings and other useful landmarks such as bridges, windmills, and so forth.

For a newcomer to a particular area, county maps can be very helpful in initial orientation and as a guide around the county. The usual scale is approximately one-half inch to the mile, and such maps are usually available from the state highway department in the capital city of the particular state.

BASIC INFORMATION FOR MINERAL ACQUISITION

State Mineral Leases

In figure 5-2, we show an example of a uranium lease, including a production royalty schedule, issued by the state of Colorado. As a word of caution, however, it should be noted that the mineral lease form (such as for uranium) varies significantly from one state to another. For example, Colorado has a sliding-scale royalty for both open-pit and underground mining, which differs from Wyoming's royalty provisions. Wyoming's current form has a fixed-percentage royalty and does not differentiate between open-pit and underground mining.

Leasing state minerals of any kind is becoming an increasingly complex matter, particularly since many states have adopted legislation which requires that requests for mineral leases must be advertised. Formerly, an explorationist could simply visit the state land office and request information or review state records pertaining to whether state land in a particular area was leased or was available for leasing. If the land was not leased, a lease application could be filed immediately, along with a check to cover the first year's rental plus the required filing fee. Such a step was all that was required to tie up the land. Usually within 30 to 60 days, the lease would be

STATE OF COLORADO

State Board of Land Commissioners
Department of Natural Resources
Denver, Colorado 80203

Uranium & Associated   M I N I N G   L E A S E   N O.   _____2613/16-S_____
          Minerals

THIS MINING LEASE, Made in duplicate and entered into this __7th__ day of

___November_____, 19_74_, by and between the State of Colorado, acting

through its STATE BOARD OF LAND COMMISSIONERS, hereinafter referred to as

Lessor, and __ABLE RESOURCES CORPORATION, 303 Petroleum Bldg.,_____

_____Casper, Wyoming 82601_____, hereinafter referred to as Lessee:

WITNESSETH: Lessor, for and in consideration of the sum of_two thousand___

____fifty-five & 00/100_____ Dollars ($2255.00), receipt of which is

hereby acknowledged as payment of filing fee in the amount of $15.00 and first

year's rental in the amount of $2240.00, and in further consideration of

Lessee's agreement to pay the following amounts annually as rental in advance

on the anniversary date of this lease so long as said lease shall remain in

effect:

LEASE YEAR	RATE PER ACRE
2nd	$1.00
3rd	1.00
4th	1.00
5th	1.00

and in further consideration of the terms and conditions hereinafter stated,

and of the payment of royalties reserved herein, to be kept and performed by

Lessee, its successors and assigns, does hereby demise and lease to Lessee

the right and privilege of exploring and prospecting for, and mining for and

taking _____uranium_____ and associated minerals of value that

can be removed in the process of mining and milling _____the same_____

from the lands hereinafter described, situate, lying and being in the County

of ___Weld_____, State of Colorado, to wit:

Figure 5-2. Uranium lease.

ACRES	SUBDIVISION	SECTION TOWNSHIP	RANGE PATENTS
640.00	All	16-17N-58W	None
640.00	All	36-16N-58W	SW¼ #5617; SE¼ #3561
320.00	E/2	34-17N-57W	N½ #6930; SW¼ #6985; SE¼ #6986
640.00	All	11-16N-59W	None

(revenues designated for School Fund)

containing 2240 acres, more or less, together with the right to use as much of the surface thereof as máy reasonably be required in the exercise of the rights and privileges herein granted, and the reasonable right to ingress and egress; the right to the use of all otherwise unappropriated water from said lands but not from surface lessee's or surface owner's water wells or reservoirs; the right to construct buildings, make excavations, stockpiles, dumps, drains, roads, railroads, power lines, pipelines, and other improvements as may be necessary; subject, however, to all existing easements and rights-of-way of third parties, and the rights of surface lessees and surface patentees, and further subject to the terms, conditions, and royalties set out in this lease.

RESERVING, However, to the State of Colorado:

A. All rights and privileges of every kind and nature, except as are herein specifically granted.

B. The right to use or lease said premises or any part thereof at any time for any purpose other than and not inconsistent or interfering with the rights and privileges herein specifically granted.

C. The right at all times during the life of this lease to go upon said premises and every part thereof, for the purpose of inspecting same, and the books of accounts and records of mineral workings therein, and of ascertaining whether or not said Lessee and those holding thereunder, by and from it, are carrying out the terms, covenants, and agreements in this lease contained.

TO HAVE AND TO HOLD The above-described premises, with the appurtenances, unto the Lessee, its heirs, successors, assigns, or legal representatives, from Twelve o'clock noon on the _7th_ day of ____November____, 19_74_, for the full term of ___five___ (5) years, and until Twelve o'clock noon on the _7th_ day of ___November___, 19_79_, and, except as hereinafter stated, as long

Figure 5-2 (continued).

thereafter as the minerals hereinabove designated are being produced in paying quantities from said premises, and the royalties and rents provided for herein, or by any extension hereof, are being paid, subject to the following terms, conditions, and agreements, to wit:

1. <u>Minimum Royalty</u> – As minimum and advance royalty, without relation to the amount of minerals mined from the leased premises, the Lessees will pay to the Lessor the following amounts:

LEASE YEAR	MINIMUM ROYALTY	LEASE YEAR	MINIMUM ROYALTY
1st thru 5th	None		

    Acreage changes resulting from surrender or partial assignment do not reduce the minimum royalty proportionately. The Board will determine the minimum royalty.

    Further, at the end of each five-year period, the Board may fix the rate of advance royalty to be paid for each succeeding five-year period of the lease.

    In the event that said Lessee does not extract from said premises said minimum amounts above specified during each lease year of the term of this lease, it is nevertheless understood that the above sums of money are royalties for the years stated, and are due and payable to the Lessor whether minerals are mined or not during such year, but that such minimum advance royalty will be credited upon the first royalties due as hereinafter provided for minerals actually produced from said premises and sold or milled during the year for which such minimum royalty was paid.

2. <u>Production Royalty</u> – Lessor hereby reserved and Lessee agrees to pay to Lessor as royalty the following:
    (a-1) For fissionable materials--see attached schedule.

Figure 5-2 (continued).

(a-2)   For non-fissionable materials--A sum equal to __10__ % of the gross
        value at the mill or buying station less reasonable transportation
        costs. Lessee will furnish evidence that prices received for material
        sold and transportation charges deducted are reasonable and fair.
        Unless otherwise agreed to in writing by the Board, deductable
        transportation charges will be from the mine to the nearest mill or
        buying station. Lessee may use the weighted average sale price
        received for all lots of ores sold during each calendar month period.

(b)   A sum of $_____ for each and every ton (2000#) of the above
        specified minerals mined from the premises. Within sixty days
        prior to the termination of each and every five-year period for so
        long as this lease remains in effect, the State Board of Land Com-
        missioners may reappraise the property herein leased and fix and
        determine the rate of production royalty to be paid during each
        year of the succeeding five-year period. Failure to comply with any
        new royalty rate set by the Board may subject this lease to can-
        cellation by thirty-day written notice by the Board.

If requested by Lessor, Lessee is to furnish proof of price received for
all minerals sold. Such royalty is due and payable on or before the twenty-
fifth (25th) day of each calendar month during the term of this lease for
minerals mined, removed, and sold by Lessee during the preceding calendar
month.

Royalty payments shall apply to payment received by the Lessee on any
mineral so sold. The minerals mined from the leased premises may be milled
in a custom or commercial plant or mill owned and operated by Lessee.
They may not be mixed or commingled with ores from other properties until
they have been crushed and sampled by the mill for the purpose of determining the hereinabove designated minerals and other minerals contained in
the crude ore that may be salvaged by milling of same. The resulting
samples are to be assayed at the laboratory used by the mill where ore is
milled and Lessor may demand an assay certificate showing the contents of
each delivery to the mill. If requested by the Lessor, the mill is to
reserve a pulp of each composite sample for testing, provided, however,
said mill will not be obliged to keep pulp for a period of more than

Figure 5-2 (continued).

ninety (90) days. In the event of a dispute as to analysis, the pulp prepared from each composite sample is to be referred to an umpire acceptable to Lessor and Lessee and the determination of the umpire will be binding upon both parties. The net weight of the crude ore on a dry-weight basis will be determined at the scales at the mill. Lessor will have the right to check the weights as often as it deems advisable. Lessee will furnish Lessor with duplicate scale weight certificates. Lessee agrees to, and will be held accountable to see that the foregoing provisions are carried out by such mill or mills as receive minerals produced from the premises herein demised.

The Lessee has the right to mine, raise, carry, and transport ores from the lands hereby leased through other lands now or hereafter owned or leased by the Lessee, and the right to raise, carry, and transport ores mined by the Lessee from other lands owned or leased by the Lessee over or through the lands covered by this lease; provided, that the ores mined from other lands are in no event mixed with ore mined from these leased premises, except as provided above.

3. Overriding Royalty Limitations - It is agreed that this lease or any subsequent assignment hereof may not be burdened with overriding royalties the aggregate of which exceeds _____. Lessor must be notified of all overriding royalties accruing to this lease. Violation of the above may subject this lease to cancellation by Lessor.

4. Weights - It is agreed that all minerals mined and taken from said premises are to be weighed and the weight thereof is to be entered in due form in weight records kept for such purposes by Lessee.

5. Reports - After operations are begun, it is agreed that on or before the 25th day of each and every month during the term of this lease the Lessee will make a sworn report on forms furnished by Lessor, in which the exact amount in weight of all minerals mined and removed from said premises and sold during the preceding calendar month are to be entered and accompanied by full payment for all royalty due for the month. Further, Lessee must furnish annually a map or blueprint of survey of all workings, with loca-

Figure 5-2 (continued).

tion of same tied to a corner established by United States surveys of some land subdivision, certified to by a licensed engineer or surveyor.

6. Inspection - It is agreed that during all proper hours and at all times during the continuance of this lease the Lessor or its duly authorized agent or agents, is authorized to check scales as to their accuracy, to go through any of the slopes, entries, shafts, openings, or workings on said premises, and to examine, inspect, survey and take measurements of the same and to examine and make extracts of copies of all books and weight sheets and records which show in any way the mineral output of the leased premises, and that all conveniences necessary for said inspection, survey, or examination are to be furnished to the Lessor.

7. Mining Methods - Failure to comply with any of the provisions of this paragraph may subject this lease to cancellation.

    In the underground and open pit workings, all shafts, inclines, and tunnels must be well timbered (when good mining requires timbering) and all parts of workings, where minerals are not exhausted, will be kept free from water and waste materials. The underground and open pit workings are to be protected against fire, floods, creeps, and squeezes. If such events do occur, they must be checked in a manner which is in keeping with good methods of mining. Such methods of mining must be used as will insure the extraction of the greatest possible amounts of minerals consistent with prevailing good mining practice.

    Lessee agrees to slope the sides of all surface pits or excavations to a ratio of not more than one foot (1') vertical for each two feet (2') of horizontal distance. Such sloping is to become a normal part of the operation. Whenever practicable, all pits or excavations are to be shaped to drain, and in no case may the pits or excavations be allowed to become a hazard to persons or livestock. All material mined and not removed from the premises will be used to fill the pits so that at the expiration, surrender, or termination of this lease the land will, as nearly as practicable, approximate its original configuration, with a minimum of permanent damage to the surface.

    Upon request from the Board, Lessee further agrees to submit plans and maps

Figure 5-2 (continued).

of proposed mining programs in advance, so that the Board may be fully
aware of the proposed operations.

8. Rights-of-way - It is agreed that the Lessor reserves the right to grant
    rights-of-way over said premises for public roads, railroads, power, tele-
    graph, telephone, ditch, and canal lines, but such grants are to be subject
    to the rights of the Lessee.

9. Notices - It is agreed that any notice required or permitted to be given
    to the Lessee under the provisions of this lease is to be sent by certified
    mail to the address set forth at the beginning of this lease or to such
    other address as Lessee may indicate in writing to Lessor and such service
    by mail will be deemed sufficient and in full compliance with the terms of
    this lease. Notice to Lessor is to be given in like manner, addressed to
    State Board of Land Commissioners, Denver, Colorado.

10. Assignment - The Lessee, with the written consent of the Lessor, will
    have the right to assign this lease as to the entire leasehold interest of
    such Lessee in all or part of the lands covered hereby, not less, however,
    than tracts of approximately forty (40) acres or Governmental lot cor-
    responding to a quarter-quarter section for any partial assignment, and for
    approval of such assignment the Lessor will make a charge of Ten Dollars
    ($10.00) for any one assignment. No assignment of undivided interest or
    retentions or reservation of overriding royalties will be recognized or
    approved by Lessor; and the effect, if any, of any such assignments or
    reservations will be strictly and only as between the parties thereto,
    and outside the terms of this lease, and no dispute between parties to
    any such assignment or reservation shall operate to relieve the Lessee
    from performance of any terms or conditions hereof or to postpone the
    time therefor. Lessor will at all times be entitled to look solely to the
    Lessee or his assignee shown on its books as being the sole owner hereof,
    and for the sending of all notices required by this lease, and for the
    performance of all terms and conditions hereof. If an assignment of this
    lease, in whole or in part, is approved, a new lease will be issued to the
    assignee for the balance of the life of the lease, covering the lands
    assigned. Said lease will be on the mining lease form in use at the time

Figure 5-2 (continued).

of assignment, and limited as to term as said lease is limited, and the assignor shall be released from all further obligations, and shall be held to have released all rights and benefits thereafter accruing with respect to the assigned land, as if the same had never been a part of the subject matter of this lease.

11. Prospecting - Lessee agrees that while using and operating any diamond, churn, or other drill on or within one-quarter mile of said premises, it will keep an accurate log on all work so done and performed, showing geological formations penetrated, the depth or thickness of each, the mineral character of each, especially mineral veins and water bearing strata, the location of same, the elevation, and tie to a corner established by U. S. surveys of some legal subdivision, and each and every thing necessary to make a complete log of the hole throughout its entire depth, a true copy of which said log will be furnished to the Lessor.

It is understood and agreed that the methods used in carrying out any program of exploration, and the rate of progress of such program may be determined by the Lessee. If the Lessee carries on any program of exploration, other than drilling, Lessee will submit to Lessor written reports showing the character and extent of prospecting being carried out on the leased premises and giving any details of mineral outcroppings, seams, and veins which may have been encountered; and Lessor agrees that during the term of this lease all such information supplied to Lessor by Lessee will remain confidential and unpublished so far as consistent with law.

12. Surrender and Relinquishment - The Lessee may at any time, by paying the State of Colorado, acting through its State Board of Land Commissioners, all amounts then due as provided herein, surrender and cancel this lease insofar as the same covers all or any portion of the lands herein leased and be relieved from further obligations or liability hereunder with respect to the lands so surrendered; provided, that no partial surrender or cancellation of this lease may be for less than tracts of approximately forty (40) acres or Governmental lot corresponding to a quarter-quarter section, the rental being reduced proportionately; provided further, that this surrender clause and option herein reserved to the Lessee will cease and

Figure 5-2 (continued).

become absolutely inoperative immediately and concurrently with the institution of any suit in any court of law by the Lessee, Lessor, or any assignee of either, to enforce this lease or any of its terms, express or implied, but in no case will surrender be effective until Lessee has made full provision for conservation of the minerals and protection of the surface rights of the leased premises.

All information in Paragraphs 5 and 11, above, must have been filed with the Board before this lease may be terminated.

13.   If lessee initiates or establishes any water right for the leased premises, the point of surface diversion or ground water withdrawal of which is on the leased premises, such right will, if the surface rights of said premises are owned by lessor, become property of lessor, without cost, at the termination of the lease.

14.   Protection against Surface Damage – Bond Requirements - Lessee has the right to utilize as much of the surface of the lands as is necessary for mining operations, and is liable and agrees to pay for all damages to livestock, growing crops, water wells, reservoirs, or improvements, caused by Lessee's operations on said lands. Further, it is understood that this lease is granted subject to surface patents, deeds, and certificates of purchase, and Lessee assumes responsibility for all claims arising from damages to the surface caused by Lessee's operations on such lands. It is agreed and understood that no operations may be commenced on the lands hereinabove described unless and until the Lessee or his assignee has filed a good and sufficient bond with the Lessor in an amount to be fixed by Lessor, to secure the payment for such damage to livestock, growing crops, water, or improvements as may be caused by Lessee or his assignee's operations on said lands. The Lessor may grant relief from the foregoing bond requirement upon application for such relief from the Lessee.

15.   Indemnification of Lessor - The Lessee further agrees to hold the Lessor harmless for any and all manner of claims arising or to arise from the said leased premises by Lessee whether from soil or surface subsidence or from any other cause or any other nature whatsoever; this paragraph is

Figure 5-2 (continued).

binding upon Lessee and upon the heirs, assigns, successors, and legal representatives of Lessee, and is a continuing obligation during and after the expiration of this lease, so long as any possibility of soil or surface subsidence remains.

16. <u>Right of Removal</u> - In the event this lease is terminated by forfeiture, surrender, or the expiration of term, and all obligations of Lessee under this lease are satisfied, Lessee may remove all his improvements and equipment from the said premises within six months from the date of such termination, and such removal must be accomplished without unnecessary waste or injury to the premises. All improvements and equipment remaining on the leased premises six months after the termination hereof will be forfeited automatically to the State of Colorado, without compensation.

17. <u>Compliance with Law</u> - Lessee further covenants and agrees that during the continuance of this lease it will fully comply with all the provisions, terms, and conditions of all laws, whether State or Federal, and orders issued thereunder which may be in effect during the continuance hereof relating to mining or other operations of Lessee hereunder.

18. <u>Forfeiture</u> - It is agreed that if for any reason the Lessee fails to keep each and every one of the covenants herein, and if such default continues for a period of thirty (30) days after service of written notice thereof by certified mail upon the Lessee by the Lessor, the Lessor has the right to declare this lease forfeited, and to enter onto the leased premises, or any part thereof, either with or without process of law, and to expel, remove, and put out the Lessee or any person occupying the premises, using such force as may be necessary to do so. In the event of the termination of this lease by reason of breach of the covenants herein contained, the Lessee will surrender and peaceably deliver up to the Lessor the above-described premises which premises are to be in good mining condition. If, upon termination of this lease for any reason, whether by surrender, forfeiture, or expiration of term, or otherwise, Lessee has not complied fully with the terms of the lease, Lessor will hold and retain possession of the property, improvements, and equipment of Lessee, as security unto Lessor for the payment of rents and royalties due it, or to protect it

Figure 5-2 (continued).

against liens, or to indemnify it against any loss or damage sustained by it by reason of the default of the Lessee, for which purpose Lessor is hereby given lien upon all such property, improvements, and equipment, which lien will attach as the same are placed upon the premises. In the event Lessor forecloses the lien in this article given to it by Lessee, Lessor may itself be a purchaser at any sale thereof under such foreclosure. Upon the termination of this lease for any cause, if the Lessee remains in possession of said premises, he will be guilty of an unlawful detainer under the statutes in such case made and provided, and he will be subject to all the conditions and provisions thereof and to eviction and removal, forcibly or otherwise, with or without process of law, as above provided.

19. Extension - If the leased premises are not producing at the end of the primary term, the Lessee may make written application to Lessor to extend this lease for an additional fixed term and the making of such extension will be at the option of the Lessor.

Lessee must pay all taxes lawfully assessed on property of Lessee located on the leased premises.

The benefits and obligations of this lease extend to and are binding upon the heirs, executors, administrators, successors, or assigns of the respective parties hereto.

IN WITNESS WHEREOF, The Lessor has caused these presents to be executed in duplicate by the State Board of Land Commissioners and sealed with the official seal of said Board, and the Lessee has hereunto set his hand and seal, all on the day and year first above written.

LESSOR:

STATE OF COLORADO

by its

STATE BOARD OF LAND COMMISSIONERS

Recommended:

By:_____

Minerals Director

_____
President

_____
Register

_____
Engineer

Figure 5-2 (continued).

LESSEE:

Attest: ABLE RESOURCES CORPORATION

                                         By: _____
_____
               Secretary   Vice President

Colorado   PRODUCTION ROYALTY -- FISSIONABLE MATERIALS   1/69

For all ores mined, saved and removed from leased premises, Lessor reserves and
Lessee agrees to pay to Lessor a royalty based on the gross purchase price as set
out below:

GROSS PURCHASE PRICE on the open market shall be the gross price received by Lessee
or his agent for ores delivered to buying station or mill including any bonus payments, transportation allowance, etc., before deducting actual transportation costs.
For determining royalty due, Lessee may use the weighted average sale price received
for all lots of ores sold to the mill or buying station during each calendar month
period.

GROSS PURCHASE PRICE (captive market). If ores are shipped to a buying station or
mill wholly or partly owned or controlled by Lessee or his agent, the price paid for
the ores shall never be less than that paid to independent operators in the same
districts for ores of like character and quality. In the event no ores are being
purchased from independent operators by the mill or buying station, then the basis
for GROSS PURCHASE PRICE shall be the market value of U_3O_8 contained in each ton of
ore (as set out in the following schedule) multiplied by the quotient obtained by
dividing the market price per lb. of such U_3O_8 in concentrate by $8.00. For deter-
mining royalty due, Lessee may use the weighted average value for all lots of ores
delivered to mill or buying station during each calendar month period.

Figure 5-2 (continued).

% U_3O_8	VALUE PER LB.	:	% U_3O_8	VALUE PER LB.	:	% U_3O_8	VALUE PER LB.	:	% U_3O_8	VALUE PER LB.
.10	$1.50	:	.18	$3.10	:	.26	$4.10	:	.34	$4.70
.11	1.70	:	.19	3.30	:	.27	4.20	:	.35	4.75
.12	1.90	:	.20	3.50	:	.28	4.30	:	.36	4.80
.13	2.10	:	.21	3.60	:	.29	4.40	:	.37	4.85
.14	2.30	:	.22	3.70	:	.30	4.50	:	.38	4.90
.15	2.50	:	.23	3.80	:	.31	4.55	:	.39	4.95
.16	2.70	:	.24	3.90	:	.32	4.60	:	.40 &	5.00
.17	2.90	:	.25	4.00	:	.33	4.65	:	Up	

On all other minerals recovered in the process of mining and milling uranium, the royalty shall be 5% of fair market value.

Transportation or other charges shall not operate to reduce the state's applicable royalty percentage. The state's proportionate share of reasonable transportation costs may be deducted from royalty payments due the state. When requested to do so, Lessee shall furnish evidence to Lessor that prices received for ore and transportation charges are reasonable and fair.

SCHEDULE A -- SURFACE OR PIT MINING ROYALTY:

GROSS PURCHASE PRICE PER TON (2000#) AT MILL	ROYALTY RATE %	:	GROSS PURCHASE PRICE PER TON (2000#) AT MILL	ROYALTY RATE %
Up to $4.00	5	:	$18.00 to $20.00	9
$4.00 to 6.00	5½	:	20.00 to 22.00	9½
6.00 to 8.00	6	:	22.00 to 24.00	10
8.00 to 10.00	6½	:	24.00 to 26.00	10½
10.00 to 12.00	7	:	26.00 to 28.00	11
12.00 to 14.00	7½	:	28.00 to 30.00	11½
14.00 to 16.00	8	:	30.00 to 32.00	12
16.00 to 18.00	8½	:	32.00 and Up	12½

SCHEDULE B -- UNDERGROUND MINING ROYALTY:

GROSS PURCHASE PRICE PER TON (2000#) AT MILL	ROYALTY RATE %	:	GROSS PURCHASE PRICE PER TON (2000#) AT MILL	ROYALTY RATE %
		:	$22.00 to $24.00	7½
$ .00 to $14.00	5	:	24.00 to 26.00	8
14.00 to 16.00	5½	:	26.00 to 28.00	8½
16.00 to 18.00	6	:	28.00 to 30.00	9
18.00 to 20.00	6½	:	30.00 to 32.00	9½
20.00 to 22.00	7	:	32.00 and Up	10

IN SITU OR HEAP LEACHING:

For U_3O_8 produced by an in situ or heap leaching process, the grade of ore being leached shall be deemed to be .12% U_3O_8 content. The royalty shall be 5% of the mill value of .12% ore.

STOCKPILES, DUMPS, TAILINGS, ETC. ON LEASED PREMISES: Unless otherwise agreed to by Lessor in writing, all material remaining on leased premises after termination of this lease shall be the property of Lessor.

Figure 5-2 (continued).

mailed for the party's signature. An authorized agent of the state would later execute the lease, and a lease on the tract or tracts was then assured for one year. Such uncomplicated procedures, accompanied by relatively low rental rates, often result in vast areas of hundreds of thousands of acres being under lease, with resultant income to the state, even though much of the land may not have any great potential for the mineral sought.

Today, some states have altered and considerably complicated the leasing procedure. In some states, if an explorationist is interested in leasing a particular tract of land, he must notify the state board of land commissioners, and they, in turn, are required to put that tract up for competitive bidding after mailing announcements throughout the industry and placing advertisements in local newspapers. The explorationist who had expressed original interest in the particular tract of land must then be prepared to make an offer against possible competitive bids. Both plus and minus factors are inherent in this system. First, the state *may* derive higher revenues from the competitive bidding; psychologically, the atmosphere of a competitive bidding situation may result in higher offers than might occur when an individual quietly files an application for a lease, unknown to competitors.

On the other hand (and as mentioned above), there are factors which may work to the detriment of the state. For example, some companies or individuals may approach the state land board (which is generally the body responsible for issuing leases) hoping to make an exceptionally large land play—say, on the order of 100,000 to 300,000 acres. Such a play might be conceived even though the company geologists could recommend, on the basis of available information, only four or five sections (say, 3,000 acres) which

appeared to be of interest for drilling or further work. If the land board was accustomed to negotiating terms for large acreage deals, an agreement might be worked out. If 200,000 acres of state lands were leased, such lands might usually carry a $1-per-acre-per-year rental, but a deal might be negotiated at 35 cents per acre for the first year, escalating to 50 cents per acre for the second year, and then to $1 per acre for the third year. The lessee, of course, almost always has the right to drop disproven lands.

If we look at the economics of such a deal from the state's position, the following considerations apply. (1) Lease on 3,200 acres at $1 per acre (assume no bidding) equals $3,200 for one year. (2) Lease on 200,000 acres at 35 cents per acre equals $70,000 for one year.

Admittedly, other factors may enter the picture, such as possible higher bids for some of the land at a later date. But on the other hand, exploration often tends to take more time than planned, and so the rentals on virtually all of the original land are paid for several years. In one large play, we made a deal for all of the state land in the center of a large county, and three years later, we had been able to drop only a few sections because of a major focus of exploration on fee lands where ore had been discovered.

As cited in these examples, states have often been able to lease huge tracts of land which would otherwise have remained unleased if such deals were not allowed to be negotiated. Some states continue to believe that they earn greater revenues from the competitive bidding system, while others maintain that their revenues are greater when the land commissioners are allowed to negotiate deals according to the circumstances. The state-owned land in some states remains totally unleased because the commissioners and others responsible cannot make up their minds about leasing procedures. This results in a significant loss of revenue for the state, as well as precluding the lands from exploration.

Federal Prospecting Limits

As we pointed out above, mining claims may not be located on federal acquired lands. Rather, an application must be filed for a federal prospecting permit. Such application is processed by the BLM and other federal agencies; if the permit is granted, it is issued for a term of two years, at a rental of 25 cents per acre per year. Each individual application may not exceed 2,560 acres; this is the maximum acreage per permit. Each company, or individual applicant, is allowed no more than 20,480 acres of federal acquired lands for prospecting purposes in any one state; this limit is reduced to 10,000 acres for leasing.

Provided certain obligations are fulfilled during the term of the permit, it may be extended for a further two-year period. At the end of the four-year period (or after the initial two-year period), the holder of the permit may apply for a preference right lease, provided he has discovered commercial mineral deposits on the lands covered by the permit.

In the case of uranium, the permit holder would have to demonstrate to the BLM and USGS that certain work had been performed and that an economically recoverable mineral deposit had been found. We have known of recent instances where preference right leases have been denied because the BLM or the USGS did not consider the discovery to be of commercial significance. Although many of these rejections have been appealed, we know of no federal ruling which has so far been reversed.

Public Domain

On other federal lands, such as public domain, mining claims may still be staked. It should be noted, however, that certain categories of public-domain lands have been withdrawn from exploration, including national park lands. In addition, in the 1950s, the Atomic Energy Commission (AEC) withdrew large tracts of known uranium lands from uranium exploration.

There are other instances of withdrawals from among federal acquired lands. For example, we experienced rejection of application for prospecting permits on an area in one western state on the grounds that such lands constituted experimental grasslands, or grasslands research area. These lands were administered by the Forest Service, which claimed that our exploratory work might interfere with their grasslands research program. It was futile to point out that grass was now poor, and had been in historical time, because of lack of rainfall. The "grasslands research" had been going on for 44 years (since 1932) with poor results, but exploration would supposedly interfere.

In the state of Utah, the minerals exploration industry is extremely dissatisfied with BLM rulings concerning withdrawn lands. Approximately 62 percent of the minerals in the state are owned by the federal government; a major portion of these lands have been withdrawn, for such purposes as park lands, scenic and wilderness areas, and so forth. The exploration industry has stated that this restrictive policy is seriously jeopardizing the economy of the state. Unfortunately, it appears that such withdrawals represent a current popular trend in the U.S. Congress, severely hampering development of resources in the western United States. Such policies do not affect the eastern United States to the same degree, since there are few federally owned minerals or public lands, as such. The Rocky Mountain region and the western states, however, suffer most from such restrictive legislation.

Public-domain lands require the staking of lode or placer mining claims as described earlier in this chapter. In the early days of uranium exploration in Wyoming, certain prospects were staked as placers. However, the AEC acknowledged lode claims as the valid claim for uranium, and lode claims have come to be the recognized method of claim staking for uranium on public-domain lands. This makes geologic sense, inasmuch as virtually all of the uranium deposits in sediments in the United States are recognized as epigenetic. Staking an extensive area of public domain for uranium exploration can become an expensive proposition (when compared to most leasing) because of the location procedures prescribed by law. This topic is discussed more fully in chapter 7; we discussed other aspects of this subject in the review of mining law in chapter 4.

Staking public-domain lands for uranium exploration in sandstone deposits must be done in accordance with the respective state and federal mining laws, and each state has slightly different requirements for the method of staking claims. In Wyoming, for example, when a lode mining claim is staked, the explorationist must drill one hole not less than 50 feet deep, or at least 50 feet of drilling must be carried out near the discovery monument in five holes 10 feet or greater in depth. Nevada, on the other hand, has no drilling requirements, but a higher filing fee is charged and a survey map (prepared by a registered surveyor) of the claims, which meets state requirements in accuracy and scale, must be filed. In actual practice, lode mining claims rarely have a valid discovery in advance of the staking of claims. Yet the law in most states requires a "valid discovery" before claims may be staked. This apparent problem is discussed more fully in chapter 4.

Special Problem of Uraniferous Lignite

With respect to uraniferous lignite, the U.S. Congress in the late 1950s passed legislation which has a direct bearing on this type of mineral. It was recognized that there was a conflict between the laws governing leasable minerals such as coal, and the laws governing the location of mining claims for metallic minerals, such as uranium. In order to resolve this conflict, legislation was passed excluding uranium associated with lignite from the locatable minerals. Consequently, the explorationist who plans to conduct exploration for uranium in association with lignite or another organic material (such as coal or a solid carbonaceous substance) should first have legal counsel undertake a careful study of the law.

Analysis of Fee Mineral Lease

Figure 5-3 is an example of a fee mineral lease form. Almost all lease forms for minerals contain provisions which are complicated and difficult for the novice to understand. Therefore we shall discuss briefly the key provisions in the sample lease, and the reasons why the lease is organized and written in the way it is.

Under the names of the lessor and lessee is paragraph 1, "Interests Leased." This paragraph describes the nature of the minerals or other interests which are being leased by the lessee under the terms of this form. This particular form is a uranium lease, although it also describes other minerals which might be spatially associated with uranium, such as vanadium and other fissionable source materials. The intent of this language is to allow the lessee to mine uranium and other valuable minerals that might be deposited in proximity to uranium and that would be removed in the uranium operation. If the other valuable minerals are not spatially associated, however, they are not included in the terms of the lease. For example, if molybdenum or vanadium were found in a bed 50 ft below the bed in which the uranium was found, lessee could not mine those minerals under this lease. This lease form is specifically intended to cover

Mining Lease

THIS LEASE is made and entered onto on _____ , 19 _____ , by and between POWER RESOURCES

CORPORATION, a Wyoming corporation, with offices at 1660 So. Albion St., Suite 827, Denver, CO. 80222,

hereinafter called "Lessee," and _____

_____

whose address is_____,

hereinafter called "Lessor."

1. **Interests Leased.** For and in consideration of the sum of Ten or more dollars paid to Lessor by Lessee, the reciept and sufficiency of which is hereby acknowledged by Lessor, the royalty herein reserved and the mutual covenants hereinafter set forth, Lessor hereby grants, leases, and lets exclusively unto Lessee, all uranium, thorium, vanadium and all other minerals (hereinafter collectively "Subject Minerals") under the lands (hereinafter "**Leased Premises**") described herein, excepting only oil, gas, casinghead gasoline, sulphur, condensates and associated hydrocarbon substances, sand, gravel and caliche. Lessee shall have and is hereby granted, the right to remove and place in waste dumps any and all sand, gravel, caliche and other unleased materials occuring in the "overburden" which is reasonably necessary to be removed in connection with Lessee's operations hereunder, but if such substances are a part of, intermingled with or associated with any of the Subject Minerals covered by this lease, Lessee shall have the right to dispose of same as Lessee may deem appropiate, without compensation to Lessor.

Lessor further grants to Lessee the right and privilege to enter upon the Leased Premises for the purpose of surveying, exploring (including without limitation, geophysical and geochemical exploration), investigating, prospecting, drilling for, developing, mining (including without limitation strip, open pit, underground, solution or any other method of recovering minerals), stockpiling, waste disposal, removing, milling, shipping and marketing any of the Subject Minerals and incident to the foregoing to construct and use buildings, roads, camp facilities, shafts, tunnels, inclines, adits, drifts, pits, mine buildings, power and communication lines, ditches, canals, pipe lines, mills and related facilities, and other improvements related to the accomplishment of the foregoing; to use so much of the surface of the Leased Premises in such manner as may be reasonably necessary, convenient or suitable for or incidental to any of the rights or privileges of Lessee hereunder or otherwise reasonably necessary to accomplish any of the foregoing; together with easements in and all rights of way for ingress and egress to and from the Leased Premises to which Lessor may be entitled, including surface and subsurface easements across other lands presently owned or hereinafter acquired by Lessor; all of the foregoing rights pertaining to access to easements across and under and use of the surface of Leased Premises and other lands above referred to being granted irrespective of the specific location and ownership of the lands on which mining or other activities are being carried on by Lessee: to the lands located in _____ County, State of _____ , to wit:

and containing _____ acres more or less. The parties further agree that any interest hereinafter acquired by Lessor to Subject Minerals underlying Leased Premises shall be deemed subject to the provisions of this lease.

SUBJECT ONLY TO: existing leases, rights and encumbrances of record; the right of Lessor to use, lease and convey the surface of the Leased Premises subject to the rights granted to Lessee hereby; and the right of Lessor to develop, explore, produce or lease for development, exploration or production of oil, gas, and associated liquid

Figure 5-3.

hydrocarbons and other unleased minerals in and under the Leased Premises insofar as the same may be developed, explored or produced without materially interfering with Lessee's operations under the terms hereof.

2. **Term of Lease.** The term of this lease shall commence as of the first day of the first entire month following the execution hereof and this lease shall continue in full force and effect for a term of ten years from such commencement date, and so long thereafter as any of the Subject Minerals are produced from the Leased Premises, provided, however, if there is no reasonable market for any of the Subject Minerals as to which reserves (including low grade mineralization not economically feasible at current market prices) have been developed on Leased Premises, or in the event production is prevented because of a force majeure within the meaning of paragraph 20, the term of this lease shall continue despite a lack of production (regardless of whether there has or has not been production of same or a cessation of production whether before or after the expiration of the primary term) of any of the Subject Minerals on the Leased Premises if, during the period for which no production occurs, the annual delay rental payments required by paragraph 3 hereof shall be made.

3. **Delay Rentals.** On or before each anniversary date of this lease, Lessee shall pay to the credit of Lessor in the _____ Bank of _____ or any successor the sum of $ _____ as a delay rental or, if in production, as a minimum royalty, the same to be reduced by the amount of any royalties received by Lessor for production from the Leased Premises during the preceding lease year and by any payments made by Lessee for purposes authorized in paragraph 13 hereof and not theretofore recovered by Lessee from payments by Lessor or credits in preceding years.

Initials

_____

_____

Page 1

4. **Royalties.** Lessor reserves as royalty ten per cent (10%) of all ores containing Subject Minerals produced and sold from the Leased Premises, based upon the value of such ores at the mine as produced in raw form before any processing or beneficiation and before such ores are transported away from the mine to a point of sale, storage, or processing. When ores containing Subject Minerals are sold by Lessee in raw form before any processing or beneficiation, the royalty shall be paid on the actual amount received by Lessee from the sale of such ores, after deducting transportation costs from the mine to the point of sale and deductions made by the purchaser of any such ores for sampling, assaying, weighing, penalties and other similar charges. When ores containing Subject Minerals are processed or beneficiated by or for the benefit of Lessee before being sold, said royalty shall be payable on the fair market value of such ores as produced in raw form at the mine, provided that, when uranium-bearing ores are processed or beneficiated for the recovery of uranium oxide (U308) contained in the ore, such value shall by determined as follows:

0.05% or less	$5.00
0.06	7.00
0.07	9.00
0.08	11.00
0.09	13.00
0.10	15.00
0.11	19.00
0.12	23.00
0.13	27.00
0.14	31.00
0.15	35.00
0.16	42.00
0.17	49.00
0.18	56.00
0.19	63.00
0.20%	70.00

Over 0.20% add $3.50 per ton for each 0.01% U308 in excess of 0.20%.

Fractional parts of a ton shall be valued on a pro rata basis to the nearest cent. All weights for purposes of thes lease shall be dry weight. The result of the above calculation shall be multiplied by X divided by $40.00 where X is the per pound exchange value for immediate delivery of U308 concentrates("yellowcake") published by Nuclear Exchange

Figure 5-3 (continued).

Corporation ("NUEXCO") for the month in which the uranium-bearing ores are actually removed from the Leased Premises for processing or beneficiating. If NUEXCO no longer publishes such exchange value, a representative comparable value for the immediate delivery of U308 in concentrates ("yellowcake") shall be utilized.

In the event uranium and other associated minerals are recovered from the Leased Premises by Lessee through in situ, leach or solution mining, Lessee shall pay Lessor a royalty in lieu of any other royalty set forth herein for each pound of U308 recovered of $1.33 multiplied by X divided by $40.00. X for the purposes of this formula shall be the per pound exchange value of uranium concentrates determined in the manner set forth in the immediately preceding paragraph above.

The royalty on any mineral recovered as a by-product shall be payable only to the extent Lessee removes and sells same or is otherwise separately compensated for same.

Payment of royalties shall be made by Lessee to Lessor, or to Lessor's credit in the Bank designated in paragraph 3 hereof. Such payment shall be made within thirty (30) days after the end of the calendar quarter within which said ore is processed, if processed by or for the benefit of Lessee or removed from the Leased Premises if sold by Lessee, and shall be accompanied by a statement showing weights, analyses and values of all ores, concentrates, minerals, and metals produced from the Leased Premises during such previous calendar quarter and the amount of the charges or costs deductible therefrom.

The mineral content of all ore mined and removed from the Leased Premises shall be determined by Lessee, or by the mill or smelter to which the ore is shipped, in accordance with standard sampling and analysis procedures. Upon request to Lessee, and Lessor's expense, Lessor shall have the right to have a representative present at the time samples are taken

In the event Lessee mills ore from the Leased Premises, Lessor shall be furnished, at the Lessor's request and at no expense, a portion of all or any samples taken for analysis. In that event, split samples shall be retained by Lessee for later analysis by an independent referee, and, in the event of a dispute concerning Lessee's assay of samples, royalty payments shall be based on the assay results of the independent referee selected by agreement of the parties. The cost of the referee shall be paid by the party whose assay shows the greatest variance from that of the referee.

All residue or tailings remaining after the initial processing or milling of the crude ores mined from the Leased Premises shall be the sole and exclusive property of Lessee; provided that, if any such tailings or residues remain on any of the Leased Premises as to which this lease is terminated for a period of one year after the effective date of such termination, then, as between Lessor and Lessee, all such tailings shall be the sole and exclusive property of Lessor.

5. **Maintainance of Operations.** Lessee agrees to maintain all roads, camps, drillsites, and mines in a good and workmanlike manner.

6. **No Drillsite near Dwelling House.** No drillsite shall be located within 200 feet of any existing well or dwelling house unless the permission of Lessor is first had and obtained.

7. **Limitation of Exploratory Work.** No work shall be done by Lessee during exploratory operations that may interfere with lambing or feeding of livestock unless the permission of Lessor is first had and obtained.

8. **Sheep-Tight Fences.** Lessee agrees to enclose with a sheep-tight fence any excavations dangerous to livestock that it may make in the course of its operations.

9. **Maintainance of Fences.** The Lessee agrees not to keep fences down at any point, to repair promptly any road damage done by vehicles and equipment of Lessee and to erect gates or cattleguards in all fences in which openings may be made for ore haulage.

Initials

———

———

Page 2

---

10. **Damages for Surface Use.** Lessee shall pay Lessor the sum of $10.00 for each exploration or development hole drilled by Lessee on the Leased Premises as a liquidated sum for any damages caused by or related to said drilling.
For all damages to the premises resulting from testing, injection and/or producing wells conducted in conjunction with in situ leaching or solution mining procedures, other than plant site damages hereinafter provided for, Lessee shall pay Lessor the sum of $5.00 per well as a liquidated sum for any damage caused by or related to the drilling and/or operation of said well.

Figure 5-3 (continued).

For each surface acre of land on which is situated any portable or non-portable facility other than a well for extracting minerals by situ leaching or solution mining procedures, and which is adversely affected by such facility insofar as the growing of planted crops, including grasses, is concerned, Lessee shall pay to Lessor, as land damages, the annual sum of $100.00 for each surface acre of land so affected. It is further agreed and provided, however, that in the event of the Lessee's use of the premises in conducting procedures for extracting all minerals recovered from the premises by in situ leaching or solution mining procedures adversely affects less than the total of two acres per year, Lessee shall pay Lessor the minimum annual sum of $200.00. However, nothing herein contained shall preclude Lessee from designating additional sites on said premises for continued in situ leaching or solution mining operations. Damage payments for such subsequent sites, if selected by Lessee, will be paid according to the terms and conditions of this Agreement, and will commence and terminate according to the terms hereof.

For all other surface lands utilized by Lessee for mining, milling, construction of buildings, construction of roads, plant sites, mine dumps, waste and production facilities and other uses which permanently or indefinitely make such lands unavailable to Lessor, Lessee shall pay Lessor 125% of the then current value of the surface area actually appropriated by Lessee based upon the value of the land for the use then being made by Lessor of such lands. Lessor upon receipt of such payment shall execute a deed conveying to Lessee the surface to such lands in fee simple determinable so long as the lands are used for one or more of the foregoing specified purposes. Notwithstanding such conveyance the rental provided for in paragraph 3 and royalty provided for in paragraph 4 shall be payable with respect to the minerals produced from such lands.

Lessee shall also pay Lessor reasonable compensation for any damage not otherwise compensated for hereunder caused by Lessee to Lessor's buildings, fences, personal property and growing crops.

Except as otherwise provided by the foregoing provisions, Lessor hereby specifically agrees that the moneys paid and to be paid as rentals and royalties pursuant to paragraphs 3 and 4 constitute a liquidated sum in full payment for all surface use, damages (including those arising from disruption of Lessor's operations) and access as hereinabove provided and hereby waives any other claim for surface use and damage which might result from the activities of Lessee, its employees, contractors or assigns. Nothing contained in this paragraph 10 shall authorize the recovery by Lessor of special or consequential damages and such damages shall in no event be allowed.

11. **Environmental and Related Considerations.** Lessee agrees to abide by all valid federal and state laws, rules and regulations governing mines, mineral exploration and environmental protection. Lessor acknowledges he is aware of the fact that his consent as surface owner may be required under existing or hereinafter enacted state or federal law to any exploration, development, mining, milling or reclamation plan and that such laws may require Lessee as operator to furnish a bond to assure compliance with the approved plan. In the event Lessee shall apply to any appropriate (federal, state or local) authority for a permit or other form of authorization to carry on exploration, development, mining, or milling operations on Leased Premises or on other premises which affect Leased Premises, Lessor shall furnish Lessee and the appropriate authority with any consent, waiver or other document required from him as a surface owner.

12. **Lesser Interest.** In the event Lessor owns an interest in the Subject Minerals which is less than the entire and undivided mineral estate therein, whether or not such lesser mineral interest is referred to herein, then the royalties and delay rentals herein provided for shall be paid to Lessor only in the proportion that Lessor's interest therein bears to the whole and undivided mineral estate. Lessor shall be entitled to retain only such proportion of the moneys previously paid to Lessor hereunder as Lessor's said interest bears to the whole and undivided mineral estate, and Lessee shall be entitled to offset all other moneys previously paid Lessor hereunder against royalties and delay rentals which thereafter become due and payable to Lessor hereunder. This paragraph shall apply with equal force and effect to reduce payments pursuant to paragraph 10 herein when Lessor owns a lesser interest in the surface estate than the entire and undivided fee simple interest therein. Notwithstanding any other provisions of this lease, with respect to that portion, if any, of the Leased Premises in which Lessor owns no right, title, or interest in the Subject Minerals, Lessee is under no obligation to pay and Lessor has no right to receive, delay rentals or royalties. In the event of any dispute as to the right of Lessor as to the payment of royalties, minimum royalties, or delay rentals hereunder or as to the amount of such payments, Lessee shall not be deemed in default hereunder if it in good faith disputes same and promptly proceeds to have the dispute resolved by a court of competent jurisdiction.

13. **Warranty.** Lessor hereby warrants and agrees to defend the title to the Leased Premises, and agrees that Lessee, at its option, may pay and discharge any taxes, mortgages, or other liens existing, levied or assessed on or against the Leased Premises; may be subrogated to the rights of any holder or holders thereof; and may reimburse itself for any such expenditures out of any royalty or rental thereafter accruing to Lessor or by enforcement of subrogated rights against Lessor in any court of competent jurisdiction.

14. **Exploration and Mining Operations.** Subject to the provisions of paragraph 11, Lessee may use and employ methods of exploration and mining as it may desire or find most profitable and economical. Lessee shall not be required to mine, preserve, or protect in its mining operations any Subject Minerals which under good mining practices cannot be mined or shipped at a profit to the Lessee at the time encountered. Lessee shall have the right and privilege at any time, during the term of this lease and so long thereafter as it may hold an interest in the minerals hereunder to use

Figure 5-3 (continued).

any and all roads or workings located at any time on or under the Leased Premises and Lessee shall have the further right of mixing, either underground or at the surface or processing plant, any ores, solutions or other products (herein called "Products") containing Subject Minerals and produced from the Leased Premises or any portion thereof with Products from any other lands, provided that the mixing is accomplished only after the Products have been sampled and after the weight or volume thereof has been determined or ascertained by sound engineering principles. An accurate record of the tonnage or volume of Products and of the analysis of Products from each property going into such mixture shall be kept and made available to Lessor at all reasonable times. The tonnage or volume of Products from each property, together with the analysis thereof, shall be used as the basis of the allocation between the properties or production royalties paid from such Products.

Initials

———

———

Page 3

15. **Assignment.** The estate of either party may be assigned in whole or in part; provided, that no change in the ownership of the Leased Premises or assignment of royalties payable hereunder shall be binding upon Lessee until Lessee has been furnished with a written transfer or assignment or a certified copy thereof. In the event this mining lease shall be assigned as to a part or parts of the Leased Premises and the Lessee, or assignee or assignees of any part or parts, shall fail or make default in the payment of the proportionate part of the royalties due from him or them or otherwise breach any covenants contained herein, such default shall not operate to defeat or affect this lease insofar as it covers any other part or parts of the Leased Premises. An Assignment of this lease shall, to the extent of such assignment, relieve and discharge the Lessee of all obligations hereunder which have not theretofore become due.

16. **Inspection.** Lessor or its duly appointed representative shall have the right, exercisable at all reasonable times and in a reasonable manner so as not to interfere with the Lessee's operations, to go upon the Leased Premises, or any part thereof, for the purpose of inspecting the workings thereon. Lessor shall hold Lessee harmless from all claims for damages arising out of any death, personal injury, or property damage sustained by Lessor or Lessor's agents or servants while in or upon the Leased Premises as herein permitted, unless such death or injury arises as a result of sole negligence or willful misconduct of the Lessee.

17. **Homestead.** Lessor hereby releases and relinquishes any right of homestead exemption which Lessor may have in the Leased Premises. If the spouse of Lessor is not named above as a Lessor but executes this Lease, said execution shall be deemed for the purpose of releasing homestead and other marital rights, if any, only.

18. **Multiple Lessors.** Whenever two or more parties are entitled to receive delay rentals or royalties hereunder, Lessee may withhold payment thereof unless and until all such parties designate in a recordable instrument an agent empowered to receive all rental or royalty payments due hereunder and to execute division and transfer orders on behalf of said parties and their respective successors in title.

19. **Taxes.** Lessor agrees to pay all general ad valorem taxes and assessments assesed against the Leased Premises and all taxes resulting from the Lessor's use thereof; provided that Lessee shall pay for that portion of such taxes which is attributable to any producing mine opened and operated on the Leased Premises by Lessee less the part thereof attributable to Lessor's royalty interest therein. Lessee shall pay all other lawful public taxes and assessments, whether general, specific or otherwise, assessed and levied upon or against the Leased Premises attributable to Lessee's operations, or upon any ores and other products thereof, or upon any prperty or improvements placed by Lessee on the Leased Premises; provided that if any tax (including ad valorem taxes) is now or hereafter levied on or measured by production, Lessor shall pay that portion of such taxes which is attributable to the royalty reserved herein. Lessee shall have the right in good faith to contest any of the above taxes, whether payable by Lessee or payable by Lessor, but shall not permit or suffer the Leased Premises or any part thereof, or any ore mined thereon, or any improvements or personal property thereon to be sold at any time for such taxes or assessments.

20. **Force Majeure and Default.** None of the parties hereto shall be liable to the other parties and no party hereto shall be deemed in default hereunder for any failure or delay to perform any of its covenants, agreements or obligations, other than the obligation to pay money, caused by or arising out or any act not within the control of the party, including, but not by way of limitation, acts of God, strikes, lockouts, or other industrial disputes, acts of the public enemy, war, riots, lightning, fire, storm, flood, explosion, governmental laws, regulations, or other governmental restraints, and inability to obtain necessary equipment or material in the open market. A party affected by such causes

Figure 5-3 (continued).

shall promptly notify the other parties in writing. No right of a party shall be affected for failure or delay of the party to meet any condition of this Agreement where such failure or delay is caused by one of the events referred to above, and at all times provided for in this Agreement the time to meet any such condition shall be extended for a period commensurate with the period of delay. Nothing contained herein shall require the settlement of strikes, lockouts, or other labor difficulties by the party involved, contrary to its wishes; the manner of handling or remedying any or all of difficulties or conditions referred to in this provision shall be entirely within the discretion of the party concerned.

Failure by Lessee to perform or comply with any of the covenants of this lease shall not automatically terminate this lease or render it null and void, but Lessor may notify Lessee in writing of any asserted default and if in default, Lessee shall have a period of sixty (60) days after receipt of such notice within which to cure, or commence to cure, such default, and, except as otherwise provided in this paragraph 20, if such default shall not have been cured within such time, Lessor may terminate this lease.

21. **Pooling** In connection with in situ leach or solution mining, Lessee is hereby granted the right to pool and combine acreage from the Leased Premises with acreage from other lands for the purpose of operating a drilling pattern set up for such mining at the boundary of said Leased Premises. In such event, Lessee shall have the location of such drilling pattern accurately surveyed to determine the position of the drilling pattern with respect to the property line or lines of the properties involved. The Lessee shall furnish the Lessors of the Leased Premises and of the other properties involved a copy of such survey. The allocation of royalties attributable to mineral containing solutions produced from the production well or wells involved in such drilling pattern shall be allocated to the respective royalty owners pro rata on the basis of surface acreage of the Leased Premises and other lands lying within such drilling pattern. It shall be conclusively presumed that the Subject Minerals which are produced are produced uniformly within the boundaries of such drilling pattern. In no event shall the drilling pattern used for the purpose of pooling under this paragraph exceed one hundred (100) feet on either side of the boundary line, or a radius of one hundred (100) feet from a point on the boundary line if the drilling pattern is irregular.

22. **Further Documents.** At the request and expense of Lessee, Lessor shall deliver to Lessee for the purpose of copying the same, any documents, abstracts, policies, or other information relating to the Leased Premises or Lessee's operations hereunder, and shall execute and deliver to Lessee any instructions, agreements, documents, or other papers reasonably required by Lessee including a recordable Memorandum of Lease to effect the purpose of this lease. Lessor shall at all times cooperate with Lessee in any reasonable way to assist Lessee in the effecting of the purpose of this lease.

23. **No Implied Covenant.** The creation of a royalty interest hereunder and other provisions hereof shall not imply an obligation on the part of Lessee to open or develop any mine on the Leased Premises, nor to mine, recover, remove or produce ores, materials, or minerals therefrom or to begin, perform, conduct, continue or resume exploration, development, mining or other operations on the Leased Premises. Any and all operations shall be commenced, conducted, terminated and resumed only if and to the extent and at such times and locations and by such methods as Lessee may in its sole discretion and from time to time elect.

Initials

———

———

Page 4

24. **Surrender.** Lessee may at any time execute and deliver to Lessor, or place of record, a release or releases covering all or any portion or portions of the Leased Premises and thereby surrender this lease as to all or such portion or portions and, from and after the date of such release, thereby terminate all obligations including a proportionate amount of delay rentals, as to acreage surrendered except obligations accrued as of the day of the surrender. If this lease is surrendered or terminated for any cause, Lessee shall have one year after the effective date of such surrender or termination in which to remove all engines, tools, machinery, buildings, structures, headframes, trailers, ore stockpiles and all other property of every nature and description erected, placed or situated on the portion of the Leased Premises surrendered hereunder by it. Upon such termination Lessee shall close off and plug any open shaft to such extent that such shaft shall not thereafter be a hazard to persons or livestock using the surface of the Leased Premises.

25. **Binding Effect.** This lease shall be binding upon the parties hereto, and upon their heirs, successors and assigns.

26. **Notice.** Any written notice provided herein shall be sent by certified mail, postage prepaid, addressed to Lessor or Lessee, as the case may be, as set forth below or at such other address as to which notice is given hereunder.

Figure 5-3 (continued).

If to Lessor: _____
_____
_____

If to Lessee: POWER RESOURCES CORPORATION
1660 So. Albion St. Suite 827
Denver, Colorado 80222

27. Other Provisions.

    IN WITNESS WHEREOF, the parties hereto have set their hands and seals as of the day and year first above written.

LESSOR: LESSEE:
POWER RESOURCES CORPORATION

_____   _____

Social Security No.:_____

_____

Social Security No.:_____

_____

Social Security No.:_____

_____

Social Security No.:_____

STATE OF_____⎫
COUNTY OF_____⎬ ss:

    On this_____ day of_____, 19_____, before me personally appeared _____ , _____, known to me to be the person(s) who is (are) described in and who executed the within and foregoing instrument and acknowledged that _____he_____ executed same for the purposes therein expressed.

    IN WITNESS WHEREOF, I have set my hand and seal.

My commission expires

My commission expires: _____

_____
Notary Public

Page 5

Figure 5-3 (continued).

uranium and associated minerals; it does not cover oil, gas, coal, and associated hydrocarbons. This particular lease form should not be used in an area where occurrences of uraniferous lignite are anticipated, since problems could arise in the interpretation of "interests leased." If the lignite were of sufficient character to justify being mined as lignite, then it could not be included in this lease. However, if the uraniferous lignite had no value as lignite, either because it was too impure or too thin, then the uranium provisions in the lease *might* be adequate. On the other hand, if the uranium occurred at the *top* of a thick lignite bed, the lease might be valid provided the value of the lignite bed was not diminished by the mining operation. Such a problem might have to be resolved in court. We recommend that the explorationist try to anticipate, and avoid, such problems by including uraniferous lignite in the lease form if there is a chance it may occur in the area. Have a different lease form printed if necessary.

The first paragraph contains wording that grants the lessee the right and privilege to explore, prospect, and so forth; to use buildings, roads, and other improvements; and to have rights of ingress and egress to the leased property. Some lawyers are of the opinion that the wording in this paragraph is less than adequate. Some lease forms provide the lessee the right to use the leased premises for mining *adjacent* properties. However, this form contains the basic provisions for entering and exploring the property and, if minerals are found, the right to develop them.

Below the interest lease provisions, there is a space for the specific description of the leased property and, directly under the description, a provision, "Subject Only To." This provision is included in the event that there is a prior lease of record. Naturally, the lessee in the prior lease would have a preferential right to the surface use,

above that of the second or third party. For example, the landowner might have leased the property for oil and gas a couple of years earlier; the following year, he may have leased the property for coal; and now he is entering into a uranium lease on the same property. Assuming that the first two leases were of record and were being maintained, the oil and gas lessee and the coal lessee would have prior claims to the use of the surface, over those of the uranium lessee.

An important point to note here is that it is the *recorded* lease which prevails in any conflicting situation. For example, an explorationist may check the records on a particular property and discover that there is no uranium lease of record. He may then enter into a uranium lease with the landowner, and immediately record that lease, even though two weeks earlier another explorationist may have taken a uranium lease on the same property but failed to place it of record. In such a case, it is usually the *recorded* lease which will prevail, unless there are extenuating circumstances. In the event of a confrontation, such a situation could be subject to the scrutiny of a court of law. A lessee has an obligation to promptly record an executed lease, but he should be allowed a reasonable time to do so. A court may determine that, in the event a landowner leased a piece of property to one party and two days later leased the same property to a second party, the first party's lease would be held valid. Regardless of the fact that the second party had immediately recorded his lease, the court may rule that the first party had not been allowed reasonable time to record his lease. On the other hand, if the first party waited several weeks before recording his lease, he should be aware that he may be leaving himself unprotected by so doing.

A landowner might occasionally lease his property for uranium exploration and, during the term (explained below) of that lease, issue another uranium lease to a second party even though both landowner and second party are aware the first lease exists. The second lease is commonly referred to in the industry as a "top lease." As long as the first lease is recorded, and the rentals are being paid, it will prevail. However, the second (top) lease would take effect as soon as the first lease is discontinued, either through default on rental payment or for other reason. Generally, since the top lease only becomes effective after the first lease becomes null and void, rental payments are not due until that time; some landowners receive double rentals as the second lessee waits for the lease of first lessee to expire, usually by the end of its term. If the original lessee should bring the property into production, the top lease holder will lose his investment. A top lease holder takes a calculated risk, since he is aware that another party has prior rights to the property in question.

On the other hand, if the original lessee has a lease with a certain term (such as 10 years) and a discovery has not been made on the property, or a discovery has been made but production has not commenced, a second party has every right to exercise a top lease and initiate development of the minerals. If the original lessee wishes to protect himself against such an event, he could possibly request a clause written into his lease stating that he has first right of refusal over any potential top lessee.

The next paragraph, reading down the lease form, relates to the term of the lease. We show here a 10-year term, which is a fairly standard duration for a uranium lease. The coal industry, on the other hand, prefers to write 20-year-term leases, since a longer lead time is usually required before the mining operation can begin, and once begun it is frequently long term.

Another item frequently included under the term of a coal lease, which is not found in a uranium lease, is a unitization clause. This permits the lessee to unitize leases within a certain mineable block; thus production from any lease in that block would hold every lease in the entire block. This clause precludes potential disruption of an elaborate mining plan, and the lessee has the right to designate units within the leased property.

The next category on the lease form is headed "Delay Rentals." This merely states that if the lessee does not bring the property into production by the end of the first year, he may delay it a further year by paying a specific rental. In effect, the minerals lease is a license to mine; thus, the delay-rental terminology acknowledges the fact that the purpose of the lease is primarily to mine, and the delay rental is a substitute payment to the landowner in lieu of a production royalty.

Landowners often raise questions concerning the amount of the delay rental and bonus payment, if any. The amount of these figures is very important to the landowner, and the amounts can fluctuate according to the demand for leases in a given area at a particular time. The explorationist has to decide what the specific land is worth to him, and must set an arbitrary figure to offer as a bonus. Depending upon his previous experience, the landowner may not automatically accept the explorationist's bonus offer. If he has been accustomed to being offered substantial sums—say, $25 per acre—he is not likely to accept $1 per acre.

Another potentially conflicting situation can arise relating to delay rentals and bonus payments: in an area where there may be several minerals, such as uranium, oil and gas, and coal, the landowner may be offered competitive bids from different lease brokers. This is common in an area where there is much activity, and landowners often play one bid against another, attempting to secure the highest bonus and rental payments. If the landowner is able to lease coal rights for $30 per acre, and oil and gas rights for $50 per acre, he is not likely to consider seriously an offer of $1 per acre for a uranium lease. For even though the uranium properties may be far more speculative than the coal or the oil and gas properties, and scarcely worth more than the $1 an acre to the explorationist according to the geologic indications, the landowner does not know (or believe) this and usually will try to maintain high rates for all minerals.

The standard, industry-wide rental on a highly speculative uranium lease has been $1 per acre bonus and $1 per acre rental. However, with the price of uranium concentrate rising so sharply, it is unlikely that this figure will remain standard for much longer, and indications are that uranium explorationists must expect to pay more in the future. In leasing properties for any type of mineral exploration, the most difficult aspect of the negotiations is establishing a bonus payment in a competitive situation. Some major oil companies have made serious mistakes in the past in bidding very high bonuses, especially in connection with oil-shale exploration and development. In leasing speculative properties for uranium exploration, it must be acknowledged that perhaps 99 percent of the acreage will have essentially no potential and will be ultimately weeded out.

The next item we come to on the lease form pertains to royalties. The figure of 5 percent quoted on the sample lease is a standard royalty rate commonly paid on ores in raw form at the mine, if the ores are sold to another party. However, if the ores are to be utilized or processed at a mill owned by the lessee, then the value

of the ore can be subject to question. To resolve this problem, the industry has designed a standard table, which is also included in this lease form. Although there are many methods used to calculate royalties, the sample shown is one of the most common approaches used in the industry.

In addition, the industry attempts to take into account the selling price of uranium concentrate derived from the ore when determining the royalty rate. Thus, language is included in the lease form to take into account the concentrate value, utilizing an $8-per-pound figure (historical AEC figure) for a base concentrate selling price. Accordingly, as the price on concentrate fluctuates up or down from the $8 figure, the royalty paid to the lessor would fluctuate similarly. Many lease calculations are being modified to reflect current higher selling prices of uranium. Consequently some royalty rates on new leases may be higher than shown. One common procedure is to increase the 5 percent figure to 10 percent base royalty.

As an illustration of how the table in the sample lease form functions, let us assume that lessee will mine uranium ore from lessor's property and process the ore through lessee's own mill. The grade of ore is 0.10 percent (2 lb per ton) and the uranium concentrate is sold for $40 per pound. The royalty per ton is calculated as follows:

Grade of ore (0.10 percent) = 2 lb per ton times $1.50 = $3 per ton.

Royalty rate (5 percent) times $3 = $.15 per ton.

Concentrate factor = $40 divided by $8 = $5.

Concentrate factor ($5) times base royalty ($.15) = $.75 per ton final royalty.

The second paragraph in the royalties section covers production from a solution mining or *in situ* process, or from ores which are brought to the surface and then leached by passing solutions through them to recover the uranium (called "heap leaching"). In these cases, the royalty is based on the uranium recovered in the solution rather than that contained in the original ore (regardless of the quantity recovered). Historically, in the United States, recovery of uranium from ore has been around 95 percent. In other words, about 5 percent of the uranium cannot be recovered from the ore. With *in situ* (solution) mining and heap leaching, less uranium is recovered from the ore. The described provision in the royalty schedule allows for payment based on actual percentage of uranium recovered.

The next few paragraphs in the royalties section are self-explanatory, down to the last paragraph, which states that tailings remaining after the initial processing or milling shall be the sole and exclusive property of the lessee; provided that if any tailings should remain on the leased premises for a period of one year after the termination of the lease, such tailings shall become the sole and exclusive property of the lessor. Lessors have been known to object to this paragraph, since they feel that they do not want the responsibility of dealing with such tailings if any do in fact remain on the premises for a year. This objection is groundless because mine and mill operators must comply with mined-land reclamation laws in any event.

Another question which landowners occasionally pose is the concern about whether all valuable minerals have been extracted from the tailings. For example, they wonder whether there could still be some valuable vanadium or molybdenum remaining in the tailings; if so, and if the explorationist should decide to return at a later date to work on the tailings, they wonder whether they would be paid another royalty. It should be clearly explained that once ore is processed through

a mill, there is no way to determine from which property the ore was derived, since ore is processed concurrently from several properties. Further, it should be explained that in milling an ore, *all* economically recoverable minerals are extracted when the ore is passed through the mill because that is the most economical procedure. Mining firms do not pass ore through a mill and deliberately avoid extracting values because of plans to hoodwink a royalty owner; yet many landowners are suspicious of this. Obviously, situations could arise where a valueless constituent of ore one year might become valuable a few years later due to price increases, but such changes in value are beyond the control of an individual company.

Paragraph 10 in this lease deals with payment for surface and other damage. This provision gives the lessee the right to have the land independently appraised, and then pay the landowner 125 percent of the appraised value for the property, thus transferring the title from the landowner to the lessee. Some landowners have objected to this clause because they fear unscrupulous lessees might misuse the right in some way, although in most cases it involves an insignificant portion of the landowner's total acreage. Language in the provision provides for the reconveyance of the land to lessor by lessee once operations are complete; thus, theoretically, a landowner could receive 125 percent of the appraised value, receive a royalty on production, and once again receive title to the land after mining is complete. Some landowners have claimed that 125 percent of the appraised value of the land was not enough and have demanded as much as 200 percent of the market value. We have encountered other instances where the land was part of a trust which specifically prohibited any sale of the property; this, of course, conflicts with the terms of the sample mining lease.

We attempt to overcome such objections by pointing out, first, that the landowner will receive a royalty during the mining operation and, second, that the ownership of the land will be returned to him after the mining operations have terminated. Since, according to the terms of the lease, the lessee has the right to enter and utilize the property, and to develop any minerals which might be discovered thereon, there is really no necessity that the surface title be deeded to the lessee. This simply gives the lessee sole possession, without any prior claims to the surface, for the duration of the mining operations.

The remaining part of the lease form is self-explanatory. It uses standard wording which appears on most lease forms.

In many instances when an explorationist enters into a uranium lease with a landowner, the landowner may request certain modifications to the lease form (often instigated by recommendations of his attorney). First, modification may be wanted in the primary term of the lease: landowners frequently object to a 10- or 20-year term because they do not wish to tie up their land for a long period of time. Depending on how much the explorationist wants the lease, he may agree to a concession on the term; but care must be exercised in shortening terms, because 6 to 10 years' time is often required between exploration and production. Therefore, a lease of less than 10 years is hazardous. A change in the term may be done by simply crossing out the stated term length and substituting the newly agreed-upon term. Whenever such a deletion is made to the lease, and new wording inserted, the lessor should always initial the change in the margin on either side, thus eliminating the possibility of the lessor claiming the lessee altered the lease after the lessor had signed it.

Generally speaking, we have found that whenever a landowner is represented by his attorney, some deletions or modifications to the lease always result. For example, we have been required on more than one occasion to include a clause to the effect that we (the explorationists) agree to pay abstract charges for any of our instruments placed of record which would increase the title instruments of record. Such a requirement was brought about by increased abstract fees after explorationists in the past had made it a practice to enter into a minerals lease and then convey the lease to numerous other parties. We believe that the inclusion of such a clause in the lease is a minor concession, especially if it satisfies the landowner and helps the explorationist to get a lease on favorable terms.

Oil, Gas, and Other Minerals Lease

This form for this lease has caused many problems in the past for those who explore for hard minerals in sedimentary basins where oil and gas are also sought, and it will probably continue to cause problems. The form was originally an oil and gas lease to which someone decided to add "other minerals." The critical wording of this document concerns (1) minerals leased:

... mining and producing therefrom oil, gas, casinghead gas, condensate, sulphur, potash and all associated salts, phosphate, vanadium, uranium, thorium, all other fissionable materials and rare earth minerals, and all other minerals whether or not similar to those mentioned, with rights of way....

(2) terms:

... and as long thereafter as either oil, gas, casinghead gas, condensate, sulphur, potash and all associated salts, phosphate, vanadium, uranium, thorium, all other fissionable materials and rare earth minerals, and all other minerals whether or not similar to those mentioned, or any of them is produced from said land hereunder or land pooled therewith. . . .

and (3) royalties:

. . . on potash and all associated salts, phosphate, vanadium, uranium, thorium, all other fissionable materials and rare earth minerals, and all other minerals whether or not similar to those mentioned, mined and marketed, one-tenth, either in kind or value at the well or mine. . . .

Most landowners (and their lawyers as well) usually fail to understand that "other minerals" are included when this lease form is executed. Some companies and individuals are apparently ignorant of the fact that this form is often harmful to the lessor's interests, but other unscrupulous companies and individuals deliberately misrepresent their intentions by using this particular lease form, and we believe that each state should have a requirement that if a party is to lease a property for "other minerals," those other minerals should be specifically named. The real harm of the "other minerals" inclusion in an oil and gas lease can be summarized as follows:

1. The landowner is often deceived by the lease, and believes the lease covers only oil and gas.
2. The landowner is denied the opportunity to lease the land to companies which might be sincerely interested in exploring for some other mineral (such as uranium). Almost without exception, companies and individuals who take oil, gas, and other minerals leases have no intention of exploring for anything other than oil and gas, and the inclusion of other minerals gives them a "free ride" speculation because they pay the same bonus and rentals for this lease as they do for a straight oil and gas lease. Not that we

are against speculation, but this is speculation by a lessee *at the landowner's expense.* In addition, a cloud on the mineral title for uranium is created.

3. The landowner receives less rental income than if the land were leased for minerals on different forms. For example, the common situation we have encountered is one in which a landowner is receiving $1 per acre rental for an oil, gas, and other minerals lease. We might like to lease the same land for uranium and pay an additional $1 per acre rental, but we cannot do so because of the "other minerals" provision.

It has been very sobering in the Wyoming-Colorado area to learn that landowners have approved as an *oil and gas lease* the lease form (Producer's 88 Revised Ruling U955) entitled "Oil, Gas and Minerals Lease," which states that the property is leased for oil, gas, casinghead gas, condensate, sulphur, potash, all associated salts, phosphate, vanadium, uranium, thorium, all other fissionable materials, and rare-earth minerals. Most lessors are probably unaware that they are leasing for any minerals other than oil and gas. As mentioned above, this type of lease is most frequently used by oil and gas companies which have no intention of ever exploring for or mining other minerals; thus the landowner's chances of having other minerals exploited on his property are quite minimal.

The courts have generally recognized that a mineral lease must spell out the royalty to be paid for production of minerals. The lease form commonly used by some oil companies provides for a one-tenth royalty in value or in kind for "other minerals," which is an extraordinarily high royalty rate for the lessee to pay on minerals mined. This provision alone could easily deter explorationists from getting seriously interested in the property.

In one instance, this particular type of lease was executed between a petroleum company and a landowner. At a later date, the landowner requested that the lessee relinquish its rights to "other minerals" so that he could lease the land for uranium exploration; the petroleum company refused to do so, even though the company had no intention of exploring for uranium. It is uncertain whether the "other minerals" portion of such a lease could be nullified by court action once the lease had been executed and recorded. To avoid such situations, we recommend that lessors, and their attorneys, include in a lease only those minerals (and associated minerals) which are actively being sought by an explorationist. Uranium should not be included in an oil and gas lease.

LANDMEN

The landman's position in the exploration organization was briefly discussed in chapter 1. The landman's function in the hard-mineral industry has evolved into a vital, highly versatile, and well-respected position. In order to explore for minerals, land must be acquired; leases must be negotiated; and numerous contacts must be made with industry and government officials, ranchers and farmers, attorneys and accountants, and countless others. Whereas originally the landman's function was merely to obtain the landowner's signature on the lease, today his role includes trading and purchasing, checking and clearing titles to land, drafting legal instruments, and, in general, performing a public relations role on behalf of the company (or individual explorationist) he is representing.

Many landmen in the mineral industry today began their careers as oil scouts, obtaining information from people they talked to in small towns and villages, farms and ranches, concerning the oil and gas leasing and drilling activities of competitive companies. Although this function is more commonly performed over the telephone today, rather than in the field, such experience has provided excellent preparation and training for the professional landman.

Landmen are required to deal with all types of people, ranging from the highest-level corporate executive to owners of large ranches, to other professionals such as attorneys and doctors, to unsophisticated farmers and landowners. Each type of person requires a distinct mode of handling, and the successful landman must learn to be flexible and accommodating in order to gain the confidence and respect of these people with whom he must do business.

To date, mineral landmen are not required to be licensed in order to buy a lease or transact negotiations. There has been some discussion among certain state legislatures about the possibility of the real estate industry restricting the activities of landmen; however, no such legislation has been enacted so far.

The explorationist who hires landmen on a contractual basis should be aware that in certain situations there could be a possible conflict of interest. Some landmen, in addition to working on a custom basis on behalf of clients, also attempt to put together acreage deals for themselves. We have had experiences with landmen who were hired to lease for uranium in a specific area on our behalf. At a later date, we dropped the leases, but had not lost interest in the area. We then discovered that the same landmen we had hired had reappeared in the area and had acquired the same leases in their own names in order to put a uranium deal together. Thus, the landman can sometimes be in competition with his client, either acquiring leases for himself or for *another* client in the same area, and this type of situation should be avoided whenever possible. We do not deny that landmen should assemble and sell their own deals, but extreme caution should be used in areas where they might compete with clients. A letter of release from the client(s) would be a good approach.

At the present time, independent landmen are usually hired at a standard daily rate of $125, although this can go as high as $200 per day in certain areas. In addition, they receive a per diem rate for personal expenses such as room, meals, transportation, and related expenses necessary in obtaining leases, such as notary and abstracter fees, copying of instruments, and so forth. The landman must be capable of performing a wide variety of duties related to obtaining leases for his clients, including checking records in an abstracter's office or courthouse to determine ownership of certain lands and mineral interests, contacting landowners, and using whatever terms may be necessary to persuade them to execute a lease. (In the early days of the petroleum industry, landmen were commonly referred to as "lease hounds" because of the "hounding" tactics which they employed in order to obtain a signed lease.)

In certain "hot" areas, it is not uncommon for landmen to be offered incentives on a per acre basis, in addition to their daily rate, for successfully leasing highly desirable acreage. This incentive could be something on the order of 10 to 25 cents for every acre leased. In some cases, landmen have been offered some type of overriding royalty, or vested interest, in the acreage, although this approach is not commonly practiced because the results are not markedly improved, and the cost could be very high.

In certain situations, landmen familiar with one particular area may be able to accomplish far more than other landmen might in the same area; thus, they may be able to demand an overriding commission in addition to their daily fee. The landman who has considerable experience in one state, and has become familiar with the records and procedures pertaining to leasing in that state, both on a state and local level, may have a temporary advantage if he is able to work faster and more efficiently than landmen foreign to the area.

In order to hire the services of a competent, trustworthy landman in a given area, we strongly suggest obtaining personal recommendations every time. The explorationist who works frequently in a particular area will no doubt become acquainted with capable landmen in that area; if these people are not available, they might be willing to recommend others. We definitely recommend against the practice of hiring landmen about whom nothing is known.

Not all landmen have the same capabilities. For instance, some are adept at record checking and know their way through courthouse and abstracters' files. Others are good talkers, best suited for handling negotiations with various landowners, especially those who might pose problems and require careful handling. We have found that some landmen who are lawyers seem to be unable to obtain leases as readily as landmen not so well versed in the law. It is vital that a landman be capable of dealing with people in the field, generally talking a layman's language. Attorneys are accustomed to talking in technical terms, to dealing with legal conflicts, and so forth; and this attitude may be counterproductive to accomplishing the goal of obtaining leases. Some major oil companies require that their landmen have a law degree, although these company landmen generally perform office work, while the company hires independent landmen on a contractual basis to carry out the field work.

The following are two important notes for explorationists who may be hiring custom landmen to work on their projects. (1) It saves time and money if the landmen are notaries so that they can notarize leases or other documents as they are executed by the lessors. (2) The landmen should keep adequate notes on landowner contacts in the field so that other landmen can later pick up where they left off if they leave the project. The landmen should routinely make descriptions and plats listing and illustrating ownership of the land, both of surface and minerals. In addition, they should make notes on attempts they have made to contact landowners, whether they made any contact, and if so, the response of the landowner. If they got the lease, fine; but if they did not get the lease, they should record a summary of the comments of the landowner so that if another landman is later sent to attempt acquisition of the same lease, or if the same landman makes a later attempt, the notes will allow an evaluation of the situation prior to the next round.

LAND DEPARTMENTS

It is incredible that at this date many mining companies, even fairly large ones, do not have land departments. We wonder how an innovative, aggressive company is able to function without the permanent services of personnel within its organization to perform the myriad of tasks associated with land leasing and record maintenance. In the petroleum industry, the land department is one of the most vital within the corporate structure; in this respect, that industry's influence on the hard-mineral industry has been favorable in demonstrating the necessity of such services on an in-house basis, aside from any independent landmen who may be hired

in addition.

Some mining companies justify not having land departments by asserting that they are specifically engineering oriented, or perhaps geologically oriented; but if they acquire and hold lands, their geologists and engineers ultimately will be obliged to spend time inefficiently performing functions that should be carried out by landmen. This unfortunate error stems from a lack of understanding by top management, and such attitudes serve only to erode the efficiency and weaken the organizational structure of an exploration or mining company.

CLAIM STAKERS

The task of staking mining claims is usually performed by a group of surveyors, hired on an hourly basis for a fee which generally includes the use of their equipment. At the instruction of the company by which they have been hired, claim stakers commonly locate claims either in their own names, in the name of the company, or in the name of a third party. They generally undertake the surveying and map-preparation work, as well as erecting necessary monuments. Depending on the wish of the company, most claim stakers will take the responsibility for the discovery work (which is frequently subcontracted) and whatever else may be necessary to complete the claims, including the related paperwork to record the claims.

Some companies prefer that the claim stakers undertake the entire operation; other organizations insist on taking care of the paperwork themselves. Claim-staking groups may sometimes bid on a job, either on a per-claim basis or, more commonly, on an hourly basis for the crew members plus a mileage allocation for the vehicles. The two- or three-member staking crew also charges for the cost of its supplies, such as the four-by-fours or other monuments which it uses in its staking program. In addition, the members expect to be paid a reasonable sum to cover their housing and meals for the duration of the staking operation.

As with hiring landmen, individuals or small companies may make a deal with the owner of a claim-staking firm whereby a slightly lower fee may be paid and the owner receives an interest in the project. This practice is rare. Explorationists and mining companies should also be aware that owners of claim-staking firms may sometimes locate mining claims on their own behalf in areas which they consider to be favorable. As is the case with landmen, this presents a potential conflict-of-interest situation.

To hire the services of a claim-staking crew in a given area, we suggest contacting persons involved in mineral exploration in that area and requesting a recommendation. Naturally, different claim-staking organizations vary in quality of work performed; some are competent surveyors, whereas we have encountered others who did not know how to establish a north-south line. Shoddy claim-staking work, caused by mistakes such as the establishment of erroneous north-south or east-west lines (off by even one or two degrees), can cause severe problems with the validity of claims in later years. Such errors sometimes require reentry, resurveying, and restaking, and often revalidating the claims. Gaps are sometimes left between claims, and discovery of these gaps by third parties can cause severe problems in the event an ore body is involved.

There is, in fact, legal certification in surveying required by most states, although this requirement is not generally enforced. It is certainly worth the effort to seek a competent claim-staking organization and to check the qualifications of its crew members before hiring them. It is not un-

common for a licensed surveyor or engineer to establish a claim-staking organization, using his license to validate the outfit, and later to hire young, inexperienced crews to perform the work in the field. There exist, however, claim-staking and surveying organizations which are extremely conscientious, utilizing only highly qualified crew members on all claim-staking jobs. The fees for such crews may be higher than for less-qualified personnel, but they are well worth the extra money.

There is also the case of an overkill in claim-staking and surveying operations. We have known of instances where the team we hired was overly meticulous and consequently took much longer than necessary to perform the location of claims. The explorationist should demand accuracy from the staking crew he hires, but the work should not be so slowly executed that valuable time is wasted and the cost becomes prohibitive.

HOLDING COSTS

Rentals

After acquiring a property, either by leasing or claim staking, rental payments or assessment work will become due on an annual basis. In addition, some contractual work may be required. The rental rate is established in the original lease, which provides a certain time frame within which the rental payment becomes due. Most contracts entered into for mineral leases, including oil and gas leases, provide for an annual rental payment. The first rental payment usually becomes due one year from the initial date on which the lease was executed. The payment is frequently made to a depository bank designated by the lessor. On occasion, the payment may be made directly to the lessor, although most individuals and companies in the mining industry prefer making payments into a bank, since the bank issues a receipt acknowledging the date of payment and the fact that the specific amount has been deposited to the lessor's account. This is helpful for accounting purposes and record keeping.

We have developed a useful practice of sending a receipt form and a stamped, self-addressed envelope, together with our rental check, to the depository bank (see figure 5-4, which shows a sample of our rental receipt). We request that an officer or employee of the bank sign and date the receipt form and return it to us. It is then filed with the respective lease for the property in question, and is thereafter available to assure that the rental payments have been made in a timely fashion. The company land clerk may be obliged to prod the bank somewhat to insure that these receipts are returned properly; some banks are negligent in their cooperation.

If the rental payment is not made on or before the due date, such default constitutes grounds for cancellation of the lease. However, if a lessee fails to make the payment through a genuine oversight, he may approach the landowner to obtain a ratification of the lease, stating that the primary term is still in effect, after having made the late payment. In most cases, the landowner will be agreeable to reinstating the lease and accepting the late payment, since he usually wants to receive money. However, there are times when perhaps there is increased activity in the area and a landowner is aware that he might now be able to obtain a higher rental than was called for in the defaulted lease. For example, in one instance in the oil and gas industry, a company failed to make a rental payment because of an oversight in that company's rental department. When it attempted to have the lease reinstated it found it was unable to do so, since there had been a large oil discovery in the area just before the lease default. Another company had immediately stepped in and leased the first company's property for a substantially higher rental. Generally speaking, the rental payment is the most vital part of insuring that a lease remains in effect, and the lessee should not knowingly fail to make a payment thereon. As explained earlier, once a property is brought into production, the royalty payment made to the lessor usually supplants the delay rental which had been paid prior to production from the property.

Assessment Work

For mining claims located on the public domain, both federal and state laws provide that $100 worth of work (called assessment work) must be performed each year in lieu of rental payments. Since most mining claims cover a 20-acre area, this means spending $5 per acre—a holding cost considerably higher than that usually required on fee leases. Most states require that assessment work be completed by September 1 of each year. Thereafter, a certain amount of time (usually 60 days) is allowed for the claim holder to file an affidavit of assessment work stating that assessment work has been completed on a certain claim or claims. This affidavit is usually placed on file in the county records, but the law in most states is somewhat ambiguous in this respect, since it does not *require* that the affidavit be filed; it simply states that it *may* be filed. Thus, if someone were interested in discovering whether assessment work had been performed on certain mining claims, and the county records failed to show an assessment affidavit thereon, this would not necessarily signify that those claims had been abandoned. It would be helpful if the law were more specific and would require that assessment affidavits be filed in the county records. There is likewise no legal

Lease Rental Remittance and Banker's Receipt

Lease No. 17005-01026   Denver, Colorado
                                                         September 9, 1976
Farmers National Bank
Denver, Colorado

Gentlemen:

Enclosed is Able Exploration Corporation check No. 1517 for $1,111.03 to be placed to
the credit of John Q. Rancher
                Rt. 1, Box 203
                Bailey, Colorado 80421
                                        Rental $1,110.03
                                        Bank Charge   1.00

To pay rental from October 1, 1976 to October 1, 1977 according to the terms of a certain
Mining Lease made by John Q. Rancher to Able Exploration Corporation on the following
1902.11 acres described as:
                Township 8S., Range 67E., 6th P.M.
                Section 14: SE 1/4
                Township 9S., Range 67W., 6th P.M.
                Section 10: W 1/2, SE 1/4
                Section 14: All
                Township 10S., Range 67E., 6th P.M.
                Section 30: Lots 1, 2, 3, 4, E1/2, W1/2, E1/2
Situated in Morgan County, Colorado, bearing date of October 1, 1976, and recorded in
Book 697 Rec. 1614287 of the records of the above County.

                                        Yours very truly,

                                        Able Exploration Corporation
                                        By: Betty L. Jones

                        BANKER'S RECEIPT
                        (DO not detach)

Deposited this day by Able Exploration Corporation in the Farmers National Bank at
Denver, Colorado $1,110.03 for credit of John Q. Rancher as above requested.

                                        Rose Waters
                                        Representative
                                        Date:

Bank please date, sign and return at once to Able Exploration Corporation, 1000 S.
Broadway, Denver, Colorado 80202.

Figure 5-4. Lease rental remittance and banker's receipt.

requirement that a locator file a notice of abandonment of claims. Such notice would facilitate the checking of mining claims to determine whether certain ones were still valid.

If the claim holder does not perform the required assessment work on mining claims, such claims will revert to the public domain. If this situation has occurred, and the original claim holder at a later date wishes to reacquire the claims, he may do so, provided that no other party has entered and staked the area in the intervening time. The relocation is done in the same manner as originally locating the claims.

As an example of how the lode claim record system functions, let us assume that in 1976 an explorationist is interested in staking a certain area of public domain and, on checking the records in the county clerk's office, discovers that one party had held mining claims in the area and had filed assessment work affidavits from 1965 through 1970. This information does not reveal if the claims continue to be held by the original locator, or if the claims have been abandoned. To determine this, the explorationist would then have to go into the field for an on-site inspection in an attempt to learn if assessment work has been done. Such an inspection still might not reveal the status of work on the claims. If drilling had been performed, it could be difficult to find evidence; yet such drilling would certainly constitute valid assessment work. One direct approach preferred by some exploration companies to avoid litigation over claims is to seek the previous holder of the claims and inquire whether the claims are abandoned or whether they are still considered valid. Some would question that this approach is a good one, because the original holder may in fact have abandoned the claims but then, on discovering that another party is interested in the area, may decide to re-

locate the claims promptly.

We suggest that if an explorationist is interested in staking a particular area, believes previous claims to be abandoned, and may possibly jeopardize his chances by contacting the previous claim holder, then he should stake new claims to the extent of placing the discovery monuments. After this has been accomplished, the previous claim holder can be contacted. If the previous holder had not, in fact, performed the necessary assessment work, then the new claims can be completed and recorded. If the previous holder stated that assessment work was done and the claims maintained, then we recommend dealing with the claim holder or abandoning the area covered by his claims. A lawsuit may thus be avoided.

Contractual Work Requirements

Contractual work requirements can also be considered a form of holding costs for mining claims. However, they differ from assessment work in that they only occur if the claims have been acquired from another party and if, as part of the contract, the second party agreed to perform specific work in order to hold the claims. In order to keep the contract valid, therefore, the work must be performed.

In the case of federal prospecting permits, required work could include the steps necessary to develop data to qualify for a preference-right lease. There is usually a requirement that the lessee drill at least one hole during the initial two-year period, or perform some equivalent exploration work which would be acceptable to the USGS conservation branch geologist or engineer supervising the project. Another requirement is that the explorationist submit logs and maps showing the exact location of the drilling and describe how the hole was plugged and abandoned. In the case of uranium exploration, the requirements to qualify for an extension are quite minimal.

On Indian lands, there is usually a requirement that a specific sum of money be spent per acre per year, and sometimes that a designation be made as to the type of exploration to be carried out. In addition, there may be further, more substantial, contractual requirements before a lease is granted.

LEASE-MAINTENANCE RECORDS

If your operation is a small one, involving few leases, it is a relatively simple matter to keep track of your own records and pay the rentals on time. One suggestion is to file the lease and rental receipts in a lease file by the month in which the anniversary date occurs and to maintain a loose-leaf notebook with pertinent information on each lease. Leases could be listed in the notebook *according to due date* on the rentals, with a separate page for each lease. When the rental becomes due, you should prepare your check, together with the rental receipt for the lessor (or the depository bank) to sign and return, and make a corresponding entry in your notebook.

As your operation expands in scope, we suggest preparing a second notebook, each page of which is a duplicate of the first, but this time filed *according to geographic location* of the leased property, with a description of the township, range, and section. This second notebook facilitates the task of seeking information rapidly whenever it should be needed.

As the organization grows still larger, such information is often computerized. If the company has its own computer, it may wish to devise its own in-house system (a computer system being a group of related programs which, as a package, performs a specific function). Other companies find it more satisfactory to utilize the services of a commercial computer

system, such as Petroleum Data Systems (PDS) in Denver, since in-house computer systems are unrelated to lease maintenance, and writing and "debugging" a new system often costs more than the fee for using an existing commercial one. Such commercial systems are virtually free of "bugs," since they have been utilized by many companies over an extensive period.

There are several custom lease-maintenance computer services available; all are relatively similar, and we do not recommend one over another. Entering the initial information into the computer system requires the services of a person familiar with lease forms who can accurately fill out a data sheet (figure 5-5 shows a sample data sheet commonly used by PDS). The data sheet, as sent to the computer company, must include the names of the lessor, lessee, date of execution, rental amount, total (gross) acreage, net acreage, name of the depository bank where the rentals are to be paid, and other miscellaneous data such as the county in which the lease was recorded, and the book, page, or a description number. A preliminary printout of this data sheet, after being prepared by the computer company, is sent back to the client company for approval. It is proofread and then returned to the computer company, where it is entered into the computer as a permanent record.

There is an initial charge of about $2.50 per lease. In addition, there is an annual fee for various services, including the 60- or 90-day notice of rentals due, check and rental receipt printing, and so forth. There is also a charge for any modifications, alterations, or changes in the original setup. For example, a certain property may change ownership, in which case the name of the lessor would change and also, probably, the name of the depository bank.

As new leases are acquired, a new data

LEASE DATA SHEET

PDS

ORIGINAL LEASE DATA ☐
CORRECTION LEASE DATA ☐

COMPANY AND DIVISION	NO. #	LEASE #	SUB #
ABLE EXPLORATION CORP. 30	1558	2075	D

COUNTRY/STATE	NO.#	COUNTY/PROVINCE	NO.#	PROSPECT, AREA OR BLOCK	NO.#	LSE NAME OR LESSOR #
S. DAKOTA	42	HARDING	31	WILLISTON	17040	JONES, JOHN

LEASE TYPE
FEE F ACQUIRED A
STATE S CROWN C
INDIAN I DOMINION D
PUBLIC DOMAIN P OTHER-ANY CHARACTER
CODE **F**

TYPE INTEREST
LEASE L PERMIT P
ORR O SURFACE LEASE S
ROYALTY R RESERVATION V
FEE MINERALS M OTHER ANY CHARACTER
CODE **L**

RENTAL RESPONSIBILITY
SELF S PAID UP U NONE N
PARTNER P H.B.P. H
COMPANY C OTHER ANY CHARACTER
CODE **S**

C*	MAP REFERENCE - SEE NOTE	LEASE DATE	RECORDING DATA	EXPIRATION DATE	RENTAL DATE	OBLIG. DATE
		MO DAY YEAR	BOOK PAGE	MO DAY YEAR	MO DAY	MO DAY
X	24 2N 15E	9 1 76	924 46	9 1 86	9 1	

GROSS LEASE ACRES	NET LEASE ACRES	YOUR NET ACRES	ORR %	LANDOWNERS ROYALTY %
680.00	124.67	124.67	——	5.00

CONSIDERATION/BONUS	OTHER CAP. COST	TOTAL LSE RENTAL	YOUR NET RENTAL	RENTAL PERIOD
124.67	——	125.67	125.67	12 MONTHS

TAX BASE	WORKING INT %	NET MIN. INT %	ACQUIRED DATE	ACQUIRED FROM
124.67	100.0000	18.3339	8 26 76	LESSOR

GENERAL INFORMATION AREA

```
05  LAND DESCRIPTION -
    T-2-N, R-15-E
    SEC. 24: S/2NW, W/2NE, SE/4, W/2SW, NESW
    SEC. 25: E/2NW, NE/4

OH  ORIG LESSOR  -JOHN JONES
OI  ORIG LESSEE  -ABLE EXPLORATION CORPORATION
OJ  SPECIAL OBLIGATIONS -

OL  ORR REMARKS -

OP  ROYALTY REMARKS -

OR  CONTRACT REFERENCE - ALPHA MINING CO. -ABLE JV-AGRMT DATED
    7-15-76. ALPHA MAY EARN UP TO 75% OF ABLE'S INTEREST.
OT  MISCELLANEOUS REMARKS -ALL LEASE RENTAL WILL BE PAID BY ABLE
    AND CHARGED AS MAINTENANCE COST UNDER JV AGRMT OF 7-15-76.
    SURFACE IS OWNED 50% BY MARY SMITH AND 50% BY THOMAS BROWN.
```

*NOTE X = SEC. TWP RNG (4,4,4) NO CODE = LON. LAT. (3,3,3,3,3)
Y = BLOCK SURVEY (4,12) (USE CODE SPACE FOR FIRST DIGIT OF LONGITUDE)

RENTAL PAID TO DATE	DEPOSITORY BANK, RENTAL RECEIVING AGENCY, PARTNER OR COMPANY PAYING RENTAL	BANK CHARLIE OR NOTICE	RENTAL AMOUNT
	FIRST NATIONAL BANK	1.00	124.67

STREET ADDRESS OR P.O. BOX NUMBER	CITY	STATE	ZIP CODE
	BOWMAN	NO. DAKOTA	58623

DEPOSIT RENTAL TO CREDIT OF: OR LESSOR LEASE NUMBER SHOW LESSOR SOCIAL SECURITY NO.
JOHN JONES
BOWMAN, NORTH DAKOTA 58623
SSN 000-00-0000

PTNR #	PARTNER NAME	PARTNER BILLING	% OWNERSHIP	N	TI
62410	ABLE EXPLORATION CORP.	$	100.0000		W
		$			
		$			
		$			
		$			
		$			
		$			
		$			

CODED BY   DATE   PDS CHECKED BY:

Figure 5-5. Lease data sheet.

sheet is filled out for each individual lease. Each original lease is retained and filed by the leaseholder. Each month, the computer system prints out a list of leases with rentals due 60 or 90 days (as selected by the company) from that date. The company must review this list and inform the system which leases to pay and which ones to drop. Using the stored data, the computer then prints out the appropriate checks and rental receipts. The company simply has to sign the checks and mail them. Thus, a substantial volume of manual clerical work is eliminated, much time is saved, and the possibility of human error is eliminated. In the case of dropped leases, however, it is the responsibility of the company to prepare the appropriate lease and file it on record in the county clerk's office, thus relinquishing its interest.

Custom computer systems may also be utilized for producing specific information on demand. For example, a custom system can give the total acreage leased in a given geographic area; it can give the capitalized costs in a given area; it can produce a full printout of all leases by expiration dates, an item which is especially useful to show the prospective buyer or investor in a deal; it can fill out Form 1099 required by the IRS; and it can provide other miscellaneous information, usually on 24-hour notice. Many individuals and companies utilizing such custom systems believe that the rental paying task which the system performs justifies the cost of the service and that the numerous additional features are simply bonuses.

Large companies usually have records departments, the primary function of which is to keep track of leases and other records. The lease numbers are generally assigned in chronological order according to the date the lease was acquired. Leases

might be given a ten-digit number, signifying that it is a mining lease, the state where the property is located, and so forth. In this way, records removed from a file can always be replaced in the file where they belong.

In order to cross-reference leases so that a particular lease may be filed by geographic location (as well as by date), a company utilizing a computer system might file one copy of the computer data sheets for that lease according to township and range, then file another sheet according to the anniversary date of each lease, by month. This is a highly recommended overall program for lease maintenance. If necessary, another set of lease records may also be filed alphabetically according to the last name of the lessor. With small and medium-sized organizations, it is unlikely that an alphabetical system would be required; however, larger companies might find the three-way cross-referencing of leases beneficial. Each company should set up lease-maintenance records according to its own specific needs—and these needs vary greatly from one organization to another, depending on the volume of leases in effect or anticipated.

Any individual or company planning to set up its own computer system (versus utilizing custom services) is well advised to visit other large companies, especially petroleum firms, which have an in-house computer system. Much can be learned from examining other systems and talking to the persons responsible for initiating and operating them. Once a company feels it can justify installing a custom computer system, there are other advantages to be gained. For example, in addition to handling lease-maintenance data, the computer can be utilized for accounting, payroll, inventory, and other purposes. The in-house system makes rapid retrieval of information possible, but custom services can usually provide information within a few days.

CONTACTS WITH LANDOWNERS

Fee Mineral and Surface Owners

When an explorationist wishes to lease a certain property, the first step should be to conduct a title search to check the ownership of the property, as described earlier in this chapter. Once the mineral owners have been established (assuming the land is owned in fee), the next step then is to contact the owner with the largest percentage (assuming there is more than one owner). Locating the owners is not always an easy task, since there is often no accurate source of information. We suggest, first, checking to see whether any oil and gas leases have previously been taken on the property; if so, those records can be checked for the address of the lessor. It should be remembered, however, that oil and gas are sometimes severed from other minerals. If no address is shown on the lease, the tax assessor's office is the next logical place to check, assuming the mineral owner also has an interest in the surface.

In the case of a property on which an oil and gas lease had been executed some time ago, and the lease contained the signature of the mineral owner but no address, it is often helpful to check the county and state in which that lease was notarized. The geographic location of the notary is often a clue as to the general location of the mineral owner. By checking the telephone information service in that area, it may be possible to locate the party. The mineral owner could have died, or moved to another location, so this method does not always produce the desired results. On other occasions, the land might not have been previously leased, and in a situation involving unidentified heirs, the name or names of the mineral owner may not be known. Contacting the surface owner (whose name may be obtained from the tax assessor's office) can sometimes help. Other times, the surface owner or lessee may direct you to a party he believes to be the mineral owner; that person may deny ownership of the minerals and, possibly, direct you to a third party. The search can become quite lengthy. Little wonder landmen are sometimes referred to as lease hounds.

If the explorationist wishes to obtain a lease strongly enough, and is unable to locate the owner, he may have the landman, through an attorney, ask a court of law to approve the lease. In such a procedure, the landman must produce evidence of a diligent, but fruitless, search for the owner. Such evidence should include title chain, records of contacts with residents in the area, and letters written in attempts to locate the owner. In order to establish a price for the lease, the landman could show evidence to the court of having leased other lands in the area. In some instances, the unlocatable person is owner of a part interest in certain mineral properties; in such an instance, it is likely that the judge would approve a lease for the part interest, even though the owner of that portion of the minerals had not been located. The bonus and rental would be paid to the court, which sum the court would hold in escrow for a number of years; thereafter, if the money had not been claimed by the rightful mineral owner or his heirs, it would be paid to the state. Problems such as these would never arise if states had laws requiring that nominal taxes be assessed on mineral ownership, as they do on surface ownership. However, when minerals become severed, the surface remains taxable even though the minerals are not subject to taxation. We seldom encourage additional taxation, but nominal taxation of mineral ownership would provide records leading directly to the owners of the minerals. As an additional step, the mineral ownership could revert to the state if the taxes were not paid in a timely fashion. In many areas, the minerals are worth more than the surface, and ownership of the minerals would be a valuable asset to the state.

It should be noted here that surface owners usually have *some* ownership or interest in the minerals, although this is not always the case. The surface owner who owns only a small portion of the minerals is usually the most resistant to leasing his interest, because he will receive a minor part of the bonus and rental, and, if exploration is successful and mining is undertaken, some of the surface could be temporarily removed from grass or crop production. However, when he purchased the land, one of the terms of the deed to which he agreed was a reservation of the minerals together with rights of ingress, egress, and mineral extraction. The purchaser of land with a mineral reservation should understand and acknowledge that the mineral owner reserved the minerals with the *hope and expectation* that exploration will be undertaken on, and perhaps production obtained from, the property.

We have found that it is often necessary to work out an agreement with surface owners whereby they receive payment for surface over federal minerals, or fee minerals owned by another. Such payment may even extend to a royalty, and a form has evolved which may be used for this purpose (fig. 5-6). A property description and royalty sche-

dule would be attached to the agreement form, but we have omitted them here. The royalty to a surface owner who owns no minerals is usually rather small, perhaps 2 percent of the ore value at the mine, and an ore-value schedule is usually included in the royalty schedule.

Once the fee mineral and surface owners have been located, they should be contacted personally, usually at their place of residence. This is frequently in a rural area, often requiring a lengthy drive from the county seat or the nearest town or city; but it is essential to talk to the owners in person to describe the lease, make an offer, and tell them of the company plans. It is rare that the landman will get a lease signed on the first visit; the process often requires as many as three or four visits before the negotiations are successfully completed.

Because of overzealous publicizing of a few bad situations (such as stip mines in Appalachia), as well as a strong national concern for protection of the environment, the mining industry is often looked upon with suspicion. Consequently, some landowners are apprehensive about leasing their land for exploration and mining purposes; they fear a devastation of the environment, damage to crops or livestock, and general disruption of their rural life-style. Also, in certain areas, the land is extremely valuable for farming. For example, wheat may sell for $5 per bushel, and a single acre may yield 300 bushels of wheat. Thus, the gross value of production from the land is $1,500 per acre per year. An owner of this type of land may not be interested in an offer of $1 per acre for exploration, especially if he has only a partial ownership in the minerals.

Some landowners are rather difficult

SURFACE AGREEMENT

This Agreement, made and entered into this _____ day of _____
_____, 19___, by and between Able Exploration Corporation, a Delaware
Corporation, whose address is_____,
(hereinafter "Able"), and John Q. Rancher, whose address is _____
_____(county)_____(state)_____
(zip)_____, (hereinafter "landowner");

WITNESSETH

WHEREAS, the surface of the lands described in Exhibit A (attached
hereto and by this reference made a part hereof) is owned by landowner.
These lands are hereinafter sometimes referred to as "Subject Properties"
or "Subject Lands;"

WHEREAS, Able desires to compensate landowner for the use of such
surface as is needed in connection with its activities in location and
discovery work, land and lease acquisitions, assessment work, exploration,
and mining and related operations;

NOW, THEREFORE, for and in consideration of the sum of $10.00 in
hand paid to landowner by Able, the receipt and sufficiency of which
is hereby acknowledged, and in further consideration of the mutual promises
hereinafter set forth, the parties hereto agree as follows:

1. Landowner hereby grants to Able access to and the right to enter
upon the surface of the lands above described and any other lands within
said county owned by landowner which might not be described hereinabove,
for the purpose of surveying, exploring (including, without limitation,
geophysical and geochemical exploration), investigating, prospecting,
drilling, developing, mining (including, without limitation, strip,
open pit, underground, solution or any other method of mining as determined by Able), stockpiling, waste disposal, removing, milling, shipping
and marketing minerals, locating mining claims, performing discovery and
assessment work thereon; to construct and use buildings, roads, camp
facilities, shafts, tunnels, inclines, adits, drifts, pits, mine buildings,
power communication lines, mill and related facilities, and other improvements related to the accomplishment of any of the foregoing; to use so
much of the surface of the lands above described in such manner as may
be reasonably convenient or suitable for or incidental to any of the
rights set forth above or otherwise reasonably necessary to accomplish
any of the foregoing purposes; together with easements and all rights
of way for ingress and egress to and from the lands above described to
which landowner may be entitled, including surface and subsurface easements across other lands presently owned or hereinafter acquired by landowner; all of the foregoing rights pertaining to access to easements
across and under and use of the surface of the lands above described
being granted irrespective of the specific location and ownership of the
lands on which the mining or other activity is being carried on by Able.

Figure 5-6.

2. Able will pay to landowner within 90 days after the location of mining claims pursuant to the rights granted herein, or upon execution if Able has completed the location of mining claims on subject lands or otherwise acquired mineral rights thereto an Initial Rental Payment in an amount equal to 25¢ for each acre of subject land covered by Able's mining claims or other mineral rights or by access roads constructed by Able. On or before each anniversary date of each separate Initial Rental Payment paid to landowner by Able, Able shall pay to the credit of land- owner in the _____ Bank of _____ or any successor the sum of <u>25¢</u> for each acre of subject land covered by Able's mining claims or other mineral rights or by access roads con- structed by Able. Provided, however, that in the event Able elects to abandon by failing to do assessment work or otherwise any mining claim or other mineral rights and gives notice of such abandonment to landowner prior to the anniversary date, Able can avoid its obligation hereunder to pay rentals with respect to such abandoned claim or claims. The rental payable by Able shall be recoupable by Able from any subsequent royalty payable to landowner under the provisions of paragraph 3 hereof.

3. In the event Able should develop a mine on lands described in Exhibit A hereto, it shall pay to landowner a royalty thereon computed and payable as provided in Exhibit B, attached hereto and by this ref- erence made a part hereof. Rentals paid by Able pursuant to the pro- visions of paragraph 2 hereof shall be recoupable by Able out of afore- said royalties.

4. In the event landowner owns less than the entire and undivided surface estate in subject lands then the rentals provided in paragraph 2 hereof and the royalties provided in paragraph 3 hereof shall be paid to landowner only in the proportion that landowner's interest in the sur- face estate bears to the whole and undivided surface estate. For purposes of determining the royalty payable to landlord his interest in the surface estate shall be deemed the interest owned by him in the surface estate pertaining to the specific lands from which minerals as to which the royalties payable are being produced. Landowner shall be entitled to retain only such proportion of the royalty and/or rentals previously paid to him as his interest bears to the whole and undivided surface estate and Able shall be entitled to offset all other monies previously paid landowner hereunder against rentals and royalties which thereafter become due and payable to the landlord.

5. All monies due the landowner from Able shall be considered paid when tendered to landowner or deposited to his credit in the Bank de- scribed in paragraph 2 of _____, which Bank shall be deemed the landowner's agent.

6. Able will use its best efforts, consistent with its authorized activities, to avoid damage to landowner's forage, crops, field, or im- provements; to avoid the creation of hazards to persons or livestock, to close all pasture gates opened by it and will plug all holes drilled by it or its employees or contractors.

Figure 5-6 (continued).

7. Subject to the provisions of paragraph 6 hereof, landowner hereby specifically agrees that the monies paid and to be paid as specified hereinabove constitute a liquidated sum in full payment for all surface use, damages (including those arising from disruption of landowner's operations) and access as hereinabove provided and hereby waives any other claim for surface use and damage which might result from the activities of Able, its employees, contractors or assigns.

8. Able agrees to abide by all valid federal and state laws, rules and regulations governing mines, mineral exploration and environmental protection. Landowner acknowledges that he is aware that the Wyoming Environmental Quality Act requires his consent to any mining and reclamation plan and provides for the furnishing of a bond by operator to assure compliance with any approved mining and reclamation plan. In the event Able shall apply to the Wyoming Department of Environmental Quality for a permit to carry on mining operation on subject lands and such application includes a mining and reclamation plan, landowner shall furnish Able and the Department of Environmental Quality with any consent, waiver or other document required under the provisions of 35-487.24(b)(x) of the Wyoming Environmental Quality Act (Enrolled Act No. 107, Senate Wyo. Session laws 1973) as amended. Landowner hereby waives bonding and other provisions of 35-487.33 of the above cited Wyoming Environmental Quality Act to the extent the same may be applicable. Landowner shall execute such further instruments as may be necessary to carry out the foregoing.

9. All notices, requests or payments which may be or are required to be given by either party under this agreement shall be given in writing and should be sent by certified mail, return receipt requested, postage prepaid, or by Western Union telegram or mailgram, charges prepaid, addressed to the parties at the addresses set forth below:

Able Exploration Corporation

_____

_____

John Q. Rancher

_____

_____

10. This agreement shall be binding upon and shall inure to the benefit of the parties hereto, their heirs, successors and assigns.

11. In the event of any dispute between the parties hereto concerning the interpretation or application of this agreement, such dispute shall be submitted to arbitration in accordance with the rules of the American Association of Arbitration.

Figure 5-6 (continued).

12. In the event the parties hereto have previously entered into a surface agreement or in the event a predecessor of Able has entered into a surface agreement which pertains in any manner to subject lands, such prior surface agreements shall be deemed to be superseded in their entirety by this agreement.

Executed as of the day and year first above written.

_____

_____

_____
Landowner

ABLE EXPLORATION CORPORATION

By_____
      Agent

STATE OF WYOMING   )
) ss.
County of   )

The foregoing instrument was acknowledged before me by _____

_____, this _____ day of _____, 197 ____.

Witness my hand and official seal.

_____
                 Notary Public

My Commission Expires:

Exhibit A (property description) and Exhibit B (royalty schedule) would be attached to this agreement. We have omitted them here. The royalty to a surface owner who owns no minerals is usually rather small, i.e. two percent of the ore value at the mine, and an ore value schedule is usually included.

Figure 5-6 (continued).

Figure 5-7. Some landowners exhibit hostility even before the landman has an opportunity to talk to them about leasing their land. The first one to great a landman when he approaches a landowner's residence is often a dog which has no intention of allowing the landman to proceed unmolested from the gate to the door of the house. In this photo, the dog, hackles bristling, can be seen approaching the photographer from the dog house. Photo by Tom Meredith.

to deal with under any circumstances (fig. 5-7). Some may even refuse to talk or discuss terms; in situations such as this, the landman may ultimately have to leave and concede defeat. However, once a lease has been granted and an exploration program is planned for the leased property, it behooves the operators to make ever possible effort to get along with the landowners and not antagonize them or give them cause for complaint. Also, the landowner's property should be treated with extreme respect. We refer here not only to the land itself, but to any machinery, fence posts, farm equipment, or other miscellaneous items which may be scattered around. In rural areas, especially where many landowners are direct descendants

of the original homesteaders, a special significance may be attached to items such as old equipment, cabins, stoves, and so forth; such items should be treated with respect by the exploration crews.

When the surface and minerals are owned by two different parties, it is the mineral owner, of course, who earns the royalties from production. Nevertheless, the surface owner must be fairly compensated for any actual damages caused to the surface of the land. In addition, the surface owner is owed the courtesy of a personal visit by a representative of the operator before drilling operations are commenced. Such personal contact minimizes any potential problems which may arise, and certainly eliminates any possible misunderstanding on the part

Figure 5-8. Access to properties is very important, and explorationists must cooperate with landowners in keeping disturbance of ranching or farming operations to a minimum. Some ranchers will post their property to keep unwanted persons out; the NO TRESPASSING—LEASED sign on this property is intended to advise hunters that the property is leased for hunting. Equipment and buildings such as those shown here are the landowner's property and must be respected by explorationists, even though they appear to have little value. Photo by Tom Meredith.

of the surface owner. Otherwise, if the operator were simply to move onto the land and start drilling, the landowner could feel justified in making accusations of trespassing (fig. 5-8).

Certain surface owners can be uncooperative and may occasionally refuse entry to property for drilling purposes even though the explorationist has the minerals under lease. Sometimes hostility exists because the landowner has no ownership in the minerals, and he therefore feels he is gaining nothing by granting entry to the property, other than whatever settlement he may be able to negotiate for damage to the land. In some instances in which the landowner owns the surface over federally owned minerals, attempts

have in fact been made to grant certain compensation from any mineral production to the surface owner. For example, there is a new proposal relating to strip-mining coal which would provide that the landowner's permission must be obtained before mining operations can begin; in addition, the government is considering granting the surface owner a portion of its royalties from production. In our opinion, this is not justified, and constitutes an unjustified disposal of assets which rightfully belong to the United States.

State Land Boards

In terms of leasing state-owned lands for mineral exploration, most states have standardized leasing procedures. They generally have a standard printed form setting forth the terms of leasing state lands, with a standard royalty and bonus rental included. In the past, some states encouraged the practice of inviting explorationists to negotiate the terms of the bonus and rental payments.

Today, with considerably more emphasis being placed on the development of energy-related resources, state legislatures are adopting more stringent policies concerning the leasing of state lands for the purpose of mineral exploration and development. The days of negotiating deals with state land boards are becoming a thing of the past as state legislators and officials come under increasing pressure from numerous environmental and other special-interest groups.

The state of Colorado requires application for a prospecting permit to be filed with the state mineral department, which is a branch of the board of land commissioners. Rental is fixed by the board, and is generally $1.00 per acre per year. The first year's rental must be paid in advance. In addition, the lessee must contact the surface owner or lessor prior to starting any drilling operations, to inform him of the nature of the proposed project.

The state of Texas has a school land board and also a land commissioner, both of which have responsibility for the leasing of state lands. The school land board acts as an executive policy-making body, with the authority to recommend leasing terms and conditions, whereas the land commissioner is involved in the day-to-day lease negotiation and management process, with the authority to reject or accept any lease proposal. State-owned lands in Texas fall into two distinct categories: state fee lands in which the state owns both the surface and minerals; and mineral

classified lands, formerly owned by the state and subsequently sold to private parties with the mineral ownership reserved by the state. On state fee lands, a prospecting permit may be obtained and, for a 25-cents-per-acre rental, the lessee may conduct exploration activities for uranium, thorium, and other associated minerals. On mineral classified lands, however, uranium, thorium, and other fissionable materials are specifically excluded from the minerals for which a prospecting permit may be issued.

State lands in Utah may be leased for mineral exploration purposes by making application to the state land board. A mineral lease is limited to a maximum of 640 acres (one section) or, "at the discretion of the director of the state land board, where the state lands are in a reasonably compact block of an area larger than 640 acres," but not to exceed 2,560 acres (four sections). Separate application must be filed for each noncontiguous tract of land to be leased, unless they fall within a six-mile radius of one another. There is an annual rental of 50 cents per acre for all metalliferous minerals (that is, everything except oil, gas, and hydrocarbon, for which the rental is $1 per acre); the first year's total rental is payable in advance. After obtaining a metalliferous-minerals lease, and prior to commencing exploration activities on the leased property, the lessee must contact the State of Utah Division of Oil, Gas and Mining and file a notice of intent to commence mining. After obtaining approval from this division, the lessee must then file a bond with the division of state lands and provide a copy of the mining plan and a separate reclamation plan.

In the state of Washington, a commissioner of public lands and a board of natural resources together have the power to negotiate the leasing of state lands for mineral exploration purposes. Upon application, a mineral-prospecting lease may be issued for a two-year term, for a minimum area of 40 acres, at an annual rental of 25 cents per acre. Under the terms of this lease, the lessee has the right to prospect for minerals by means of core drilling, trenching, and so forth. The lessee may remove adequate ore for assay purposes. This lease may then be converted into a mining contract, as long as the initial two-year lease term has not expired. However, the lessee must show proof that $1.25 per acre per year in development work has been carried out. The mining contract is then issued for a period of 20 years; there is a rental of 25 cents per acre per year for the first and second years, and 50 cents per acre per year for the third and fourth years. To hold the contract beyond the fourth year, the lessee must again show proof of having completed $1.25 per acre per year in development work. Beginning with the fifth year, the rental or minimum royalty is $2.50 per acre per year; in addition, $2.50 per acre per year in development work must be conducted. Once production begins, the usual production royalty to be paid to the state is 3 percent of the gross income.

All state lands in Wyoming are under the jurisdiction and control of the board of land commissioners, including the leasing for minerals. On one application, an offer may be made to lease up to 1,280 acres (two sections), provided such tracts of land are contiguous or fall within a six-mile radius of each other. The lease is issued for a ten-year term, with a preferential right provision for a succeeding ten years at the discretion of the board. The rental is 50 cents per acre per year for the first five years of the lease; $1 per acre per year after commercial discovery; and $1 per acre per year for the sixth through the tenth year of the lease, or any renewal thereafter. The total first year's rental payment is due in advance.

The responsibilities of the various state land boards can sometimes be seen as conflicting. First, they must try to earn as much revenue for the state as possible, through rentals and bonus payments. Second, they must encourage resource development in the state, insuring income for the state as well as job opportunities for the citizens. Ultimately, the major income to the state results from the development of resources, not simply from leasing the land, and exploration must precede development. On the other hand, state land boards are constantly under political pressure from various environmental groups to limit exploration and development of resources, to protect the land, and to preserve the wilderness areas. Pressure against resource development also comes from agricultural sources. In North Dakota, for example, agricultural interests are very strong, and attempts have been made in the past to enact legislation which would withdraw state lands from mineral exploration and development. The issue has not been resolved, and at this writing the state lands in North Dakota lie essentially unleased.

In most states, the board of land commissioners is appointed by the governor of the state. Typically, once these officials have been appointed, they tend to remain in office for many years, surviving several different administrations.

6
Continuing Detailed Geologic Work

GENERAL DISCUSSION

Continuing geologic work before a land play is made can be extremely interesting: there is excitement and fascination in examining a new or different area to determine if it really appears to have potential for the discovery of mineable deposits. In fact, some explorationists have stated that "the fun is in prospecting and initial discovery; later development work is drudgery." We have found development work to be just as fascinating if the working environment is good—more about that in chapter 12.

Geologic work before a play is made is critical to the process of arriving at a decision concerning whether to make a play, and if so, where. In large plays, millions of dollars may be spent evaluating an area recommended by the geologist, so his observations and conclusions must be sound. In the case of an independent

explorationist financing his own plays, the wisdom of his decision to proceed, based on his own observations, may literally make or break his career as an independent. Such risks lend an additional air of excitement to the mineral exploration business. But now back to the focus of the chapter.

If the regional surveys outlined in chapter 3 indicate that the prospect area has reasonably good potential for commercial uranium deposits, and if preliminary land work shows that acceptable exploration and mining rights can be acquired in the area, then detailed studies such as those discussed in this chapter may be carried out to provide data for further decision making. In large prospect areas with substantial acquisition costs or in areas with large acquisition costs per unit of area (such as a small, high-cost area), detailed geologic work should be planned and carried out to help guide acquisition.

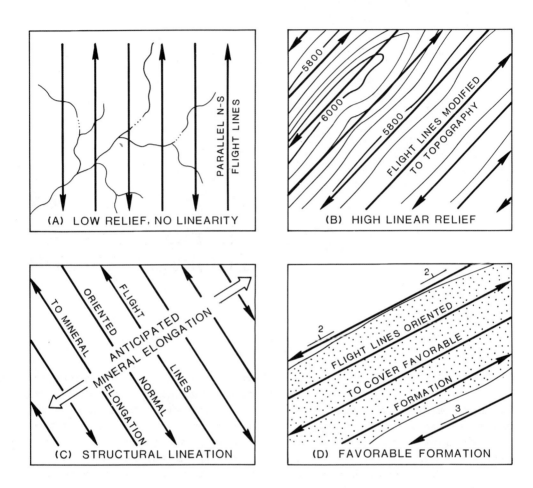

Figure 6-1. Idealized maps showing flight patterns for detailed airborne radiometric surveys.

It is frequently possible, and desirable, to acquire low-cost leases scattered through the area of interest immediately after the early regional surveys are completed, provided the geology and general land situation appear favorable. Later studies will help guide acquisition of more costly properties, if they are acquired at all.

Geologic work described in this chapter is usually a continuation of the geological, geochemical, and geophysical studies started or carried out in the regional or preliminary program, and is usually designed to evaluate and improve the definition of targets which were broadly or vaguely defined by the regional surveys. Much of the work described in this chapter may be carried on during the drilling programs, which last for years in large and successful projects.

AERORADIOMETRY

One of the exploration procedures which should provide data for continuing evaluation of mineralization and alteration and should help locate possible drilling targets is aeroradiometry, which is the airborne search for radioactive anomalies. In regional aeroradiometric surveys, flight lines are widely spaced (usually 1 to 4 km). Such surveys will sometimes define large anomalies or relatively large areas of high background, but many anomalies of small areal extent are likely to escape detection. More specifically directed radiometric surveys, using a small, fixed-wing aircraft or a helicopter, may be carried out over areas indicated to be favorable by the earlier regional work. These areas may be selected on the basis of surface lithologies, outcrops of certain formations, structural projections, regional geochemical anomalies, or high-background radiometric anomalies.

In areas of low topographic relief and complex (or poorly understood) geology, where no preferred orientation of mineralization is known, the pattern of detailed radiometric surveying might consist of simple north- and/or east-oriented flight lines. The lines may be spaced at 200 to 1,000 m. If topographic relief is pronounced, it might be necessary to use a flight pattern modified to fit topography. In some areas, topography and geology are related, so that flight lines are better oriented parallel to topography or even across the topographic grain (fig. 6-1). Detailed aeroradiometric surveys often require preliminary ground surveys with flight-line markers which can be identified from the air. These markers may be placed at the ends of flight lines or at intersections of flight lines and a road for convenience. In most cases, the measurement of distance between flight-line markers can be accomplished with a vehicle and its odometer. The markers may consist simply of large numbers drawn out on the ground with white lime or other readily visible material.

Surveys by helicopter are significantly more costly than are surveys by fixed-wing aircraft, but in many areas, the cost differential is small compared to the advantages of the helicopter. With the helicopter, detailed surveys of anomalies can be easily performed at the time the anomalies are first detected; indeed, in many cases the helicopter can land nearby to allow sampling or other surface work. A further advantage is that in the aeroradiometric survey, the helicopter can maintain more nearly uniform terrain clearance and thereby facilitate detection and interpretation of anomalies.

The exploration program which resulted in the discovery of the large Jabiluka uranium deposits of northern Australia included a detailed helicopter radiometric survey (Rowntree and Mosher, 1976). This discovery is discussed below in this chapter.

SURFACE MAPPING AND GROUND RADIOMETRIC SURVEYS

The priorities of field work and, indeed, the extent and effectiveness of the work depend on the geologic setting and the quality of exposures. In some areas, no surface geologic mapping would be done. For example, surface mapping would be omitted where the exploration targets are identified as specific stratigraphic units known to be confined to the deep subsurface, and where surface expressions are of no significance to mineralization. In some of the producing areas of New Mexico and Wyoming, uranium ore bodies have been found in sedimentary host rocks at locations so remote from surface exposures of the rocks that it is highly improbable that surface mapping would be of benefit in providing data to help guide drilling programs. This problem can be readily recognized if one acknowledges that geologists continue to argue over the meaning of lithologic characteristics of host-rock outcrops in the producing districts. Even in an area of relatively simple

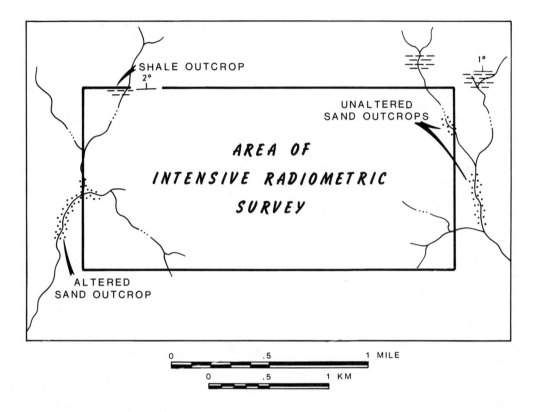

Figure 6-2. Map illustrating a basis for selecting an area for intensive surface radiometric survey. Geologic mapping sometimes establishes favorable areas between scattered limited outcrops.

geology and one as well documented as Shirley Basin, Wyoming, not one explorationist has come forward with a technique (mapping, geochemical, or other) to claim accurate prediction of ore at depth as a result of surface studies.

It is generally recognized that data for surface geologic mapping can best be gathered in areas of good exposures where the mineralized or altered rocks can be studied. Where this is the case, surface work should be meticulously carried out to provide information for later courses of action such as land acquisition or drilling. In some areas where the targets are shallow, surface work may be hampered by covers of soil, wind-blown sand, or glacial deposits. Such complications may limit the amount of data that surface mapping can provide in these areas, but radiometric or geochemical methods which "see" through or below such cover may be employed successfully.

Topographic maps at a scale of 1: 24,000 are adequate for assembling most of the data from field work. If topographic maps are not available, it is best to acquire aerial photographs and possibly have controlled mosaics prepared. Controlled mosaics are assembled by laying the centers of the photographs along linear flight lines and adding ground control from maps or surveys for definite picture points (Low, 1952, p. 266).

Aerial photographs in natural color may simplify and significantly improve the mapping. Often the important geologic formations or structures can be readily identified and mapped on color photographs so that field work can be focused on areas of interest.

Strata-controlled Deposits

In most cases, strata-controlled deposits occur in specific stratigraphic units which can be identified early in a program. If these favorable units are exposed around the basin margin or along incised drainages, field work may be concentrated on such outcrops. Potential host rocks should be mapped individually.

In areas where roll fronts are anticipated and where good exposures exist, detailed study and mapping of sandstones might establish targets for the initial phase of drilling. Mapping the altered sandstones and their unaltered equivalents sometimes gives the approximate loci and projections of roll fronts in the subsurface. Several drilling programs in the Powder River Basin, Wyoming, were enhanced by the mapping of surface exposures of altered and unaltered sandstones. In many other areas, however, the hosts are not adequately exposed to carry out detailed mapping in advance of drilling.

As the mapping is being carried out, possible mineralized zones within the sandstone (such as carbonaceous intervals) and sandstone-shale contacts should be carefully surveyed with a scintillation counter. Unexposed projections of the favorable outcrops may be examined through surface radiometric surveys (fig. 6-2). If the area to be covered is large, a vehicle-mounted total gamma or gamma-ray spectrometer survey may be considered. For easy relocation, anomalies may be marked on the ground, by flagging for example, or they may be evaluated as they are encountered.

If an anomaly is detected where bedrock is not exposed, the anomalous area should be surveyed in detail on the ground on foot to determine the locus of highest gamma count. A hole or holes should then be dug to see if the radioactivity increases with depth below the surface. If so, and if the source cannot be reached by manual digging, a backhoe or auger drill might be used to help locate the source. In some cases, a backhoe is useful to expose a vertical section of the mineralization. We have used this method on several occasions with success (figs. 6-3a, 6-3b, 6-3c).

Figure 6-3a. Backhoe excavating trench as co-author M. O. Childers looks on, scintillometer in hand. Trenches such as this may be dug to expose certain critical stratigraphic or mineralized intervals, thereby providing faces from which more detailed information may be gathered. Care must be exercised to assure that the walls of the trench will not fall in on persons in the trench.

It should be remembered that digging and auger drilling at the site of radioactive

Figure 6-3b. After the trench is completed, a fence should be promptly erected to prevent livestock from being injured. A detailed examination of the rocks exposed in the trench walls may be made by the explorationists. The potassium ferrocyanide test discussed in chapter 3 is valuable to precisely define uranium concentrations in the walls of the trench.

anomalies are simply low-cost, preliminary steps in an evaluation. They may expose some limb mineralization associated with an altered sandstone, or they may even expose a shallow extension of a subsidiary roll or roll front. But most of the mineralization on typical roll fronts occurs in porous and permeable sandstone which, in a near-surface situation, will allow most of the uranium to be leached by influent water from the weathering sandstone. For this reason, the explorationist is normally obliged to drill well below the water table when evaluating roll fronts. Ordinarily such drilling must exceed 30 m in depth in the western United States. Exceptions to this are many of the deposits in the Colorado Plateau, which occur above the water table; indeed they rarely are totally destroyed

by chemical weathering. This is partly due to their occurrence in relatively impermeable sandstone and partly due to the common association of uranium and vanadium in fairly stable oxides. In the late 1950s and early 1960s, geologists misinterpreted shallow, weathered mineralization in Wyoming because they were making an analogy between those deposits and deposits which had been previously described in the Colorado Plateau. Consequently, they failed to recognize the large potential in Wyoming for ore bodies below the water table.

We have seen evidence of many early attempts to evaluate surface mineralization before the significance of altered sandstone was recognized. In some cases, it is obvious that the explorationists grossly underestimated the depths required to determine

Figure 6-3c. Photography is an important part of accurately recording observations made in the course of geologic field work. Here mineralization in a trench wall is being photographed. When the trench is no longer needed, it may easily be refilled and the surface reseeded.

if ore existed nearby. For example, in one location on the west flank of the Powder River Basin, high-grade uranium-vanadium mineralization occurs in a thin clayey and carbonaceous zone near the top of a porous, permeable, and altered sandstone 30 m thick. During the middle 1950s, prospectors attempted to follow this mineralization downdip by driving an inclined shaft. The shaft bottomed in altered sandstone at 20 m depth. In the late 1960s, an exploration company drilled farther downdip and encountered the altered sandstone at depths of approximately 50 to 75 m without locating unaltered sandstone (or roll front). In 1969, a company with which we were associated acquired the prospect and located the roll front in the sandstone at a depth of about 360 m and about 2 km northeast of the original prospect shaft (fig. 6-4). This roll front has not yet been evaluated by

close-spaced drilling, but preliminary drilling results have been very encouraging.

These same prospectors had established a camp about 200 m east of the inclined shaft, and geologists working in the area in 1969 noted that a sandstone at the camp site changed from pinkish to gray along strike. Careful work with a scintillometer, followed by digging a pit with pick and shovel, revealed uranium mineralization exceeding 1 percent in grade. Without realizing it, the prospectors had camped on higher-grade mineralization than anything they exposed in the inclined shaft.

In the Colorado Plateau, numerous ore bodies have been discovered directly by surface work, and many drilling programs have been initiated on the basis of surface mapping. Mineralization in the Salt Wash Member of the Morrison Formation and in the Mossback and Shinarump Members

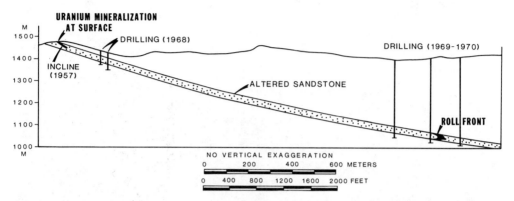

Figure 6-4. Cross section through a roll front in a thick sandstone on the western flank of the Powder River Basin, Wyoming, illustrating the progressive evaluation of mineralized and altered outcrops. Early exploration efforts based on experiences gained in the Colorado Plateau were ill suited to Wyoming roll fronts.

of the Chinle Formation frequently occurs in hard, impermeable parts of the sandstone bodies. Also, chemical weathering undoubtedly proceeds more slowly in the arid climate of the Colorado Plateau than in most humid regions.

Much of the initial exploration in the Colorado Plateau consisted of surveying the outcrops of favorable sandstones and conglomerates with Geiger counters or scintillometers and sampling mineralized rock so that chemical analyses could be made. The successful prospectors learned to recognize favorable host rocks and concentrated their efforts on them. Aerial radioactivity surveys were also a popular prospecting technique.

Geologists later began to define favorable trends along which ore deposits were most likely to occur. To some extent, these trends have influenced exploration activities; indeed, in the case of the Uravan mineral belt, it appears that exploration might have been overly influenced by the "mineral belt." Published information suggests the Uravan mineral belt is vaguely defined and inadequately explained.

On the other hand, some trends can be well explained on the basis of geologic

mapping. For example, mapping in northeast Arizona and southeast Utah has shown that uranium ore occurs in sandstones and conglomerates which record Triassic (Chinle) paleodrainages. It was possible to map these channel deposits and their paleocurrent features and project favorable trends into the subsurface (Malan, 1968). Such surface geologic mapping has obvious value in exploration. It establishes a valid control on mineralization.

In our own present-day programs in sedimentary basins, however, we find that we seldom assemble, or even attempt to assemble, maps of detailed geology. The reason is that we cannot afford the time and expense required to assemble a surface map when we know that a few drill holes will provide better data faster and usually at lower cost. A recent example should help illustrate this approach. We carried out regional exploration in an area where literature studies had suggested that a uranium play in Paleocene sediments might be made. Topography in the area was gently rolling and outcrops were quite scarce, with only an occasional sandstone or shale exposed in banks of dry streams. We flew the area both in fixed-wing aircraft and in helicopters and carried out several weeks' work on the ground, examining surface geology and doing a lot of pick and shovel work to expose critical geologic features such as altered sandstone. In the course of our work, we found an outcrop of strongly mineralized sandstone (locally exceeding 2 percent U_3O_8) and had a backhoe dig a trench through the best anomaly. Mineralization exceeding 0.10 percent U_3O_8 in thicknesses of more than 2.5 m was found in the trench. This might seem an ideal spot to begin surface mapping in an attempt to decipher the geology of the area, and yet we elected to do *no* additional surface mapping for the reason mentioned earlier. We will get more good information in less time and at lower cost from drilling than from surface mapping, and the prospect will have to be drilled regardless of whether we surface map or not. If we do *any* detailed surface mapping on that prospect, it will be *after* we have the information from the drilling.

This brings up another thought on field mapping. Geologists mapping for the U.S. Geological Survey almost always do an excellent job, but we have seen many mapping projects go on for several years as the geologists cover every metre of the surface trying to find decent outcrops. When they do find one, they must try to make the most of it and may spend a great deal of time (and money through wages and other expenses) gleaning every shred of information from rocks exposed. Most of us in industry cannot afford the time and expense of such a procedure when evaluation by drilling is so quick and relatively inexpensive and, on top of that, provides better data in most cases. We have seen geologists from the Survey spend three to four years mapping the surface of an area an industry geologist may map in three or four months or less. The difference, aside from the Survey's strict attention to detail, usually does not reflect on the geologists involved, but rather on the procedure. Industry often will carry out exploratory drilling early in the program, and the industry geologist will use data from the drilling to help interpret the surface geology. The Survey seldom conducts drilling in conjunction with mapping, to the detriment of the project.

Structure-controlled Deposits

Lithology is often a strong secondary controlling influence on mineralization in structure-controlled deposits, and so it is useful to identify and map both the structure and the favorable lithologic units. If favorable areas for mineralization can be mapped reliably, the radiometric and geochemical surveys may be concentrated in them. Unfortunately, exposures are often inadequate to precisely map or delineate favorable or unfavorable areas, but it is often possible to outline areas generally and either carry out further work there or eliminate them from the surveys.

Radiometric surveys have been important to all of the exploration programs which have located significant structure-controlled uranium deposits. In most of these programs, aeroradiometry first detected the anomalies, and subsequent ground surveys pinpointed sources of mineralization. In some cases, these ground radiometric surveys have been extremely detailed. The Jabiluka deposit in Northern Territory, Australia (fig. 6-5), was first detected by a ground radiometric survey which consisted of a grid with 30-m line spacing (Rowntree and Mosher, 1976).

The Jabiluka program utilized several survey teams, each of which included one geologist and one or more field technicians equipped with gamma-ray spectrometers.

The mineralization at Jabiluka occurs in a schist (Koolpin Formation) which is uncomformably overlain by middle

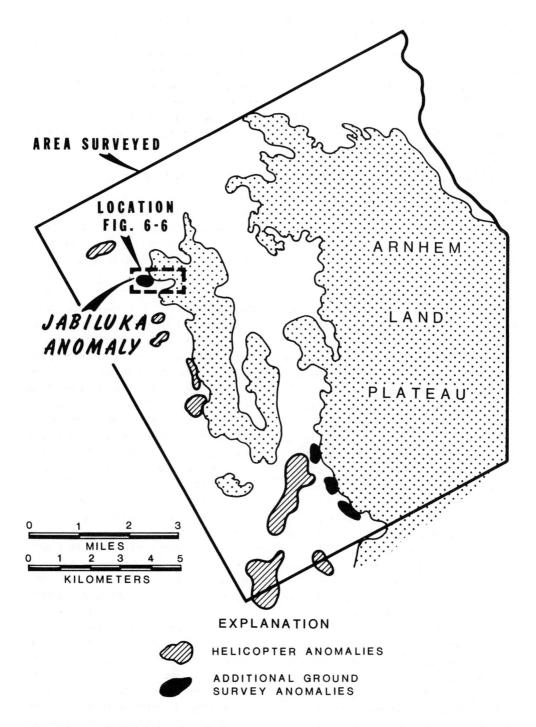

AREA SURVEYED

LOCATION FIG. 6-6

JABILUKA ANOMALY

ARNHEM

LAND

PLATEAU

0   1   2   3
MILES

0   1   2   3   4   5
KILOMETERS

EXPLANATION

HELICOPTER ANOMALIES

ADDITIONAL GROUND
SURVEY ANOMALIES

Figure 6-5. Map of Jabiluka uranium district, Northern Territory, Australia, showing radiometric anomalies from helicopter surveys and later ground surveys. Modified from Rowntree and Mosher (1976).

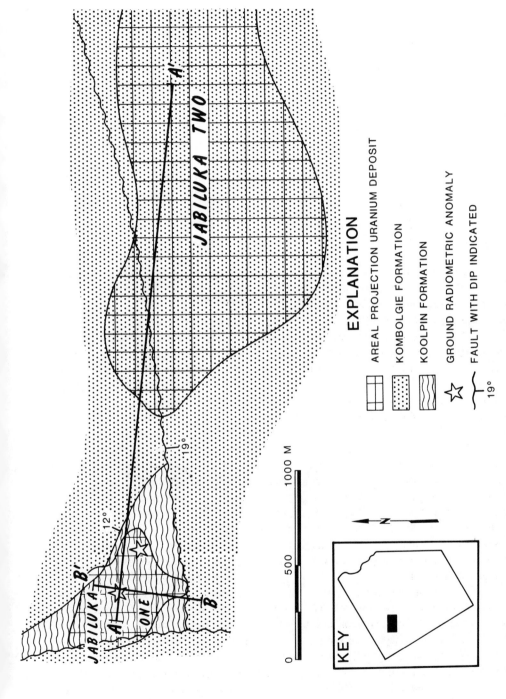

EXPLANATION

AREAL PROJECTION URANIUM DEPOSIT

KOMBOLGIE FORMATION

KOOLPIN FORMATION

GROUND RADIOMETRIC ANOMALY

FAULT WITH DIP INDICATED

Figure 6-6. Map of Jabiluka uranium district, Northern Territory, Australia, showing surface geology, radiometric anomalies, and projected outline of uranium deposits. Cross sections A-A' and B-B' are shown in figs. 6-7a and 6-7b, respectively. Modified from Rowntree and Mosher (1976).

Proterozoic sandstone of the Kombolgie Formation immediately east of the discovery point (fig. 6-6). The host rock is not exposed in the vicinity of the surface anomaly due to thick regolith and talus from the Kombolgie. The 30-m grid survey detected a gamma anomaly of 2x background. A more detailed examination of the anomalous area revealed a fragment of mineralized schist and an anthill with a gamma anomaly of 50x background.

Two anomalies were discovered at Jabiluka, and pits were hand excavated at both places. A cobble of quartz schist 1 m below the surface in one pit assayed 8 percent U_3O_8. Next a backhoe was used to cut trenches across the anomalous area where regolith was thought to be thin. These trenches cut into bedrock and exposed the ore bodies. This was followed by shallow drilling with an auger drill to establish the controls on mineralization in advance of the deep diamond drill program (figs. 6-7a, 6-7b). Rows of closely spaced holes (3 to 10 m apart) were drilled to depths of 15 to 25 m (Rowntree and Mosher, 1976).

The Jabiluka case emphasizes the importance of detail in radiometric surveying and, indeed, the importance of thorough evaluation of anomalies. As pointed out by Rowntree and Mosher, it is also very important to maintain flexibility in a prospecting program. Rigid adherence to a priority program based on size and strength of anomalies detected from the air would have lessened the chances of discovering Jabiluka, one of the world's largest known high-grade uranium deposits.

Most of the uranium deposits in northern Saskatchewan were first detected by airborne radiometric surveys. Ground surveys next located and evaluated the radioactive anomalies, which frequently were found to be mineralized boulders in glacial drift (Lintott and others, 1976). The explorationists studied the mineralized boulders to identify the bedrock lithology which constituted the exploration targets. At the same time, detailed ground radiometric surveys were carried out. Anomalies were excavated in an effort to locate mineralization in bedrock. Finally, drilling was carried out to locate and define bedrock mineralization.

The water table is very shallow in northern Saskatchewan, and there are numerous lakes. The Canadians have developed instrumentation and field procedures for underwater radiometric surveys. This has been an important method of exploration in recent years.

Radon and helium surveys are also used in these northern Saskatchewan exploration programs, and these methods are discussed in following sections of this chapter.

Intrusive-controlled Deposits

Mineralization in intrusive-controlled deposits is primarily controlled by lithology, and usually the uranium is concentrated in a specific phase of the intrusive host rock. For this reason, detailed geologic mapping is important in establishing ore controls. As the host rock is determined and mapped, progressively more detailed radiometric surveys can be executed. In this manner, the surface geometry of the mineralized rock can often be outlined.

The following description of the field work that formed the basis for the drilling program at the Rössing uranium deposit, South West Africa, is based on the excellent paper by Berning and others (1976). Early in the program, geologic mapping was completed that showed the general structure and formations of the area, including the broad distribution of the intrusive rock which hosts uranium. Also, early reconnaissance radiometric surveys located the main area of anomalies.

A detailed ground radiometric survey

229

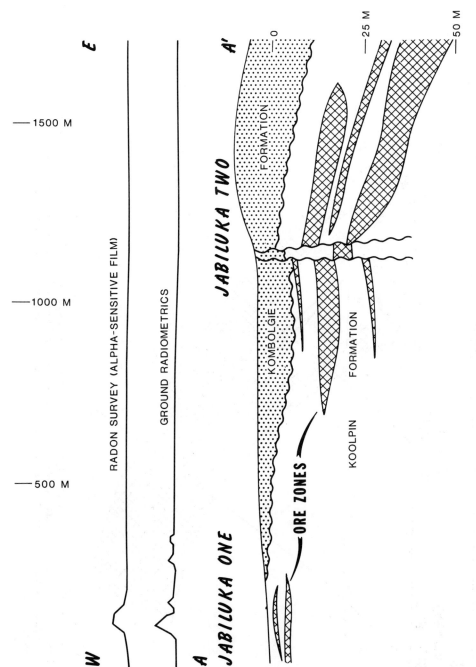

Figure 6-7a. East-west cross section through Jabiluka area, showing relationships between uranium deposits, and profiles from radon and ground radiometric surveys. Radon survey utilizing alpha-sensitive film did not detect the large high-grade Jabiluka 2 deposit at 25-m depth. Modified from Rowntree and Mosher (1976).

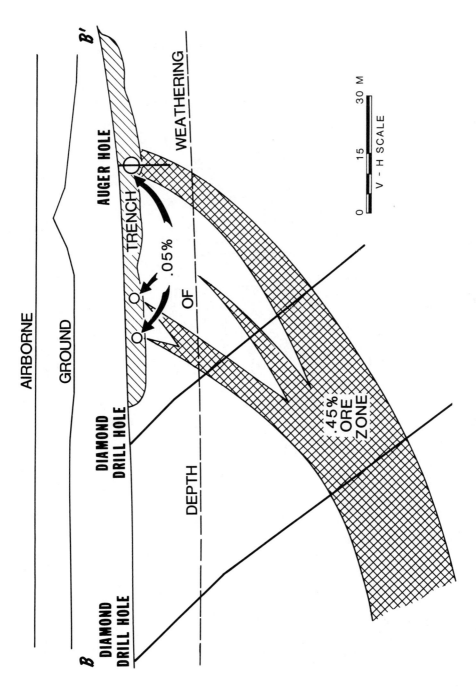

Figure 6-7b. North-south cross section through Jabiluka 1 deposit showing shallow auger drilling, trenching, and later deep drilling. Auger drilling and trenching provided data for planning later diamond drilling. Two profiles above the cross section show results of airborne and ground radiometric surveys. Note that detailed ground radiometric survey detected the surface extension of the ore body. After Rowntree and Mosher (1976).

was carried out over the area which contained the main anomalies, covering 200 km². This survey consisted of plotting scintillometer readings at 30 m (100 ft) intervals along traverse lines spaced 90 m (300 ft) apart, with the scintillometer held 45 cm (18 in.) above the gound. This detailed survey successfully delineated the exposed part of the Rössing uranium deposit.

The surface geology of the main uranium deposit was mapped by plane table at a scale of 1:1,000, and a detailed topographic map was prepared. This detailed geologic and topographic map formed the basis for planning and interpreting the drilling of the ore deposit.

Initial drilling consisted of percussion angle holes averaging about 40 m in depth. This was followed by deeper incline core drilling which defined the ore body at depth.

RADON SURVEYS

Radon Migration

Radon-222 anomalies have been measured in soil gas, ground water, and surface water. Radon in water is sometimes measured as a supplement to uranium in reconnaissance surveys (described in chapter 3). However, tests have indicated that radon-222 does not migrate far from its parent, radium-226 (Dyck, 1975). Radon-222 has a half-life of 3.8 days, and it is a dense gas which does not diffuse rapidly. In an excellent paper, Tanner (1964) reviewed the data pertinent to radon migration in soil and ground water and estimated that radon-222 will not migrate more than about 10 m in dry soil or 20 cm in wet soil. He indicates a maximum migration of radon in one-dimensional moving ground water of 2,300 m. Tanner (1964, p. 178) suggested that "migration of several most mobile members of the uranium series,

uranium-238, uranium-234, radium-226, and radon-222, is probably responsible for many anomalies, particularly those which imply movement through distances of many meters."

Tanner's data suggest that substantial variations in radon in soil gas might simply correlate with variations in moisture content of the soil. If maximum radon migration in soil varies from 10 to .2 m, depending on the moisture contuent, it is to be expected that substantial variations in measured radon will result from variable moisture profiles. Moisture content varies with clay and sand content and topography. Caneer and Saum (1974) reported large variations in the amount of radon monitored at a sample site that were related to varying soil-moisture contents. They measured alpha intensities (counts per minute) as high as 679 in moist soil, whereas in relatively dry soil 1.2 m away the alpha intensity was 548 cpm. They found that in one site when the soil was water saturated the alpha intensity was only 154 cpm, but 10 days later, when the soil was only slightly moist and they cleaned the hole out and deepened it 5 to 8 cm, the alpha intensity increased to 326 cpm. They concluded that this twofold increase was caused by the reduction of moisture in the soil. Further, they concluded that an *optimum* moisture content lower than the saturation point yields highest radon values.

Dyck (1972, p. 217) found that radon was lost from water samples contained in plastic bottles, but he also found that freezing the sample preserved the radon. Radon diffusion can be almost halted by freezing, and radon diffusion is enhanced by increase in temperature. These are extremely important factors for the explorationist to keep in mind.

Caneer and Saum (1974) reported that "factors such as variations in barometric

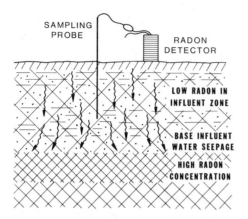
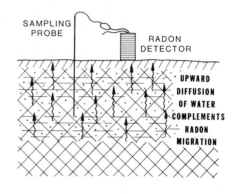

Figure 6-8. Hypothetical cross section showing how radon in soil might be inhibited in upward migration by influent seepage of water during and after heavy precipitation or when spring thaw is taking place.

pressure, minor changes in hole depth, seasonal change, and repumping of open holes after extended time lapses have little or no effect on the radon content of specific sample sites." Kraner and others (1964) showed that changes in barometric pressure and wind affect radon measurements in soil gas slightly, but not enough to be of concern to the explorationist.

Tanner's data suggest a simple inverse relationship between the mobility of radon in soil and the amount of moisture in the soil. The data reported by Caneer and Saum would be more meaningful if additional information were included relating to the moisture changes taking place in the soil for a few days before the measurements were made. Temperature variations before sampling might also be significant. We would suggest that very local changes in soil character might also affect the readings.

Assuming that a sample site for monitoring the radon content of soil gas is located a few metres (or less) above a radium-226 source, and that radon typically diffuses upward past the sample point, the radon measurements probably

would be higher when the soil is losing moisture and soil moisture is diffusing upward. Conversely, if the soil is gaining moisture from influent seepage (for example, during a rain storm), the radon would be carried downward, and radon measurements would be lower. We believe that the movement of water in the soil for a period of time before sampling is important, and this would have to be considered in any extended radon survey which might be continued during and after weather changes (fig. 6-8).

Surface radiometric surveys are capable of detecting bismuth-214, which may occur within 30 cm of the ground surface. Radon soil gas surveys may detect radium-226, which occurs less than about 10 m below the surface, according to Tanner (1964). This may be decreased to less than a metre below the sample point if moisture has been moving downward through the soil opposing the upward diffusion of radon. We suggest that radon surveys be conducted when upward diffusion of radon is enhanced by upward diffusion of soil moisture.

There have been reports of radon

anomalies being defined by soil-gas surveys over uranium ore bodies which occur as much as 120 m below the surface (Caneer and Saum, 1974; Gingrich and Fisher, 1976). Gingrich and Fisher (unpub. preprint) suggested that radon might migrate upward from deep uranium mineralization to near-surface sample points by the following mechanisms: "(1) gas movement across geothermal gradients to cooler and hence lower pressure areas; this is predominantly an upward movement; (2) earth tidal pumping effects, the opening and closing of fracture and pore spaces; (3) pore pressure changes due to seismic stresses; (4) deep effects of barometric pressure changes; (5) diffusion pressures of more concentrated soil gases (O_2, N_2, CO_2, Ar, He and other gases); (6) upward aqueous transport vectors; (7) soil and air temperature differentials; and (8) wind speed, direction and turbulence effects on the soil gas." Earth tidal pumping effects and deep effects of barometric pressure changes are exotic theories with little evidence to support them. Gases with higher constants of diffusion than radon are effectively trapped in the subsurface under a variety of conditions. Seismic stresses associated with earthquake activity are known to result in increased radon migration to the surface, but this is confined to areas of seismic activity. "Upward aqueous transport vectors" are effective, in some places, in bringing radium and radon up to the surface. But these are restricted to open fracture systems or similar phenomena which in special cases result in significant upward movement of water (transporting radon and/or radium). The upward diffusion of water in soil, mentioned earlier herein, enhances upward radon migration, but not more than a few metres of migration distance

are involved. The other mechanisms cited by Gingrich and Fisher have insignificant effects relative to long-range migration of radon.

The few case histories where radon anomalies have been measured in soil gas over uranium deposits 50 to 120 m in depth have been reported by people who are selling radon-surveying services. We are obliged to conclude that the possible alternatives to long-range radon migration have not been adequately discussed. The reports of success could also be placed in better perspective for the general explorationist if those reporting case histories would also cite the larger number of cases where no radon anomalies can be detected (by any method) over large high-grade uranium deposits which occur at less than 100 m in depth. Industry geologists could help if they would report on unsuccessful cases.

The importance of more than superficial evaluation of radon in soil gas is exemplified by a case history of a survey in Karnes County, Texas, where Tanner (1964, p. 178) "found a halo of enhanced radon concentration in soil gas at all sampling points within about 30 m of a uranium ore body. Without further knowledge, the halo might have been ascribed to radon migration. However, the vertical distribution of radon in the ground was inconsistent with upward radon migration of more than a few centimeters, and the horizontal gradient of radon concentration was inadequate to account for the anomaly. Displacement of the anomaly in the downhill direction, correspondence between the gamma-ray intensity and radon concentrations in holes near each other, and uranium-series disequilibrium at the prospect (J. N. Rosholt, Jr., written commun.) indicated that the anomaly resulted from migration of the precursors of radon."

Dyck (1972, 1975) has studied radon diffusion in still lake waters and determined that radon will not move more than 6 m in these waters.

Soil-Gas Radon Survey Methods

There are two competing techniques for measuring radon in soil gas: the alpha-sensitive film and the silver-activated zinc sulphide counter (radon emanometers which utilize ionization chambers are also available). An earlier method of passing soil gas through activated charcoal has been largely abandoned.

The alpha-sensitive film method has been widely advertised in the industry. The film is an acetate variety which is not sensitive to light but may be sensitive to very high temperatures. Alpha particles which strike the film create "damage tracks" which can be etched and microscopically counted. Film for radon detection purposes is available at very low cost (less than $.10 a strip) from Eastman Kodak Company, Rochester, New York, together with instructions for etching. The film may be ordered in sheets or as a roll and must be cut for placement in containers for use. At least one service company has established a practice in supplying film and containers and later etching and counting the alpha particle tracks for a fee. Many operators have found this service to be prohibitively expensive.

The following steps summarize the alpha-sensitive film method of measuring radon in soil gas (fig. 6-9a):

1. A small strip of alpha-sensitive film is attached to the inside base of an inexpensive container such as a plastic cup. The containers are stored and transported in plastic bags to prevent exposure to dust and radon gas.
2. The container with the film strip is placed in a shallow hole 0.4 to 1 m in depth with the open end down to prevent rock or soil particles from falling on the film strip (fig. 6-9a).
3. The hole is covered or backfilled and marked for relocation.
4. After the film has been in the hole for a period of three or four weeks, it is removed and taken to the laboratory for a simple procedure of etching with a sodium hydroxide solution and counting the etched alpha particle damage tracks under a microscope.

The zinc sulphide counter method of measuring radon in soil gas is carried out in the following steps:

1. To sample the soil gas, a hole may be drilled into the ground to a depth of about 75 cm with equipment, as shown in figure 6-10, or a hollow rod may be driven into the ground. We prefer the latter sampling procedure because it reduces the effects of atmospheric contamination.
2. The soil gas should be passed through a filter to remove dust and moisture before it enters the counting chamber.
3. The soil gas is pumped into the counting chamber, and a reading is recorded in counts per minute. This process is repeated until a maximum (or plateau) reading is recorded; then the sample is held in the chamber for two or three minutes to determine if radon-220 is significant in the sample. If radon-220 is contributing significantly to the alpha count, the reading will drop proportionately in this time frame, because radon-220 has a short half-life of 54.5 seconds (figs. 6-9b, 6-9c, 6-9d, 6-10).

The alpha-sensitive film method of radon measurement reputedly integrates the radon concentrations in soil gas over an extended period of time (usually 2 to 3 weeks), and supposedly will assure that

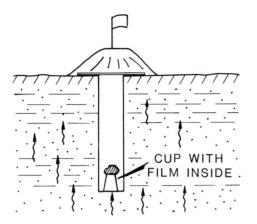

Figure 6-9a. Diagram illustrating the method of emplacing the alpha-sensitive film for soil-gas radon measurement. A hole must be drilled to a depth of about 75 to 100 cm. A cup with the film attached to its interior is placed on the bottom of the hole, and the hole is covered and marked.

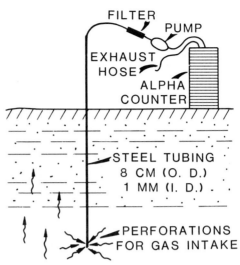

Figure 6-9b. Diagram showing the probe method of sampling soil gas for radon analysis using a zinc sulphide alpha counter.

measurements will not be erratic due to barometric variations or temporary unusual wind conditions. However, in most cases, these factors apparently do not result in significant changes in measurements made by the zinc sulphide method. Disadvantages of the film method include its high cost (when the service company is used), the delay between placing the film and getting a track quantity, and the lab setup required. In some cases, it is desirable to distinguish radon-222 from radon-220, and this cannot be done with the film method.

We have found that immediate determinations provided by the zinc sulphide method give the explorationist more flexibility in planning and executing a program. If an anomaly is indicated by a measurement during a survey, the sample pattern can be modified to get additional measurements, or the anomaly can be marked for later evaluation. If additional soil-gas measurements are desirable, they can be made immediately (before atmospheric and ground conditions may change).

The important variable which influences the movement of radon in soil appears to be water. In the form of optimum soil moisture, water may enhance radon anomalies. In periods of downward seepage or upward diffusion of water, results of a radon survey may be greatly different than in times when the soil is simply moist. It follows that soil-gas radon surveys should not be conducted immediately following heavy or extended rainfall or during the spring thaw. However, periods of rainfall in which influent seepage does not extend to the depth where soil gas is sampled will not adversely affect the survey.

The alpha-sensitive film method requires that the measurement proceed once the films are emplaced. If the period of measurement coincides with extended rainy weather, persistent downward seepage might effectively prevent radon from reaching the points of measurement. In this case, we recommend that the period of measurement be extended.

Radon surveys are usually carried out as grid patterns, and the density of spacing

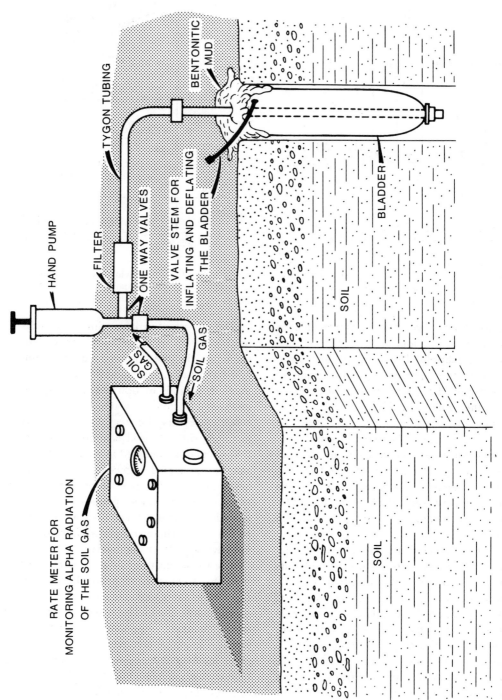

BENTONITIC
MUD

TYGON TUBING

BLADDER

HAND PUMP

FILTER

ONE WAY VALVES

VALVE STEM FOR
INFLATING AND DEFLATING
THE BLADDER

SOIL GAS

SOIL GAS

SOIL

SOIL

RATE METER FOR
MONITORING ALPHA RADIATION
OF THE SOIL GAS

Figure 6-9c. Diagram illustrating the method of measuring radon in soil gas in which a hole is drilled and the sampling probe is sealed beneath an inflated bladder and bentonitic mud.

Figure 6-9d. Radon detector attached to probe ready to begin pumping soil gas for analysis.

depends on the program objectives and anticipated mineralization. If the anticipated mineralization is projected to be elongate, the sample stations may be laid out in rows normal to the trend, with 20 to 100 m spacing within rows. The rows may be spaced 100 to 300 m apart. If no anticipated elongation of mineralized trends exists, a simple grid on 50 to 300 m spacing might be used.

Interpreting and Evaluating Radon Anomalies

Most of the radon surveys we have seen show some anomalies, and usually the most difficult part of a program is that of interpreting and evaluating the anomalies. In some areas, a background value is very difficult to establish due to complex variations in radon determinations. These variations may relate to complex bedrock geology or to complex variations in soil or alluvium.

It is helpful if the radon data are plotted on a detailed topographic map to facilitate correlation of anomalies and topography. Information which might be useful in making an interpretation of water movement in the soil and bedrock should also be plotted on the map. This includes springs, bogs or seeps, sandy or clayey soil, salt or alkali concentrations, or phreatophytes.

Bedrock geology might be important to the interpretation of radon anomalies, because lithologies and geological structures influence soil development, moisture, ground-water movement, and fluid diffusion. A radon anomaly which supplements independent favorability based on geologic mapping is a good basis for drilling or excavation.

Figure 6-10. Soil-gas pump and hoses being tested before an auger hole is drilled preparatory to taking a soil-gas sample. The radon content of the sample will be determined by a silver-activated zinc sulphide analyzer. We recommend against the procedure of drilling holes to obtain soil-gas samples.

Strong radon anomalies should be evaluated by drilling or other means. Many weak anomalies will relate to moisture, soil, or topographic variables, but in some cases, a weak anomaly may be a clue to uranium ore at depth. Each anomaly must be evaluated in light of its geologic context.

Some areas contain such abundant strong anomalies that interpretation becomes almost impossible. In one program in 1967 in Wyoming, we decided to run a radon survey in the Great Divide Basin south of Crook's Gap, where uranium was actively being mined. In order to obtain some radon value over ore, we first deployed the "sniffer" unit to the Gas Hills uranium district to obtain some readings from drill holes which penetrated known ore bodies. We were using an activated-charcoal-type unit at the time, and in that process, soil gas is passed through activated charcoal which traps the radon; radioactivity readings of the charcoal are made later. Some relatively high values were obtained. The unit was then sent to the Great Divide Basin, where sniffing in old seismograph drill holes and auger holes produced some tremendous anomalies, much higher than those recorded over known ore bodies. Claims were staked and drilling was carried out, with some holes drilled to 300 m, but no good uranium mineralization was found. Although the radiometric background was generally high and a few streaks of mineralization almost reached 0.01 percent U_3O_8, the findings did not correlate with the radon anomalies. We finally concluded that the scattered weak uranium mineralization and the porous sandstone section from the surface downward combined to produce strong

Figure 6-11a. Explorationist drives metal tube into the soil so that soil-gas sample may be taken. The device in the explorationist's right hand is being used as a hammer; with an up-and-down motion, he hits the striking ring which is visible near the ground surface. This tube has been driven about 120 cm into the soil. A sampling syringe is visible on the surface about 1 m to the left of the tube.

radon anomalies even though no concentrations of uranium approaching commercial grade were present.

In another situation, we were able to get good radon definition of a shallow ore body in the Powder River Basin of Wyoming. A roll-front ore body existed above the water table under a mesa capped by a shale 10 m in thickness. The radon unit did not find the ore body, but it was used later to see if definition could be obtained (it could). It is questionable if the radon unit could have detected the ore body if the shale layer had not been penetrated by drilling.

Some explorationists who have tried radon surveys off and on for several years in the Colorado Plateau have concluded that once drill holes penetrate an ore body, radon anomalies can be found over the ore body. Before the ore body is drilled, however, no radon anomalies will be found unless the ore body is within about 30 m of the surface.

Numerous radon anomalies have been drilled in many different areas with negative results. In fact, some have never been explained. Strong anomalies sometimes occur over shallow weathered sandstones which contain very weak uranium mineralization or relatively insoluble daughters of the radium group. Conversely, there are numerous cases where radon surveys have not detected anomalies over shallow, large, high-grade uranium deposits. A classic example of this is the Jabiluka 2 occurrence in northern Australia, where an alpha-sensitive film survey was carried out over the deposit with no resultant anomaly (Rowntree and Mosher, 1976). In fact, Rowntree and Mosher stated that "the alpha particle survey did not detect Jabiluka Two which is overlain by a minimum of 20 meters of Kombolgie Formation Sandstone." If an alpha particle survey will not detect the largest high-grade uranium deposit ever discovered, and at shallow depth, we must admit that the technique has severe limitations.

HELIUM SURVEYS

Soil-gas helium surveys are now being perfected by several groups, but case histories are still lacking. Preliminary work by the U.S. Geological Survey has been encouraging, and we believe that this technique may prove to be extremely valuable in locating deposits that are very deep or otherwise undetectable as a direct radon anomaly.

Because helium is a stable isotope with no time restriction (such as the 3.8-day half-life of radon) on its migration, and because it is very light, with a much higher constant of diffusion than radon, the helium anomaly around a uranium ore body should extend much farther than the radon anomaly. Radon has been used much more extensively than helium in past exploration projects because radon is easier to sample and measure. However, with the new technology for measuring helium, we believe that it will be more useful in exploration than radon.

The application to reconnaissance work of helium surveys in ground water is discussed in chapter 3, and that discussion includes a brief description of the equipment used.

When sampling soil gas for helium analysis, it is extremely important to avoid contamination by atmospheric gas or exposure of the sample point to the atmosphere. Helium diffuses rapidly into the

Figure 6-11b. A soil-gas sample is drawn from the metal tube by means of a sampling syringe. A sample of soil gas can easily be taken from the same setup and processed through a radon detector, thereby providing additional exploration data.

atmosphere. The best method of sampling soil gas for helium analysis is to use a thin, hollow metal probe (tube) which can be driven into the ground to a depth of about 120 cm (fig. 6-11a). The inside diameter of the probe should be very small to minimize the volume of air within the probe and to give the probe maximum strength in relation to its outside diameter.

If there is a significant thickness of loose dry soil, silt, sand, or other material on the ground surface where a sample is to be taken, this should be removed to permit maximum penetration into the moist soil or relatively dense packed soil. After the probe has been driven into the ground, the sampling syringe is inserted through an airtight septum on the top of

Figure 6-11c. Geochemist with a soil-gas sample in a sampling syringe prepares to enter the vehicle in which the mass spectrometer is contained.

the probe. First the probe is purged by withdrawing a full syringe (10 cc) of gas and discharging it to the atmosphere. The syringe is then reinserted and a sample of soil gas is withdrawn for analysis (figs. 6-11b, 6-11c, 6-11d).

G. M. Reimer (1976, personal commun.) has found that, in most soils, there is a rapid and somewhat erratic increase in helium concentration from the surface (5.2 ppm) down to a "breaking point" at about 50 cm depth, below which a relatively uniform low gradient is usually developed (fig. 6-12). More data are needed to establish this critical depth under varying soil and weather conditions. In some areas it might be necessary to develop a more sophisticated sampling technique to get uncontaminated samples from below the zone of erratic variation.

Reimer (1976a) has tested the plastic syringes used for sampling soil gas and found that they "retain helium very well. Several hours can pass before a significant reduction in an increased helium content would be noticed. This sampling technique would allow remote areas to be explored. The only equipment that the sampling crew needs is a probe or two, a cleaning wire, and a belt pouch of syringes." It would be good to make quantitative determinations of rates of helium loss at various concentrations if a program is undertaken in remote areas where much time is required for transport of samples to the mass spectrometer.

We recommend that both helium and radon be sampled and measured in soil-gas surveys. If a helium anomaly can be supplemental with a radon anomaly, the prospect

Figure 6-11d. Truck-mounted mass spectrometer and recorder. Syringe in which soil-gas sample was collected is visible just above and to the right of center of the photograph. This mass spectrometer is tuned to ^{4}He. The use of portable mass spectrometers in the field is in an early stage of development, and considerable improvement in both equipment and technique can be expected.

is enhanced substantially.

SOIL SAMPLING

Soil sampling for uranium has not been carried out extensively in the past. Explorationists have instead relied heavily on radiometric surveys, radon surveys, and uranium hydrogeochemical surveys.

Soil geochemical surveys for uranium have weaknesses similar to those outlined in chapter 3 for sediment surveys. Uranium is fixed by humates in the soil, but it is readily leached from soils without humates. If humates are fairly uniformly distributed through the soils of an area, and if the area is not covered by glacial deposits, a soil geochemical survey might be

successful in locating a shallow uranium deposit. A soil geochemical survey might be considered when soil cover is too thin for a soil-gas survey. If a soil-sampling geochemical survey is planned, other pathfinder elements should be considered in addition to uranium. Selenium, molybdenum, copper, vanadium, and other elements frequently show positive correlation with uranium, depending on the province.

Soil geochemical surveys are usually conceived and executed in an effort to detect anomalous concentrations of trace elements which have migrated by solution and/or diffusion *away* from an ore body (fig. 6-13). It is interesting to note that the converse of this might also form the basis for

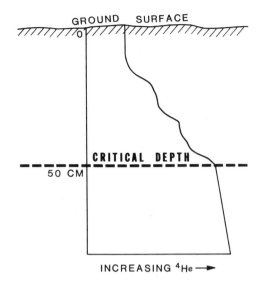

GROUND SURFACE

CRITICAL DEPTH

50 CM

INCREASING ⁴He ➝

Figure 6-12. Profile showing typical increase of helium in soil gas with depth. Based on oral communication with G. M. Reimer, 1976.

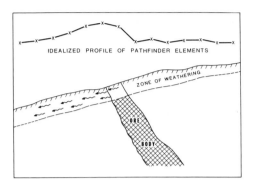

IDEALIZED PROFILE OF PATHFINDER ELEMENTS

ZONE OF WEATHERING

ORE BODY

Figure 6-13. Idealized cross section and profile of concentration of pathfinder elements in soil over an ore body where weathering processes transport elements away from source.

soil or surface bedrock geochemical surveys where supergene structure-controlled deposits or roll-front deposits are anticipated. The object of such a survey might be to detect anomalous concentrations of trace elements *left behind* in altered rock by mineralizing solutions (fig. 6-14).

Lee (1975) found that chloritic minerals replaced montmorillonite in sandstones of the Westwater Canyon Member of the Morrison Formation in the Grants uranium region, New Mexico. These authigenic chloritic minerals were enriched in vanadium and iron and probably were formed at the time of uranium mineralization (142 to 144 ± 5 m.y. ago). Limited work has indicated that altered sandstones in Wyoming are sometimes enriched in selenium, vanadium, and uranium. With newly developed techniques for rapid analysis and computer programs for processing the data, soil and bedrock geochemical surveys may become useful in identifying altered outcrops (fig. 6-15).

Soil sampling for uranium has been used

in the state of Washington, where soil cover is extensive. The results of these surveys have not all been released, but uranium anomalies have been found where no surface radioactivity can be detected.

In our exploration efforts in the Rocky Mountain region, we have yet to find a locality where we believed soil sampling would be of significant value in determining the potential of the area or prospect, and we are not aware of significant results being obtained by others as a result of soil sampling. Nevertheless, it is one of the techniques which must be kept in mind as a source of data to help guide exploration. Advances in the knowledge of uranium geochemistry may one day send us scurrying back to some of the questionable prospects where previous insufficient data kept us from moving ahead with an exploration program.

Geobotanical prospecting methods are often described in texts pertaining to exploration. Our studies of these procedures, as well as attempts at field usage, have led us to the conclusion that the value of this approach is highly questionable. As a result, we do not use geobotanical techniques

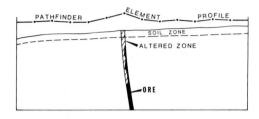

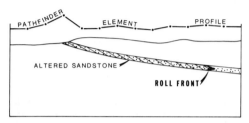

Figure 6-14. Idealized cross sections and profiles illustrating how pathfinder elements left behind by mineralizing solutions (which formed supergene deposits at depth) may be detected by soil or surface bedrock geochemical surveys.

in any of our programs at the present time.

NONRADIOMETRIC GEOPHYSICAL METHODS

Smith and others (1976) have reviewed the case for "other" geophysical methods, and they have suggested that these "methods have not been extensively used in uranium exploration for two general reasons: (1) uranium minerals do not have a distinctive physical property which produces the kind of geophysical anomalies commonly associated with sulphide or iron ore bodies, and (2) at least in the United States, many ore deposits have been found at shallow depths in sedimentary environments where drilling is relatively inexpensive." We agree with these observations and also with the authors' conclusion that increasing depths and costs will force uranium explorationists to incorporate more geophysical methods into their exploration programs. Certainly the efficiency of many uranium exploration programs could be improved by judicious use of selected geophysical methods.

Several exploration companies have attempted to use induced polarization (IP) surveys to help guide drilling programs aimed at delineating roll fronts. The concept behind such surveys is that altered sandstones with highly oxidized iron minerals such as hematite, goethite, and limonite would have lower chargeabilities than pyrite-rich unaltered or ore-bearing sandstones. This difference might also be enhanced by alteration of the clay minerals, because altered clay minerals such as kaolinite often are less chargeable than unaltered clays such as montmorillonite.

Most of the proprietary data from these past IP surveys have not been made available to us, but the results which we have seen have been inconclusive. None of the IP surveys which we have seen have disclosed distinct strong anomalies on the fronts. The U.S. Geological Survey is currently evaluating the use of IP in uranium environments. The Survey's methods of downhole IP and surface IP surveys utilizing new and improved equipment and processing are more encouraging than the past attempts we have seen.

Limited use of high-resolution seismic surveys to locate paleochannels in unconformities has resulted in some reported success. South of the Gas Hills district in central Wyoming, altered and mineralized sandstones at depths of 200 to 450 m are restricted in some areas to Eocene erosional valleys cut into the Precambrian basement rocks. Seismic surveys were useful in defining these old valleys to guide drilling programs.

Other geophysical methods having potential for future exploration programs are discussed in chapter 13.

PROSPECT EVALUATION

The objective of the detailed work outlined in this chapter is to define favorable areas for land acquisition purposes and possibly for drilling. In some instances,

Figure 6-15. Clearing soil from the surface so that a detailed examination of bedrock, in this case uraniferous lignite, may be made.

specific areas for land acquisition can be sharply defined by geochemical, geophysical, or geological methods, but in most cases, acquisition, if undertaken at all, must be planned in only broadly outlined acquisition areas. Occasionally the work described in this chapter results in an ore discovery at the surface (such as Jabiluka), but that is exceptional.

When the detailed surveys are completed on an anomalous area, the data should be integrated so that plans for leasing or staking may be formulated. In most of our projects, the geologic map forms the basis for the evaluation, and any geochemical or geophysical anomalies are superimposed in color or on transparent overlays.

If interest in the area continues, data from this part of the evaluation may be (and should be) used for planning some of the initial drilling. When target areas have been only broadly outlined by surface work, early drilling may be planned on topographic maps at a scale of 1:24,000. However, it might be more efficient to prepare maps at a larger scale (such as 1:12,000) if sharply defined targets have been discovered.

The scale of maps to be prepared during various phases of an exploration program, which starts with a large area and progressively refines targets, is an extremely important consideration. We have observed several programs where the operators were working with maps which were too small to accommodate the data.

Geologic maps and cross sections which illustrate the concepts of mineralization and anticipated depths should be prepared. We have witnessed drilling programs in which the geologists arbitrarily established drilling depths without a basic knowledge of the depths required to penetrate all of the potential zones. In one such program, the holes were all too shallow to have any possibility of encountering mineralization. This could have been avoided if surface work had been properly completed and the data integrated into the initial drilling program.

In some instances, the data available are insufficient to permit the geologists to plan drilling depths with confidence. In such a situation, a few of the initial drill holes should be drilled to depths adequate to test deeper potential host rocks within the prospect. Such drilling, of course, must fit into the budget and must be timely.

Recently we were involved in a uranium project in a basin where we felt confident there was little chance for mineralization below 100 m. Yet, when we began drilling, a few of the holes were drilled to 300 m so that more of the stratigraphic section could be examined. If there is mineralization at depth it is best to discover it early in the game.

7
The Decision To Make A Play

MARKET CONSIDERATIONS

In the previous chapter, we discussed steps which may be taken in a continuing evaluation of the geology of a prospect. As the studies progress, the explorationist must also continue to make decisions concerning disposition of the prospect from a business standpoint (fig. 7-1). An important consideration is the projected exploration timetable, which is closely tied to the market for the mineral in question and to forecasts of sales for that commodity. Unfortunately, forecasts of this type are often difficult or impossible to make, even for the short term. Consequently, serious difficulties arise in attempting to plan exploration for a particular mineral when results of successful exploration may not appear on the market for eight to ten years.

When embarking on a raw exploration program, it is reasonable to assume that four or five years will be required *before* an ore discovery can be made. Some discoveries may be made sooner, but others may require even more time. Mine production usually does not commence for five to eight years *after* a discovery is made, and even longer lead times are required in environmentally sensitive areas (which now may include the entire United States).

When we review past exploration efforts of several major companies, which have staffs analyzing the future markets for various mineral commodities, we find that they seem to have fared no better than the individual who uses plain common sense to predict demand. For example, many of the companies which became involved in copper exploration did not anticipate the low copper prices of mid-1976. Some producers have been obliged to curtail mining and exploration activities until copper prices become more favorable; others will continue copper exploration in anticipation of better future prices.

247

248

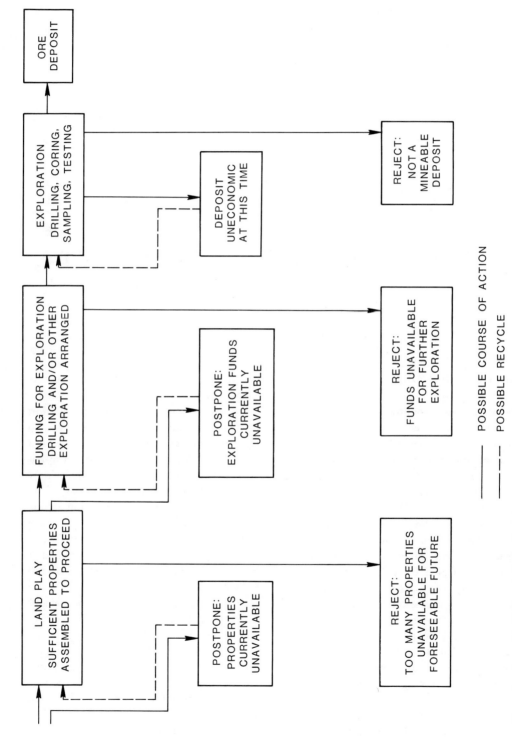

ORE DEPOSIT

EXPLORATION DRILLING, CORING, SAMPLING, TESTING

DEPOSIT UNECONOMIC AT THIS TIME

REJECT: NOT A MINEABLE DEPOSIT

FUNDING FOR EXPLORATION DRILLING AND/OR OTHER EXPLORATION ARRANGED

POSTPONE: EXPLORATION FUNDS CURRENTLY UNAVAILABLE

REJECT: FUNDS UNAVAILABLE FOR FURTHER EXPLORATION

LAND PLAY SUFFICIENT PROPERTIES ASSEMBLED TO PROCEED

POSTPONE: PROPERTIES CURRENTLY UNAVAILABLE

REJECT: TOO MANY PROPERTIES UNAVAILABLE FOR FORESEEABLE FUTURE

—— POSSIBLE COURSE OF ACTION

--- POSSIBLE RECYCLE

Figure 7-1. Second half of exploration flow sheet.

Uranium exploration was slow and relatively uneventful from 1959 to 1965, because large reserves were available and the $17.60 per kilogram ($8 per pound) price was not particularly attractive. In 1965, many companies began to forecast better future prices and recommenced exploration. Figure 7-2 shows footage drilled by years. The market did not materialize, however, and many companies that were engaged in exploration failed to find the quick bonanza they had expected. As a result, in 1970 expenditures for exploration drilling fell once again, but they began a recovery in 1974. Some companies were able to survive these ups and downs and are still involved in uranium exploration and development today, but many had to rely on income from other sources because their uranium exploration activities resulted in no discoveries.

In the late 1960s, some small operators fell by the wayside. Many of these failures resulted from the following sequence: (1) anticipated good market for uranium (and thus uranium properties), (2) substantial expenditures for property acquisition (much with borrowed capital), (3) failure of a good market for uranium (and uranium properties) to develop, (4) inability to sell deals (properties), (5) inability to regain spent capital, and (6) loss of properties and capital. As we explained in chapter 5, acquisition and retention of mineral exploration properties is expensive. A person or company acquiring properties must either (a) have the financial ability to hold or explore the properties, (b) turn them over to another party or otherwise obtain financing, or (c) relinquish them. Many small operators went under when the market failed to materialize, never to return to prospecting or mining.

In the mid-1960s, one prominent energy-mineral–oriented company spent several million dollars drilling out uranium ore bodies in the Powder River Basin of Wyoming in anticipation of a sharply increased uranium market (and prices) in the late 1960s. The market did not develop, and ten years later, by 1976, no ore was yet being produced. Someone erred in a forecast. Had the money been invested elsewhere, at a reasonable return on investment, the company would have benefited significantly. The delineation of reserves could easily have been delayed eight or nine years.

Many independent explorationists and small companies are able to function effectively by making plays (an industry expression for establishing a land position) in certain minerals which are "popular" at a given time. The plays usually are followed by geologic evaluation or exploration, although not necessarily. If gold deals are currently in demand, a few gold prospects may be originated or renovated, and likewise with uranium, silver, copper, and so forth. By such fast footwork these groups can often generate enough income to stay in business during slow periods and can sometimes obtain substantial income during periods when competition for deals is brisk. As might be expected, however, errors in timing are not uncommon among independent operators.

We are acquainted with a small group of prospectors who were successful in selling miscellaneous uranium properties in the mid-1960s to the late 1960s. In 1968, utilizing profits from earlier property sales plus borrowed capital, they decided to make one more large play: they staked approximately 20,000 new uranium mining claims in a remote but relatively flat area. Costs for staking the claims were about $35 per claim, or roughly $700,000 total. The staking crews had to camp out for periods of two or more weeks at a time to reduce driving time. Since the claims were staked in the summer, no assessment work was due until September of the following

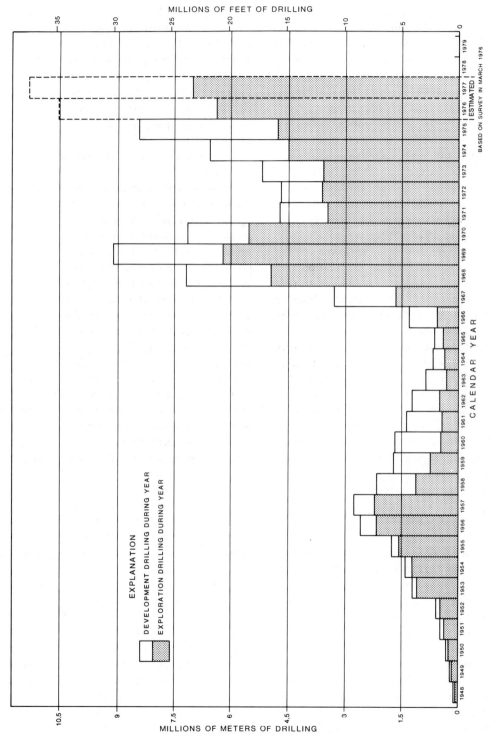

Figure 7-2. Footage of development and exploration drilling, 1948-1977.

year (1969).

Due largely to a declining interest in uranium properties from 1968 to 1971, the prospectors were unable to sell the properties. In 1969, assessment work totaling $2 million was due on the claims ($100 per claim for 20,000 claims), the bank was concerned about its loan, and no prospective buyers had been found. A public offering scheme was attempted, to no avail, and the prospectors finally abandoned the ground. They somehow managed to raise the money to satisfy the bank, but their financial loss was staggering. The error was a direct result of poor timing in the market. Undoubtedly they will restake some of the same ground now that the uranium market has recovered and will again try to make a sale.

Some companies and individuals do not worry about future demand. They simply explore whatever good prospects are available, contending that it is futile to attempt to forecast demands. If they have a gold prospect, or fluorspar, or molybdenum, or whatever, they explore on the assumption that if they make a good enough strike the market will take care of itself. Such extreme optimism is rarely found in other professions.

On the other side are the pessimists. They, too, do not believe in forecasts. They wait until a good market already exists before undertaking expenditures for exploration and land acquisition. Pessimists are also the first to fade when the market drops or shows sustained weakness, even though they may have hired an exploration staff, opened several offices, and made other commitments. Those of us who have been involved in mineral exploration for more than ten years have seen such organizations come and go. Frequently they are offshoots of businesses other than mining. Some petroleum companies, with their spasmodic efforts in uranium, are good ex-

amples of this. They hire a staff which they may lay off in two years' time—hardly a recommended practice for establishing a stable group of explorationists with expertise in uranium.

In contrast to the market forecasters with a seller's viewpoint are those who forecast the market from a buyer's standpoint. We have seen some significant errors there, too. A recent classic example is the manufacturer of nuclear reactors which contracted to supply uranium at very low prices (less than $22 per kilogram or $10 per pound) based on a forecast that uranium prices would continue low for many years. From the buyer's standpoint, that was an optimistic forecast, but from a seller's standpoint, it was pessimistic. Some in the uranium industry believe that the low uranium price quoted by this manufacturer kept the price artificially low for too long. When the price finally rose, it rocketed to about $88 per kilogram ($40 per pound). This manufacturer has become involved in uranium exploration and appears now to have become, at least in part, a pessimistic buyer (the company's explorationists must be optimistic, however).

An important policy we have developed that seems to work successfully is to *avoid mineral exploration situations where the economics, environmental considerations, or politics appear to be questionable.* In other words, it makes good sense to devote time, money, and effort to those mineral situations where the economic, environmental, and political factors are positive; by so doing, chances of success greatly increase. For example, we consider hard-mineral exploration in the entire state of Alaska to be marginal at the present time. Our reasons are (1) unsettled and disturbing land situation (native claims, withdrawn areas, "study" areas, tentative state selections, land disputes between federal and state), (2) extremely high operating costs,

(3) short season, (4) inaccessibility of much of the state, and (5) environmental red tape and restrictions which boost costs even higher. Southeast Alaska does not necessarily have all of these problems, but that area has a very high rainfall.

For these reasons we are not exploring in Alaska. Others *are* exploring there, and we wish them success, but we have more plays or new concepts in the "lower 48"— with none of Alaska's problems. One day, Alaska may become more appealing, but it probably will not be in the immediate future.

How does one forecast demand, then, so that exploration expenditures can be made in a timely fashion? There seems to be no safe way, but some generalities can be made. Energy will be in short supply for the foreseeable future; thus, energy-mineral exploration should be a rewarding endeavor. Oil and gas will see increasing demand (and, hopefully, increasing price); coal will be a valuable commodity *if* well-located deposits of suitable quality can be acquired or discovered; uranium should see continued strong demand, at least through the year 2000.

Forecasts by the U.S. Bureau of Mines indicate serious shortages of many metals in coming years, but the bureau cannot predict downturns in the market of any mineral commodity because of the complex economic and political factors involved. We occasionally look at mineral prospects other than energy minerals, but in the last analysis, we always return to energy minerals because we believe that the economics of the other minerals or metals do not compare favorably. Our specialty, of course, is uranium; and we are particularly pleased with market developments and forecasts for this mineral. None of us has a crystal ball, however, and perhaps by 1978 we will once again see uranium concentrate selling for $33 per kilogram ($15

per pound) or less, in which case we are all wrong, and the industry will suffer a severe recession with associated layoffs and company failures. We are optimistic sellers, though, and we forecast prices continuing to exceed $77 per kilogram ($35 per pound) for the rest of this century.

We wish to stress here that the explorationist should exercise good judgment before embarking on an exploration program for any particular mineral commodity. In addition to investigating the market for selling a deal or selling a mineral product, the explorationist must consider grade, size, location, and other factors which we discussed more fully in chapters 3 and 6. Once these factors are taken into account (including financing, which is discussed below in this chapter), and the time has come to make a play, certain steps should then be taken in order to carry out the program efficiently.

DEFINE BOUNDARIES; DETERMINE PRIORITIES

As geologic work in an area progresses, the explorationist can observe where the favorable areas lie, and thus can begin to assess the size of the project or prospect and determine which ground should be considered for acquisition.

As additional data are gathered and the picture becomes more complete, the explorationist must then decide on the approximate configuration of the area which might be desirable to acquire, later to be explored or sold. We have often found that the area of geologic interest in an exploration project is quite extensive, and we have therefore found it necessary to assign first, second, and even third priorities to prospects within one area. Aside from economic considerations, assigning priorities facilitates concentration on the more favorable prospects first.

The question of a large play versus a

small play invariably arises. By small, we mean anywhere from 5 to 50 mining claims (40 to 400 ha or 100 to 1,000 acres); a large play would be from 100 to 10,000 mining claims (800 to 80,000 ha or 2,000 to 200,000 acres). Today there are deals being put together consisting of 120,000 ha (300,000 acres) or more as an initial position, with additional land to be acquired later. Some areas of interest for specific plays are 60 by 240 km (40 by 130 mi); other projects have a nationwide or even a worldwide scope. Of course, larger plays require larger investments, and this is a major factor the explorationist must consider when deciding what size play to make.

The boundaries of the area to be acquired must be defined as precisely as possible, and this task is best accomplished by the explorationist working on the particular project. He is most familiar with the geology of the area, and it has been his responsibility to map trends, strikes and dips, mineral occurrences, and other information which might indicate the location of favorable ground so that property acquisition can be knowledgeably guided. Even early plans must be formulated within certain guidelines dictated by the company's or the individual's goals and financial capabilities.

ASSEMBLE LAND TITLE INFORMATION; ESTIMATE ACQUISITION COSTS

Within the areas selected (first priority, second priority, and so forth), all land title information must now be assembled and posted on maps, in order to estimate acquisition costs. For example, the first priority area could well consist of three types of lands: (1) federal mineral lands (public domain), on which mining claims must be staked; (2) fee mineral lands, where uranium leases must be taken; and (3) state

lands in the area, which can (and usually should) be leased if they are available. Once this information is posted on maps, showing areas where acquisition based on geology is contemplated, the explorationist can begin to anticipate his acquisition costs. Topographic as well as ownership maps should be studied in this connection.

Custom surveyors are the best source for obtaining estimates of staking costs on public-domain lands. Staking costs are usually directly related to several factors, including the topography of the area, access roads or trails, and remoteness from commercially available food and lodging. Vegetation in the area should be considered, as well as any possible environmental disturbance. Experience gained by the explorationist in traveling in the area will be valuable in the determination. By analyzing these factors, the average cost of a mining claim on federal lands within the area can be estimated. The hostility or friendliness of landowners in the area will have an effect on the costs and even on the acquisition itself, as the following example illustrates.

We were looking at a prospect near a village in Alaska. The prospect was on a steep hillside within sight of a paved highway, and partly within sight of an Indian cabin at the base of the hill about 200 m (660 ft) below the prospect. We had obtained property maps from the Bureau of Land Management, which showed that although the cabin was on land which had been deeded to the Indians, the prospect was on open federal land. In addition, several kilometres from the prospect northward was open federal land, including access to the paved highway. The small tract deeded to the Indians had recently been surveyed, brass caps had been set, and trees had been cleared along the property line, so there could be no mistaking which land was privately owned. Carefully skirting the

Indian land through the trees, we arrived at the prospect at about 7:00 a.m. We could see the cabin below us through the trees, and it was evident that no one was up yet. Soon we were busy chipping away with rock hammers and examining rock types. Before long, dogs began barking and a woman shouted up to us to get off "her mountain." We politely replied that this was land owned by the United States government and that we had a right to be there. She was obviously perturbed and, amid cursing, disappeared into the cabin. The dogs continued to bark, and we went about our work.

Later, a field assistant carried down a load of samples to the truck. On the return trip, about halfway across an opening, a shot was fired. We took cover among the rocks, looked down at the cabin, and saw the Indian woman with a rifle aimed in our direction. She was again cursing, and threatening to call the Alaska state troopers. We heartily welcomed this suggestion, and we told her so before she drove away.

When the woman returned, more shouting ensued, although we were now careful to keep sizeable rocks between us and the cabin. She then went into the cabin, and in a few minutes, a very close shot rang out, the bullet whistling through the trees just over our heads. Soon a wild-eyed boy carrying a rifle appeared from the direction where the shot had originated. He had obviously been sent by his mother to chase us off the mountain even if it meant shooting at us, and we were unarmed.

The boy calmed down, and we explained the meaning of the survey line— that we were on land belonging to the United States, and that he and his mother were breaking the law. He returned to the cabin, but by then we were so unnerved by the shooting that we decided to leave the mountain and call the state troopers. We left and never returned to that prospect,

only partly because the collected samples did not show much promise. Although the Indians were indeed breaking the law, our legal position would have been of little consolation if a bullet had hit someone.

The cost of acquiring the fee mineral lands should be estimated as part of the expense of making a play. Discussions with landmen or other explorationists should give an indication of the current rates for uranium leases on fee lands. For instance, lands may have been leased in a given area for $5 per ha ($2 per acre) per year; to this must be added the fees and expenses of landmen hired to acquire the leases. This might add another $2.50 per ha, for a total of $7.50 per ha.

Costs for state-owned land, if such land is available in the area, can readily be obtained from the state land board, unless there is a requirement that the land be put up for competitive bid (see chapter 5). These costs can range from $1.25 to perhaps several hundred dollars per ha. If you are dealing in a "cold" area, then you probably will be able to acquire state land at a reasonable cost.

FINANCING THE PLAY

After gathering the above information, it should now be possible to estimate all of your land acquisition costs. The total budget must now be considered to insure that adequate funds are available to carry your plan forward; otherwise, alternative financial arrangements must be made. Perhaps a partner with additional funds could be brought in, or just a portion of the total area could be acquired. Many different solutions can be found when an explorationist has the determination and the expertise, but less than adequate funds, to carry a project forward.

Here are a few examples of arrangements which have recently been worked out.

1. The first party (originator) contributes one-third of the initial acquisition money for half of the play; the second party (investor) contributes two-thirds of the funds for the other half of the play. Thereafter, each party shares expenses on a 50-50 basis. The first party usually has the responsibility for all work involved, on a fixed-fee basis, and is also responsible for selling the project. Because the first party has invested acquisition funds, he has a strong incentive to sell the project or move it forward.

2. The first party contributes 25 percent of the money for half of the play (and is usually the operator); the second party contributes 75 percent of the money for the other half of the play.

3. The second party contributes the total funding (within certain guidelines), and then recovers his investment from revenues derived from a sale of the property before any other funds are distributed between the parties. This arrangement is most successful when the second party is able to make a selling effort on his own behalf, since the first party's incentive to sell is somewhat diminished.

4. The first party acquires a substantial land position and then brings in a second party. The second party pays the first party a larger sum than the land originally cost. An area of interest is defined, and the first party, as operator, proceeds with a land acquisition and exploration program, utilizing funds supplied by the second party. After certain expenditures, all interests become working interests. Funds may also be advanced by the second party on behalf of the first party for the opening of a mine or the construction of a mill; in such event, these funds are usually recovered out of 50 percent of the first party's share of initial production receivables. This arrangement calls for a high degree of confidence in the first party by the second,

but it has proved successful in several deals within recent years.

5. The first party acquires a property, or an option on a property, and then raises money for exploration and additional land acquisition through a public offering, whereby a corporation is formed and the necessary steps are taken to offer corporate stock to the public. This can be accomplished through a "Regulation A" (commonly termed "Reg A") or "short form" filing with the United States Securities and Exchange Commission (SEC). Such stock offerings must also be approved by the secretary of state (colloquially known as the "blue sky commissioner") of the state in which such an offering is to be made. We will not discuss public stock offerings in detail in this book; if the explorationist should decide that a public offering appears to be a desirable alternative, then a competent lawyer who specializes in securities law should be consulted.

In brief, here are some of the *advantages* of a public offering:

1. It is often a viable means of raising funds for exploration and acquisition. At this writing, up to $500,000 can be raised in a Reg A offering (from this sum the organizers net about $400,000 after legal fees and broker commissions have been paid).

2. The organizers can usually keep a substantial portion of the stock—sometimes a majority.

3. Once the shares start trading in the market, a trading price will be established for the stock. This may be of value in that the company could perhaps make acquisitions through the issuance of stock (within the law, of course).

4. If the company is successful, there may be certain tax advantages to the organizers (offerers) when their stock appreciates in value, since taxes are

paid only on the capital gains value of stock sold in a particular year. Consequently, taxes on income from stock sales can be more readily managed than taxes on sudden income from bonuses, option money, or purchase payments from selling a property. In some cases, successful public companies are merged into other companies on a tax-free stock exchange basis; this could be a favorable situation for the organizers and shareholders alike.

Some *disadvantages* of a public offering include:

1. An inordinate amount of time must be spent in meetings with attorneys (both the explorationist's and the underwriter's), underwriters, accountants, brokerage firms handling the issue, the SEC (the local office and perhaps also the headquarters in Washington, D.C.), bankers, and personnel from the secretary of state's office. This is time taken away from exploration.

2. A significant expense may be incurred even if only minimal steps are taken toward Reg A or full registration, or if the unexpected occurs—such as worsening stock market conditions. Some explorationists have endured the entire process of a planned stock offering, only to discover that the market was so depressed at the time of the offering that the issue did not sell and had to be withdrawn. Others have been obliged to buy part of their own issue in order to meet the minimum requirement.

3. The explorationist who decides on a public offering will be obliged to carry the shareholders (who have made a one-time investment) on every subsequent deal he makes for

as long as he remains with that company.

4. Any public company is subjected to the scrutiny of the SEC and other regulatory agencies, and a massive volume of paperwork must be submitted frequently and on a timely basis. Rather than addressing himself to matters geologic, such as new prospects, alteration, and mineralization, the explorationist finds much of his time is concerned with SEC forms 10K, 10Q, and 8K.

A word of caution: in the early days of mining in the United States, it was not uncommon for an individual or group to form a company and sell shares in that company to finance work at a mine or prospect. Many people were swindled by charlatans who took advantage of the times, situations, and the greed of other individuals. Consequently, United States securities laws extend even to the sale of part interest in a deal, if certain steps are taken.

For example, let's assume that a prospector in Grand Junction, Colorado, discovered an outcrop of uranium-bearing sandstone. After staking claims and performing some work to expose additional mineralization, he decided to drive an adit underground on the mineralization and to drill holes from the top of the bluff 150 m above. He estimated that the cost, plus a modest management fee to himself, would be $115,000.

After telephoning and visiting acquaintances in Salt Lake City, Albuquerque, Denver, Grand Junction, and some small sales efforts; these monues are advanced by $100,000 from 20 persons, or $5,000 each. Letters were sent to all 20 stating that they each owned one-thirtieth of the mining claims, for a total of two-thirds share of the prospect (with the prospector retaining a one-third interest).

The work proceeded until the funds

were spent; no ore was produced, and none was even in sight. Three of the investors decided they wanted their money returned, but the prospector was unable to repay them. A lawyer was contacted, and a letter was sent to the SEC. The lawyer filed suit on behalf of the investors, and within a few weeks, the prospector was arrested by the county sheriff. He was charged with fraud, selling unregistered securities, and selling securities without a license.

In financing arrangements, it is often undesirable for the organizer (first party) to have an investor (second party) retain an interest for the duration of the project simply because he made an early investment. For example, perhaps a first party (whom we shall call Western Exploration) has discovered an area which appears to have good potential for uranium deposits. Costs for acquiring a commanding land position in the area are estimated at $30,000 for the first year. Holding costs for the same properties for the second year are estimated at $20,000 (this will not become a serious concern until later).

We shall assume further that Western Exploration has insufficient funds for land acquisition costs in this project. Therefore, Western contacts a second party (whom we shall call Resource Investors), and they reach an agreement whereby $30,000 will be spent for land acquisition on a one-third/two-thirds basis, with Resource Investors expending the larger portion. They allocate $3,000 for report and map preparation, and $2,000 for subsequent sales efforts; these monies are advanced by Resource Investors. Any profits or losses from the venture are to be shared equally.

Land acquisition is supervised daily by Western Exploration, which also furnishes Resource Investors with monthly progress reports. Western Exploration prepares maps and a report, which require several weeks' time, for presentation to potential

third-party investors. The presentations are made intermittently over a period of several months (with considerable time and money expended for telephone calls, travel, and presentation of the "show and tell"). The companies find they must each advance $5,000 to the project to cover additional expenses.

Finally, the properties are successfully placed with a mining company under an option arrangement. The deal is negotiated by Western Exploration over several weeks' time. The mining company pays $30,000 for a one-year option, which is divided between the parties.

After ten months have passed, the mining company returns the properties, along with logs of 25 drill holes and a map showing location of same. Western Exploration personnel examine the logs and decide the properties still look good, and perhaps even better than before the drilling was done. Once again, interpretive maps and cross sections must be laboriously prepared, utilizing the best talent Western Exploration has; and in the meantime, $10,000 in rentals and assessment work comes due and must be paid. The amount invested to date is:

	$30,000	Initial land costs
	5,000	Maps, reports, sales effort
	10,000	Sales effort
	10,000	Rentals and assessment work
	$55,000	
less	30,000	(Mining company's investment)
	$25,000	($5,000 net was spent by Western and $20,000 by Resource Investors)

Another sales effort is launched by Western, and after several months' hard work, and the expenditure of another $10,000, a second mining company that will take an option on the properties is found. This time the option payment is $50,000, and after the deal is painfully negotiated, Western begins to wonder if, because of the time and grief involved, *it should have borrowed or otherwise raised the original money for this project.* The option payment is divided for a net return of $15,000 to Western, and nothing to Resource Investors. Resource Investors may wonder if the investment will ever pay off.

After another year, the option is exercised for $200,000, which is divided between Western and Resource Investors. The net income to Western is $115,000, and to Resource Investors it is $100,000. In addition, in the event uranium is produced from the property at a later date, royalties from production will be divided by the parties.

This may seem like a good deal for both parties—but is it, in fact? Resource Investors made one moderate and several smaller expenditures, put in virtually no time or effort, netted $100,000, and holds a royalty interest. Western, on the other hand, invested substantial time and effort over a two-year period, utilizing geological and negotiating talent which might have been spent on other, perhaps better, deals, for a net return of $115,000 plus a royalty interest. If Western had borrowed the initial $40,000 which Resource Investors provided, rather than bringing in a partner, they could have netted $215,000 (minus interest) and held the entire royalty interest. In so doing, however, they would have run a substantial risk of being unable to sell the properties or otherwise repay the loan.

This example illustrates how important it is for explorationists to exercise extreme care when considering the participation of investors who have a passive role in a project. When an investor merely supplies modest funds at an early stage of the program, the general result is an inequitable partnership. If the prospect has good geologic potential (and competent explorationists should not be wasting time on properties which do *not* have good potential), then it is probably worthy of an investment, particularly if the explorationist is a successful promoter and can sell his ideas and projects.

We heard of an interesting account of one scheme for obtaining financing in New Mexico, although it was *from* a mining property rather than *for* it. An unscrupulous individual bought two patented mining claims, each 20 acres (or a total of 871,000 sq ft), for approximately $10,000. After having copies of an inexpensive deed made, he advertised in a national publication: GET YOUR FOOT IN THE MINING BUSINESS, ONLY $1.00. For each dollar he received, he mailed a deed to one square foot of the property. Obviously, his profit potential was enormous. His scheme apparently went well until postal officials learned of it.

To return to more serious matters: the explorationist must accurately estimate the time required for assembling his project to the stage where exploration may commence. Acquisition and exploration of the properties, or a sales effort to interest another party in the properties, must be carried out rapidly before lease rentals or assessment work becomes due. As we have pointed out, the holding costs for mining claims and leases can be high—as much as several hundred thousand dollars per year for a large project. It behooves the explorationist to progress quickly, in order to avoid paying holding costs over an extended period of time.

An ideal course of action is to acquire the properties and explore them as soon as possible. Thus, the apparently worthless properties can be returned to the owner or relinquished, whereas the more favorable ones may be retained and explored further. Funds are often spent unnecessarily for holding costs on properties which could have been "disproven" promptly with more aggressive, early exploration. It is extremely difficult to disprove *totally* the potential for a given property; but at some point, the explorationist must make a judgment when the available evidence weighs against the existence of an ore deposit on the property.

ACQUIRE THE EASY, INEXPENSIVE LAND FIRST

Once the land ownership picture has been assembled, land acquisition costs have been estimated, and the project appears favorable in all respects, land acquisition comes next. We recommend acquiring the easy, inexpensive land immediately. In many western states, the fastest initial step is to acquire or file on state-owned lands within the area. For example, in Wyoming (and some other states), it is a quick and simple procedure to file, through the offices of the state land board, applications for any open state land which has not been previously filed upon by others. Forms must be completed, and payment of bonuses, if any, and first year's rental must be made. This immediately withdraws the land from other filings. The cost is reasonable, usually ranging from $1.25 to $2.50 per hectare ($.50 to $1.00 per acre).

In the case of federal acquired lands, the explorationist must file an application with the Bureau of Land Management for a prospecting permit for these lands, and this should be done at a very early stage in his leasing program. Issuance of prospecting permits may take months, or even years. No exploration may be conducted until they are issued, but when an application is pending, the lands are not available to anyone else either. In the meantime, the adjacent lands can be drilled, provided the explorationist has obtained leases on them. If the drilling proves favorable, the explorationist will bide his time until the federal prospecting permits issue (and he may try to encourage their issuance). If the drilling is unfavorable, the applications for prospecting permits may be withdrawn, with the only penalty being the loss of filing fees, which are nominal. Acreage limitations *do apply* to federal acquired lands within states.

In western states, MIN plats should be ordered from the Bureau of Land Management office responsible for the area. To expedite the initial phase of the program, the explorationist might also arrange for mobilization of the necessary crews, such as landmen and claim stakers, to carry out the field work. From county records and MIN plats, a determination must be made of the ownership of all lands within the area of interest, such as federal acquired lands, Indian lands, withdrawn lands, and so forth, in addition to fee, state, and public-domain lands.

The landman should check the county records immediately for two reasons: first, if there are federal lands in the area, to learn whether any mining claims have been located on these lands and, if so, whether such claims are being maintained as evidenced by recorded assessment work (bear in mind that in most states assessment work does not have to be recorded); and second, to determine where the fee lands are located in the area and the names and addresses of the owners.

This information usually should be gathered concurrently with ongoing geologic studies carried out in the area, as described in chapter 6. Even though the size of the area, and the areas of priority, may not yet have been determined, it is a minor cost for the landman to check records while the explorationist gathers additional geological information. If the land picture appears unfavorable, the explorationist has the option to abort the entire effort so that exploration may be focused on a different, perhaps distant, area.

MOBILIZE CREWS: LANDMEN, CLAIM STAKERS

Mobilization of landmen and claim stakers can usually be accomplished through a few telephone calls and subsequent meetings. In addition to checking ownership records, the landmen are usually responsible for contacting landowners and attempting to obtain mining leases. They might also contact surface owners (those with no mineral ownership) to arrange for access to the properties.

The number of landmen required depends upon the size of the project, difficulty of access, timing, and budget considerations. Sometimes as many as 10 or 12 landmen are utilized on large projects, with one landman supervising the entire operation. On such sizeable projects, substantial sums of money can be spent in a hurry, particularly in "cold" or "open" areas (those in which there is no other exploration being conducted) where thousands of acres may be acquired in a few weeks.

The claim stakers are responsible for locating or staking claims by physically placing monuments and performing other work as prescribed by law. If requested by the explorationist, stakers usually agree to perform discovery work, at additional cost. As with landmen, the number of claim stakers required is directly related to the timing of a project and the size of the area to be staked. Entire townships have been staked by some companies in acquisition efforts, which required several months to complete.

The usual staking procedure entails the use of 10 x 10 x 120–cm (4 in. x 4 in. x 4–

ft) wooden (pine) claim posts. Power augers are often used to dig holes so that the posts may be set into the ground (figs. 7-3, 7-4, 7-5). (Some prospectors have suggested that if western United States mining law required planting a shrub or tree, instead of "substantial monuments" such as four-by-fours, the West would now be far greener.) Some claim stakers use half-inch steel reinforcing rod (rebar) for setting claim posts. In the field, a piece of pipe with a welded cap on one end is placed over a 60-cm (2-ft) length of rebar, and the rebar is driven vertically about 30 cm (12 in.) into the ground. A 10 x 10 x 120–cm post with a 30 x 1.25–cm (12 x .5-in.) hole drilled into one end is then placed over the protruding end of the rebar. Livestock rubbing against the posts have been known to bend the rebars, but they seldom knock them over.

Four-wheel-drive vehicles are frequently required for travel in rough terrain. In some instances, helicopters can facilitate the staking effort.

As part of the procedure for locating mining claims on public domain, federal and state mining laws usually provide a 60-day protection period after placement of a discovery monument, during which time the claim may be perfected and recorded. In other words, placement of the discovery monument will hold the ground for 60 days against other locaters. After placing the discovery monument, the explorationist has the right to continue performing work in the area, such as excavating or drilling, to determine whether there is potential for mineable deposits. If no potential is found, the claim can be abandoned without completing all the steps necessary for recording the claim (such as placing the additional corner and side center posts or other monuments), thus saving considerable time. Unfortunately, however, drilling to evaluate the property usually

Figure 7-3. Claim posts being unloaded from a vehicle in the area where claims are being staked. The posts leaning against the tailgate of the truck have a hole drilled in each of them and will be used as discovery monuments, or "DMs," as they are called. A location notice or card is immediately fastened to each DM, and later a location certificate is prepared, folded, rolled up, and placed in the hole previously drilled. Photo by Bob Carrier.

cannot be completed within the 60-day period due to lack of availability of equipment, complex nature of the geology, shortages of trained personnel, and so forth. Consequently, the explorationist must usually complete work on the claims and record them, thus allowing more time at a later date to evaluate the property in detail before making a decision as to potential value of the claims.

The explorationist who acquires claims for later sale is unlikely to drill deep holes on the claims, or otherwise incur substantial expenditures, while he has only the discovery monuments to hold the ground. Many sellers, in fact, do not expend large sums on claims at any stage—they prefer to avoid such high risks. Prospective sellers often complete the claims and attempt to

sell them later. This does not mean that a seller cannot place the discovery monuments and then attempt to sell the claims immediately; this is a common procedure. Prospective buyers, however, expect the claims to be properly completed and filed of record in the appropriate court house, unless otherwise stated by the seller.

Claim staking has evolved from the relatively simple task of making a discovery, posting a notice on a discovery monument, erecting corner posts or monuments on a few claims, and recording their locations, to today's staking efforts which sometimes involve areas of 100 sq km (36 sq mi) or more. An important requirement of the old mining law was that a discovery of valuable minerals be made on each claim. Yet today, even with our advanced under-

Figure 7-4. Surveyor gives chainman directions for the spotting of the next claim post. Previously placed posts are visible behind surveyor. Photo by Bob Carrier.

standing of complex stratigraphy, geochemistry, and mineralization habits, a good evaluation of properties, whether leases or claims, requires months or even years. It has been suggested that the legal requirement of "discovery of valuable minerals" should now be supplanted by "geologic evidence that deposits of value may exist within the boundaries of the claim, although not exposed at the time of staking." Current exploration geology techniques enable explorationists to forecast favorable conditions for mineral deposits to depths of several thousand feet in many cases. Yet present mining law still requires a discovery before a claim may be staked.

If explorationists operating in the western states adhered strictly to the law, very few claims would be staked because of the "valuable discovery" requirement.

More than 60 days, the period of pro-

tection to a newly located claim, may be required to drill one hole of 900 m (3,000 ft) or, in tough drilling conditions, even 120 m (400 ft) in order to make a deep discovery. In Wyoming, New Mexico, Utah, and other states, a significant percentage of most claims currently being staked probably do not have a valuable discovery according to strict interpretation of the law. In most situations, those holding the claims intend to confirm a discovery at a later date; nonetheless, a shallow discovery may be difficult to prove at the time of staking.

You may well ask how this type of procedure can function effectively in a competitive field such as uranium exploration. It does so only because those involved in exploration recognize the *claimed* rights of others to a tract of land on which a claim has been staked, regardless of whether a valid discovery has been made. Throughout most of Wyoming a code of honor exists among most explorationists. In New Mexico, however, the same code apparently does not apply; there, claims have been jumped by persons suspecting that the claims do not meet the strict requirements of the law.

COMMON PROBLEMS

Fee Mineral Holdouts

In any land acquisition effort, there are certain common problems which occur repeatedly. For instance, "fee mineral holdouts" are common in some areas where land is being leased. The holdout is one who owns fee minerals (either an individual, a corporation, or other entity) and refuses to lease the property. The reasons can be diverse: perhaps the owner objects to language in the lease; the bonus or rental offer may be inadequate; the term of the lease may be unacceptable; or the owner may be willing to grant permission to drill, but not to mine.

For example, let's assume that you have defined an area where you believe the geology to be favorable, and the going rate for the execution of a uranium lease by fee mineral owners in that area has been around $5 per hectare ($2 per acre) bonus and $5 per hectare rental per year. Your landmen begin contacting the mineral owners and discover that most of them are demanding $17.50 per hectare bonus and $5 per hectare rental. Perhaps higher bonuses recently have been paid by companies buying oil leases, and the owners see no reason why uranium should be leased for any less. (As commonly used in the industry, the term "bonus," meaning the fee paid to a mineral owner for executing a lease, includes the first year's rental; thus, a $5 per hectare bonus and $5 per hectare rental means a payment of $5 per hectare for the first year and subsequent years. A

Figure 7-5. Field laborer starts up a power auger in preparation for augering holes in which claim posts will be placed. Photo by Bob Carrier.

$10 per hectare bonus and $5 per hectare rental means that $5 per hectare was paid as a consideration for the execution of the lease *in addition to* the rental.)

If enough fee mineral holdouts exist in an area because terms mutually satisfactory to both the mineral owner and the explorationist cannot be agreed upon, the entire play may fall by the wayside. However, if there are just one or two small holdouts out of 10 or 15 fee mineral owners, it may still be possible to assemble a workable land package and explore the area. From the land position attained, concrete exploration evidence may be gathered before attempts are made to deal with the remaining fee owners. If the exploration data proves negative, no further contact need be made with the holdouts. In the long run, it is usually possible to work out a satisfactory arrangement with even the most stubborn holdouts. Time has a mellowing effect, and many mineral owners will ultimately decide to welcome the income from leasing.

In a "cold" area, a procedure frequently followed by exploration companies and individuals is to instruct the landman to make a pass through the area, leasing all land which can be acquired for a reasonable rate under a standard lease form. However, if any public-domain lands exist in the area, consideration must be given to staking that ground at the same time the fee mineral owners are contacted—literally within hours. Ranchers approached for a fee mineral lease have been known to rush into staking adjoining public domain lands the very next day, before those offering to lease could do so. In some of these cases, the lands were not leased, nor were the claims sold, and so the ranchers' gambling was to no avail.

The most favorable areas in which to conduct a fee leasing program are often nonirrigated grasslands or desert lands, where individual landowners have moderate to large holdings (say, 240 to 2,400 ha, or 600 to 6,000 acres) with simple title chains. Once scattered acreage is leased in an initial leasing pass, the explorationist frequently reviews the land situation to determine where the holdout lands are situated, and what terms the mineral owners are demanding as a leasing concession. If the explorationist-lessee has sufficient coverage, he may elect to drill in the area, or otherwise perform sufficient exploration, to determine whether the geologic concept behind the play is sound. If the concept is disproven, the area will probably be abandoned, the claims dropped, and the leases relinquished. In such event, those landowners who leased property received the first year's bonus and rental payment in any case, whereas the holdouts received nothing. Since most mineral prospects turn out to be worthless, just as most wildcat oil tests turn out to be dry holes, the fee mineral owners who hold out simply forego the opportunity of receiving some bonus and rental payments.

On the other hand, if preliminary exploration produces encouraging results, the lessee may be obliged to review the original offer for the fee leases. Frequently the same offer is made again, or the offer may be increased, but landmen almost always have specific financial limits in dealings with landowners. This is referred to in the industry as "dollar limit" or "dollar tops" (as in "$10 tops"), which means that the landman has the authority to offer a *maximum* of $10 per acre, hectare, or other unit of area.

The first limit to be increased is usually that of the bonus payment; for example, an initial offer of $2 per acre may be raised to $3 per acre. The greater the desire of the explorationist-lessee, and the more stubborn the landowner, the higher the figure may rise. However, there has to be positive geologic evidence before a company will pay high bonuses and rentals —say, in excess of $25 per hectare ($10 per acre) bonus and $12.50 per hectare rental for wildcat or unproven acreage. Higher prices, of course, may be paid for lands with known reserves, or which are otherwise considered to be very good prospects for the occurrence of mineable reserves.

Aside from bonus and rental payments, other aspects of the lease may be unacceptable to some landowners. Royalty provisions, for example, may pose a problem: landowners may demand a higher royalty, or require a different basis for the royalty. They may demand a five-year lease term, rather than the generally accepted ten-year term; and they may request a surface damage payment in addition to their bonus and rentals.

Occasionally, a landowner with substantial acreage may insist that all of his land in a given area be leased; if the explorationist is unwilling to lease the total area, the landowner may refuse to lease any portion of it. In other instances, landowners may own the surface and mineral rights of some lands, and only the surface on others; yet they may insist on payment for the surface and mineral estate alike or refuse to lease at all.

These are just a few examples of problems which a landman may encounter, and he must have the ability to find solutions which fit within the guidelines and limitations set by the explorationist or company by whom he is employed.

A "hot" exploration area is one where a discovery is rumored to have been made, or where other indications point to excellent potential for a discovery. When a rumor is first heard, companies often dispatch crews of landmen to the area to initiate record checking and leasing. Since land records are available in limited quantity, the first group of landmen to enter the area will be able to establish priority in utilizing them, while others must wait their turns. If an area is deemed to have very good potential and an active play is under way, some companies may instruct their landmen to acquire leases without first checking ownership records. This can be accomplished by approaching landowners in the area and directly asking them which lands they own. This information is entered on the lease according to the landowner's oral claim of ownership, and a sight draft (usually 30 days) is written for the bonus. The company relies on the "lesser interest" clause to protect it in such an event, and an effort is made to check the records before the draft is to be paid. Many such drafts are paid without benefit of record checking if the play is sufficiently active.

Hot areas produce other phenomena in leasing arrangements. A landowner may be approached by 10 or 15 lease brokers during one week, each offering different bonuses, rentals, royalties, and so forth. Before long, many landowners become holdouts, playing brokers against one another to see who will make the best offer.

Active oil and gas leasing in an area can also generate mineral lease holdouts, especially if discoveries of oil and gas have been made. Such discoveries create a hot petroleum leasing area, and bidding for oil and gas leases by competing firms tends to inflate prices, sometimes in excess of $250 per hectare bonus and $62.50 to $125 per hectare rental. Land leased for oil and gas is usually available for concurrent leasing for the exploration of uranium or other minerals (unless the "oil, gas, and other minerals" situation arises, as discussed in chapter 5). A landowner who has leased his oil rights for $250 per hectare ($100 an acre) is unlikely to want to deal with a uranium landman offering $5 a hectare. We have encountered landowners who have leased oil rights for $7.50 a hectare and claim uranium leases should therefore be worth at least $12.50 a hectare. In such situations, the landman has to use persuasive skills in negotiating the purchase of important leases at a reasonable price.

Access to Property by Stakers

In traveling to and from federal lands, claim stakers often must travel across property owned by a holdout or other uncooperative party. The explorationist should be aware of the potential problems this can cause. If the owner will not permit the staking crew to cross his or her land, and no other access route is available, it might be necessary to use a helicopter to gain access to the property. This is an expensive alternative, and it may be preferable to reconsider the original fee requested by the holdout.

The surface owner who possesses no mineral rights can present a further problem for the explorationist. This situation has been encountered frequently in many areas in the Powder River Basin of Wyoming, where the United States sold or otherwise conveyed the surface and retained the mineral rights, including rights of ingress and egress by prospectors and miners. Many ranchers now living in the area have virtually assumed control of these tracts, regardless of the mineral ownership and reservations in the deed. In such situations, the explorationist has the legal right to enter the property to explore, and to mine if a discovery is made. Yet when the surface of such lands is privately owned, surface owners frequently attempt to prevent other parties from entering the property, unless a prior agreement has been made between the owner and the party in question. Even though the surface owners are not legally entitled to a production royalty, they have been known to cause such difficulty that the explorationist often grants them a small production royalty, or a small per-acre payment plus a royalty, despite the fact that the minerals are owned by the federal government.

Problems can also arise when a rancher or other entity owns the surface and minerals on certain lands, but owns only the surface on other lands. If he is approached for a lease on the mineral lands, he may agree to lease those lands only if the explorationist agrees to pay a production royalty on *all* lands, including those where only the surface is owned. This scheme should be considered a form of blackmail, but it has been used frequently and successfully by some ranchers to extract a royalty concession from the explorationist for lands where the minerals are not owned by the rancher.

Nonresident ranch owners often employ a foreman to run the ranch. Such foremen are sometimes hostile and more difficult to deal with than the owners themselves, particularly when the explorationist is seeking access to land or supplies of water for the exploration program. Undue hostility on the part of the foreman should be brought to the attention of the ranch owner at once. For example, we encountered one foreman who met every visitor to the ranch house with a shotgun. He was finally fired when the owner learned of his behavior, and he probably would have been dismissed sooner if others had told the ranch owner about his actions.

The above examples are some typical challenges which the explorationist must meet in his efforts to acquire land in a given area. Unless the land is successfully acquired and controlled by the explorationist, no play can be made.

Accuracy Considerations

In claim staking, we remind the explorationist that the accuracy of surveying usually falls off in direct proportion to increased haste in the staking operation. In areas which are not rough and which have few trees, one staking crew can locate 40 or 50 claims per day. At such speed, they would not set *all* the corners—only the discovery monuments and the corners between them. As with any surveying activity, staking is slowed down significantly if the crew members have to work around obstacles such as trees, or have to set claim posts in hard ground, or correct for slope distances.

Regardless of the character of the terrain, the faster the staking crews work, the greater likelihood there is for error. Of course, accuracy is also dependent upon the proficiency of each staking crew. Some crews are able to stake fast and accurately, particularly if they are ambitious, have an able surveyor, and have been working together as a team for at least several weeks.

Inexperienced and perhaps incompetent crews can be expected to make some serious mistakes, regardless of the speed of their work. Such work must frequently be redone.

On the other hand, accuracy can be over-emphasized. In one particular staking job, where we had planned simply to set discovery monuments and follow up immediately with a drill, a registered U.S. mineral surveyor was hired for the task. Explaining that we required speedy work and were not so concerned about a high degree of accuracy, we pointed out that if a discovery were made we could resurvey and correct any possible defects in the discovery monument locations at a later time. Despite our instructions, we found that the surveyor was only setting four discovery monuments a day because extreme accuracy had always been inherent in his work. Despite our urging, he was unable to change his *modus operandi* in order to stake a substantial quantity of claims in one day. Consequently he was transferred to another project where greater accuracy was required in surveying drill-hole locations; a less experienced and certainly less competent crew was hired to complete the rapid task of placing the discovery monuments. (As it transpired, when we later drilled the area, the geologic concept was disproven, and the discovery monuments were removed for use elsewhere.)

CONTROLLING AN AREA

The question of controlling an area and having confidence in showing it to others sometimes places the explorationist in a quandary. From our own experience, we prefer to control those portions of an area which are most favorable from a geologic standpoint. If the area of interest is exceptionally large, then real control of it is probably not feasible. In such a case, the explorationist might actually have under lease or claim only scattered tracts of land, with efforts under way to acquire other favorable portions of the area. Often the explorationist has no immediate intention of leasing or staking even half of the available acreage.

With a large area of interest, it is best to carefully select and screen prospective purchasers or investors before showing the prospect, rather than to attempt to control the entire area on the ground by acquisition. We have found that most major companies will not enter into leasing competition if the seller controls part of the ground through leases or claims and has acquisition efforts under way in the area at the time the play is shown to them. In such situations, other companies are unlikely to attempt to go around you and begin leasing or staking claims in the area independently, particularly if the seller controls some of the more favorable-appearing ground. If they have a genuine interest in the area, such companies usually prefer to deal with the explorationist, attempting to work out an agreement, rather than secretly trying to attain their own position within the area. However, a significant new discovery, or rumors of one, can cause the most benign company officials to adopt predatory characteristics.

The question of adequate control is closely related to the party to whom you wish to show the prospect. It depends, to a large extent, on the nature of the prospective buyer's business. Most utility companies, for example, with little expertise and virtually no exploration experience, are safe to show prospects where land acquisition is incomplete. Such companies are beginning to hire personnel with some exploration knowledge, but most of them at this time do not have complete exploration staffs. If you have assembled a prospect, or have an area where you are acquiring lands, it *may* be safe to arrange an appointment with the appropriate personnel of a company you consider to be trustworthy and explain your concept and your property position, but there is a certain amount of risk until you control all of the prospective lands.

Showing a prospect to a mining or exploration company, as opposed to one not involved in exploration, involves potential risk, especially if that company is a competitor in exploration for the same mineral. In such cases, the explorationist should have his land position essentially completed before showing the prospect. Although it is unlikely that such companies would enter the area in competition with the seller, rumors or casual comments between acquaintances could leak critical information to other parties who might enter the area and begin acquiring land in competition with the seller. Therefore, we stress the importance of having a firm land position before showing a prospect to potential competitors.

It is important to understand that a potential investor will express far more interest in your prospect if you have some (or most) of the land already under control. A psychological factor is involved here: by expending risk money for land acquisition, you have displayed confidence in your own idea.

We have also discovered that buyers often like to have other companies, especially substantial ones, in the area. It gives them a feeling of security to know that the area has attracted large companies.

SOME ADVICE TO NONEXPLORATIONISTS

A nonexplorationist who wishes to enter the mineral exploration business, either in the hope of earning a profit or to obtain a source of raw material, faces some special problems. Our discussion will be restricted to uranium because therein lies our experi-

ence and expertise.

Once a decision is made to become involved in uranium exploration, there are many courses of action which may be followed, including—

1. Hiring geologists and forming an exploration group. This approach should be taken with a view to staying in the business for at least several years. New prospects can be generated and submittals of deals can be reviewed. Drilling and other evaluation can be carried out.
2. Acquiring an existing uranium exploration firm, either by merger or purchase.
3. Hiring consultants to locate and pass judgment on deals.
4. Forming joint ventures with companies possessing the expertise to carry out exploration and evaluation.

Combinations of these and other approaches can be worked out. The one essential factor, however, is the technical competence of the explorationists conducting the programs. In our opinion, exploration is no different from most other professions: there is a serious shortage of competent individuals. In uranium, particularly, the recent surge of exploration activity and entry into the field by many new companies and ventures has seriously depleted all available experienced personnel, both competent and otherwise. Although many recent graduates are being hired for exploration programs, several years must pass before they can gain sufficient experience to be of real value.

From our many years of experience, we have concluded that few people can be considered truly competent in this specialized field of uranium exploration. When we are asked for recommendations, or are seeking a new employee ourselves, it is difficult to compile many names. Hopefully, many new members of uranium exploration teams will prove to be competent; unfortunately, experience suggests that this is unlikely.

We believe there are ample localities where significant new deposits can be discovered. Possible shortages of uranium will not result from the lack of a resource; rather, shortages may result from: (1) insufficient areas to explore (too many withdrawn, populated, or otherwise unavailable areas), (2) exploration activities curtailed by environmental restriction, (3) shortage of adequate risk capital, and (4) shortage of competent exploration personnel.

The nonexplorationist, then, in becoming involved in uranium exploration, should direct primary effort at obtaining competent explorationists for the program; this will be his single greatest problem. Given adequate financing over a reasonable number of years, ore discoveries followed by production will surely result.

8
Selling A Deal

PREPARING FOR SALE

Selling a deal, or play, can be one of the most interesting and rewarding, but also one of the most frustrating, aspects of the mineral exploration business.

Let's assume that you have progressed to the stage of acquiring a property, or properties; you have obtained some analytical data, and perhaps have on hand some maps and some written material, such as geological publications concerning the area. You must now be prepared to assemble the prospect into a form which would be saleable to another party in the industry, either as an outright sale or in the form of a joint venture. No matter whether you are employed by a company and desire to sell an idea or project to someone else within your company, or whether you are an individual operator, planning to approach several business organizations—regardless of whom you wish to interest in your project, there are certain steps to be taken in order to make a good presentation. We should point out that although much of the following information is directed toward small operators or individuals, others, such as the exploration and land staffs of larger companies, may find it of interest. Sellers must frequently seek out buyers, and selling goes on constantly, even within companies.

At the outset, we should recognize that different individuals have different personalities and differing objectives; thus, the fact that a technique or procedure has been successfully used by one person or group does not necessarily mean it will work for someone else. The techniques we describe herein have been used successfully in our sales efforts, and it is important to note that the foundation for each of the plays we have assembled and sold is good basic geology. If we develop a concept, follow through the steps described in earlier chapters to evaluate the concept and land situation, and *then* decide to move ahead and make the play, we are confident we can sell it. However, our expertise is in uranium, and we would certainly seek outside expert advice if we were to consider making a play in copper, silver, or some other mineral in which we are not specialists.

We recommend others who are not specialists in any particular mineral to follow this advice in dealing with mineral properties: Be sure good basic geology forms the foundation for your play. If the geology is questionable, have one or more consultants study the prospect; perhaps with some minor expenditures data can be gathered which will substantiate the concept.

Concerning sales techniques, we are acquainted with one prospector from Wyoming who went to New York in the mid-1960s and obtained more than $1 million from private placements of stock in a small publicly traded company of which he was president. This person had no formal geological training (which he liked to emphasize), but had sold used cars, so he knew how to sell. To prepare for the selling, he simply cut the gray cardboard backing from some tablets of paper, and on each piece of cardboard hand-lettered such headings as "GOLD—WYOMING—10 patented claims" (and their names), "SILVER—IDAHO—5 patented claims" (and names), "URANIUM—WYOMING—200 unpatented claims" (and name of group), and so forth. When he met with various investors and brokers in New York, he would place the 10 or 12 cards around the room (some even on the floor), and would relate a glowing tale of potential along with numerous human-interest or humorous stories about each prospect or project. Everyone in Wyoming was astounded when he returned in a short time having successfully placed the stock. This approach contrasts sharply with the approach we use, but success is what

counts, and this prospector was successful.

Outline of Deal

The complexity of the deal you have in mind will have a direct bearing on the type of information you must prepare for your prospective buyers. Most of us try to make the deal seem simple, even though it may be quite complex. On big exploration deals, which are usually phased, we have found that it does no good to break down the program into details. It is better to describe only broad categories in which money will be spent during the program—categories such as land acquisition and rentals; exploration drilling; and, in phase one, any front-end payment to the seller. Breaking it into further detail is pointless, because these plans always change as the program progresses anyway. But it is important to remember that if you actually reach the negotiation stage, initial material you supply to prospective buyers will probably be an important part of your early discussions.

To begin with, an outline of the deal is important when contacting prospective buyers or joint venturers. This should be limited, if at all possible, to not more than two letter-size pages, wherein the seller should establish his or her credibility and convince the buyer that the terms asked for are realistic and justified. The seller should not adopt a defensive attitude; rather, the rationale for the terms should be explained. He should also be assertive (a popular term) but not aggressive.

Most deals involving mineral exploration programs are structured to fit the size and geology of the area, with the exception of the simple situation where a cash payment is made to the seller and the seller retains a production royalty. In order to show how a relatively complex deal can be outlined effectively, we present below an

example of an outline similar to one successfully used in placing a sizeable sandstone uranium deal into a joint venture.
The area involved also had significant coal
potential, and this was included in the outline. The name of the corporation is hypothetical, as is the name of the basin.

SHAPELY BASIN
URANIUM PROJECT
by Spruce Exploration Corp.

****

Outline of Deal

1. An area of interest will be defined.
   Within this area, Spruce Exploration
   Corporation (Spruce) has leases,
   permits, and claims on approximately 250,000 acres; has filed for 35,000
   additional acres; and is continuing its
   leasing efforts.
2. Investor will have the right to earn
   up to 70 percent interest in the project by expending exploration money
   in six phases, as follows:

Phase	Expenditure	Cumulative	Interest Earned
I	$2,000,000	$ 2,000,000	10%
II	2,000,000	4,000,000	20%
III	2,000,000	6,000,000	30%
IV	2,000,000	8,000,000	40%
V	2,000,000	10,000,000	50%
VI	2,000,000	12,000,000	70%

1) Phase I expenditure includes $2-per-
   acre management and consultation
   fee paid to Spruce for all acreage
   held under lease or prospecting
   permit at the time of signing. Said
   payment shall apply retroactively to
   acreage filed before signing and
   which issues after signing.
2) Phase II expenditure includes additional one-time-only $2-per-acre
   management and consultation fee to
   be paid to Spruce if, prior to the end

of Phase III, more than 1,000,000 lb
economically recoverable U_3O_8 in
ore is discovered on lands controlled
by the joint venturers. Ore is defined
as material containing not less than
0.04 percent U_3O_8 with a grade-
thickness of not less than 0.16 ft
percent.

3) A phase will end when the funds
   budgeted for that phase are spent.
4) After the completion of Phase VI,
   each party must pay its share of exploration and development costs, except as provided in item 8 hereunder.
5) Investor shall have the right to drop
   out of the program at the end of
   any phase, with no further financial
   obligation.
6) Spruce shall be exploration operator
   of the project on a cost plus 15 percent basis (applied to expensed items
   only) for at least the first three
   phases. Thereafter, investor shall
   have an option to take over exploration as operator. It is envisioned that
   a separate entity, owned jointly by
   the joint venturers, shall be established to manage mining and production activities. An operating committee composed of representatives of
   Spruce and investor shall be formed.
   Operator shall furnish monthly operating reports to nonoperator.
7) In the event early U_3O_8 production
   is obtained from joint venture properties before investor has earned
   a full 70 percent interest, said production shall be treated as though
   investor had earned 70 percent in
   those properties provided investor
   continued with the exploration program through all six phases.
8) Investor shall agree to advance up to
   $5 million of Spruce's share of sub-
   stantial production facilities and
   equipment, and shall recover same

from 50 percent of Spruce's share of first production. Spruce shall have the right to advance its share of such expenditures, if it so desires.

9) Investor and Spruce shall each be free to do as it sees fit with its respective share of uranium concentrates produced by the venture.

This project is subject to prior sale or commitment.

Spruce Exploration Corporation
February 20, 1976

This outline is set forth in an assertive manner. It contains sufficient information for the prospective buyer to see what kind of deal the seller has in mind. If it appears to include terms which the buyer finds unacceptable, he can immediately advise the seller. A six-page summary of the geology, uranium potential, and exploration plan was sent with the outline.

We firmly believe that an investor who participates in an attractive uranium exploration project should pay a premium for the right to get into such a program. After all, the explorationist has probably spent many years reviewing hundreds of prospects and ideas and also spent many thousands of dollars before one prospect is found which really fits all the favorable criteria. The premium is especially justified if the prospective area has the potential for large reserves—say, more than 9,000,000 kg (20 million lb) U_3O_8 in ore. There are deals being made wherein the investor puts up 75 percent of the money to get 50 percent of the deal, and pays only a token amount for the property position. In such deals the explorationist puts up 25 percent of the money for 50 percent of the deal and, as operator, usually gets 5 to 10 percent on expensed items as a management fee.

In other deals, the investor may put up 90 percent of the money for a 50 percent

earned interest; however, in such instances, the explorationist often must contribute a substantial land position as part of his own cost.

As inferred in the above outline, it is our philosophy that when the geologic concepts are good, and it is a seller's market, the investor should pay a premium to get into the play. He then should pay *all* of the expenses, through several phases of the exploration program, in order to earn a substantial interest in the project. The explorationist has, in effect, a carried interest through to production (see explanation of the term "carried interest" later in this chapter), and we believe that he deserves such a reward if the project is successful. Explorationists selling deals for other minerals or for uranium in a buyer's market will have to determine what the market will bear and structure their terms accordingly. The type of deal which can be made, however, depends to a large extent on the quality of the play.

If the exploration project is not successful, or only marginally so, the investor has nonetheless had the opportunity, and gained the experience, of participating in an exploration program which might have provided a handsome reward, or provided an assured supply of yellow-cake for conversion to fuel for nuclear power plants. The risk here is similar to drilling an oil test on an anticline which has been mapped at the surface: you don't really know what is down there until it is drilled; and in order for it to be drilled, someone has to be willing to invest some relatively high-risk dollars. If no one will risk the dollars to drill, nothing will be found.

Historically, in the United States at least, oil companies large and small have carried out exploration drilling with dollars derived from revenue ultimately paid by consumers of petroleum products, such as gasoline. In the uranium business, electricity generated by nuclear power plants is utilized by consumers. It is these consumers who not only should, but *must,* bear the cost of exploring for and developing new reserves of uranium for use as fuel in these plants. In our opinion, utility companies can increase profits and insure fuel supplies in a coming era of increasing fuel *shortages* by financing and retaining an interest in exploration and mining. Consequently, utility companies should include in their annual budgets an allocation for exploration for, and development of, future fuel supplies. A substantial part of this allocation should necessarily be high-risk money, which should be invested in good-quality exploration ventures.

The uranium exploration industry, and the rest of the mineral exploration industry as well, has almost always lacked adequate high-risk capital, and the present time is no exception. There are many good exploration projects to be assembled, but it is no easy task to sell a project after it has been put together. "Old-guard" mining companies are generally too conservative to consider participating in many of these projects. Oil companies are somewhat less conservative, but they still want to leave the basic explorationist or discoverer with only a very small interest, since they prefer to acquire and drill their own company-generated deals.

Funds of the type put together for drilling in many oil deals are not satisfactory for most mineral exploration ventures because of the long time lag between exploration, discovery, development drilling, mine development, mill construction, and production. Thus, from our viewpoint, the most logical means of generating uranium exploration funds is through the consumer. These funds should be collected by the utility company serving a particular area and then be expended by the utility company in exploration projects. It makes very

little sense in an era of possible fuel shortages that a utility company will spend $800 million for construction of a nuclear power plant but not one penny to explore for a raw-material source of fuel for that plant. In the past, the conservative boards of directors of most utility companies disregarded the need for exploration. Hopefully, this situation is changing, and the very necessary exploration funds may be forthcoming as utility companies realize the value of expending exploration dollars.

Maps and Cross Sections

Almost all information packages relating to mineral exploration projects to be sold must contain adequate maps showing the location, size, and potential of the prospect to a prospective partner, purchaser, or supervisor. Generally, the maps to be prepared should include—

1. A small-scale map showing the regional location, and perhaps even the location on a national scale, displaying its relationship to prominent geographic features or to other mining or exploration areas. The map scale might be 1:6,000,000.
2. A district map showing, on a larger scale, where the area is located with respect to other significant nearby features. It should include the geology of the area. The scale of this map might be 1:126,720.
3. Topographic maps, which are usually on a scale of 1:24,000. These enable you to show greater detail about the prospect, such as the location of claims and leases. Some geologic features can be shown in relation to the topography of the areas. On these maps, it is important to display and emphasize any mineralization which might be present on or near the prospect.
4. Additional detailed maps, perhaps to

a scale of 1:576; also, cross sections showing the geologic relationships in the area. The maps should be keyed to one another, showing as much detail as is necessary to accomplish the purpose you have in mind.

All of these maps should be of professional quality and appearance; they should be drafted by a professional draftsman, and presented in orderly fashion. Color should be used, wherever possible, to emphasize highlights.

If considerable coloring of the maps is required, we suggest having this done by a professional coloring service. The coloring services prepare an overlay from the map which is to be colored; they cut out various color areas, and then spray the color onto an underlying map. For example, if you have prepared a map to be colored in red, yellow, and green, the coloring service would have one print made for each of the colors. On one sheet, they cut out the red colors and spray these on an underlying map; on a second map, they cut and spray the yellow; and on the third map, the same procedure is followed for the green areas. Coloring by spraying is effective for large areas only; the smaller areas usually have to be colored by hand. Utilizing professional coloring services is a tremendous timesaver, and it produces a neat, accurate, and attractive map.

The same comments apply for cross sections. If you wish to have cross sections to accompany your maps, these should be accurately made by a professional draftsman (and professionally colored in the manner described above for map coloring). Cross sections are frequently shown on the sheets prepared for the maps, although in uranium exploration projects, the cross sections are usually prepared separately.

On some of the projects we have sold in the past, we did not have the cross sections drafted; instead, we took the original probe records and had them reduced to a scale of about 1:576, and had prints made from these reductions. The reduced prints were then put on cross-section paper, and correlations were made showing the roll-front geology which we believed to exist in the prospect area; but no prints were made from these cross sections. We simply carried the originals with us, and, when showing the cross sections, used the maps to show where the cross sections had been made.

This method has worked very well for us, and it has saved considerable time and expense in attempting to get cross sections drafted. In any case, cross sections are usually changed after more detailed drilling is completed, so no real benefit is accomplished by drafting the cross sections in advance.

You should determine how many maps you will need to make your presentation. If you are making a large-scale presentation, contacting numerous companies, then it is a good idea to prepare 12, 15, or even 20 sets of maps. At a later stage, you might want to put these maps into a brochure format, or perhaps into a rolled group, to show to the prospective buyer.

Prospective investors or joint venturers frequently ask if they may retain some of the maps and cross sections. You should be willing to leave them for short periods of time—say, one or two days; but it is quite unnecessary to leave maps or cross sections with prospective investors for a longer period unless they are on the verge of making a deal. *Under no circumstances* do we leave property maps in early contacts with prospective investors, unless all the desired property is under control.

Written Material

The next item to be considered in preparing your information package for selling a deal is the written material describing the prospect to be submitted to various companies. The nature and extent of this material depends entirely upon the type of prospect you are selling and how you plan to present that prospect to a prospective investor or purchaser.

We are firmly convinced that the direct mailing approach—assembling a comprehensive package of information and mailing this to a list of 15 or 20 companies as possibly interested parties—is not a satisfactory means of enticing a prospective investor. Many companies fail to respond to mailing-campaign submittals; the maps, cross sections, and written description of the geology of the prospect will simply go into their files.

Rather, as mentioned earlier, we suggest either sending a brief summary, not more than one or two pages, to various companies; or simply telephoning the person responsible for reviewing submittals in each company and inquiring about their possible interest in the project. If there is a genuine interest on the part of a particular company, we arrange an appointment to visit the company, making a "show and tell" presentation and answering any questions which the prospective purchaser might have. We have found that it is far more efficient to first determine a preliminary interest by a brief letter or telephone call; if an interest is established, then follow up with a personal contact.

There are, however, specific instances when it is desirable to prepare some written material on your prospect. For example, you may have a small deal where perhaps you might want simply to sell the claims outright and retain a small royalty interest. In this instance, it might be desirable to put together a map and a written description of the geology of the prospect, detailing its potential and possibly its past production. Such information could be sent to potential purchasers. As a general

rule, however, we maintain that it is not advisable to assemble descriptive information and send it as an unsolicited submittal to various companies.

DECIDING ON TERMS

Once you have assembled a prospect and begun readying your maps and cross sections for display to prospective purchasers, you must determine what kind of terms you intend asking for the property. There are three important factors to remember in selling a mineral deal, and they can be effectively shown as a triangle (fig. 8-1).

All exploration projects involve these three factors. If you, as a prospect owner, seek a direct sale of properties to a mining company, you will not be responsible for exploration expertise after the deal is made; the purchaser of the properties will assume that responsibility. On the other hand, if you are attempting to sell a deal to a company which has no exploration expertise, then you may have to assure them that this important function will be handled competently, possibly by consulting firms. If you and your associates can carry out the exploration, then a joint venture might be formed with the investor advancing most (or all) of the money and with you providing the properties and exploration expertise.

As in any other aspect of the mineral business, the terms you ask depend on several interrelated considerations. Some of the more important factors are (not necessarily in this order)—

1. Reserves on the property, if any.
2. Size and quality of prospect (geology compared to known deposits).
3. The amount of time and money you have invested in the property, and your own monetary needs at the time.
4. The amount of optimism the seller has with respect to the ultimate po-

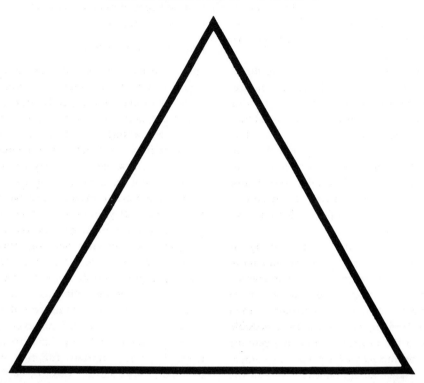

EXPLORATION FUNDS

PROPERTY   **EXPERTISE**

Figure 8-1. Factors in selling a mineral deal.

tential of the property for production, and the expected timing of production.

5. Location in relation to known deposits and milling facilities.

6. Local or regional mineral exploration activity in the area.

7. Mining costs, either from past production or estimated mining costs, in the event of an ore discovery.

8. The value you believe the property to have in relation to other mineral properties that have been sold.

9. What the market will bear: is it a buyer's or seller's market?

All of the above factors combine to give the seller some idea of what price he might be able to obtain for a certain property, or what kind of deal he might make. For example, if a company or individual as seller is willing to accept some kind of cash payment and a simple royalty in consideration for some properties, this is a simple and often relatively fast type of feasible deal (unless, of course, the asking price is too high).

In deciding the terms, the seller must first determine his goals and objectives in the sale. He should formulate a primary plan and put a time limit on pursuing the sale on those terms; he should have a secondary plan for use if time on the first one runs out. Ultimately, of course, the buyer's desires must be taken into consideration. In this respect, the seller should consider the alternatives discussed below.

Cash or stock payment. This is also referred to as bonus or front-end money. It is a payment which is made to the seller, usually at the time the agreement is executed. It may be a one-time payment, or it may be staged payments arranged in conjunction with an option for the property. The payment might be in the form of cash or stock (either registered or unregistered). The cash payment, of course, is directly dependent upon the potential value of the property; it is common for the seller to believe that his property has a higher value than the buyer is prepared to believe. Some median ground must be reached in negotiations, which we discuss later in this chapter. Basically, however, it is not uncommon for a cash or bonus payment to be paid to the seller by the buyer at the time of execution of the agreement, and it is certainly a significant factor to be considered by the seller when he is determining what kind of a deal he can make for his properties.

It must also be recognized that nearly all companies operate on a fixed annual budget. Many deals compete for parts of that budget—including ones generated by the company's own personnel. The money spent for bonuses cannot be spent on exploration on the property in question. Many companies are very sensitive about this and do not like to make bonus payments. Nevertheless, they will if the proposed play is sufficiently attractive.

In the case of stock: if the seller is dealing with a small company, the stock is probably unregistered. Unregistered stock may be sold only under SEC Rule 144, which requires, among other things, a certain holding period for the person who has acquired this type of stock. Therefore it is common for the seller to request a larger amount of stock (commonly referred to as a discount) because of the time delays in disposing of it.

Deciding what size of payment to ask for a property or deal is a complex matter. The seller should certainly take into account the factors described earlier. Bonus payments may range from zero to several million dollars. The initial price the seller should ask is probably one which is on the high side, but not unrealistic; if it is too high, it may discourage some potential buyers.

In respect to the cash payments, as well as other items discussed below, the seller should almost always be willing to negotiate and should not take an absolutely firm position concerning the price he is asking for the property. In a "cold" area, the seller must use balanced optimism in deciding on terms. By balanced optimism, we mean he must be optimistic about the prospect and be able to convey this feeling to others in order to generate enthusiasm for the play, but he should not be so optimistic that he overprices the deal. There is a very delicate balance between holding relatively firm to the prices you are asking and yet being willing to compromise in some areas. It is not uncommon for the seller to take a lower (or no) bonus payment if he is to retain a higher royalty and, conversely, to ask for a higher bonus payment if he is to take a lower royalty than he had requested. If the bonus payment involves a substantial sum of money—for example, more than $100,000— it may be desirable (from the seller's standpoint) to ask for staged option payments rather than a bonus, because of tax consequences. This is explained below. Taxes are frequently an important consideration in pricing a deal.

Retained royalty. This is also called

overriding royalty or ORR. It is a royalty to be paid from production and sale of ore (or concentrates produced from ore) derived from the property. This is in addition to the landowner's, or base, royalty in the case of lands owned in fee or by the state. The retained royalty is an obligation by the buyer of the ore, or the producer of ore from the properties, to pay to the person retaining the royalty a certain amount as set forth in the contract.

The amount of royalty is usually based on a certain percentage of the gross value of the ore at the mine. In uranium deals, the royalty is sometimes based on the selling price of the uranium concentrate produced from ores which are mined from the properties. A concentrate royalty differs from an ore-value royalty in that the concentrate royalty is based on a product that has been produced from a raw material (the ore) and consequently has "added value" because of that processing. A royalty on ore value has to be quite high in order to equal a concentrate royalty.

The seller generally tries to retain the highest royalty he can negotiate with the buyer. In numerous deals over the past few years, sellers have retained a royalty of 5 percent of the concentrate sales price; such a royalty is strongly resisted by some buyers, even to the extent of refusal to negotiate unless it is agreed the royalty will be based on ore value.

An illustration may help clarify the difference between the two approaches. Let us assume that two prospectors have side-by-side uranium claim groups of similar size in New Mexico. They are each contacted by separate mining companies who wish to acquire the claims. Both deals are made, with identical cash payments, but Prospector A retains a 5 percent ore-value royalty, while Prospector B retains a 5 percent concentrate royalty. Ore is discovered and production initiated on both

properties; then the difference in royalties becomes apparent.

In A's case, the mill buying the 0.20 percent average-grade ore from the mine was paying the mining company $22 per contained kilogram ($10 per contained pound), or $44 per metric ton ($40 per short ton), at the mine (0.20 percent ore contains 2 kg per metric ton, 4 lb per short ton). A's royalty therefore was $2.20 per metric ton ($2 per short ton), which is equivalent to $1.10 per kilogram ($.50 per pound). In B's case, he was not concerned about ore grade as long as the operator mined a substantial quantity of the ore. However, he was pleased to learn that the company operating the mill, which was processing ore from the claims he had sold, had made a deal to sell the uranium concentrate for $77 per kilogram ($35 per pound). B's royalty checks reflected five percent of $77, or $3.85 per kilogram ($1.75 per pound), which is $7.70 per metric ton ($7 per short ton) of 0.20 percent ore. B's royalty, then, was 250 percent more than A's.

As the above example points out, royalties may vary considerably. To a large extent, they are related to any other outstanding royalties which may exist on particular properties. If the properties are mining claims located on the public domain (as were the claims of the above prospectors), then there is no other outstanding royalty, since the federal government receives no royalty payment from production on mining claims. In selling claims, the seller can ask for a royalty which might run as high as 5 to 7 or even 10 percent of the concentrate selling price. On the other hand, if the property is a fee lease, or state lease, which already has a base royalty attached, then the seller must ask for a smaller royalty so that the sum of the two royalties does not present an unrealistic burden on the buyer-producer.

In the 1960s, a commonly used uranium fee lease specified a landowner's royalty of 5 percent based on ore values, and a sliding scale was incorporated to account for ore grade. If a seller included such a fee lease or leases in a deal, he frequently added 3 percent ore value as his royalty. The resulting total royalty was then 8 percent of the ore value. With mining claims, it was common for the seller to ask for 5 to 7 percent as an ore-value royalty, assuming there were no other outstanding royalties, which was slightly less than the total royalty for fee or state lands.

In the mid-1970s, with the advent of higher uranium prices, we are seeing the royalties creeping upward. It is not uncommon for a landowner's royalty to be 10 percent, if the ore-value schedule is used. In special situations, landowners are now demanding and receiving 5 percent of the concentrate value as a royalty.

Another type of royalty the seller might request is an advance royalty, to commence a year or two after the agreement is signed. An advance royalty, as described in this text, is a royalty paid to the seller (usually on an annual basis) and later recovered from his share of royalties once the property goes into production. Of course, if the property fails to go into production at a later date, then the buyer (who has paid the advance royalty) must write off that amount of money as a loss, as it is not recoverable.

An advance royalty provides a source of income for the seller during those years when there is no production; thus he receives income from the properties whether they are producing or not. An advance royalty is written into an agreement to assure that some work will be done on the property, and it helps to avoid the situation where the buyer simply acquires the properties and holds them without attempting to commence production. Such

an incentive may not be necessary where substantial lease rentals must be paid annually, usually assuring that the buyer will either develop the properties or return them to the seller. Rentals alone may not provide such incentive, however; therefore, the seller may require an advance royalty or an annual work commitment.

Advance royalty, then, is one of the terms that a seller may request from a prospective buyer. Although it adds to the complexity of the deal, it may be a worthwhile goal.

Work commitment. A work commitment is an arrangement whereby the buyer agrees to perform certain work on a property within a specified period of time. The commitment may require a specific monetary expenditure, but more commonly it specifies work, such as drilling or tunneling, to be carried out on a property. A work commitment is rarely requested when the seller has negotiated an advance royalty; both generally accomplish the same purpose, which is to insure that the buyer is performing work on the properties. A work commitment is often a desirable factor from the seller's standpoint if he wishes to expedite exploration on the properties, which will ultimately bring them into production.

A work commitment can require the buyer to return the properties to the seller if the buyer does not perform the amount of work specified in the agreement. The commitment can be to the seller's advantage, because if properties are returned he may be able to resell them promptly to another party on favorable terms.

Carried interest. In some agreements, particularly those between companies, the party which controls the properties sometimes retains what is called a carried interest. This is a retained interest which costs the owner nothing up to a certain stage of the exploration program as established in the agreement.

For example, if company A retains a 30 percent carried interest in a deal with company B, B agrees to finance the project to a certain point, either defined by the amount of money spent, by passage of time, or by the accomplishment of a certain goal. In mining deals, it is usually defined by the amount of money spent. Once B has spent that amount, say $5 million, A must begin paying 30 percent of costs as defined in the agreement; at that point A's interest becomes a *working interest,* because A is now paying its way. In such agreements, B may recover funds (plus interest) advanced on behalf of A's carried interest out of 50 percent of A's share of first production. B will usually want to recover the money from 100 percent of A's share, but A may require some income during the payout period.

When both parties have working interests, most agreements contain a clause which spells out a decrease in interest if one party wishes to proceed and the other does not or cannot afford its share of the necessary costs. This is called an *interest-reduction clause* and is an important part of working-interest agreements.

Continuing with the example described above, assume that the interest-reduction clause specifies that, after the initial $5 million is spent by B, each party must pay its share or suffer an interest reduction of 1 percent for each $50,000 spent on its behalf by the other party. Thus, if B proposes to spend $2 million in the next phase and A does not provide its share of the money, A will suffer an interest reduction of 12 percent, leaving it with only 18 percent (30 percent of $2 million is $600,000, which at $50,000 per percent equals 12 percent).

In such situations, a well-financed company which wants total control of a property can quickly spend its small partner out

of the project, and this very thing occasionally happened in the oil business. Later a different approach evolved, and now agreements are written to specify that if the interest of a partner is reduced to a certain level (say, 5 percent), that interest would automatically be converted to a royalty and could be reduced no further.

Net-profits interest. Another approach which is not uncommon in deals between companies is a net-profits interest in lieu of a royalty. A company which retains a net-profits interest receives that percentage of the profits after all costs are deducted. It differs from a royalty in that a royalty is usually paid on the value of the ore at the mine, or at some other point, with no deductions, as explained earlier.

As an example of a net-profits interest, let's assume that company A retains 15 percent net-profits interest on a property in a deal with B. The agreement provides that B can recoup all expenses (including reasonable general and administrative costs, and interest) before A begins to share in profits. B spends $1.4 million in four years before the start of production; then production begins, generating gross income of $350,000 per year, with operating costs of $100,000 per year. It is now evident to A that several years must pass before any income will result from this production, and 5 percent royalty now appears more attractive than 15 percent net profits.

From the seller's viewpoint, net-profits interest may be preferable to a royalty if (a) the operating costs of the project are low, (b) production can be obtained rather quickly, and (c) the cash outlay is low or there is a fast payout. It is common for the percentage retained in a net-profits deal to exceed the royalty retained for a similar deal. For example, depending on the type of project, a 5 percent gross-ore-value royalty might equal a 20 percent net-profits interest.

Joint venture. Mineral exploration joint ventures are usually entered into by companies to carry out exploration in a given area or on given properties. Reasons joint ventures are undertaken include the facts that: (1) the financial risk is spread out; (2) one or both (or more) of the participants may have expertise, property, or financing needed for the project; and (3) politics may be involved, particularly in foreign countries. Uranium exploration joint ventures are common in the United States and usually are between companies which have expertise and perhaps property and companies which have financial strength (and may need the uranium). There are many different ways in which joint ventures can be structured, but usually the financially strong party advances most or all of the money, and the other party puts in properties and manages the program.

In order to enter into a joint venture and manage or operate, the seller must possess expertise in conducting such a program effectively and efficiently and must have a staff capable of handling the scope of the project (Bloomenthal, 1976).

Discussion

Frequently, individuals and small companies engaged in putting together uranium prospects strive to work a mixture of deals concurrently. Their objective is to generate enough cash to cover ongoing overhead costs; at the same time, however, they should have some long-term goals and criteria.

A deal with so-called front-end money plus a royalty provides an immediate cash flow, whereas a joint venture *may* provide more significant long-term income further down the road. It is generally acknowledged that, for every dollar of front-end money asked for, the seller may forfeit ten dollars in possible production income. For example, if in selling a deal you ask for a payment of $5 per acre plus 25 percent carried working interest, you might make a deal on the same property for no front-end money and 50 percent carried working interest. Established mining companies usually prefer a straightforward lease or purchase arrangement, with a small royalty retained by the seller, while utility companies generally are interested in a joint venture arrangement, since they rarely wish to operate an exploration program. We again point out that tax considerations play a large part in the explorationist's decision as to what type of deal he should try to make.

There is probably no limit to the diverse combinations of property deals which can be made for mineral prospects. Some examples are discussed later in this chapter in order to illustrate how a particular deal is made and to show the variety of deals which have been arranged between parties.

Table 8-1 has been prepared to show what types of deals are usually feasible between parties. In this table, the buyer is assumed to be a mining company with substantial venture capital. Three types of sellers are used as examples: individual prospector, small company, and large company. Five types of mineral prospects are used as examples, and eight types of payments or commitments which are common in the industry. The overall assumption is that the prospect area where claims and leases are held is in a reasonably favorable geologic situation with at least a modest chance of discovery of one or more uranium-ore bodies, but not in a situation where there appears to be an excellent chance of ore discovery. The possiblity of a deal being made between the buyer with substantial venture capital and one of the three types of sellers is rated from 0 (improbable) to 6 (very commonly done).

Those who embark on a course of selling a mineral deal should be aware of the

TABLE 8-1. SOME TYPES AND PROBABILITIES OF URANIUM DEALS

	Bonus payment		Advance royalty*		Royalty on ore produced		Royalty on concentrate		Work commitment		Carried interest		Area of interest		Joint venture†	
	Seller is	Possibility	Seller is	Possibility	Seller is	Possibility	Seller is	Possibility	Seller is	Possibility	Seller is	Possibility	Seller is	Possibility	Seller is	Possibility
Seller has small block of claims or leases	A	6	A	5	A	5	A	3	A	5	A	2	A	1	A	1
	B	6	B	5	B	5	B	3	B	5	B	4	B	1	B	4
	C	6	C	4	C	5	C	3	C	5	C	6	C	1	C	6
Seller has scattered claims and/or leases in an area where prospective buyer has no properties; some other properties available within area	A	6	A	5	A	6	A	3	A	5	A	2	A	5	A	1
	B	6	B	5	B	6	B	3	B	5	B	4	B	5	B	4
	C	4	C	4	C	6	C	3	C	5	C	6	C	2	C	6
Seller has scattered claims or leases in an area where prospective buyer already has properties	A	6	A	5	A	6	A	3	A	4	A	2	A	1	A	1
	B	6	B	5	B	6	B	3	B	4	B	4	B	1	B	4
	C	6	C	4	C	6	C	3	C	4	C	6	C	1	C	6
Seller has large block of claims or leases	A	6	A	5	A	6	A	3	A	6	A	2	A	6	A	1
	B	6	B	5	B	6	B	3	B	6	B	4	B	6	B	4
	C	6	C	4	C	6	C	3	C	6	C	6	C	6	C	6
Seller has a geologic concept for a play in a new area where properties are availabe, but owns no property there	A	1	A	3	A	5	A	2	A	0	A	2	A	3	A	1
	B	1	B	2	B	5	B	2	B	0	B	2	B	2	B	2
	C	0	C	0	C	0	C	0	C	0	C	0	C	0	C	1

Note: Buyer is assumed to be a mining company with substantial venture capital. It is assumed that the prospect area where claims or leases are held is in a reasonably favorable geologic situation with at least a modest chance of discovery of one or more uranium-ore bodies, but not in a situation where there appears to be an excellent chance of ore discovery.

Key: A: Individual prospector or group of prospectors; short risk on capital B: Small company; can afford to pay modest way later
C: Large company; can afford to pay way later

0: Improbable that this would happen	6: Very commonly done
1: Usually not feasible	
2: Extremely difficult	
3: Difficult	
4: May be done, but uncommon	
5: Good possibility	

†Assumes no royalty on production to participants

*After two or three years have passed

realities of the exploration business, and should realize that a series of false starts and dead ends usually awaits them. A few turn-downs or poor receptions should be expected. The expected adversities include: (1) persons incompetent to review submittals, (2) managements with poorly conceived and arbitrary guidelines for what constitutes a good prospect, (3) pessimists who regard every prospect as a poor one, (4) those who want to see the play and hear your presentation simply for their own benefit, with no intention of buying, and (5) little or no good communication between local company offices and the main office. If you were to call thirty companies interested in uranium exploration and advise them that a good prospect was available in a producing district with sizeable ore potential at shallow depth, chances are that only about five of those companies would be seriously interested. Only two or three would proceed to the point of considering a counteroffer, and negotiations might be held with one or two. Thus it is to be expected that most submittals go nowhere; but after all, it only takes one buyer to close a deal. If your project is sound geologically, persistence will result in success.

An important principle to keep in mind in deciding on terms, and a consideration which should be passed on to prospective buyers (if they are not already aware of it), is that the greater the risk is, the greater the potential return will be; conversely, the better the guarantee, the less the potential return. For example, assume Company P has a property with 1 million kg (2.2 million lb) proven uranium reserves but essentially no potential for additional reserves in an existing mining district. Utility W makes a deal to buy those reserves in the ground for $8 million. Utility W has a good guarantee the reserves are there, but the potential return is low.

Company Q, on the other hand, has some good exploration properties but no reserves in any category. The *potential* on Q's property is high, however, and if ore bodies are present, it could exceed 20 million kg (44 million lb). Utility X opts for a phased program to invest up to $8 million with Q. Thus X is willing to take a greater risk but the potential return is great.

WHICH COMPANIES TO APPROACH WITH YOUR DEAL

Let's assume that you have progressed with your uranium prospect to the stage where you have a property, or properties, and you have gathered some data, including maps, assays, and perhaps some written material concerning the geology. You are now ready to begin contacting prospective purchasers or joint venturers with your deal. But who and where are they? If you have no specific leads on companies which might be interested in your prospect, there are several avenues you can follow to reach companies known to participate in uranium exploration and/or development.

The regional office of the Energy Research and Development Administration (ERDA) in Grand Junction, Colorado, maintains lists of companies which are producers of uranium, through either mining or milling. ERDA also maintains a list of companies and individuals currently engaged in exploring for uranium or who have expressed an interest in uranium. Attendance lists are prepared by ERDA for each uranium seminar held in Grand Junction. ERDA also maintains and publishes lists of utility companies which have built, or are planning to construct, nuclear power plants. This is a prime source for seeking prospective purchasers; in addition to reviewing the lists, we recommend talking to ERDA personnel in the Grand Junction office—they frequently can provide some insight into who might be interested in the acquisition of uranium properties, or where a prospective seller might make submittals.

In addition to ERDA's published lists, Moody's Investors Service, Inc., New York, annually publishes *Moody's Public Utility Manual,* which contains descriptive information about all publicly owned utility companies in the United States. This manual is available at most libraries; however, we suggest that if you intend to approach utility companies with deals of a fairly substantial size, purchasing Moody's manual would be a worthwhile investment. It is available from Moody's Investor Service, Inc., 99 Church Street, New York, NY 10007. We have not found it necessary to purchase the manual more than once every five or six years, since much of the information does not change significantly over that period.

Combining data from ERDA reports and Moody's manual gives a good overall picture of utility companies in the United States, including their use of nuclear power plants, current or projected, for generating electrical power.

The Atomic Industrial Forum (AIF), headquartered in Washington, D.C., frequently holds meetings concerning raw-material sources for nuclear fuel. Attending such meetings gives an excellent opportunity to become acquainted with persons potentially interested in exploration for uranium. The AIF also publishes lists of individuals attending these meetings, which are available upon request.

Among other sources are trade magazines, which sometimes publish advertisements wherein companies request uranium prospect submittals. In addition, companies advertising for geologists or engineers to work on uranium projects are usually interested in receiving submittals. Telephone companies' Yellow Pages in exploration centers, such as Denver, are also recommended for seeking companies interested in receiving submittals.

In the past, companies frequently advertised their interest in uranium prospect submittals. Today, however, such advertising is less commonly found—perhaps due to the sophistication the industry has now acquired. Although they may have an active exploration program, many companies prefer to maintain a low profile and consequently do not advertise. Such companies may prove to be relatively difficult to locate; however, ERDA is generally aware of their existence and their activities and can be contacted for the relevant information. Inquiries via telephone calls to consulting geologists who work in uranium also may be fruitful because they frequently have clients interested in acquisitions.

How do we determine which type of company might be the most likely prospect for a particular uranium submittal? This knowledge and insight comes with extensive experience in the industry; however, we will describe a few procedures which have proved successful. First, concerning utility companies in general: although they will indeed require large amounts of uranium as fuel for nuclear power plants, they have not in the past been willing to participate in exploration programs. With few exceptions, the utilities generally have not yet been inclined to invest risk capital in exploration projects.

Take the case of a deal with no proven reserves proposed to a utility company for $5 million. Would this utility be likely to accelerate its activity in negotiating to buy the project if the price were cut in half when suddenly there were 450,000 kg (1 million lb) of U_3O_8 in ore found on the property? No, they would not be likely to accelerate their effort; they have always moved slowly and will probably continue to do so. Consequently, as an industry, they have not been high on the list of companies to whom properties might be submitted.

A prime exception among utilities is the Tennessee Valley Authority (TVA), which has formed several joint ventures with existing mining companies; until very recently, however, TVA was virtually alone. TVA's exploration activities have expanded to the point where the company now has an exploration group. TVA also purchased the Susquehanna uranium mill at Edgemont, South Dakota.

Since 1975, when Westinghouse Electric Corporation notified utility companies that it might not be able to fulfill existing delivery contracts for U_3O_8, more utilities have shown an interest in exploration programs. Thus, many should now be considered as potential purchasers for a deal, provided that the seller is in a position to operate the program.

As a general rule, large mining or energy-related companies will not be interested in small deals—that is, properties which have a rule-of-thumb potential for less than 3 million lb of U_3O_8 in ore (unless it is near their existing mine-mill operations). Large companies are usually interested in a property which might be capable of supporting its own uranium mine-mill complex, and most of their efforts are devoted to properties with such potential. If they already own properties in a particular area, these companies may well look at other properties in the same area which could contribute to a mill they might construct.

By the same token, large companies are more likely to be interested in a very large deal than are the small companies. Some small companies have a total annual exploration budget of $2 million or less, and the allocation for uranium exploration may be only part of that figure. Some uranium deals require an annual expenditure of $2 million on the one project alone; therefore, we do not recommend talking to small

companies about such a deal, unless it is for a part interest.

It is generally easier to sell a deal to a company whose personnel or consultants really understand the geology and potential of the play. This is particularly true if the seller has proposed some tough terms. The consultant or company geologist who goes on a field trip to examine a prospect and doesn't understand what he is shown will not build up enough enthusiasm to sell company management on the potential of your play.

It is also true that it is difficult to interest certain companies, which have expertise in exploration for a certain mineral, in a joint venture for that mineral where they put up most or all of the money. They prefer their own deals.

INITIAL CONTACT

We suggest that your initial contact with a company which you believe to be a prospective buyer be in the form of a telephone call to the exploration manager, chief geologist, or whoever is responsible for reviewing submittals. In submitting uranium deals to utility companies, the person to contact is usually the nuclear fuel purchasing agent (sometimes referred to as "buyer of nuclear fuels," or other similar titles).

When telephoning these companies, we suggest asking for the person *by title* if the name of the individual is not known. When you are connected with the person responsible for reviewing submittals, you should immediately identify yourself, the company or organization you represent (if any), your location, and the nature of your call. Then ask, and make a note of, the name and title of the person to whom you are speaking.

In the interest of saving time for both parties, inquire immediately whether that company is interested in mineral exploration projects of the type you have assembled. If there appears to be an interest, you can then briefly give a few pertinent facts about your deal, providing the party at the other end with a concise summary of your project. If there still appears to be sufficient *genuine interest,* some written material may be sent.

We emphasize genuine interest because there are a few individuals responsible for reviewing submittals who are lacking in scruples. They will request written information on every deal that comes along, simply for their own edification and to accrue more data for their files, although they may have no serious intention of considering your offer. Other companies adopt the attitude that if a prospect is submitted they should look at it as a courtesy, even if they are fully aware that the idea does not fit their requirements. Such "courtesies" are a waste of your time and theirs. With experience, you should be able to screen out most of these nonserious buyers (referred to as "tire-kickers" by some explorationists, apparently because persons looking at used cars in a lot with no intention of buying often walk around idly kicking the tires). Some of the clues to a nonserious buyer can be gleaned from the first telephone contact. Remarks such as "We're short on exploration money right now, but if it looks good, perhaps we can work it in next year"; "We're not taking any new prospects, but bring it in and if it has good potential we'll ask for more money"; "The potential sounds too limited, but maybe we should look at it anyway"; "Our management doesn't like Cretaceous plays, but if we like it we'll recommend it anyway"; and the classic, "I doubt that we'd be interested, but if you wish, bring it in and we'll look at it," are all clues to a nonserious buyer. We recommend politely declining at this time and suggesting to them that you may be

back in touch at a later date. After all, later on they *may* become interested.

Companies which receive submittals should promptly advise the seller of interest or lack thereof as a matter of courtesy. The notification of lack of interest is a positive response in that it allows the seller to concentrate on other potential buyers, making his selling effort more efficient.

The question frequently arises whether it is better to write or to telephone when submitting a mineral deal. Our experience, and that of many of our colleagues, is that it is preferable to telephone the company prior to sending any written material. To illustrate this point: In 1973, we had assembled a uranium project consisting of slightly over 100,000 acres of fee and state lands. A fact sheet was prepared and mailed to a large list of companies, including utilities. We received responses from only about 5 percent of those to whom we wrote. We do not believe this weak response was due to an inadequate submittal letter and supporting data. Rather, it was a matter of our letter becoming eclipsed by issues of more immediate importance. Within less than one year from the time this deal was assembled, we formed a joint venture on the project under favorable terms, resulting from a telephone call followed by a show-and-tell presentation.

We do not intend to imply that submittal letters should never be written, but that they are far less effective when not preceded by a telephone call. Regardless of the potentiality of the company we are contacting, a station-to-station telephone call is faster and actually less costly than a letter, figuring general and administrative costs. Another important factor: unless you telephone, you may not know to whom you should address your correspondence. Simply addressing the information package to "Buyer of Nuclear Fuels," or "Chief Geologist," will not command the same attention as a specifically addressed envelope.

FURTHER COMMUNICATION

After contacting a potential buyer who has expressed interest in your project, the next usual step is to send written information describing your proposal in more detail. This should be sent within a few days of your telephone conversation. The correspondence should include certain important information, such as:

- Reference to the telephone conversation (date, etc.).
- Brief description of you and your company (if any).
- Location of the properties, including proximity to other mineral mining and milling facilities, if applicable.
- Size and ownership of the properties.
- Brief description of the geology (remember: many of the individuals who may read your submittal may not be geologists; avoid a lot of geological detail).
- A summary of work (if any) which you have carried out on the properties.
- Reasons why you believe the properties to have potential, and what that potential might be in terms of tons of ore and grade of the ore.
- Probable means of mining and milling the ore and reasonable timetable.
- Brief financial terms of the deal.
- A small map may be included, although this is optional and not considered necessary for large projects.

In 1974, we put together a uranium deal which included 150,000 acres of leases. We proposed that the investor provide $8 million in a six-phase program. After several telephone conversations, we sent written information consisting of the following:

- Cover letter (one page).
- Outline of deal (two pages).
- Review of geology (four pages).
- Mining and milling possibilities, including timing thereof (two pages).

This was followed, a few days later, with a copy of our company's annual report and photocopies of recent newspaper clippings, to acquaint the prospective buyer with our company. Due to complications caused by some properties in the area which were not under our control, negotiations over this deal lasted almost one year, but we were finally able to conclude a good agreement; by "good," we mean good for both sides.

It should be remembered that the purpose of the initial correspondence is to determine whether a particular company has an interest in your deal. No one will buy the properties simply because of your telephone call or your letter; but you may be able to ascertain whether there is any interest.

It is important that the terms you set forth in your letter, although of a "ballpark" nature, are reasonably accurate. It would be poor business practice to state one set of terms in your letter and then request different terms at a later date, unless you had a valid reason for so doing (this is discussed more fully under "Negotiations").

If the company is still interested after receiving your information package, a representative may then contact you to receive additional information about the property. (If you hear no further word, you may assume that the company is not interested.) The person contacting you may be a consulting geologist, or he may be a company employee or an officer of the company.

In some companies, especially utilities, it might take three, four, or even five months for your letter to find its way through the hierarchy system and reach someone who can give you a response. Even then, the response may not be definitive; the company may simply indicate that your proposal is under review. If you have received no response from a "high potential" company within, say, 45 days, we suggest writing a second letter (enclosing a photocopy of your original letter), asking why you have not heard from them. Occasionally, the person responsible for answering might be on vacation, or your original letter may have been pigeonholed for some other reason. In such event, a second letter may generate some action.

Once you have received a response from one company expressing some interest in the property, you should be sure to follow up promptly, either with telephone calls or additional correspondence. It is essential that you continue to maintain contact with the interested party.

A major problem often encountered in submitting prospects to companies relates to the ratio of price to size. For example, the seller may have a small property which he values highly, and which therefore has a rather high asking price attached (which the larger companies can afford). Yet large companies may claim: "Although this prospect might fit our budget, the property is too small to fit our criteria—therefore we don't think it has the potential to interest us." On the other hand, the smaller companies might respond: "We like the property, including the size, but we cannot afford that amount of money." It follows that smaller companies are also ruled out as potential investors in a large property with a large asking price, since they rarely have the necessary financing.

KEEPING RECORDS OF CONTACTS

We have found, through experience, that it is vital to keep accurate records of companies to which we submit a deal. One effi-

cient method is to maintain a small, loose-leaf notebook (about 10 x 17 cm [4 x 6.5 in.] —similar to an engineer's field notebook), with alphabetical tabs. Enter names, addresses, and telephone numbers of companies to which you have submitted a deal, together with brief details of any subsequent contacts you may have had with them; include personnel, dates of contacts, and the nature of the companies' responses. You may also make notes concerning the information you have submitted to each particular company. It is important that the notebook be loose-leaf because this allows flexibility in adding needed pages, and companies added to the group can be inserted in proper alphabetical order. Once the deal is sold, the notebook pages can be removed as a group and will be of significant value as a reference in selling another deal.

The small notebook is especially handy to take along when traveling. There are times when you need such information at your fingertips (maybe an unexpected contact with a prospective buyer); the notebook enables you to check quickly on your last contact and on where the discussions stand at the present time.

THE PRESENTATION

Let's assume, then, that you have gone through all the necessary preparations of gathering your data and contacting several companies. One company has expressed an interest in seeing your presentation, and an appointment has been arranged. If possible, at the time the appointment is set up, you should find out who is to represent the company at the meeting, and get some details of this individual's background. Such information is most helpful in determining what kind of presentation you should make.

If you are to make a presentation to someone with a geological background, then you may wish to explain in detail the geological aspects of the prospect. On the other hand, if your presentation is to be made to, say, a buyer of nuclear fuels for a utility company—someone with an electrical engineering background, for example —he will, naturally, be interested in data concerning economics and timing of the project, but it is unlikely that he will be interested in more than just a summary of the geological considerations.

If a company is interested in looking into your prospect further but lacks the necessary geological expertise, it may well hire a consulting geologist for this task. You may then explain the geology of the prospect to the consultant.

As you enter the meeting, by way of personal introduction, you should mention the name of the company you represent (if any), your title, and, most importantly, the ownership of the properties in question. Incredible as it may seem, persons have been known to make presentations attempting to sell properties which are owned by another party—sometimes without the owner's knowledge.

After the initial introductions, you should request some information about the individuals to whom you are addressing your presentation (if you have not found out earlier). It is important to remember that your audience has probably allocated a specific amount of time for your presentation; therefore, little time should be wasted in idle chat. In fact, we recommend asking, at the beginning of the meeting, how much time is available for your presentation and the discussion which may follow. You must then work within this time frame, unless your listeners choose to extend it.

Regardless of their professional backgrounds, the listeners will be making several observations about your presentation, including the following:

- Your verbal presentation: how you come across; how well you *appear* to know your subject.
- The conciseness of your presentation (which ties in with your knowledge of the material).
- The visual material which you display as part of your presentation, and how you handle this material.
- Your responses to their questions.

The prospector with no formal geological education need have no serious concern about his inability to discuss geology in detail. If the listeners are aware of the seller's background, they will usually avoid questions which the seller may not be able to answer. Instead, they may concentrate on such questions as: Where does the best mineralization occur? What is the history of mining or exploration in the area? What recommendations are you making for additional work on the properties?

Selling a deal or idea is not the sole province of persons outside a company or organization. Indeed, selling, or convincing others that your proposals are valid and worth investment, is carried on more or less continually within companies as well. For example, district geologists representing three separate exploration areas must each try to convince the chief geologist or exploration manager that the projects they propose represent not just good investments, but the *best* investments available to the company. The exploration manager must then sell a vice-president on the idea, and so on up the organizational ladder.

Every exploration budget has limitations and there are almost always more places to spend money than there is money available. Consequently, the geologist who can convince or sell the exploration manager on his ideas will have funds made available for the projects, and those who fail usually receive minimal funds.

In the case of a diversified company

Figure 8-2. The presentation made by the explorationist is extremely important in the process of selling. Good maps and other illustrations, good knowledge of the geology of the area, and a good knowledge of the geology of known deposits of the mineral being sought are essential.

which may deal in several commodities such as petroleum, mining, plastics, manufacturing, and so forth, the head of each division, having been sold himself on certain projects by persons under his direction, must present and sell to a budget committee projects representing the entire division effort. This must be done successfully in competition with other division heads who all have their own projects and monetary requirements. Accordingly, our suggestions concerning presentations apply to intracompany selling as well as selling to a company by outsiders.

Some exploration companies, before reviewing your data, may ask you to sign a form stating that you are aware of and acknowledge that this company may have some properties, or may be conducting exploration, in the area of your prospect. Since this is a relatively common practice, we have shown in chapter 9 under "Finder's Fees" a typical form (fig. 9-1) which is often used by companies as a statement sheet, to be signed by the seller before the company will review a submittal. This is a form of legal protection for the exploration company; the need for such protection evolved through instances in which an explorationist approached the company with a mineral prospect that the company declined to purchase at the time, in an area where the company acquired an interest at a later date, for reasons unrelated to the original submittal. Since the explorationist with the original submittal could claim that the company became involved in the area because of knowledge gained from his presentation, this could result in litigation. Therefore, such protection is justifiable. We know of no cases where adverse circumstances have resulted through use of this form; but, certainly, some problems have arisen when a form of this type has not been used. (The finder's fee problem is discussed more fully in chapter 9.)

When making a presentation (sometimes referred to as show and tell) to a small group representing a company, at least some of whom are knowledgeable geologists, maps and cross sections as described earlier in this chapter, as well as written material, should be the components of your presentation (fig. 8-2). Your accompanying descriptions should be presented in a well-informed and systematic fashion. Your audience may raise questions now and then as you proceed. Indeed, questions should be invited, so that you are sure everything is clear as you progress from one map to another. Questions also give you an insight into your audience's reaction to your presentation.

You should discuss your reasons for believing the properties to have potential for the discovery of mineable mineral deposits. The more information you can offer on this subject, the better. Some tabulations might be prepared showing analyses or identification of minerals, location of samples, or other factual information of this nature. Water analyses may be tabulated or shown on a map to indicate anomalous characteristics of water you have found in the area.

Some of the questions commonly asked at this stage are:

- On what portion of the property do you expect mineable deposits to be found, if any exist?
- At what depth are they likely to be found?
- What is the anticipated grade of the ore?
- What minerals might be expected to be found?
- How many tons of ore do you expect to be found on the properties?

These questions are some of the most crucial to be addressed by the seller in a show and tell. Another particularly disarming question we have been asked was: "You appear to be well acquainted with our company; tell us why acquiring this property would be of benefit to us." If you are talking to a company which insists on having at least 4,550,000 kg (10 million lb) U_3O_8 on the property, then you have little chance of making a deal. Obviously, a basic objection to this type should be brought to light on the telephone prior to the interview being set up. Make the most of *any* meeting, however, because if they don't buy this one, they may buy the next one.

It is a good idea to carry along samples of rocks and mineral specimens from the properties to present during your show and tell. Scientists of many disciplines are interested in looking at mineral specimens, particularly geologists and mining engineers who may be in the group. Not many are needed, just a few representative specimens.

We are frequently asked to leave maps and other pertinent data with a prospective buyer. If ever you are obliged to leave maps with a company, it should be clearly understood that you expect the maps to be returned within a certain minimal period. It is actually a fairly common practice in the industry for brochures to be prepared and either mailed or carried to the meeting and left with prospective buyers. Yet we continue to question the value of this approach, unless all available properties in the area are under our own control or the control of another party. Industry ethics dictate that a company which has reviewed a submittal has an obligation to stay out of the area for a reasonable length of time (at least three years) unless it already has property there. If the prospective buyer does have properties in the area, the seller should be told of this immediately before he discloses his own full property situation and his ideas concerning mineralization. Companies with a reputation for acquiring land in a seller's area after reviewing a submittal lose future submittals because their reputation becomes known in the industry.

Although we require several other sets of work maps, it is our policy to prepare only two sets of final maps on a prospect for use in a presentation. These include the geologic maps, the property maps, and the cross sections. As stated earlier, we sometimes use only our original cross sections, in which case there are no others available. We have found it quite satisfactory to initially show the basic geologic maps without leaving any for the company to retain. If there is a desire on the part of the company to find out more about your prospect, that company may send its geologists or engineers to your office to acquire additional information, or you may return to their offices.

A frequently encountered problem area when making submittals is with company employees who are interested in discovering all they can about the geology of an area—what you and others are doing there, and so forth—yet who have absolutely no intention of making a deal for your prospect. This can become a time-consuming and time-wasting process for the seller as he finds himself being used as a mineral training instructor and his properties as mineral training grounds for the personnel of other companies. Consequently, a real effort must be made by the seller to determine which companies are genuinely interested before ever making an oral presentation, and certainly prior to going to the field.

In the initial telephone contact we reveal the size of the area (but not the location), what we believe its potential to be, and the kind of deal we are asking. It is often possible to determine immediately whether your prospect is interested in them. As soon as your general terms for the deal are stated, most companies will advise you if they are not interested—maybe they cannot handle it financially, or perhaps it does not fit their exploration requirements for some other reason.

Some potential buyers may indicate that *if* the prospect appears to be a good one, they would discuss further the terms you are asking, or they might be willing to make a counteroffer. In any event, the potential buyer usually requests a meeting, and a presentation should be made as requested.

We recall one occasion in Casper, Wyoming, several years ago, when we were helping sell a joint venture on a uranium prospect in the Powder River Basin. The proposal involved a property on which con-

siderable drilling had been carried out. It was a very good prospect, with thick altered sands, ore-grade mineralization in drill holes, and some ore-grade exposures near the surface. The property was under the control of a small company at the time, and a joint venture partner was needed to provide financing in order for the project to proceed.

After several telephone calls, a show and tell meeting was arranged with a substantial mining company at its offices in Casper. We assumed that one or two geologists might be present, together with the land manager, to review the submittal. When we arrived, we discovered 30 to 35 company employees assembled in the meeting room to listen to our presentation. Personnel had been flown in from Denver, Albuquerque, and other locations, and our presentation was obviously intended as an educational seminar for the company personnel.

We proceeded to make our presentation to the group in that instance; however, we have subsequently established a policy of refusing to make a presentation should we find ourselves in a similar situation. We later determined that this company had no interest whatsoever in our prospect; they merely wanted to benefit from the educational procedure. From such experiences, we now recognize certain indications of a lack of interest. For example, hesitancy shown on the part of the geologist (or project manager), or an accommodating attitude—perhaps they appear willing to look at your prospect simply as a matter of convenience—cause us to lose interest promptly in such companies, since we are wasting everyone's time. We much prefer a straightforward "No" to a devious response.

However, if the seller is able to convince the prospective buyer that his properties do have good potential, and the buyer is sincerely interested, then arranging a field trip is the next probable step.

FIELD EXAMINATION
BY THE PROSPECTIVE BUYER

Following a favorable show-and-tell presentation, the company generally indicates an interest in looking at the prospect "on the ground," so to speak. A mutually agreeable date is arranged for the field trip. It is the seller's responsibility to estimate the length of time required to adequately review the prospect, bearing in mind that the field trip must be designed to fit within the prospective buyer's parameters and criteria. Sometimes the company representative simply wants a brief glance at the area; other times, companies require a finely detailed review of the prospect. The nature of the company's interest should be kept in mind when arranging a field trip.

The seller should be sufficiently well acquainted with the prospect that he is able to point out the basic geologic features of the property, including particular outcrops or exposures which are especially representative or of importance. In some areas such as deeper mineralized zones, where the prospective host rock or zone might be at depths of more than 1,000 ft, there may be no significant outcrops or exposures visible on the property. In such cases, the property can be reviewed on the surface, and perhaps some nearby adjoining properties can also be visited. If the company representative is still interested, it might be worthwhile to visit outcrops of the host rock wherever they occur within the district, perhaps even as far as 25 km (15 mi) away. This decision should be made by the company representative.

Field examinations are interesting exercises, enabling the seller to gain considerable information concerning the competence of the prospective buyer's representative. On occasion, we have taken geologists representing prospective buyers to look at certain mineral prospects, and have returned convinced that they understood very little about the geology of the mineral deposits in question. In such situations, we concluded that these geologists would be unable to write competent reports on their examinations. If this impression is clearly conveyed, then the seller would be unwise to spend much time on the property with such representatives.

Other times, company representatives are alert, competent geologists with whom the seller can discuss the geology of a prospect in meaningful terms, perhaps comparing it with other known areas. Such a visit to the prospect area can be a valuable experience for both parties.

Likewise, company representatives may be sent to review prospects in cases where the seller lacks the expertise to show or discuss the property and may not even understand the geology of the prospect. In such situations, the company representative may be unable to obtain adequate information from the seller, and he may be obliged to terminate the prospect examination prematurely.

When conducting a field examination, we recommend showing certain features of the prospect which demonstrate its favorable potential. In showing a roll-front uranium prospect, for example, if there are good sandstone outcroppings (perhaps altered) on the prospect, these should be examined. Solution banding at the surface is also impressive, if it is visible. We suggest emphasizing radioactive anomalies, particularly if they exist at the contact between an altered sand and shale. High-grade mineralization is always of interest (figs. 8-3, 8-4).

If we have carried out some water sampling and have found uranium anomalies present in the water, we point out the sample locations. The topography of the area can also be examined, the vegetation should be reviewed, and wildlife can be

Figure 8-3. Preparatory to showing a property to a prospective buyer, it may be necessary to make excavations at critical locations, particularly if ore can be exposed. Here the authors complete a trench dug to expose mineralized rock. A multimillion-dollar deal was concluded shortly after buyer's representatives examined the prospect, including a close study of mineralization in the trench.

surveyed in the course of looking at the prospect. The locations of any holes which have been drilled and which produced favorable results can also be examined, if time permits.

Although we have rarely found this to be necessary, it is sometimes a wise investment to charter a small airplane to fly over a particular area. For example, if you are looking at a rather large prospect, or a large area around a prospect, and time is limited, it may be desirable to spend a few hours flying over the property. This would enable the company representative to see the overall setting, as well as some specific prospect locations, if any are definable. In an aerial survey, the seller should have sufficient familiarity with the prospect that he can readily locate certain significant features and point them out to the company representative.

Company representatives may frequently express skepticism concerning certain information or data presented by the seller. This is to be expected, and it is perhaps a healthy sign. Sellers tend to be optimistic, whereas prospective buyers are generally pessimistic. We have been told that one of the geophysical supply companies plans to market a new scintillation counter in two models: the 121S (for sellers) and the 121B (for buyers). The S model will be equipped with a loud alarm and flashing red light which will be activated when a radioactive anomaly of a preselected intensity is encountered. It will have a large crystal for good sensitivity. The B model, on the other hand, will have no alarms or lights and will have a small crystal for low sensitivity.

Figure 8-4. Buyer's representatives examine an outcrop in the company of the seller's representative.

Seriously, however, some prejudices will be encountered on both sides. We recall showing a Cretaceous sandstone uranium prospect in Colorado to a representative of a major company which had open-pit uranium operations in Texas. For reasons unknown to us, the company's exploration managers had decided that the company was only interested in first-cycle sediments as a host for uranium deposits in sandstones. This policy did not become known to us until we had almost completed the day of showing this third- or fourth-cycle sandstone prospect to the company's geologist. Regardless of other favorable features which were present—even ore-grade mineralization at the surface in the sandstones—the company had no further interest in our prospect. This same company had operations in Texas in sandstones which were not first cycle; but for the Wyoming, Colorado, and Montana area,

they had established criteria whereby the sediments had to be first cycle. This company should not have sent its representative to review a prospect which was already ruled out before he left the office.

Prospective buyers frequently request mineral samples; this practice should certainly be encouraged, but it is very disturbing when the company representative collects samples of altered sandstone with the anticipation of getting some favorable uranium assays from them back in the lab. The seller should assist the company representative in selecting locations from which samples can be taken. This should be done with care, because the company representative might possibly assume that the seller is trying to steer him toward samples with a higher value. This is a rare problem with uranium, however, since radioactivity is an excellent sampling guide.

If the seller is a prospector, or if the

buyer's representative is inexperienced, it is also a good idea for the buyer's representative to request that the seller collect a sample, which will be marked as picked by the seller. This procedure precludes the seller from later complaining that the buyer's representative, through incompetence, chose poor samples. This procedure also protects the buyer's representative from the problem of a disgruntled seller who complains to the buyer that the geologist did a poor job of evaluating the property.

The seller should request copies of the sample analyses, since the samples were obtained from the seller's property. There is usually no valid reason why the buyer should object to sending assay results to the seller and, as a general rule, it is a good idea for both parties to have the results of the analyses.

We recommend that the buyer's representative (such as a consulting geologist) who carries out the examination of an offered deal should gather as much information as time allows in the field and from publications. There is, however, an additional important source of information that is frequently overlooked. That source is other people, particularly geologists, who may have done some work on the property at some time in the past. For example, for several years we were closely associated with a small company carrying out uranium exploration on a property which we regarded very highly for specific geologic reasons. Later, when we were no longer associated with the company (which had no geologists on its staff), they began looking for a joint venture partner. One of the companies interested in the deal hired a prestigious consulting firm to look at the project. Despite urging by the small company, the consultants did not visit or call us to discuss the geology of the prospect. The prospective buyer turned the project down, and the deal was taken by someone

else. Months later, we were meeting with the consultants about another matter, and that project happened to come up. They expressed surprise at our knowledge of the area, and we sharply disagreed with their estimate of potential reserves on the properties. They were far too low because of inadequate comprehension of the stratigraphy and mineralization. Why the client declined to enter the venture was readily apparent.

At the time of the field examination, the seller should have on hand adequate maps to show the company representative exactly where the property is located, the access roads, any adjoining properties which might be owned by others, location of samples and results of assays, and other pertinent information. Here, again, the company representative might request some prints of these maps for his use in preparing a report to the company. We still maintain that it is generally not necessary to make maps available at this stage. The company representative should by now have adequate information to prepare his report; he may even wish to use your maps to facilitate his note-taking on the location of samples. At a later date, if a genuine interest has developed, copies of maps may be made available to the prospective buyer.

Many novices in the exploration field have heard of, or read about, the practice of "salting" a claim so that the property will appear to have more mineral value than it actually has. In the gold rush days, claims were salted with appropriately placed gold dust, nuggets, or ore to mislead a prospective buyer. Gold chloride solutions have even been injected with hypodermic needles into locked leather pouches containing samples. This practice still goes on today, although on a diminished scale, and buyer's representatives must be wary of salted properties. Uranium ore has been hauled from mines to prospects, and we are aware of one instance where altered sandstone was hauled from a producing district and dumped over an outcrop on the prospect. The most incredible case, though, involved the dumping of man-made radioisotopes down a shallow drill hole so that good radioactivity would show up on the probe record. We understand the FBI entered the latter case but was not successful in learning the origin of the isotopes.

NEGOTIATIONS

Let's assume, now, that the company representative has visited the properties (several visits are common), reviewed the maps and other data available from the seller, and conducted independent research in other areas (such as the literature) concerning this prospect submittal. He may have requested further information, or copies of maps, in order to carry out additional research, and he may wish to ask more questions. The seller, meanwhile, may begin to form some idea of the buyer's thinking through questions he has asked or data he has requested.

It can be expected that the prospective buyer will contact the seller within a few weeks after the field examination, indicating whether his company is interested in commencing negotiations or in discussing the proposal further.

If interest continues, and the prospective buyer is satisfied with his findings, the next step is to arrange a meeting in an effort to negotiate a mutually agreeable deal. Prior to such a meeting, of course, the seller will have presented the terms he is asking for the prospect, although he may not have received any response from the prospective buyer as to their acceptability.

At the time of the meeting, the company representatives can be expected to review the results of their studies of the prospect, pointing out first the prospect's weaknesses. They may mention a few positive aspects of the prospect, but they will tend to emphasize weaknesses because they will want to make the best deal possible. The property situation may be reviewed, after which the prospective buyer might make a counterproposal—usually considerably lower than the price and royalty the seller is asking. Some buyers like to make an initial offer *shockingly below* what the seller had in mind; they theorize that after the shock has worn off, any figure above that will look good to the seller. It is common for the prospective buyer to request an option period, and often a free option is requested, so that the buyer may carry out some additional examinations including, perhaps, some drilling on the prospect to confirm whether it has merit.

The seller should bear in mind that the prospective buyer has certain objectives in mind. The buyer certainly seeks to make a deal for the property, or a further meeting would not have been requested. However, the company probably has certain guidelines and policies within which it is trying to operate in working out a deal with the seller. Further, the seller should be aware that the buyer's representative is not just *interested* in making a deal at the lowest possible price, he is probably *determined* to do so. The buyer is usually well financed, and another deal one way or the other isn't going to make or break him. The seller may have stretched available finances to assemble the properties and may be anxious to make a deal. If the seller is in a financial bind and the buyer becomes aware of it, it is likely that the buyer will try to use the seller's eagerness to sell as a tool for leveraging the terms down in favor of the buyer. We have seen this happen in several mineral deals where the seller was essentially forced to allow the buyer to dictate terms of the deal because the seller was under financial pressure. There is nothing wrong or unethical about such tactics, but sellers should be aware that many well-financed buyers have a predatory instinct, and if they detect a weakness, they will move in for the kill. This is a fact of business life in general, not just exploration.

As we stated earlier, in negotiations the seller has to consider his own objectives. On occasion, the objectives of the prospective buyer and the seller are so far apart as to be irreconcilable; once these positions are established, there may be no further negotiating sessions. Often a buyer will telephone the seller prior to planning negotiations to see whether there is any flexibility on the part of the seller. If the seller appears inflexible, and the terms are too tough for the prospective buyer, then there can be no meeting. In almost all cases, however, there is some room for compromise on both sides, and so the discussions continue.

Once the prospective buyer has stated his position, the seller is then in a position to compare or contrast this with his own original proposal. For example, if he had not considered selling an option, but proposed selling the properties outright and retaining a royalty, he might balance this against an option. Hypothetically, if the seller were asking $100,000 for the properties, and the prospective buyer requested an option to explore the properties prior to making a financial commitment, the seller might suggest that the buyer pay $50,000 for a short-term option (six months, for example) and then an additional $100,000 to exercise the option. This sort of approach is a prime example of compromise; it allows the buyer time to examine the property at lower cost than the $100,000 originally asked. The seller runs a certain amount of risk, because he would receive less money initially than if the entire amount were paid, but there exists the opportunity for a further gain of $50,000 if the property has sufficient value.

Several details of a deal are subject to negotiation. Royalties usually must be negotiated. The buyer generally wishes to pay a lower royalty than that asked by the seller. A proposed area-of-interest concept (discussed later in this chapter) may be objectionable to the buyer, and it may be modified, or perhaps the buyer might attempt to eliminate it completely from the discussions. Timing is a vital factor; the seller may wish to have the property explored promptly, whereas the buyer may request a year or two in which to adequately explore the property. Such differences of opinion should be negotiated.

If the seller is experienced and competent in uranium geology and believes strongly in the promise of his prospect (or has a competent representative whom he trusts), then he should be able to hold a firm position while negotiating terms. If the seller is asking a certain price and the prospective buyer will not or cannot meet that price, then another buyer can undoubtedly be found. Belief that the prospect is valuable places the seller in a far stronger negotiating position.

On the other hand, if the seller is uncertain about the value of his property and has no technically competent representative to confirm the value, then he is in a weaker negotiating position in demanding tough terms. Technical competence is extremely valuable in negotiations from the standpoint of both parties, since the opposing side cannot possibly use erroneous information to downplay the value of the prospect.

Surprising as it may seem, using erroneous information is not an uncommon practice. Buyer's representatives sometimes tend to downplay certain aspects of a prospect through the use of technically incorrect data. If caught at it, they may feign ignorance or oversight. If the seller is not adequately versed in the technical details of the prospect, he may be unable to defend his position and will therefore lose ground in the negotiations. Again, technical competence and the ability to utilize this competence in negotiations is an enormous asset to the seller.

It is also a common practice in negotiations for mineral deals to use a third party as a scapegoat who demands specific, perhaps unreasonable, terms and who can be blamed for an inability to reach a compromise on certain terms. This practice is particularly prevalent in large companies where the seller might be obliged to deal with a landman or with a district or regional geologist. These individuals (who, of course, wear white hats) will often use a company vice-president, a board member, or another uninvolved party (who wears a black hat) as their reason for being unable to meet, or having to demand, certain terms.

Similarly, the seller can utilize this same approach to his advantage, unless he is an individual prospector or the sole owner of the property. If the property has more than one owner, the seller can state that he is negotiating on behalf of the partner or partners (provided they do not attend the meetings), and consequently may demand certain terms, or deny certain concessions, because they would be "unacceptable to the partners."

One of the most frustrating aspects of dealing with a large company, especially if the company is undergoing significant personnel changes during your negotiations, is that you may be obliged to negotiate with different persons within the company, rather than with one party. For example, in one particularly extended negotiating session over the course of a nine-month period, we found ourselves dealing with seven or eight different negotiators representing the buyer. With each personnel change, we were obliged to brief the new individual on what had transpired during our previous sessions and to begin many arguments anew. The deal was finally concluded, but only after repeated exhausting and nerve-wracking episodes.

SUGGESTIONS TO BUYERS

While most of this chapter has dealt with problems and approaches from the seller's standpoint, a few comments from the buyer's standpoint may be of value. In the first instance, the buyer must have confidence in the personnel who have conducted the examination of the property in order to proceed with negotiations on a sound basis. In the case of large companies particularly, negotiations are often attempted by individuals within the company who have no first-hand knowledge of the prospect. They have on hand the seller's data, along with a report and recommendation by their own personnel, but the two sets of information probably do not agree on significant points such as potential reserves, possible grade, costs to evaluate, timing, and so forth. If the buyer has confidence in the representative who looked at the prospect, then an offer can be structured around criteria set forth by the representative. With little confidence in the representative, the buyer might be inclined to make no offer at all, or to make an offer so low that regardless of how poorly the prospect may turn out, little loss will result.

We prefer to deal with a buyer's representative who is an exploration geologist himself, preferably with a good background in uranium. If such a person is in an upper management position, then we probably showed the prospect earlier to a geologist under his supervision. Before beginning negotiations, we encourage the management person to go to the field to look at the prospect, even if only for a short period. Such a visit can lay the groundwork for the negotiations to follow as the negotiators become better acquainted and discuss the positive and negative aspects of the play.

Recently, we concluded a multimillion-dollar deal within 90 days from the time a representative of a particular mining company looked at our data. The sequence of events was as follows: (1) we made a phone call to the chief geologist, describing the prospect and terms; (2) the chief geologist visited our office to look at the data and to discuss the concept; (3) the chief geologist visited the prospect in the field; (4) we prepared written information and supplied maps so that the chief geologist could give a show and tell for company management; (5) we took the exploration vice-president and the chief geologist to the prospect (the vice-president, in addition to being the negotiator, was also an exploration geologist); (6) the very evening after completing a three-day look at the property, the vice-president made a counteroffer which was low but not *too* low, and negotiations were under way. By 2 p.m. the following day, oral agreement had been essentially reached. A trip to the home office was necessary in order for the chairman and president to review a few points (we were accompanied by our lawyer, and theirs attended as well), and then the preparation of an agreement went to the lawyers. Ours prepared the first draft and stunned us all by having a 40-page agreement draft and a 15-page operating agreement draft on their lawyer's desk in less than 10 days. From there, things moved quickly (thanks to prodding of their lawyers by the vice-president), and in what must be record time for a deal of that size, the agreement was signed. The deal we accepted was not what we originally had in mind, but we were willing to compromise because of the fast action by the buyer.

The buyer's representative should remember that it is usually very important for the seller to get an offer as soon as pos-sible. In mineral exploration, substantial costs are often incurred as properties are assembled through staking and leasing, and it is important for the seller to move the deal promptly. As a result, most sellers are willing to accept a lower price or commitment if the buyer will move quickly. We strongly recommend to buyers that if you receive submittal of an attractive prospect that meets your criteria for exploration, you will be well advised to make an offer, particularly in a seller's market. We advise against simply asking the seller if he would consider a lower price, or asking him to "let you know" if the terms of the deal change. Very often sellers will consider that a last-ditch kind of suggestion and will try to sell the deal to others before finally advising you that the terms have changed.

Perhaps an example would help clarify this point. Let's assume a seller has assembled an attractive group of properties for uranium exploration. He is asking $200,000 for a two-year exploration option and $200,000 as an option-exercise price along with a 5 percent concentrate royalty. Companies A and B both look at the deal and like it, but both would like a smaller option payment and exercise price. Company A's representative meets with the seller and advises him that the company is interested, but the price is too high; he asks that the seller contact Company A when he is ready to talk about "a lower price." Seller has no idea what "lower price" Company A has in mind: $10,000; $50,000? Company B, on the other hand, takes the initiative and *makes an offer* in the form of a three-stage option rather than a two-stage one. The company offers $100,000 for the first two years, $150,000 for the second two-year period, and $200,000 to exercise the option at the end of four years, for a total of $450,000, which is $50,000 more than seller was asking. Company B believes that such a deal benefits them in several ways: (a) the immediate cash outlay is reduced, providing more money for exploring the property; (b) within two years, they should know whether to keep the option for the second two-year period; (c) if they keep the property four years and then exercise the option, they will know if it is a good deal or not. Company B further explains to the seller that if he *really* has confidence in the properties, he shouldn't mind allowing the property to be explored for a lower initial payment when the company is willing to pay more for it in the long run. The company also requests a prompt decision.

It is probable that such an offer would be taken by the seller, but what about Company A? It would have been happy with a deal such as that negotiated by Company B; the difference was that it wouldn't make an offer. By the time the seller could get back in touch with Company A, the deal would probably be made—and why should there be any hurry? After all, Company A was only modestly interested.

KEEPING NOTES

Beginning with the first serious response from a buyer suggesting sufficient interest that negotiations may result at a later date, it is recommended that the seller begin keeping notes of all correspondence and meetings between the parties. Such a procedure is particularly valuable if the negotiations prove to be very lengthy and complex, involving many facets of a deal where give-and-take discussions take place. It is often impossible to tell ahead of time how long and involved negotiations might be, so a standardized note-taking procedure is desirable. For these purposes, we prefer a secretarial spiral notebook (bound at the top), and a *different notebook for each company* with whom negotiations are carried out. Usually, negotiations for a particular deal are made with just one com-

pany, so one notebook will be used in most instances. Very detailed note-taking is unnecessary, but the meeting dates, attendees, important points discussed, conclusions reached, and so forth, should all be included. In addition, telephone calls and other items (even newspaper clippings) should be added. These data will prove extremely valuable as negotiations go on. You may find weaknesses in the other party's position, or you may discover a comment they made at an early meeting that suggests a compromise in your favor.

LETTER OF INTENT

At the point in the negotiations where some general agreement is reached, some explorationists insist on preparing a letter of intent setting forth those terms on which both parties agree. Through experience, we have found that the time involved in preparing a letter of intent, if it is to be of any value, may as well be spent in preparing an agreement in principle.

Theoretically, the difference between a letter of intent and an agreement in principle (frequently referred to as AIP) is one of depth. A letter of intent usually describes in a few pages the salient points on which the parties agree. An AIP describes such points in greater detail, including important provisions, and consequently is a lengthier document. In a complicated deal, the AIP may contain 20 or 30 pages, and it may include exhibits, such as a listing of properties owned by the seller or on which the seller might have an option.

The AIP should be executed and notarized by the parties, and is a legal, binding instrument. It is usually stated in the AIP that a more definitive agreement is being prepared. Because of the rather complex nature of an AIP, which is a step toward a final agreement, attorneys usually require a few weeks for preparation of the initial draft, which is then followed by the final draft.

When negotiating sessions are drawing to a close, and the parties have reached agreement on certain items, it is good practice to make a list of the essential parts of the agreement. The list should be read to all parties present, so that any questionable points may be discussed and more definitively agreed upon prior to the legal document's preparation. In this regard, we strongly recommend that all parties be represented by legal counsel—preferably the same counsel who is to draw up both the AIP and the definitive agreement.

In chapter 9, "Legal and Accounting Aspects," we discuss more fully the use of lawyers in various negotiating sessions. We also outline some techniques which can be used to expedite the preparation of written material, such as the AIP or the definitive agreement.

THE AGREEMENT

Who Should Draft?

Whenever the time arrives for the AIP or the final agreement to be drafted, a disagreement invariably arises between the parties as to which party should prepare the first draft of the document. Typically, a large company will attempt to prepare the first draft of the definitive agreement, since the company feels it may have an added advantage in the inclusion of certain wording. Also, large companies usually like to include specific language in all agreements they make, most of which is acceptable, and which is usually referred to as "boiler plate." A major problem with this procedure, however, is that there are often extensive delays, sometimes running into weeks or even months, because corporate lawyers are generally busy with many other projects and can scarcely find the time to put an agreement into writing. This situation varies, depending on the character of the persons other than lawyers

involved in making the deal, and how much pressure they can exert on their lawyers.

We always opt for preparing the first draft of an agreement ourselves. We can usually get it done in about one-quarter or one-third of the time it takes a large company to prepare the draft; however, this probably reflects our experience, as well as the competence of our legal counsel. We contend that corporate lawyers can insert their specific language after they receive our first draft.

What may constitute a major deal for a small company, or a small group of individuals, may be a very minor deal for a large company. While the smaller company might tend to concentrate its efforts on getting the deal written promptly, the larger company might place it behind an accumulated backlog of paperwork. More about this aspect is discussed in the following chapter.

We have found that it often takes 10 or 12 drafts, over a period of several months, before a complex written agreement which is satisfactory to all parties can be completed. A similar exchange of drafts also takes place during the preparation of an agreement in principle. It generally begins by one of the parties preparing the first draft of the final agreement (this is usually prepared by the legal counsel representing that party). The draft is reviewed by the first party, and any necessary revisions are made before it is sent to the other party. We suggest that the very first draft of the AIP, as well as the final agreement, be placed on a magnetic card typewriter, or a similar memory typewriter, so that subsequent revisions can be made rapidly and efficiently.

Once the agreement is drafted, it is submitted to the other party and that party's legal counsel for review and comment. The other party makes such notations and comments as he feels are desirable and necessary to meet his objectives. Such comments can be made either on the draft itself (or on a photocopy), or on additional sheets which may be inserted into the agreement. A complete retyping may be required if there are extensive changes. The amended draft is then returned to the originating party. Matters can be expedited if changes are noted on the returned draft, with marginal notations indicating where changes have been made. Any new wording inserted into the draft should be underlined if the draft has been retyped. The second party's lawyers may also wish to record their draft of the agreement on magnetic cards, so that later draft preparations may be expedited.

When the amended draft is received by the first party, he and his legal counsel may review the comments to see how it fits in with their concept of the deal, and how the comments may affect their position. The first party prepares yet another draft, incorporating the acceptable changes suggested by the second party, and perhaps some more of his own.

At this stage, it may be considered necessary to hold another negotiating session so that any problem areas, for example, tax considerations or a work commitment, can be agreed upon. The exchange of drafts back and forth, and additional negotiating sessions, may last for weeks, or perhaps even months, before a final agreement is prepared. If the deal is a complex one, involving large expenditures, then it is even more likely that an extended period will be required to produce a final document which is acceptable to both parties.

For example, a uranium exploration and development deal signed in 1974 between the Tennessee Valley Authority and American Nuclear Corporation required 15 months of negotiations. A deal involving a uranium joint venture in Weld County, Colorado, between Power Resources Corporation of Denver and Wyoming Mineral Corporation (a Westinghouse subsidiary) required 10 months of negotiations.

The need for legal counsel's presence in discussions is determined by the topics to be discussed and whether they actually involve some legal aspects of the deal. The representatives of the two negotiating parties must make this decision. Some companies hesitate to meet at all with a seller unless he is represented by a lawyer, since they do not wish to appear to be "taking advantage." Need for a lawyer may depend on the competence of the negotiating representatives to handle the discussions, and their ability to carry negotiations forward without the aid of legal counsel.

The final written agreement is a document with which both the buyer and the seller may have to live for many years. It is well worth the considerable time spent, and the numerous exchanges of drafts, to insure that all parties are satisfied that the agreement provides the opportunities and protection they feel they need, and adequately reflects the terms agreed upon during the negotiating sessions. Both parties and their legal representatives must be sure that the final agreement expresses clearly all points upon which agreement has been reached, and any tendency toward ambiguity should be removed from the agreement.

Some Basic Provisions

Area of Interest

An area of interest is the area within which the parties agree to perform certain functions, such as land acquisition and/or exploration and mining; it is defined by legal subdivision, such as section, township, or range. The area-of-interest concept is necessary in certain instances and is not workable in other instances.

The definition of an area of interest in

an agreement can work to the benefit of both parties in a situation where both parties wish to continue land acquisition and other functions in a cooperative arrangement rather than in competition with one another. As an example, let's say that Company Q has scattered leases on 7,680 acres (12 sections of 640 acres each) in two adjoining townships, and it is continuing its leasing efforts. Company R then becomes interested in joining Company Q in the project. Q and R write an agreement specifing an *area of interest* within which either Q or R may acquire leases, but any leases so acquired become a part of the joint venture. In such arrangements, it is common for R to become the operator, since R usually has the responsibility of financing the project. Company Q will probably require (and should insist upon) the right to continue acquiring leases in the area. Q's right to acquire leases is insurance for Q that leases which appear favorable may in fact be leased (or attempts may be made to lease them) and will not be at risk of falling into the hands of a competitor because of inaction by R; this is very important. However, if R is the operator, Q will have to spend its own money to acquire leases, because R will not be likely to advance funds to Q for this purpose.

This problem is worked out by agreement between the companies that Q can acquire leases in the area utilizing its own funds, but it must then promptly offer said leases to R. R shall then have 30 days in which to decide whether to accept the leases for the venture, and consequently reimburse Q for the actual monies spent on acquiring the leases. If R chooses not to accept the leases, then Q is at liberty to retain the leases for its own account outside of the venture.

Buyers frequently do not want an area-of-interest concept included in a deal; yet this may be an extremely important factor for the seller, because it can provide him with larger area of influence should commercial production be established from his properties or from nearby properties. The area-of-interest concept is one of the most important items to be considered in the preparation of a joint venture agreement, and while it may appear to be a one-way street in favor of the seller, this is not necessarily so. If the seller is an exploration operator, or if the seller is a good explorationist active in the particular area, then the buyer may feel this concept provides protection in the event that the seller comes up with a good property or properties.

Phased Program Concept

In virtually every deal short of one in which the reserves are pretty well blocked out, most buyers prefer to invest in a property in increments or phases, with the right to drop out if results of the exploratory work are unsatisfactory. Such an approach was included in the Spruce Exploration proposal described above in this chapter. Under such a concept, the buyer agrees to provide financing for a project one phase at a time and to earn a certain interest by completing each phase. Such an approach may be satisfactory to both seller and buyer. The seller will have funds available to explore the property, including later funds if encouragement is found, but will not yield substantial interest in the deal *unless* substantial expenditures are made by the buyer. From the buyer's standpoint, an opportunity is provided to earn a substantial interest in the deal, but large expenditures will be necessary only if encouragement is found; thus the immediate commitment is reduced, but long-range rewards are possible.

We have made several deals involving the phased concept, and it works out very well. Typically, we prefer to have the buyer earn *no interest* in the first phase. Then, if the buyer is discouraged and drops out, the seller has 100 percent of the deal back again for possible resale. In one joint venture, the buyer was a mining company and insisted on operating the exploration program; as the company's management put it, there was "no way" someone else was going to spend their money. The agreement was structured so that by spending approximately $6 million in five phases, they would earn 70 percent interest. However, they earned no interest in phase one. They spent approximately $500,000 in that phase, but, because of what we considered a poorly run program, nearly $400,000 was essentially wasted. When the meeting was held wherein they advised us they were dropping out because of insufficient encouragement, they expected us to be very unhappy. Not so—we were delighted to be disassociated with such poor operators, and, in addition, we were confident that we could do a minor amount of drilling (based mostly on their results) and find some ore. If we could do this, perhaps an even better deal could be made with another party (eternal optimists again).

That is exactly what we did. We completed $40,000 in drilling on our own, acquired a few more leases, and sold the deal again for a roughly $8 million commitment to earn 60 percent interest (10 percent better than the previous deal). The wisdom of pursuing this project has been proven, because significant uranium reserves have now developed within the defined area of interest.

In another instance, the potential for an area did not look very good after $750,000 had been spent in three phases, and so we suggested to the buyer that phase four (in which, according to the agreement, $500,000 was to be spent) be divided into two stages. Thus, additional data could be gathered to prove or disprove the prospect,

but we felt it might be a waste of money to force the buyer to continue to finance exploration when the project might be disproven with just one-half of the funds specified in the agreement.

Cash or Stock Payments

It is common practice in mineral deals for the seller to receive some kind of payment (often referred to as a bonus payment) in return for executing an agreement with the buyer. This payment can take many forms, but it is most frequently in cash; sometimes it is paid as corporate stock. The timing of these payments from the buyer to the seller can be an important factor to both parties, but it is generally of greater importance to the seller, if the seller is a small company, an individual, or small group of individuals.

The buyer often attempts to obtain the properties without any bonus payments to the seller. If a bonus payment must be made, the buyer may wish to make several small payments over an extended period rather than one larger payment. This method can have certain advantages for the seller. For example, perhaps the buyer had agreed to pay the seller $100,000 as a bonus for executing an agreement on a group of properties. Because of the tax consequences to the seller, he would realize little income if the payment were received in one year, and so he may prefer to have the payments spread over two years, or even three, rather than receiving one payment.

The Internal Revenue Code treats payments of this type as one payment, all of which would be taxable in the year the deal is made and in which the initial payment is made, unless the buyer has an option on the properties and can drop the properties within a certain time frame. In order for the seller to receive a $50,000 payment during the first year and $50,000 during the second year without an adverse tax impact, the agreement must be written as an option, giving the buyer the right to drop the properties before the second payment is made if he so wishes. This procedure may be unacceptable to the seller, because it gives the buyer the right to drop the properties if conditions should change, or if some unfavorable geologic evidence should be discovered during exploration activities. This is a risk the seller must take if he wishes to reduce tax obligations for a particular year.

Buyers often propose to make bonus payments in the form of *advance royalties,* since this method allows the buyer to deduct the bonus as an expense item, rather than as a capitalized item, in tax computations. The seller must decide whether this is an advantage or a disadvantage to him in the transaction. An advance royalty is recovered from the royalty due to the seller from ores or concentrates produced from the property. If the seller is convinced that production will be obtained from the property, he probably should insist that a bonus not be treated as advance royalty. On the other hand, if there is some doubt about the existence of ore on the property, then in order to close the deal, the seller may agree to regarding the bonus as an advance royalty.

Sometimes—especially in cases where a small publicly held company may be acquiring properties from a seller—the form of payment may be corporate stock in that company. If such payment is proposed, the seller should certainly seek the advice of competent legal counsel, because it is probable that the stock he would receive might be unregistered or "restricted" stock (also called "lettered" stock) which he could not immediately sell on the market. In such a transaction, the unregistered stock is often "discounted," so the seller should demand more shares than would be the case with registered stock. Sales of unregistered stock

which he might subsequently make would be made under SEC Rule 144. The phrase "lettered stock" originated when it became mandatory to place a legend on the back of each certificate representing unregistered stock. The legend pertains to the fact that the stock is not registered and may not be sold or traded.

The seller should consider carefully any tax implications if he is to accept stock as a bonus payment. Such transactions are likely to be viewed by the IRS as a payment at the quoted values of the stock at the time the transaction is made, and therefore the seller would have a tax obligation identical to that imposed if he had been paid in cash.

If the stock issued as payment to the seller is registered or "unrestricted" stock—that is, readily saleable (if there is a market)—then the seller is less likely to be justified in demanding a discount on the stock value. But if, as is usually the case, the stock is restricted stock, then it is not uncommon for the seller to receive a discount of perhaps as much as 50 percent from the current market selling price of the stock.

For example, if the stock of Company G is traded on the over-the-counter market for $0.20 bid/$0.25 asked, and Company G proposes to issue to the seller 100,000 shares of unregistered stock in return for an assignment of mining claims, there is no immediate value to the seller in such a transaction, because the stock cannot be sold on the market; there is, however, a tax liability. The seller might be satisfied with the stock-payment concept, but he may negotiate for a larger number of shares (130,000, for example) to compensate for the lack of an immediately liquid bonus because he cannot sell any of the shares for two years.

Another recommendation for the seller who is considering accepting a stock payment is that his lawyer should review the background of the company and the history of the stock activity and value. Such an investigation should be carried out as a forecast to satisfy the seller that he *may* get a fair cash value for the stock when and if he sells it. A conservative lawyer will almost always recommend against such a trade; yet, some good profits have been made by those willing to take a chance.

It would be unwise for the seller to accept a relatively small number of shares of stock for the property if the stock offered were grossly overinflated. Even with a 25 percent discount on the current market value, he would be running the risk that by the time the stock could be sold it would have returned to a low price. If these events occurred, the seller would have to accept the low price for the stock, resulting in a low price for his property. Ideally, from a seller's standpoint, stock accepted for property would significantly appreciate in value by the time he could sell it.

It should be emphasized that attorneys who are not familiar with stocks, the stock market, and laws administered by the Securities and Exchange Commission may not be qualified to advise a client on making a deal which includes payments in the form of stock. Prospective sellers considering stock payments are advised to obtain the services of an attorney who is well acquainted with stock transactions, securities laws, and SEC requirements, or to ask his own attorney to obtain additional counsel for the deal.

There are other types of payments which are very important in many exploration deals and should be mentioned. One is the *management fee*. Once again, a management fee was mentioned in the Spruce Exploration proposal at the beginning of the chapter, and in that case, a fee of 10 percent of expensed (not capitalized) items was suggested. It is common for the operator of an exploration project to charge the project a management fee ranging from 5 to 15 percent; such a fee is to cover miscellaneous home office general and administrative expenses associated with running the project. Of course, the project itself would have to rent space, obtain office equipment and furniture, and (probably) hire people; such costs must be paid for from the exploration budget. Land acquisition costs are almost entirely capitalized, but drilling and similar work is an expense item. No management fee is paid on land acquisition costs (or other capital items), because such a procedure removes any doubt that the operator (manager) may be acquiring land in order to collect the 10 percent fee.

Another fee that is common in our joint venture agreements is what is called a *consultation fee*. This is a fee agreed upon by buyer and seller as compensation to the seller for assembling the project and for acting in consultation on the conduct of the program. This fee is usually paid *in addition to* the management fee described above, and it may amount to $1 million or more in payments keyed to phases in programs costing $5 million to $10 million.

Options

As we mentioned above, it may be desirable for the buyer, and sometimes for the seller, to have the agreement set up as an option, which gives the buyer an opportunity to carry out investigations on the property before additional payments are made, or before additional commitments go into effect. The option is either (1) included as a sort of preamble to the agreement itself; or (2) it is set apart and labeled as an option and agreement; or else (3) a short separate option (two or three pages) is prepared, wherein an agreement is attached to, and referenced in, the option. The length of the option may run from a few days (for a "hot" deal) to several months; a staged option may even run into years, depending on the complexity of the deal. Options in uranium deals usually do not extend for more than two years.

A typical two-year option may provide that on or before the end of the first year, the buyer has the right to make a payment to the seller and/or perform work on the property, and then proceed into the second phase of the option. Finally, prior to the end of the total option period, the buyer has the right to exercise the option or to return the properties to the seller. The seller should *always* insist on the inclusion of a clause which provides that he will receive copies of factual engineering and geologic data developed on the property if the option is not exercised.

As explained earlier, the seller with a high degree of confidence in his properties rarely objects to giving the buyer an option; this offers the buyer an opportunity to examine the property and to determine what the real value might be. In addition, options may be advantageous for the seller, because he might command a higher price paid in cash through the options than he would otherwise be able to obtain if he simply asked for a one-time cash payment at the beginning.

Royalties and Participation Interests

From the standpoint of both parties, the royalty provisions are among the most important aspects of the agreement. Basically, a royalty is a consideration paid to the seller, or lessor, for allowing the buyer, or lessee, to mine, remove, and sell ore from the property. Royalties are usually paid in some direct proportion to the value of ore removed and sold from properties described in the agreement and controlled by the seller or from production from properties within the area of interest.

For example, let us assume that uranium ore was mined from a certain property

and sold to a uranium mill. Let us also assume that the seller had retained a 5 percent ore-value royalty in the agreement, with payment to be made based on the value of ore at the mill. In this case, if we assume $5 per ton shipping charges and $20 per ton payment for the ore by the mill, the net value of the ore at the mill would be $15 per ton. The seller, with a 5 percent royalty, would receive 5 percent of $15 per ton, or 75 cents per ton, as a production royalty payment.

Advance royalty. Another consideration which may be included in an agreement, and which is certainly to the seller's advantage, is the advance royalty provision. This, of course, provides that the seller will receive a royalty payment regardless of whether the property is in production. The advance royalty is usually paid to the seller on an annual basis. An advance royalty, if of a significant amount, can create an incentive for the buyer to put the properties into production, and it provides a source of income for the seller during a time of nonproduction.

An advance royalty is recoverable out of the regular royalty paid when production commences. For example, the seller might receive $50,000 in advance royalties over a three-year period prior to production; in the fourth year, production commences. If we assume that the seller's royalty in that year would have been $75,000, he would receive only $25,000 for the first year of production, since the operator would be recovering monies paid earlier as advance royalties.

Concentrate royalty. Another type of royalty which is becoming more common in uranium deals (and which was discussed earlier in this chapter) is the concentrate (also called "yellow-cake") royalty. This royalty is based on the selling price of the uranium concentrate produced by the mill from ores mined from the seller's properties. Some buyers will not deal with sellers or lessors if the concentrate royalty is higher than 3 percent; other buyers refuse to consider a concentrate royalty under any circumstances. They point out, with strong justification, that the seller is entitled to a reasonable royalty based on production of raw ore, but not to a royalty based on a product produced through costly processing of the ore. A comparison can be made between a concentrate royalty and royalty on gasoline produced by a refiner of crude oil. Royalties extending to a refined product are unknown in the oil business.

In-kind royalty. Another important feature which the seller might consider including in the agreement is the right to take royalties in kind. This provides the seller the right to take his share of the production as either ore or concentrate, depending on the language of the agreement. In the case of ore in kind, the provision is less important to the seller; if the seller takes his share of the ore produced, he will then have to sell it, with the concomitant problems of hauling, loading, and so forth.

In the case of a uranium-concentrate royalty, however, it may be extremely important for the seller to have the right to obtain his share of the concentrate produced. For example, the seller might retain a 5 percent concentrate royalty; ore produced from his property might be milled to produce 500,000 lb of uranium concentrate per year. If the seller has the right to take 5 percent of 500,000 lb of concentrate production in kind, and he exercises that right, he would receive 25,000 lb of uranium concentrate and may sell the concentrate to whomever he chooses. The seller may allow the buyer to sell his share together with the other concentrates produced by the mill, or, if he disagrees with the price that the buyer is asking for concentrate, he may then go into the market

and seek his own purchasers, assuming that there is a sufficient quantity of concentrate to interest them.

Retaining the right to take the royalty in kind, then, gives the seller a greater degree of flexibility in the type of deal he can make with concentrate produced from ores mined from the properties. Another important factor is the tax consideration: by receiving royalty in kind, the seller would have no tax consideration; he would also have no tax obligations from the sale of the concentrate until such a time as a sale was made.

It is important that the buyer have the obligation to package the concentrate, so that the seller can acquire his share of concentrate already packaged and then dispose of it as he sees fit.

Participation interests. Carried interest, net profits, and joint ventures may be referred to as participation interests. These approaches have been discussed above. It is not likely that any of these types of deal will be made between an individual seller and a company, but such deals are common between companies.

Work Requirements

In some agreements, it behooves the seller to include work requirements to be performed on properties optioned or sold to the buyer. Work requirements are usually not a necessary inclusion in the agreement if the seller is to receive an advance royalty, because both the advance royalty and the work requirement tend to encourage the buyer to proceed promptly with exploration and development of the properties.

There are situations, however, where it might be desirable, from the seller's standpoint, to insure that the buyer proceeds promptly with exploration work on the properties. Perhaps the seller has some work commitments which must be completed in order to comply with terms of an agreement written when he acquired the properties. If so, these requirements must be satisfied.

A timely opportunity to acquire adjacent or nearby properties may also dictate the prompt commencement of work on a property. In many existing mining districts, those who carried out early exploration were among the first to recognize the potential of the area, and consequently, they had the opportunity to begin *aggressive* property acquisition under terms which were favorable to the buyer. Of course, having an opportunity and making prompt use of such opportunity are two different things. Companies or persons responding sluggishly to opportunities often are left out; hence we emphasize the word "aggressive."

One of the legal requirements to hold mining claims is annual assessment work. Assessment work is required by statute on each mining claim in an amount specified by law, usually $100 per claim per year. A mining claim consists generally of 20 acres, and so assessment work usually costs $5 per acre per year, which is a relatively high holding cost when compared to a fee lease, which might cost $1 per acre.

In any agreement, the buyer should certainly assume the obligation of performing assessment work on claims, or else he should return the claims to the seller at least 60 days prior to the end of the assessment year. In most states, the assessment year ends August 31 at midnight. Theoretically, the seller can perform the necessary assessment work on returned properties in 60 days.

In addition to the assessment work, it may be desirable for the seller to require that additional work be performed by the buyer. This could include environmental studies, road building, drilling, or underground work such as drifting, assaying, or amenability studies in order to satisfy

either an obligation which the seller may already have incurred or to satisfy a certain time schedule established by the seller as one of the requirements of the deal.

Another obligation to be assumed by the buyer is to make rental payments for leases in a timely fashion. If, for any reason, a rental payment cannot or will not be made, the buyer should advise the seller 60 days prior to the date the rental payment is due. If the buyer fails to timely notify the seller, the buyer should be obligated to make the rental payments regardless of whether he plans to retain the properties or relinquish them.

Lawsuits have arisen over the loss of leases because of a buyer's failure to pay rentals. For this reason, many assignments include a provision that, in the event of inadvertent loss of a lease because of nonpayment of rental, liability of the buyer is limited to the original bonus price plus rentals paid for the lease. It is easy to see how a seller might file suit against a buyer if the buyer lost a lease through nonpayment of rental and later an ore body was found on the property. Such situations are not uncommon in the petroleum industry.

Renewal of Leases

The seller should include in the agreement a provision specifying that if the buyer renews or extends any leases previously held by the seller, the seller shall have an interest in these leases as set forth in the agreement. For example, one of the leases held by the seller might lapse during the time the buyer is carrying out exploration, and the buyer may then proceed to acquire a new lease for the properties. If this occurs, the seller should be assigned a royalty for the new lease, as though the lease had been acquired by the seller. In addition, any leases or claims assigned or reassigned to the seller by the buyer should not be reassigned by means of an instrument which reserves a royalty to the buyer.

The above provisions should apply to mining claims which may be relocated within the area by the buyer; that is to say, the seller should retain a royalty on production from such relocated claims.

Right To Relinquish Properties

From the buyer's standpoint, it is important that he have the right to relinquish properties or to return to the seller those properties which he believes to have been disproven or which, for any other reason, are no longer of value to the buyer. This is an important provision, and one to which the seller should not object. After all, if the buyer returns properties which the seller believes to be of value, the seller has the choice of exploring, developing, reselling, or dropping the properties. It is pointless for the seller to require the buyer to continue spending money, carrying out assessment work, or maintaining leases on properties which the buyer is satisfied have been disproven. By the same token, the buyer should not object to timely reassignment, allowing the seller time to perform work on, or pay rentals for, certain properties.

We know of one such agreement written for a uranium exploration program in Wyoming, wherein the seller insisted upon a provision that *both* buyer and seller must agree that properties were disproved before they could be relinquished. In this instance, the seller proved to be completely unreasonable in deciding when properties had been disproven, with the result that the buyer was obligated to continue spending time and money on properties which the buyer's geologists were convinced had no potential for the production of commercial uranium. The buyer eventually had to make the seller a substantial cash payment (approximately $200,000) to persuade him to acquiesce and allow the buyer to relinquish the properties.

A proven, workable arrangement for a reassignment provision might state that when the buyer plans to relinquish or reassign a property he shall make timely notification to the seller and, at the same time, shall provide the seller with factual data (as discussed below). The seller shall then have 30 days in which to request assignment or reassignment from the buyer. If the seller fails to notify the buyer accordingly, then the buyer shall have the right to relinquish the properties (that is to say, pay no further lease rentals or perform no further assessment work). This is a fair provision for both parties. It is a waste of time and money for the buyer to reassign properties to the seller if the seller does not want them.

Access to Data and Data Furnished upon Relinquishing Properties

It is important for the seller that the agreement provide him or his authorized representative the right of access to the properties to examine the work in progress, including a review of the production or exploration data. (This provision may be placed in operating principles.) In some cases, particularly concerning production from the properties, the seller may require that the buyer furnish actual receipts of ore sales or concentrate sales to verify royalties paid to the seller.

In addition, for properties returned to the seller by the buyer, or for properties the buyer plans to relinquish, the seller should require that the buyer furnish copies of all factual engineering and geologic data resulting from the buyer's work on the properties. This should include maps, drill-hole logs, copies of assays, and any other factual information developed by the buyer. Such a provision need not include interpretations made by the buyer or by his representatives. The buyer should have no objection to these provisions.

The Value of Illustrative Calculations

We have found it extremely valuable in all mineral deals to include written examples illustrating the various complicated aspects of a deal—particularly pertaining to royalties or to working-interest relationships. For example, if the seller were to retain an ore-value royalty, it would be an excellent idea to include in the agreement an example of a situation where a certain grade of ore is produced from the seller's properties and specific costs are assumed in transporting the ore to a mill or to a buyer, thus demonstrating how a purchase price is determined.

The value of sample illustrations in the text of an agreement increases with the complexity of the deal. If sliding-scale royalty rates are to be considered, or even an ore-value sliding-scale royalty with a multiplication factor for the concentrate selling price, examples of such computations should be illustrated to avoid later problems in determining the exact intent of the agreement. In addition to their use in uranium agreements, we frequently find similar examples written into deals involving coal, copper, and other minerals. Such practice significantly reduces the possibility of conflicting interpretations.

Operating Principles and Accounting Procedure

Along with exhibits describing area of interest and so forth, a joint venture agreement should have an exhibit describing operating principles. This exhibit (sometimes called an operating agreement) contains such provisions as (or procedures for) insurance coverage, record maintenance, payment of taxes, geological information and access, operator's duties, title procedures, technical committee functions and voting, payment of costs, and so forth. The accounting procedure contains provisions for preparation of accounting reports and functions including advances and payments, audits, budgetary control, and details of chargeable costs and expenditures. Such documents obviously are not needed except in joint venture or partnership arrangements. They should be prepared by legal counsel in coordination with accountants knowledgeable in mining and processing the mineral or minerals sought.

9

Legal and Accounting Aspects

WHY LEGAL ADVICE IS NEEDED

Complexity of Laws

As is the case with most industries today, laws pertaining to exploration and mining are becoming increasingly complex. Those of us involved in mineral exploration are obliged to comply with federal as well as state mining laws, including laws relating to locatable lands and nonlocatable lands. In most situations, there are certain requirements to fulfill, including discovery work and assessment work on mining claims, and performance requirements for mineral prospecting permits on federal acquired lands.

The Environmental Protection Agency (EPA) has stringent regulations relating to the environmental impact of exploration and mining operations. Potential problems exist when funds for exploration from public investors are sought, and laws and regulations of the Securities and Exchange Commission (SEC) must be observed. Other federal agencies, such as the Forest Service, Bureau of Land Management (BLM), U.S. Geological Survey (USGS), Bureau of Mines, and Internal Revenue Service, also must be dealt with from time to time during the course of exploration activities.

At the state level, the state archaeologist, soil conservation service, land use commission, planning boards, state geologist, tax commission, and other regulatory bodies all tend to play a role, to a greater or lesser extent, in determining the path exploration may take. States are beginning to enact more restrictive regulations pertaining to exploration and mining, some of which are so stringent that they can and do have an adverse effect upon an exploration program. We firmly believe that exploration *can* be carried out with minimal environmental damage; yet, because of certain restrictive or otherwise adverse legislation in certain states, the explorationist often finds it economically necessary to explore for minerals in states with more reasonable regulations. We shall provide only one specific example here, but the explorationist is encouraged (perhaps "warned" is a better word) to gather information concerning exploration and mining in any specific state before undertaking an extensive exploration program.

We have observed recently that counties within states are beginning to clamor for a voice in the regulation of exploration (and, of course, mining) activities. For example, a law was recently passed in Colorado giving authority to the various counties to designate mining areas and to regulate activities associated therewith, including exploration drilling. This means that in addition to federal and state bureaucracies, the explorationist must deal with county planning commissions, county zoning boards, county commissioners, and even the county health department if a question about contamination of ground water is raised. The county agricultural agent may also become involved if grass or crops are disturbed by exploration activities. We would much rather see state-wide regulation instead of county-by-county ordinances which likely will produce such a jumble of laws that it will become almost impossible to function effectively.

Tax regulations pertaining to mineral exploration and development have become more complex in recent years as a result of new Internal Revenue Service codes and rulings concerning deductible items, expense items, capitalized items, depletion, and so forth. Considerable expertise in tax law is required to comprehend all of the ramifications, many of which can have an important bearing on the conduct of an exploration and development program. Accountants can provide tax advice, but the attorney representing an exploration firm

should have a firm grasp of tax laws pertaining to mineral exploration and production, since he is usually the immediate source of advice. In addition, the attorney usually has a broader perspective of the entire program than does the accountant.

Another reason for legal advice is the ever-present possibility of lawsuits. In mineral exploration (as in any other business), litigation should be avoided at almost all costs; in exploration, a good working knowledge of mining law by the explorationists themselves can be invaluable in this respect. The old saying "a gram of prevention is worth a kilogram of cure" is very applicable here. Lawsuits produce many adverse effects, perhaps the worst of which is the loss of time. That, along with the fact that lawsuits can be very costly, is often an incentive for the parties to settle out of court if possible. Because of the nature of exploration activities, and the diverse, sometimes conflicting, personalities which can be involved, disagreements or misunderstandings tend to arise which may result in a lawsuit. Later in this chapter some aspects of lawsuits and specific case histories will be cited to illustrate this point.

Pitfalls

Most pitfalls in the mineral exploration industry are related to field work. Persons and vehicles are often on the ground in a given area performing various kinds of work, and this is usually where complaints for damages or misunderstandings with landowners and others occur. We mention here just a few items which can lead to problems in the field.

Drilling wrong location. When carrying out exploration drilling in the field, the explorationist occasionally may become disoriented, or perhaps mistake a distance or direction, and thus drill a hole in the wrong location. This is not a serious legal problem if the hole is drilled on property controlled by the explorationist. However, if the hole should be drilled on fee mineral land not held under lease, or on other uncontrolled property, then some adverse consequences can result. For example, the explorationist may have attempted, unsuccessfully, to lease a certain property. A hole might later by error be drilled and probed on that unleased property. The fee mineral owner, discovering the hole, can charge the exploration company with trespass and can sue for physical damages which might have resulted to the property. In addition, he can sue for damages to the mineral *potential* of the property, charging that the company which drilled the hole illegally evaluated minerals on the property to the detriment of the mineral potential of that property.

We experienced a situation of this type in Wyoming, where we had located mining claims in an area consisting of public domain except for one small, narrow strip of fee land across a small valley within the federal mineral area. Unfortunately, there was some confusion at the time the staking crews were given instructions to stake the area, and they staked mining claims across the public domain *and* the fee land. Under Wyoming law, and in accordance with federal law, it is required that a discovery be made on each claim. In Wyoming, 15 m (50 ft) of drilling (with each hole a minimum of 3 m [10 ft] in depth) may constitute validation. After the discovery monuments were erroneously located on the fee land, a drilling rig followed the staking crew and drilled one 15-m (50-ft) hole on each claim near the discovery monument. Four of the claims were on the fee mineral land, and four holes were drilled on this property. The landowner discovered that the drilling had been carried out on his property before we learned of the error. He immediately contacted his lawyer, and we were

placed on notice that unless some kind of settlement could be reached promptly, a lawsuit would be filed. There was no denying that we had done the drilling on the property, and even though we had attempted earlier to obtain a lease, we had nevertheless performed work on his land without his consent. After the landowner's lawyer had lectured us about our "lust and greed in acquiring land," we were able to settle out of court for $1,500. This figure was probably less than the legal fees would have been had the case gone to court, and we undoubtedly would have lost the case. As it turned out, we later disproved the potential of the area and abandoned our exploration efforts there. The landowner held no hard feelings over the incident, nor did we; we later leased his land for oil and gas exploration.

Accusations of damages to wells or aquifers. Another risk the explorationist runs when conducting field work is that a rancher or farmer may claim that drilling activities in the general vicinity of a water well or spring resulted in damage to that well or spring. Most leases used in the uranium industry state that no drilling may be done within a certain distance of a dwelling (usually 60 m [200 ft]). In the course of a drilling program, exploration holes are sometimes drilled in an area where there is a well for watering livestock; other times, drilling may be carried out within 150 to 300 m (500 to 1,000 ft) of a dwelling which has a nearby well. Most ranchers and farmers object to the use of explosives in drill holes (a practice frequently followed in seismograph operations) close to an existing water well or dwelling, but most of them also realize that simply drilling a hole (which is later to be plugged) near an existing water well does not have any impact on that well.

Once, when we were exploring for uranium in an area in Wyoming where the surface was quite dry, there was a rancher living several miles away who had a surface grazing lease on a section (256 ha, or 640 acres) of state-owned land. A water well had been drilled in the center of this section, and a windmill on the well was being used to pump water for cattle which grazed here. With the permission of the state (since we had a state uranium lease), and after advising the rancher, we drilled a hole near the south quarter corner (center of the south line) of the section, 0.8 km (.5 mi) south of the water well. In four hours' time, we drilled to a depth of about 240 m (800 ft), probed the hole for the usual electric log (resistance, SP, and gamma), logged the samples, and then plugged the hole. The dip was flat in this area, and it was apparent from the results of our drill hole that there were no good aquifers above a depth of 180 m (600 ft).

Three weeks after we had drilled, plugged, and abandoned the hole, we received a call from the rancher claiming that his well had begun pumping mud and then had become almost dry immediately after we had drilled our hole. He stated that the well had produced 90 to 110 l (25 to 30 gal) per minute before our drilling, but now produced only 4 to 11 l (1 to 3 gal) per minute. He insisted that our drilling had damaged his well (he was unable to explain how), and he demanded either that we pay damages or that we drill a new, deeper well. When we investigated the situation, we found that his well had been drilled to a depth of only 105 m (350 ft) in a stratigraphic section which was almost entirely shale. Further, we learned that the well location had been selected by a "water witch" using a willow branch. We suspected, after talking to the driller who drilled the well, that it had *never* functioned satisfactorily. We concluded that the rancher was simply attempting to have a deeper well drilled at our expense in order to

reach the lower, better quality sands which might produce more water. We gave him copies of the logs from the hole we had drilled, and advised him of our interpretation. Dissatisfied, the rancher hired a lawyer who advised us that, if necessary, a lawsuit would be filed over the matter for what he claimed were just damages. The rancher also filed a complaint with the state board of land commissioners, claiming that his well had operated perfectly before we had drilled and was no longer functioning satisfactorily.

We believed the rancher's claim to be entirely fabricated, and thus unjustified, and we refused to yield to the pressure. Drilling and casing a new well down to a reasonably satisfactory aquifer would have cost about $6,000. We presented our own case to the state board, whose members tended to agree with us. Wishing to be fair, we advised the rancher that he should contract the services of the original well driller to pull the rods from the well and reenter it with a drill or swab in an attempt to discover the problems, and we agreed to underwrite this cost, *provided* we were not held responsible. We suspected that the thin casing might have collapsed from age prior to our drilling. The rancher refused this approach, and the matter is closed as far as we are concerned. However, litigation may yet result. In a situation of this nature, we often prefer to go to court rather than to yield to pressure from someone attempting to receive damage payments which, in our opinion, are not justified.

On the other hand, one must be realistic in a situation involving a threatened or actual lawsuit. Defending oneself in a lawsuit can be costly—partly through legal fees but, more importantly, through time lost in dealing with the case in terms of preparation, testimony, responses, and so forth. Consequently, many disagreements are settled before a suit is filed; others are settled before a suit comes to trial. The explorationist must balance the cost of a lawsuit or settlement in terms of both money and time.

For example, let's assume that a landowner, or other party with a valid complaint, has threatened to sue or has filed suit claiming damages amounting to $15,000 plus legal costs. The complaint charges that heavy drilling equipment and water trucks utilized by the explorationist have weakened or damaged a privately owned bridge on the landowner's property. If an inspection reveals some damage has, or may have, occurred, the explorationist is well advised to meet with the landowner and attempt to work out mutually agreeable terms, perhaps in the form of a cash settlement, or repairs to the bridge at the explorationist's expense.

If the explorationist notices damage which has occurred to the landowner's property and reports that damage to the landowner before the landowner has discovered it, there is less likelihood of any legal action being brought against the explorationist. For example, some gates to fields or pastures are barely wide enough for drilling equipment to pass through. We once inadvertently tore down a gate post as we were moving a drilling rig from one field to another. We immediately dispatched an assistant driller to purchase a larger creosoted post and replaced the broken one, which was quite old. The landowner was most pleased with our considerate action.

Attempts to reach negotiated settlements with surface owners may sometimes backfire. One company exploring in Colorado encountered a particular problem with a surface owner: this company had executed a mineral lease with the mineral owner, and thus was operating entirely within the law in carrying out exploration drilling on the property. The exploration area consisted of gently rolling land with very sparse grass cover, and the only visible disturbance after the drilling was completed was the outline of mud pits dug with a backhoe and later refilled after the hole was drilled. Nevertheless, the surface owner began to harass the company; he complained to such an extent that the company decided to settle on mutually agreeable terms for "grass damage." The surface owner demanded $2,000; the company disagreed with this sum, and made a counteroffer of $500. The two parties finally decided to call upon the county agricultural agent to assess the damage, and they agreed to abide by the agent's decision. The agent visited the property and prepared a memorandum describing what he believed to be the actual damage in terms of dollars; this amounted to $14.20. The agent cited initial poor quality grass, minimal areas disturbed, and prompt grass regrowth. The company immediately sent the surface owner a check for $14.20 and considered the matter closed. The enraged surface owner claimed that the agent was favoring the company, and returned the check in shreds.

Several years have passed since this incident, and other companies are now operating on the same property, but these companies have agreed in advance to pay the surface owner $5 for each hole drilled. As a result, the harassment factor is minimized, although not eliminated. This particular surface owner happens to be an extremely difficult individual, who has even declared that a shotgun is ready if employees of the original exploration company should ever return to the property.

Livestock injury. Whenever persons and equipment are working in a pasture area where livestock are kept, there always exists the possibility of injury to the livestock. With mineral exploration, the most likely situation for livestock injury to occur is one in which drill holes or excavations are left open. Unplugged drill holes are particularly hazardous, because a horse or cow can easily step in such a hole and break a leg. To avoid this situation, explorationists should plug all drill holes promptly and adequately. A cement plug is most satisfactory; however, other types of plugs may also be utilized. (This subject is discussed in more detail in chapter 10.)

Another potential hazard to livestock is excavations or pits into which the animals might fall. In exploration, such excavations are sometimes made to provide rock samples for testing; mud pits are also frequently dug prior to drilling a deep hole. Excavations are also made in the course of mining ore bodies by open pit. Most states require erection of a fence around any kind of excavation into which livestock might fall and become injured, and we recommend this as a precautionary measure in any case. Erecting a fence is a relatively simple, inexpensive matter, and it can avoid many problems with landowners.

In the Rocky Mountain region particularly, a backhoe is commonly used to dig a pit for use in mixing mud and settling cuttings during the drilling of exploration holes. Occasionally these pits are left open overnight, either because the driller did not have time to move the drilling rig onto the hole location, or perhaps because a problem arose which demanded his attention.

An empty pit is hazardous enough, but a pit full of mud is even worse. We recall one instance in the Powder River Basin of Wyoming: we were conducting a widely spaced drilling program in a fenced field where some black angus steers were pastured. We began drilling a hole late one afternoon at a site where a mud pit approximately 1.5 m (5 ft) deep, 3 m (10 ft) long, and 1.2 m (4 ft) wide had been dug. The driller was unable to complete the hole before sundown, so the drill hole and the adjacent full mud pit were abandoned for the night. There were no cattle around the rig at the time, and no precautions were taken to fence the pit.

When we returned the following morning to begin drilling, an assistant driller happened to notice an animal's head sticking out of the mud in the pit (fig. 9-1). One of the steers from the pasture had fallen into the pit during the night. Its body was submerged in the mud, and only its head was visible. Since the animal weighed between 320 and 365 kg (700 and 800 lb) and was almost wedged in the pit, we were presented with quite a removal problem. Eventually, we decided to extend the pit, utilizing the backhoe to make an incline in front of the steer so that it might be able to walk out. This was accomplished, and after a few unsteady moments, the animal appeared to be recovered from the ordeal and soon rejoined the herd. In this instance, the rancher was on hand to watch the rescue. He was cooperative and appreciated our concern for the animal. Later on, he remarked that if he had wanted one of the steers in the pit there was practically no way he could have gotten one in there.

Animals are naturally curious, and when equipment of any kind is brought into a pasture, they can be counted on to investigate. Also, in the wintertime, they are accustomed to being fed from vehicles such as trucks. Bran, cottonseed hulls, ground-up paper, and other materials are often carried in paper or cloth sacks on the drillers' water trucks as lost-circulation material (see chapter 10). Care should be taken to insure that these sacks are protected, since cattle have been known to eat some of these materials if they have access to them, occasionally resulting in sickness and even death.

Explorationists usually are willing to compensate ranchers for any livestock

Figure 9-1. View from above of a black angus steer which fell or was pushed into a mud pit near a completed drill hole. Drilling mud is shoulder-deep in the pit. The animal was removed without injury by means of a ramp dug with a backhoe.

injured or killed as a result of their work an area. On the other hand, situations have occurred in which a rancher has hauled a dead horse or cow from another location and placed it in an excavation or mud pit while the drillers were not present. Demand has then been made for reimbursement for the animal; of course, the deceased animal was always "the best on the ranch." It is often difficult for the explorationist to prove that the death was not his responsibility.

Gates left open or fences torn down. In most areas where exploration is carried out, explorationists are obliged to pass through fences erected by ranchers or farmers in order to gain access to the lands to be explored. Most companies make a concerted effort to insure that gates are closed properly and that no fences are torn down, although errors occasionally happen. It is the explorationist's responsibility to see that all workers close gates if they are found closed, and leave open those which are found open (fig. 9-2).

A gate left open in error can pose a serious problem for the rancher: livestock can move into an adjoining pasture, where they might mingle with other animals. Even worse, the livestock could stray onto a highway, possibly causing the death of the animals or traffic accidents and injuries. On the other hand, if a gate is found open it should be left open, unless it is known that the landowner wants it closed. Sometimes a rancher may have grass available for livestock in two pastures but water in only one, and thus prefers the gate between the two to remain open.

In situations where a cattle guard or

Figure 9-2. Field personnel must exercise care in opening and closing gates and in traveling across land to keep damage to a minimum.

other structure is damaged, it should be repaired or replaced, perhaps even left in better condition than originally. This is part of maintaining good relations with landowners, and these efforts are usually greatly appreciated.

Wet ground. During exploration programs, particularly in a spring thaw or after a heavy snowfall or rain, the ground often becomes so wet that travel in the exploration area may cause severe rutting of the ground surface, especially in fields, but also on dirt or gravel roads (fig. 9-3). Explorationists should make special efforts to cease operations when ground which might be susceptible to damage becomes excessively wet. Naturally, there are times when traveling over wet ground cannot be avoided—for example, it may be necessary to move a drilling rig just a short

distance onto a paved or gravel road for travel to a distant location, or a member of the field crew may have to return to town. However, as a general rule, every effort should be made to keep travel to an absolute minimum under wet conditions; indeed, many environmental regulations now require this. Provisions relating to travel on wet ground are included routinely in regulations attached to prospecting permits for federal acquired lands in the United States.

Damage to crops. The explorationist occasionally must conduct exploration drilling in an area where crops are growing. Most leases include the stipulation that the landowner be paid for any damage to growing crops. In the case of severed minerals, such damages, of course, should be paid to the surface owner or the person owning

Figure 9-3. Wet ground must be taken into consideration in planning and executing exploration programs, because surface damage occurs much more easily under wet conditions. Much of the water in the left foreground of this photograph resulted from drilling a hole (which is being probed), but the standing water in the trail at right from a brief spring snowstorm suggests that ground conditions are marginal and that a shutdown may be necessary until drier conditions prevail.

the crops. Payment for damage to crops is also applicable in situations where the federal government owns the minerals and an individual owns the surface.

It is important that arrangements for crop damage payment be made *before* the drilling begins. For example, if the explorationist plans to drill in part of a wheat field, he should make contact with the person who owns the wheat and agree to pay for any damage to the crop. Let's assume that four different locations will be drilled in an area of about 8 ha (20 acres) in one corner of a section of wheat land. Even though damage would probably occur in only four small areas within the 8-ha tract, the explorationist might pay *in advance* for the total estimated harvest from the 8-ha. Most growers cooperate with ex-

ploration programs, especially since the explorationist can give assurances that only a very small portion of their crop will be disturbed. They may still harvest 80 or 90 percent of the crop from that particular tract, and, because of the advance payment, receive double payment thereon.

Concerning damage to growing crops, it is essential that a payment agreed upon by all parties be determined ahead of time. This should be a fair payment, based on the estimated market value of the crops which might be destroyed. Whenever possible, the explorationist should attempt to plan the drilling program after the crops are harvested or, in the case of wheat, in stubble strips lying fallow. Such procedures will avert most damage claims by growers.

Incorrect title. On occasion, a landman

may make an error when checking county records to verify mineral title on a particular tract of land. Thus, an explorationist might be found unwittingly drilling on unleased land—property on which no right to explore has been obtained. In such a situation, a settlement would have to be worked out with the rightful mineral owners, once their identity had been determined. Perhaps a mineral lease could even be obtained from the owners, if they are agreeable.

Finders' fees. Most companies engaged in exploration are obliged to pay a finder's fee on occasion. An individual or group may become aware of what is believed to be favorable mineral potential in a particular area, or may hear about an available mineral deal. This information may then be offered, either in written form or through a personal visit, to a company or an individual. The person initiating the information may propose that a certain deal be made, or may suggest that a certain area is a likely one for mineral exploration and development. If the recipient of the idea works out a deal with the suggested party or conducts exploration or development in the suggested area, the so-called finder might claim an interest in the deal or demand a cash payment, asserting that the activity, or the conclusion of the deal, resulted from his efforts as a finder. The courts have sometimes recognized the legitimacy of some finders' fees, depending on the circumstances.

Sometimes finders perform a valuable service. However, it should always be spelled out clearly in the *initial contact* with a potential finder whether the company has an interest in hearing about a possible deal which may be available through a finder. Some companies have established a practice whereby any party submitting a deal must sign an agreement stating that he is aware the company is conducting mineral exploration in various areas and that the company is under no obligation to pay any fee to any individual in return for bringing a deal to the company's attention. This agreement is significant in that it protects companies or individuals against the possibility of paying unjustified finders' fees. Figure 9-4 is a sample agreement form, called a disclaimer, which is often used when property submittals or ideas from possible finders are received. We suggest that all companies and individuals consider the use of this type of agreement when receiving submittals. (Such forms are not commonly used in the petroleum industry.)

Although there are legitimate circumstances for paying finders' fees, it should be understood that there are certain unscrupulous persons who attempt to use finders' fees as a means of collecting revenue or gaining an interest in a deal when such revenue or interest is not justified. A written agreement similar to the one shown in figure 9-4 should be signed before the recipient reviews a submittal. This procedure protects companies and individuals against the dishonorable finder.

One might justifiably wish to know the difference between a person who might be considered a finder and one who might be selling a deal (as discussed in chapter 8). The key factor lies in the control or ownership of property. The person who has an idea but has no property under control would be seeking a finder's fee in attempting to sell the idea. The person who controls property would be attempting to sell a deal, even though such a person generally has plans concerning the exploration or development of that property.

Avoiding Problems

We suggest paying special attention to three important aspects of exploration in the interest of avoiding legal problems.

ABLE EXPLORATION CORPORATION
(company name)
(company address)

LETTER OF TRUST (DISCLAIMER)

To whom it may concern:

I have information which may interest Able Exploration Corporation, its divisions
and subsidiaries (hereinafter collectively referred to as "Able"), relating to certain lands located in _____(sec., T., R.)_____
_____(county)_____(state)_____.

I submit the information voluntarily and freely state that Able did not request it
or in any way influence or induce me to divulge it. I realize that Able may have no
interest in the proposal or that they may already be aware of the data I have. I
understand that Able does not usually accept such information from persons not in its
employ and that it is not obligated to pay a fee, commission, or any compensation for
such information. Recognizing this fact, I now state that Able will not be obligated
to make any payment of any kind whatsoever, now or in the future, to me or to any
other person or firm for the receipt, evaluation and use of the information I am dis-
closing with respect to the lands referred to above. I have placed no limitations of
any kind, nor do I do so now, on what use, if any, Able may make of such information.
Accordingly, I understand and intend that Able shall have absolute discretion in the
evaluation and use of such information, and that by such evaluation and use no partnership nor joint venture agreement between Able and myself is implied. I do request,
however, that Able keep confidential for at least six (6) months from the date hereof,
the information submitted by me which was not previously known to it, or is not in the
public domain, and I acknowledge that Able shall have no liability to me if it fails
to comply with this request through inadvertence of its officers or employees.

No confidential relationship exists between Able and me and it is my intention and
understanding that no such relationship will be established by the submission of this
information.

This letter states my intention and the entire understanding between Able and me
regarding the above-mentioned lands and any information I may have given them, either
orally or written. I hereby agree to hold Able harmless from liability to me or to
any other person or firm for any problem resulting from my disclosure of said information to Able.

                              Yours truly,

Date: _____   _____
                                            On My Own Behalf/On Behalf Of
Address: _____

         _____   _____
                                            A Company Of Which I Am

         _____   _____
                                            And Duly Authorized to Represent

Figure 9-4. Letter of trust (disclaimer).

Careful planning. The exploration program should be carefully planned in the office, taking into consideration all possible problems which might arise. Persons responsible for the program should be well versed in the scope and purpose of the program and procedures to be used. The land situation should be carefully reviewed and, if field work is planned, surface owners should be contacted for permission to enter their lands. Appropriate state and federal agencies in the area where work is to be conducted should be advised of the program (if such notification is required). Attention paid to the land situation during the planning stages can avoid many later problems.

Adequate number of competent persons. Despite the most meticulous planning, unanticipated events arise or plans are changed in every exploration program, thus necessitating that certain decisions be made on the spot. Making proper decisions requires, first, that there be sufficient crew members in the field to oversee all operations, so that problem situations do not become unmanageable. Second, field crew members should be competent to make such on-the-spot decisions as are required in a field program. As plans change and events occur in the field, well-versed, competent supervisors should have the ability to make correct decisions, thus avoiding problems which might otherwise arise through careless, incompetent procedures or work.

Good legal advice. We stress many times throughout this text how vital it is to obtain and adhere to sound legal advice on any questionable aspects of a planned exploration program. If the program manager fully understands all of the legal aspects, then it is usually unnecessary to consult a mineral lawyer. Frequently, however, there are some areas where the explorationist may not fully understand the legal ramifications, or he may have a specific question relating thereto. In such a situation, to avoid problems which might waste time and money, a competent lawyer should be consulted before work is initiated. A reputable, experienced lawyer frequently has the answers to such questions immediately at hand, and one or two telephone calls to such a lawyer can often resolve uncertainties. The explorationist should take the time to see that any questions are answered satisfactorily; it is far simpler to obtain advice in advance than to correct a problem later.

For example, some old mining claims may exist in a specific area where an exploration group is interested in the mineral potential and the acquisition of properties. County records may indicate that regular assessment work was recorded several years ago, but none for the past three. A ground check reveals no evidence of recent activity. The following questions should be asked of the lawyer: (1) Should the ground be considered "open," and new claims staked as if former claims never existed? If so, what specific procedures should be followed? (2) If new claims are staked, what are the liabilities if the former claim owners make accusations of "claim jumping"? (3) If the former claim owners are approached before their original claims are jumped, what is to prevent them from hastening to the field to recommence work, once they know that another party has an interest in the claims?

Additional questions could be asked, but these are some of the more important ones. Experience is a useful guiding factor in such situations. Our inclination would be to contact the previous claim owners and ask whether they had abandoned the claims. If they promptly initiate work on the claims once again, so be it; perhaps we might try to work out a deal with them. It is almost impossible to prove that a claim holder has not performed the required annual assessment work.

In 1975, a case was settled out of court after approximately two weeks of testimony in U.S. District Court in Cheyenne, Wyoming. The case involved certain mining claims held by an individual prospector which had been jumped by a major mining company. The claims, which were in several groups in the Powder River Basin in Wyoming, had been staked in the 1950s by the prospector, who also carried out mining in the area. The prospector continued to visit and work on the claims each year, and affidavits of assessment work were filed each year for the claim groups. In the late 1960s, the mining company embarked upon a large-scale staking program which covered an area of many square kilometres, including the claims held by the prospector. The company believed the prospector had not performed the necessary assessment work (although he asserted that he had); further, the company believed the prospector's claims to be invalid because of lack of valid discovery.

Early in its program, this company could have made a favorable deal with the prospector by paying him a relatively small cash amount and assigning a royalty; instead, as time went by, the two parties perpetuated their disagreement. The company then made a discovery of a roll-front complex at a depth of about 180 m (600 ft) on the disputed property. A second mining company entered the scene and made a deal with the prospector; a suit was filed, and a large amount of time and money was subsequently spent. The litigation resulted in an out-of-court settlement whereby the overstaking company was obliged to convey some prospectively valuable mineral properties to the second company or risk relinquishing half of an ore body.

FINDING A GOOD MINERAL LAWYER

General Characteristics

Ideally, a mineral lawyer should have a solid educational background in mining law, extensive experience in mineral deals of various kinds, and a working knowledge of taxation pertaining to exploration and mining. As with other specialty legal fields, a mineral attorney should have the ability to use the English language skillfully, and should be adept at negotiation. He should be able to function efficiently as a member of a team which includes a team leader (company management), landmen, geologists, accountants, and all others involved in exploration activities. He must have the facility to influence decisions skillfully, while at the same time maintaining harmonious working relationships among the members of the group.

It would be unusual for an explorationist to delve into a lawyer's educational background before retaining him to work on a mineral deal. It is important, however, for a lawyer who practices in the area of mining law to have included mining law courses as part of his training. Many law schools in the Rocky Mountain states offer such courses, although not all law schools include this area of interest as part of their curricula. Obviously, any law student intending to pursue a career in mining law should attend a law school which offers specialized courses in this field.

It is important that a mineral lawyer have a reasonable working knowledge of mineral taxation. This knowledge is valuable for giving proper advice to clients in structuring deals, so that all parties can take advantage of tax treatment pertaining to the deal. Knowledge of tax laws also is important in negotiations. Some lawyers are also certified public accountants which, given the required experience, is a useful combination.

Just as a recent law school graduate could not be considered a good trial lawyer, a graduate with little experience cannot be a good mineral lawyer. There are many diverse facets inherent in the exploration industry which must be learned through practical experience. Ideally, a mineral lawyer's experience should include participation in negotiations; writing exploration and development agreements (and observing and evaluating how the agreements function); field trips to observe uranium exploration and mining and milling programs, and familiarity with each of these activities; litigation (although good trial lawyers are specialists in this field); mineral title examinations; and a working knowledge of the Internal Revenue Code pertaining to exploration, mining, and milling. Thus a lawyer recently admitted to the bar would most likely be inadequately prepared to establish an independent mineral law practice, unless he happened to be a former mining engineer or geologist who had gained the necessary experience before attaining a law degree. Otherwise, we recommend that a lawyer join an established firm engaged in this facet of legal practice in order to acquire mineral-related experience.

In our discussion of the assets and abilities of a good mineral lawyer, we would be remiss not to include a few words about the client-lawyer relationship from the lawyer's standpoint. A good working relationship can be established and maintained between client and lawyer as long as the client does not withhold information from the lawyer. In an effort to state the specific case to the lawyer, a client may sometimes reveal only that information which tends to indicate that his position is the correct one and that the other party is at fault. This practice can lead to wasted effort and extra expense, not to mention embarassment to the lawyer when it is revealed that the client presented only a partial or distorted description. It should be remembered that lawyers are bound by the Code of Professional Responsibility to keep in confidence any matter which a client may relate in a lawyer-client conversation.

Sources of Information

For the explorationist interested in finding a good mineral lawyer, there are several sources to consult for information. An initial possibility might be business associates. Generally speaking, if you are engaged in mineral exploration, you undoubtedly have some associates who have worked in a particular area, and these associates should be able to recommend attorneys to you. If you have no associates in the given area, we suggest telephoning a mining company in that area and asking to speak with the legal counsel. After you have explained your needs, the corporate attorney will no doubt be glad to make a recommendation, and one lawyer recommended by another is frequently a good choice.

An excellent source of information on lawyers in the Rocky Mountain region is the Rocky Mountain Mineral Law Foundation, a professional organization with headquarters in Boulder, Colorado, whose members practice throughout the Rocky Mountain region. The RMMLF holds several meetings each year, and publishes books and portfolios consisting mostly of mineral-oriented papers presented at its meeting by lawyers and others (Bloomenthal, 1976). Many of these publications contain valuable information about many different aspects of mineral law, and each lawyer-author tends to write about his own specialty. By reading through some of the papers, you can discover quite a bit about the various lawyers before contacting one you might select to represent you. For example, in March 1976, the RMMLF held a meeting in Phoenix, Arizona, concerning minerals on Indian lands. Papers were presented by lawyers, Indians, and industry representatives. If you were seeking legal advice for problems relating to Indian mineral lands, the publication resulting from that meeting would be an excellent source of information about lawyers knowledgeable in Indian affairs, not to mention the information concerning Indian lands.

The bar association of the state in which legal help is required is not a useful source of information. Bar associations are not permitted to recommend lawyers, nor may they provide specific information about a particular lawyer's area of interest.

Law schools in the region where you are working can be reliable sources of information. For example, if you required the services of a uranium lawyer in the Denver area, you might contact the University of Denver College of Law or the University of Colorado College of Law. If you speak with a faculty member or professor who teaches mineral law and tell him that you are seeking a specific type of legal help, he may be able to comment on certain lawyers who are practicing in that field.

Information for the final evaluation of a specific lawyer must come from the lawyer himself. Once you have received recommendations, it is perfectly acceptable for you to contact the attorney, arrange an appointment to discuss your situation, and at that time inquire about his background and qualifications. He should not object to giving you this type of information and should welcome the opportunity to discuss your position. There are certain instances, however, where lawyers may decline to undertake a particular case. For example, we have found that mineral lawyers working primarily on mining and exploration

deals or other aspects of the mineral industry, often avoid cases involving trial work where they must spend much time and effort representing a client in court. This is perfectly understandable, because there are trial lawyers who specialize in this type of work. Thus, a meeting with your prospective lawyer enables you to exchange ideas and become acquainted; there is no obligation on either part at this stage and, if it should appear that he does not fit your needs, you should advise him promptly and continue your search.

It is generally felt, and it is our experience, that better lawyers are to be found in large cities. They tend as a rule to gravitate to metropolitan areas, where they are better able to keep abreast of events and can usually establish a more substantial business. However, we recommend, whenever possible, hiring a lawyer who is familiar with the laws and regulations of the state in which you are working. For example, if you had your headquarters in Denver and had an exploration project in Texas, you probably should seek legal counsel in Texas, although such counsel could work in coordination with your Denver attorney. Lawyers working with mineral laws in a particular state are usually familiar with the procedures in that area. Also, they can often make recommendations if a particular type of help is needed; for example, lawyers are often acquainted with reputable landmen in their state.

LAWYERS' FEES

A lawyer's professional knowledge and ability are usually the result of an expensive education combined with continuing study, work, and experience in his career. There is no simple answer to the question of how a lawyer arrives at the fee he charges. A discussion of legal fees was included in volume 1 of *What Everyone*

Needs to Know About Law (Newman, 1975); the following paragraphs are quoted from this publication:

Setting a fee is more a matter of judgment than of formula. Too many subjective factors enter into the fee-setting process for it to be susceptible to precise analysis. The Code of Professional Responsibility lists several factors as proper for consideration in fee-setting. These include:

Time, labor, and skill involved in a particular case. Time is simply the number of hours a lawyer spent working on a case. Labor and skill are closely related to each other, and the lawyer's fee will reflect the amount of each required of him by the particular case. And, of course, an attorney with thirty years' experience and a national reputation will charge more than a young lawyer six months out of law school.

The sum of money involved and the benefit to the client. If a lawyer spent ten hours of office time and one day in court to save his client $30,000, he will charge more than if he had spent the same amount of time to save his client $3,000.

Customary charges of the bar. This element is particularly important in the more "routine" types of cases, such as drafting an uncomplicated will, or handling an uncontested divorce, a no-asset bankruptcy, or an adoption. If the case is more complicated, the lawyer may use the customary fee as a minimum and charge an additional amount for the extra time and skill required of him.

The contingency or certainty of the compensation. Generally, if a lawyer is plaintiff's counsel in a negligence suit, he gets paid only if he wins.

Naturally, he will charge more in such a case than if he got paid win or lose.

The regularity of the employment. Like many shopowners, a lawyer may give a better deal to a steady customer.

Another factor which contributes to the size of a lawyer's fee is his overhead. One estimate holds that operating expenses account for about 40 percent of a lawyer's gross income. To achieve a before-taxes income of $24,000 a year, a lawyer would have to make a gross of about $50,000 annually. Since lawyers average about 1,300 fee-earning hours (hours of work chargeable to a client) per year, a lawyer would have to charge his clients about $30 an hour to net $24,000 in a year. Most successful lawyers, especially in large cities, have a net annual income greater than $24,000.

Lawyers usually do not come cheaply, although the cost must be considered in light of the benefits rendered by the lawyer's services. In any case, the client should be sure to discuss the matter of fees with his lawyer before any legal advice is given. A full and honest discussion about fees beforehand will prevent needless disagreements and hard feelings later.

Most lawyers who are working on contracts or negotiations charge an hourly fee. In the Rocky Mountain region, this fee runs from a low of about $35 per hour to a maximum of about $120 per hour.

Trial lawyers, on the other hand, often base their fees on the difficulty of the trial, the time required, the amount of money involved (if any), and the ability of the client to pay the fee. Some trial lawyers require an advance payment based on an estimate of their total fee for handling the case. Such an advance may not be required if the client is a large company or person with substantial assets, but in the case of an individual with modest assets, or a small company, payment in advance is frequently required. If the trial or its preparation is expected to be lengthy, the prepayment could amount to many thousands of dollars. Presumably, prepayment is required so that, in the event of an adverse ruling by a judge or jury, the lawyer will not be obliged to sue a client for payment of fees.

In our dealings with lawyers over several years in the business, we have found that it is more costly in the long run to hire an inexperienced lawyer who might charge a lower fee than to hire a well-experienced attorney whose fee is higher. The more expensive experienced lawyer may prove more economical because he usually operates more efficiently, can work faster, and is likely to have a better organizational setup. Overall, you should seek a lawyer who is well trained, well experienced, and competent in his field.

HELPING YOUR LAWYER HELP YOU

Assuming that you have gone through the search and selection process and have retained a lawyer in the field of mineral exploration, you may now wish to discuss a pending deal. One way in which you can facilitate your lawyer's work is to prepare a typewritten draft outlining the basic aspects of the deal. In fact, if the deal is similar to a previous one, you might photocopy the earlier agreement, make appropriate changes on the copy to reflect details of the new deal, and present it to your lawyer as a prototype for the new agreement. After reviewing this draft, the attorney can then make suggestions for changes or deletions in accordance with your wishes.

We have found that when we take the initiative on this first step, most lawyers are able to complete the job faster and more accurately than if they had been obliged to prepare the agreement from scratch. We certainly do not wish to imply that the client should dictate the path which his lawyer must follow; rather, there should be a meeting of the minds and agreement on a course of action.

When submitting written comments to your lawyer, be sure to include all pertinent material, such as copies of correspondence to and from the parties involved in the deal (or the dispute), together with a memorandum containing your views on the matter. You should also describe the other party's position as fully as you can, so that your lawyer may understand both sides of the story. Again, you should not depend on your lawyer to do the work from scratch; do some of the preparatory work yourself, and the process will move forward more rapidly and efficiently. This procedure can save time and money. You are usually in the position of knowing many details which may not come to light unless the lawyer questions you intensively. Obviously, you need not write to your lawyer every time you require some advice; many matters can be handled briefly and effectively over the telephone. However, significant topics such as litigation, a pending deal, negotiations, and so forth benefit from accurate, complete written descriptions prepared by the exploration-ist-client.

If you should become involved in litigation, you can assist your lawyer by preparing a chronology—that is, an itemized list of all events relating to the particular case, from the earliest to the most recent. The chronology should include dates, names of persons involved, details of what transpired, and any other data—pro or con. In complex litigation, where events may have taken place over a lengthy period, the chronology can become a major undertaking, consisting of perhaps 75 to 100 typewritten pages.

Preparation of a chronology can be greatly facilitated if a diary is maintained with daily entries. We strongly recommend that all persons engaged in exploration activities maintain a diary in which all relevant events are documented, including the exact location, the nature of work performed, and names of persons with whom they talked, giving a reasonable amount of detail. A record of expenses may also be kept in the diary, perhaps at the bottom of each page.

There are convenient commercial diaries designed with one page for each day of the year, allowing adequate space for recording details of each day's activities. A popular model is the *Day at a Glance* diary, published by Eaton Paper Division of Textron, Inc., in the United States, and available at most stationery stores. The pages of these diaries, which may also be used for daily appointments and memoranda, are divided into quarter-hour segments in the margins (although we seldom use the time designations). These handy booklets measure 13 x 20 x 2.5 cm (5½ x 8 x 1 in.) and have a flexible cover.

Although daily recording is often possible, the explorationist may sometimes find that three or four days pass without an opportunity to record the daily events. Such lapses are tolerable, provided they are kept to a minimum—no longer than, say, a week. If events are not promptly documented, loss of detail results. Information recorded in a diary can be extremely valuable for recalling mineral exploration events, especially where litigation is involved. The diary enables you to review where you were, what you were doing, and to whom you talked. Although not every detail may be recorded, it is possible to determine the general course of events and thus reconstruct the daily happenings. We

again stress that all persons involved in mineral exploration should maintain an adequate diary on a daily basis.

Some explorationists use a second booklet, similar to the diary, as a planner in which they note upcoming appointments, meetings, field trips, and other future plans. The diary and the planner can be used in conjunction, one for recording future activities and one for recording past events.

SOME PROBLEMS WITH LAWYERS

Lack of Education or Experience

Occasionally, lawyers representing persons or companies with whom you are dealing are ill prepared for the task. Their knowledge of the subject matter may be insufficiently broad, due to lack of education or experience or both. In situations of this nature, it is helpful if your lawyer can explain the subject matter to the other party's legal representatives. For example, the other party's attorney may fail to comprehend the practical aspects of an exploration project, may not be well versed in mining and reclamation activities, or may lack an understanding of IRS codes for mineral taxation. Such inadequacy can be offset, to a large extent, if your lawyer is familiar with and can explain the subject matter. If your lawyer is unable to apprise the other party's attorneys sufficiently for them to understand the matter at hand, a substantial delay may result while the subject matter is researched.

Despite the requirement of the bar examination, incompetence does exist in the legal profession. You may find some lawyers unable to perform adequately those services for which they have been hired. Large mining companies are usually represented by competent legal counsel, but such corporate attorneys are not always knowledgeable. We have also had dealings with lawyers working for smaller companies whom we found totally incompetent to handle the subject matter. We fail to understand why a well-informed company or individual might hire a lawyer to work on mineral exploration and development when that lawyer had no knowledge of the industry, did not understand this type of deal, and was unfamiliar with tax laws pertaining to the deal. It might be considered unethical to criticize the legal representative of the other party; however, such criticism is sometimes called for. If the client is at all perceptive, he should be aware that his legal counsel is not capable of performing the functions for which he was hired. An incompetent lawyer can create an unfavorable situation for a client, since incompetent work may result in the executing of a poor or inadequate agreement. Certain items may be included which do not properly belong and which should have been omitted; other points may be omitted when they should have been included. If the lawyer is incapable of handling a given situation, and the client is not sufficiently aware to know that the lawyer cannot handle it, adverse consequences are bound to result.

Delays

Another potential problem which explorationists may experience with lawyers relates to delays in the course of putting an agreement into final form. Such delays can occur when attorneys propose putting an oral agreement into writing, or agree to modify or act upon a draft which has been presented. The delays can attain absurd proportions: for example, a draft promised within a week to ten days may not be ready for examination until two or three months have passed. Lawyers representing large corporations tend to have numerous duties to perform concurrently, and any mineral deal (including yours) is just one among many items demanding their attention. Seldom does any one deal have high priority.

On the other hand, lawyers representing individuals or small corporations usually accomplish things more rapidly, especially since a mineral deal is likely to be the client's most important project. Thus, whenever possible, we suggest that the attorney representing the small corporation or individual prepare the initial draft of an agreement. This would then be submitted to the lawyers representing the other party, for their review and comments. Lawyers representing large corporations may not accept this procedure, because they usually prefer to prepare the initial draft, including much so-called boiler-plate language from other agreements. Also, corporate attorneys generally believe that they have a certain competitive advantage in writing the initial draft of an agreement because, in theory, they will be able to include certain phraseology which might come under close scrutiny if inserted later. We recommend that the lawyer representing the smaller party prepare the initial draft of the agreement with particular attention paid to the important points of the agreement. Anything inserted later by the corporate counsel, including "boiler plate," could (and should) then be carefully scrutinized by the smaller party and its legal representative.

In most mineral deals, unless there are extreme complications, a competent mineral lawyer should be able to draft an agreement to submit to the other party within 10 to 14 days from the time an oral agreement is reached. This would allow for a moderately complex agreement of perhaps 30 to 35 double-spaced legal-size pages, and it assumes, of course, that the lawyer does not have too many concurrent matters of a pressing nature. There are certain procedures which can be utilized to accelerate the process of preparing agreements. The effectiveness of any of the following tools, however, is largely dependent upon the strength of the negotiators.

1. A *monetary penalty* is charged if a draft or agreement is not prepared within a given time. For example, in negotiations with a prospective buyer involving legal work to prepare an exploration and mining agreement, a short, separate agreement could be prepared specifying that if the final agreement is ready for signature within 60 days there will be no monetary penalty. However, if it is not ready by that time, the buyer must commence purchasing options in 30-day increments, at a cost of $10,000 (or other figure) per increment, payable in advance. The buyer may wish to add to the 60-day period all those days during which the draft is in the hands of the seller, and therefore out of the buyer's control. A deadline such as this can often expedite considerably the preparation of an agreement.

2. A *time penalty* can also be used as an incentive to speed up preparation of the agreement. For example, if you are negotiating with a company which requires a two-year option on your property to conduct exploration, the option period should begin the day the agreement is being discussed. In such an arrangement, the time the lawyers take to work out the agreement is part of the period the buyers are granted to carry out drilling or other exploratory work on the property. It is not uncommon for some companies or individuals negotiating for (or buying) properties to use the draft and agreement preparation time as an additional free option period. If property can effectively be removed from the market for several months while an agreement is being prepared, that time can be used to advantage by the buyers to carry out additional studies, exclusive of physical work on the property which would require the owner's permission.

3. So-called *earnest money* could be paid at the outset to the seller. Such payment would remove the deal from the market for a specified period while negotiations are carried out and agreements prepared. The seller usually must have an extremely attractive property to accomplish this.

4. If the seller has a deal which is of interest to several possible buyers, he can *refuse to remove it from the market* until a final agreement is signed.

Negotiating by Lawyers

Before negotiations on an agreement commence, the negotiating role played by lawyers, if any, should be agreed upon in advance between lawyer and client. To a large extent, this determination should be dictated by the capabilities of both the client and the lawyer. If the client is capable of conducting and wishes to conduct negotiations on his own behalf, then the lawyer should simply make comments at the appropriate times, or answer questions; above all, the lawyer must not make any comments which might weaken or jeopardize a client's position. Some individuals prefer that their attorneys conduct all negotiating, within general guidelines, making only occasional comments on the direction of the deal.

During negotiations, it is especially valuable to be represented by a lawyer who is well versed in all aspects of exploration, mining, and milling for the particular mineral under discussion. The other party may raise certain points, or suggest reasons why he requires specific concepts included in the agreement; he may object to a particular point, and if you are not sufficiently knowledgeable to refute that point, your lawyer should be able to do so by explaining why a proposed provision has adverse ramifications. (This is discussed in more detail below.)

INSTANCES WHEN LAWYERS
ARE COMMONLY NEEDED

Exploratory work of a general reconnaissance nature, undertaken in an area where no land is under control, is sometimes referred to as a "raw material survey" or "prospecting" for accounting purposes. Prior to or during such a survey, you might consult your lawyer concerning the exploration and mining laws of the state or other political subdivision; you can then be aware of any potential problems or restrictions. For example, if you are considering mineral exploration in Mexico, you should arrange an appointment with your lawyer to discuss the matter, allowing him time to research the laws and regulations pertaining to that area. Additional conferences may be necessary as the raw material survey progresses.

Before undertaking land acquisition in a new region, you should confer with legal counsel on several items. First, you should inquire about laws governing mineral ownership in the area (or country) under consideration. If you are considering acquisition of land, the laws must be examined in greater detail. Regulations pertaining to exploration and development must be further reviewed, particularly if you are considering a foreign country. Procedures for land acquisition and exploration must be studied, including claim location procedures (if any), lease forms, concessions, surface-damage agreements, exploration plans required by law, environmental restraints, and export or import restrictions on machinery and commodities. Needless to say, competent legal assistance in the area where exploration is planned is essential.

As an exploration program progresses, your lawyer should be kept informed of developments. If a discovery is made, you or your lawyer may wish to initiate a title examination of properties where the discovery was made. The title examination is usually not performed by your lawyer, although he may supervise the preparation and approve the final report. If curative work is required, your lawyer should advise you and suggest the steps to be followed.

If you are undertaking an exploration project and might later wish to sell the deal or bring in a partner, you should discuss with your lawyer at an early date the kind of deal you have in mind. For example, if you are considering selling the properties, you might review the terms, including price, royalties, timing, options, and so forth, and ask for your lawyer's comments. He may be able to point out potential problem areas to be attended to, or a different approach to the deal may be suggested, especially concerning tax aspects.

Mineral exploration ventures frequently are lacking in venture capital. The explorationist, especially a small company or an individual, must often attempt to raise money. In the past, unscrupulous mining operators have been known to raise funds illegally or to abscond with the capital. Consequently, many types of financing arrangements, particularly those involving the public, have now come under the jurisdiction of the SEC. If you are interested in the public approach for venture capital, a competent lawyer should either be able to advise you concerning securities laws or to direct you to one who can. In addition, your lawyer might be able to suggest other avenues to follow in attempts to raise venture capital for an exploration project. The early capital carries the highest risk in most exploration deals; consequently, it is the most difficult to find. Once a project has progressed to the stage where there are some reserves, or where data indicate it to be more than a mineral prospect, investment capital can be more readily found.

As exploration moves forward, one of

the lawyer's most important functions is to avoid lawsuits; frequently, it is the advice of a competent lawyer which prevents a lawsuit from occurring. In properly assisting an exploration company, the successful lawyer does not go to court to defend his client; rather, he prevents his client from getting into court.

The following comments were made by Maurice Mitchell, chancellor of the University of Denver, in an article which appeared in *Colorado Business* in 1975; they are appropriate in this connection.

Few businessmen remember, but there was a time long ago when being involved in a lawsuit was viewed as something of a social error. You just didn't "go to court."

Corporate lawyers handled routine incorporation papers and contracts, sat in on real estate closings and wrote occasional unpleasant letters to disagreeable people. As an added service, they handled wills and estates for top company officials. The fees were a minor factor in the corporation overhead. Institutional lawyers were not much busier.

Today's business enterprise pays dues to many segments of the world around it, but among the fastest growing is the cost of legal services. Some entrepreneurs wonder whether we aren't drifting into a phase of business existence in which the cost of legal advice and protection may be more than we can afford.

Here are just a few of the facets of business life in which expensive legal backup is now a built-in factor:

Government: Today's businesses exist in a rat's nest of laws, guidelines, regulatory agency requirements—local, state, regional, national and international. Most are written by lawyers to cover

a seemingly endless array of contingencies. Lawyers dominate the legislatures and express their concern for problems by writing laws and more laws.

Taxes: Although competent auditors can tell you what the tax advantages may be, a prudent businessman has his lawyer on hand, too. It is more common today to retain and use both than one alone.

Social Pressures: The cost of social change is felt more sharply by businesses than most other institutions. Anyone who has reviewed the requirements of the government for federal contractors knows this well: Safety procedures and equipment, affirmative action programs and pension plans all call for legal guidance on a scale that is mammoth compared to such needs three decades ago.

Unions: The contractual relationships between business and labor, often handled by trained negotiators, almost always end up being completed by lawyers for both sides.

Lawsuits: Going to court may have been gauche years ago, but it is a national pastime now. A businessman or professional can expect to be sued by anyone and everyone: a stockholder who lost a few dollars per share in the market, a competitor who thinks you have too big a slice of the market, a trust account customer whose portfolio didn't double his estate.

When a college student sues for the return of tuition because she didn't think she learned anything, it is fair to assume that the courts are today's sweepstakes opportunity for a certain segment of our society.

Perhaps the most troublesome of these few categories is the matter of lawsuits, especially the use of the courts by those who often seem to be engaged in fishing expeditions. Many feel that encouragement to harass is provided by the contingency-fee arrangement, which permits lawyers to engage in speculative litigation, often paying the expenses of the case in return for a high percentage of the amount awarded. The business-man-defendant, caught between the hazards of a courtroom showdown and his mounting legal costs, is all too ready to settle.

It may be concluded that foreign individuals and companies will often settle for terms leading to harmony, but those of us in the United States insist on confrontation and turmoil in an effort to achieve victory.

ACCOUNTING ASPECTS

As tax reporting and accounting procedures become more complicated, it is little wonder that the explorationist may look with bewilderment at these problems which must be dealt with in addition to complex geological and legal problems. We have found, however, that it is possible to function effectively in the mineral exploration field without getting overly involved in accounting aspects. Indeed, we would caution explorationists not to become so involved in accounting problems and procedures that geological work is neglected. Accounting is a specialty field, just as law is, but our experience has shown that the explorationist must have a good working knowledge of law because of legal problems and situations in the field. The working knowledge of accounting and taxes need not be in such depth, provided a specialist is available to perform the accounting functions.

All major mineral companies have accounting departments which handle not only payrolls, receipts, and disbursements, but also are involved in tax planning for exploration and development. The small company or individual will most likely call upon an outside accounting firm to prepare its tax returns, financial statements, and so forth. Because these services are not available internally, such firms may not be familiar with all the activities of a small company or an individual. They are not in a position to render tax planning advice unless a specific situation is presented for their study.

Consequently, the following summary of tax considerations for exploration and mining is presented as a brief review for the explorationist. These data have been summarized by Stanley Hallman of Elmer Fox, Westheimer & Co., Certified Public Accountants.

Deduction of Mining Costs for Tax Purposes

Most tax controversies have arisen over the distinction between *mining exploration* costs and *mine development* costs. The difference between the tax treatment of these items is described under the headings "Exploration" and "Development" below. One of the tests for determining the classification of an expenditure is the purpose for which the expenditure was authorized. Thus, expenditures paid or incurred in connection with drilling, where the purpose is to delineate the extent and location of the existing commercially marketable deposits and facilitate their development, are treated as development expenditures.

Mining activity falls within five functional groups: prospecting, acquisition, exploration, development, and production. Each of these categories is explained in further detail below (see also table 9-1).

Prospecting. Prospecting is the search for geological indications leading to acquisition of exploration rights in areas of further interest. The activity can range from general observation of industry activity and broad surveillance to extensive and

TABLE 9-1. SUMMARY OF EXPENDITURES INCURRED AND THEIR TREATMENT FOR TAX PURPOSES

Description of expenditure	Treatment for tax purposes
Prospecting activities (also called raw material survey)	
a. General geological and geophysical survey	a. Capitalize costs applicable to project and allocate ratably to the property units acquired
b. Options to lease or buy property	b. Capitalize, if property acquired
c. Rights to conduct geophysical work	c. Capitalize as part of geological and geophysical work
d. Salaries	d. Most companies consider this overhead and expense
e. Equipment	e. Capitalize and depreciate over useful life
f. Supplies for prospecting	f. Capitalize as part of geological and geophysical work
g. Geologists	g. If on staff, probably expense; if independent, capitalize
h. Geophysical crews	h. Capitalize as part of geological survey
	Expense all above items (except item e) if no property acquired
Acquisition activities	
a. Lease bonus	a. Capitalize
b. Delay rentals	b. Under IRC Section 266, may be capitalized or expensed
c. Lease broker commission	c. Capitalize and allocate ratably to property acquired
d. Abstract and recording fees	d. Capitalize
e. Legal fees	e. Depending on nature of work done, probably capitalize
f. Title search fees	f. Capitalize if property acquired
g. Legal costs of title defense	g. Capitalize
h. Ad valorem taxes	h. Under IRC Section 266, may be capitalized or expensed
i. Minimum royalties	i. Depending on nature of payment, probably can be treated as IRC Section 266 carrying charge and expensed or capitalized
Exploration activities	
a. Drilling of noncore holes	a-g. Exploration expenses may be expensed or capitalized and amortized pursuant to the provisions of IRC Section 617.
b. Drilling of core holes	
c. Tunneling	
d. Trenching	Above activities will incur costs for labor, administrative overhead, third-party charges, depreciation or rent of drilling or mining equipment, mining supplies, and access or support facilities for crews in remote areas.
e. Removal of overburden	
f. Sinking shafts	
g. Driving tunnels, drifts, or crosscuts	

TABLE 9-1 (*continued*).

Development activities

a. Expenditures are similar to those for exploration activities. The development function begins when the exploration activity reveals with reasonable certainty with respect to a given deposit the existence of commercially marketable quantities of ore.

a. Development expenditures are deducted or capitalized under IRC Section 616 at the taxpayer's option. If an election to capitalize is made, the costs are amortized over the productive life of the property. The costs are in addition to cost or statutory depletion allowable under Section 612 or Section 613.

Production activities

a. Utilization of facilities for extraction of ore in commercial quantities developed in exploration and development phase. Additional costs include:

   Royalties
   Depreciation of production machinery
   Maintenance and repairs
   Labor
   Overhead
   Supplies
   Restoration
   Professional fees
   Fuel, light, water, and power
   Interest

a. Expenditures matched against revenue from sale of minerals extracted

Public offering

a. Underwriting discounts and commissions
b. Registration fees
c. Transfer agent fees
d. Printing
e. Accounting fees
f. "Blue sky" qualification fees and expenses
g. Travel
h. Salaries
i. Other
j. Legal fees

a-i. No tax authority for expensing or capitalizing most costs; little litigation on this matter

j. Legal fees for organizing partnership have been held to be capital costs.

costly physical tests in particular geographical areas. The intent of prospecting is to narrow the search for minerals to areas of greatest promise by obtaining geological evidence that indicates the presence of minerals.

Costs associated with this function would include options to lease or buy property rights to lands for geophysical work, salaries of the prospecting group, and equipment with which to do the work. The salaries and supplies for prospectors in the field, as well as salaries and expenses for geologists and geophysical groups, are additional expenditures which may be incurred. For tax purposes, these costs are primarily capitalized as costs of an area study. If properties are acquired as a result of the work done, the costs are allocated ratably to the property acquired. If properties are not acquired, the costs are expensed.

Acquisition. After the prospecting function determines that a particular area has interest and the decision is reached to obtain property rights, the acquisition phase is entered. In this activity, the right to explore a property and recover any minerals discovered is obtained. Such right is attained by a lease or purchase of land and mineral rights or of mineral rights alone. Usually, a lease bonus for signing and a royalty on production is incurred at this time. The agreement would also include a provision for the payment of delay rentals which would preserve the rights in the lease over a period of years as long as the other provisions of the agreement are being maintained.

Expenditures incurred in this phase could include lease broker commissions, abstract and recording fees, legal fees, title search fees, legal costs for title defense, and ad valorem taxes or minimum royalties. For tax purposes, these expenditures are capitalized except that (under Section 266 of the 1954 Internal Revenue Code) carrying costs of the property may be expensed or capitalized at the taxpayer's preference.

Exploration. The exploration function is similar to prospecting. The objective is to find minerals, but exploration is distinguishable from prospecting because prospecting seeks an area of probable mineralization. The exploration activity probes the area of interest for specific deposits.

A favorable prospect is proven only by some means of physical access to the expected mineral body. Initial indications of minerals lead to an expansion of effort to define the prospect further to determine the likelihood of its being commercially productive. Exploration methods include the drilling of noncore or core holes, sinking of underground shafts, excavation of drifts or crosscuts, and the removal of overburden. Costs include labor, administrative overhead, depreciation or rent for drilling and mining equipment, mining supplies, and access or support facilities for crews in remote areas. For tax purposes, these expenditures are treated in accordance with Section 617 of the 1954 Internal Revenue Code as amended in 1969.

Section 617, as amended, provides that post-1969 exploration costs incurred in foreign areas are deductible up to $400,000. Foreign exploration costs in excess of $400,000 must be capitalized, and expenditures deducted are subject to recapture. Domestic exploration expenditures are deductible without limitation but must be "recaptured," that is, added back to the taxpayer's income under certain conditions. Generally, when income from production on a property has been attained, the recapture may be accomplished either immediately, by including in income all previously claimed exploration expenditures, or ratably, in which case the deduction for depletion with respect to the property in question will be lost to the extent that exploration expenditures have been previously claimed. In the event of a sale, the unrecovered expenditures are treated as ordinary income rather than as a capital gain.

Development. The development phase of the mining activity arises when exploration expenditures define a commercially productive ore body. Expenditures for commercial production require construction of access and material-handling facilities. The mineral body is opened to further mining by drilling, removing overburden, sinking of shafts, or driving of tunnels. Roads, dikes, primary cleaning or processing equipment, and field storage are required to move, separate, store, and prepare minerals for shipment. Support facilities (camps and so forth) may be required for the work force, particularly in remote areas.

Development expenditures are not completely distinguishable from either exploration or production costs. All three activities are frequently carried on at the same time. In some mines, the shafts and drifts required for exploration frequently provide a means of access for development of the ore body.

Much of the development expenditures are not recoverable in the event mining is not continued, because the expenditures consist largely of labor and construction equipment. Often, the costs of moving equipment exceed its salvage value and the equipment is simply abandoned.

The development phase of a mine continues until production on a commercial scale is obtained. Development expenditures can continue after commercial production is begun, because successive stages follow in a logical sequence—development, production, and then development of additional reserves. The staging of the expenditures avoids long periods of delay between

the outlay for tunnels and shafts and their subsequent use as production facilities. Usually development precedes production by a year or two.

Some development expenditures are made only as production progresses. Additional tracks, lighting, ventilation, and roof supports in the form of timbers, roof bolts, or pillars of minerals are examples of progressive development expenditures. Development expenditures such as wells, shafts, tunnels, and drifts are frequent extensions or augmentations of similar facilities that were provided by exploration. Such facilities can be equally useful for subsequent development and production.

The risk of failure on development expenditures is considerably less than that incurred in prospecting or exploration because a mineral deposit is known to exist. However, errors in original estimates of reserves and changes in market price or competitive conditions may make future expenditures uneconomical.

Under Section 616, expenditures incurred after the existence of ores or other minerals in commercially marketable quantities have been disclosed are classified as development expenditures. These expenditures are deductible on a current basis or may be deferred and amortized over the productive life of the property. The method chosen by the taxpayer would most likely depend on his other taxable income and the effect the choice would have on his allowable percentage depletion.

Production. Production is the commercial extraction of the ore outlined by the exploration and development expenditures. Facilities include tunnels, shafts, wells, and equipment or other facilities constructed during the development process, along with mining or processing equipment. Labor is a substantial part of production costs, especially in hard-rock mining, strip mining, and quarry operations. Production methods vary, depending on the mineral involved and its depositional location. Mining methods frequently require blasting or mechanical digging, which results in quantities of broken ore in a pit or underground mine that must then be transported from the mining area for further refinement and processing.

Elements of production cost include royalties, depletion or amortization of capital facilities provided during exploratory and development phases, depreciation of production equipment, maintenance of mines, facilities, and properties, direct and indirect labor, supplies, supervision and administrative costs, overhead, and the restoration of the landscape destroyed by mining. Costs of production are a charge against the income from the minerals being sold and are deducted on a current basis.

Planning Expenditures

Tax advantages resulting from the choice of particular alternatives are stressed in the subsequent discussion of particular problems. A few general fundamentals that will serve as a guide to many of the more common situations are set forth below.

1. Ordinarily it is desirable to a party receiving income to have it taxed as a long-term capital gain rather than as ordinary or depletable income. The reason is that capital-gain deduction is equal to 50 percent of the excess of net long-term gain over net short-term loss. Therefore, only about one-half of net long-term capital gain is taxable. In addition, the capital gains tax is limited to 25 percent of the first $50,000 of long-term capital gain. If the gain were not a long-term capital gain, the entire amount would be taxable at ordinary rates.

2. As a corollary to the foregoing proposition, it is ordinarily desirable to the "vendor" to have a transaction regarded as a sale rather than a lease or sublease, because any consideration received by the "vendor" in a lease or sublease transaction must be treated as depletable income. The capital-gain deduction of 50 percent will probably result in a lower gross income, and lower taxes than the depletion deductions, which are limited to 50 percent of taxable income from the property, that is, 50 percent of income from the sale of oil and gas or minerals. Treatment as a capital gain results in an immediate deduction of 50 percent. The depletion deduction, on the other hand, depends on the amount of minerals that is actually extracted and sold.

3. If income cannot be considered as capital gain, it is desirable for it to fall within the depletable income classification. The reason is that the recipient of depletable income can take a deduction for cost or statutory depletion, whichever is the greater, within the limitation that the statutory depletion deduction cannot exceed 50 percent of the taxable (net) income from the property. Although depletable income results in higher taxes than capital-gains income, it is still better than ordinary income, which is subject to the highest tax rates.

4. It is desirable to the taxpayer incurring expenditures for the development of mineral properties to have as small an amount as possible charged to capital expenditures amortizable through the depletion allowance. If such charges are capitalized, the taxpayer frequently realizes no tax benefit therefrom because statutory depletion can be taken in any event and does not depend upon the cost basis of the property. Statutory depletion is calculated on the basis of income produced by a property, not the cost of the property or production. There is no guarantee that capitalized costs will be recovered through depletion, because there may never be any income. It is advantageous to get an immediate deduction by expensing development expenditures rather than capitalizing them.

5. It is desirable for the taxpayer financing the development of mineral properties to be in a position to deduct the development costs (intangibles in the case of oil and gas; exploratory and development costs in the case of other minerals) as current expenses or as expenses prorated against production and to recover expenditures on physical equipment through depreciation. Otherwise, such expenditures must be capitalized by the taxpayer as part of the acquisition costs of the mineral interest and amortized through the depletion deduction. The deduction of development costs insures the investor of an immediate tax saving. If he capitalizes the costs, his only tax saving is the depletion deduction, which is not directly related to those costs.

6. It is desirable with respect to each separate property for 50 percent of the taxpayer's taxable (net) income (after deductions for all expenses other than depletion but including development expenses) to be equal to or in excess of the amount obtained by multiplying the taxpayer's gross income from that property by the appropriate statutory depletion rate. The reason is that the statutory depletion deduction for each property cannot exceed 50 percent of the taxpayer's (net) income from that property. If the statutory depletion exceeds that 50 percent figure, the excess cannot be deducted and is lost.

7. It is ordinarily desirable to taxpayers investing in mineral operations to be taxed as individuals (co-owners) or a partnership and not as a corporation. If taxed as a corporation, part of the income from the mineral operations will be subject to double taxation, and in addition, the taxpayer will be deprived of part of the benefit he would otherwise have received from the statutory depletion deduction. A corporation is a separate taxable entity. Income tax is imposed at the corporate level, and then the shareholder pays a tax on the corporate earnings that are distributed to him. Thus, there is a double taxation on corporate income distributed to shareholders. In addition, all dividends distributed to stockholders are taxed as ordinary income, even if those distributions include capital-gains income or depletable income.

A partnership is not taxed as an individual entity. All income is distributed to the partners, and then taxed at the individual level. Income distributed through a partnership also retains its character. That is, the income will be taxed at the individual level as ordinary, capital gains, or depletable income.

8. It is ordinarily desirable to have a transaction regarded as a tax-exempt exchange or sharing arrangement rather than a sale. The reason is the obvious one that no tax is paid with respect to such transactions, whereas a gain from a sale is subject to taxation. No gain is recognized on the exchange of "like kind" property held for productive use or investment.

We suggest the explorationist hire a good accountant or accounting firm to handle most of the accounting aspects. Those with experience in exploration and mining will be able to do the best job.

10
The Drilling Program

GENERAL DISCUSSION

Drilling is an extremely important part of almost any mineral exploration program. Drilling is the procedure we use to reach down into the earth to obtain samples and make various recordings in an attempt to decipher part of the geologic history of an area. The samples we obtain from this drilling, and the recordings we make of certain phenomena in the drill holes, are usually the only clues we have to enable us to analyze and determine what happened hundreds or thousands of metres below the present surface in rocks which may be 2 billion years old. Those of us in mineral exploration must put all the clues together to find that phenomenon of nature called a mineable ore deposit. We can ill afford not to glean every bit of information from every hole we drill. This chapter deals with the practical aspects of gathering that information.

At the outset, we wish to make it clear that our discussions will be concerned primarily with rotary, noncore, nondiamond drilling. Many, perhaps the majority, of exploration holes being drilled worldwide are rotary noncore holes. Our discussions further, of course, will be from a geological standpoint and consequently will not go into great detail about the mechanics of drilling, which in itself could require several volumes. For further information concerning drilling, particularly so-called diamond drilling, we recommend to the reader *Diamond Drill Handbook* by James D. Cumming and A. Percy Wicklund (3rd ed., 1975, published by J. K. Smit and Sons, Diamond Products Limited, 81 Tycos Drive, Toronto 19, Ontario). The book is 547 pages long and well worth the $20 it cost in 1976. The text treats the following subjects of particular interest to the geologist: mechanical features of a diamond drill, drilling bits, fluids (mud) for drilling, directional drilling, and deviation (including surveying of drill holes).

In any mineral exploration project, the drilling program must always be a compromise between the necessity of cost control and the desire for acquiring quality data. Administrators and exploration managers who lack a thorough understanding of geologic concepts are commonly inclined to stress the cost control aspect of a project; as a result, that project can suffer from poor data quality. Likewise, an exploration manager who is weak in geologic concepts can sometimes become carried away with "blind science," and the result can be excessive use of certain techniques. Thus, some projects are overburdened with expensive and excessive coring followed by unnecessary detailed petrographic work and numerous chemical analyses which do not focus on the problem of finding ore.

Many exploration programs are neglected once a budget has been provided. Important day-to-day decisions controlling financial expenditures are sometimes made quite casually by persons lacking experience and good judgment. We suggest that the status of the project geologist supervising the field geologists in charge of the drilling programs on a day-to-day basis should be upgraded throughout the industry; this would assure a better return on the overall exploration investment.

Another problem which persists in the uranium exploration industry, and maybe a common problem in exploration for other minerals as well, relates to the number of geologists who supervise drilling programs. Drilling can and often does get ahead of geologic planning; when this happens, wasteful drilling programs are the result. Exploration drill holes may be drilled too deep, too shallow, in the wrong location, too close together, and so forth. The exploration manager who tries to trim his costs by operating with an inadequate number of field geologists is making a mistake which will result in less than good geologic data being gathered from the drilling, and drilling funds will be wasted.

THE FIELD GEOLOGIST'S EQUIPMENT

In most exploration drilling programs, the field geologists supervise drill crews (industry terminology is "drill crew" and "drill rig," rather than "drilling crew" or "drilling rig"), probe operators, and, from time to time, survey crews. In order to carry out those functions, field geologists must be able to travel to drill holes in the field in a timely fashion in order to describe samples, examine mechanical logs, and make interpretations; all of these are necessary for planning new drill holes. Field geologists must also be prepared to do a certain amount of surveying, which is necessary to locate drill holes in some instances.

Vehicles

The field geologist should have a vehicle capable of traversing the terrains of the project area under the varying conditions which can be anticipated during different seasons. In most areas where uranium exploration is being undertaken, this means a four-wheel-drive vehicle. The geologist should have adequate protection for his maps and other equipment. Some field geologists use pickup trucks with a weatherproof box located in the forward part of the bed. However, we have found that enclosed vehicles are more suitable than pickup trucks for field geologists, because the maps and equipment are better protected from snow, rain, and dust and are more accessible to the geologists. Further, four to five persons can ride comfortably in an enclosed vehicle, while three is the maximum in a pickup (fig. 10-1).

Accessory Equipment

In most areas, it is advisable that the geolo-

Figure 10-1. Field geologist with four-wheel-drive vehicle and some of the recommended equipment: shovel, tow chain, and long extension jack which can also be used in conjunction with a chain and tree, post, or stake to pull a vehicle that is stuck.

gist's vehicle be equipped with such accessory items as a large jack (sometimes called a handyman jack), a tow chain or two, rope, a tool kit with a wide selection of wrenches and screwdrivers, tire chains, a shovel, jumper cables, flashlight, and first aid kit.

Radios

On many projects, two-way radios, capable of providing communication between geologists and the probe operator for distances of 20 to 30 km, can greatly enhance the efficiency of the operation. Such radios save a great deal of time, allowing the geologist to concentrate his efforts on locating drill holes and logging samples. On a project where several drills are operating, two-way radios consistently save substantial mileage on the part of the probe truck as

well as the geologist. Ideally, on a project involving four drilling rigs, two probe trucks, and five geologists, there should be seven radios on the job: one in each of the geologists' vehicles and one in each probe truck. In addition, of course, a base station radio should be installed at the exploration base with a range adequate to cover the field operation.

Other Equipment

Other equipment generally needed by field geologists to locate drill holes accurately includes a measuring tape about 30 m (100 ft) long, a Brunton compass, a hammer and an axe, brightly colored flagging, wooden laths, 5 x 5 x 45–cm (2 x 2 x 18–in.) wooden stakes, and stainless-steel-tape label makers (fig. 10-2).

Figure 10-2. Geologists verifying a bearing with a compass preparatory to staking a new drill-hole location between two holes which have already been drilled. The distance will be measured with the steel tape lying on ground.

It is important that field geologists be supplied with the best maps available for locating themselves and for plotting drill holes and geologic data. For the first phase of widely spaced exploration drilling, topographic maps at a scale of 1:24,000 are preferable. If these maps are not available in the areas being drilled, then the next best map available should be used. For lack of better maps, field geologists must sometimes used generalized planimetric highway maps at a scale of 1:126,720 for preliminary widely spaced drilling. Unfortunately, this may lead to inaccurate location of drill holes and errors in elevations of the drill holes. Intermediate phases, in which closely spaced drilling is carried out, should be plotted on maps of larger scales. If such maps are unavailable, and the prospect appears to have considerable promise, it may be necessary to have custom mapping companies prepare topographic maps from aerial photographs. Such services are quite expensive, however.

Emergency Supplies

When field geologists are working in rough and remote areas, they should be equipped with emergency supplies of food and water. If the weather conditions are potentially inclement, they should have foul weather gear for emergencies, including down-filled sleeping bags.

Field personnel should always be aware of emergency procedures to follow in the event an accident or unexpected problem arises. For example, in the Rocky Mountain states, a geologist or other person who is in a vehicle which gets stuck while a snowstorm is raging should *always* stay with the vehicle. Most deaths in such a situation result when someone decides to walk for help.

DRILLERS AND EQUIPMENT

In most uranium exploration projects within the sedimentary environment, drilling is done with a rotary drilling rig; the drilling medium is water, or water and additives.

The Drill Crew

In a typical program, the drill crew consists of a driller, who operates the rig, and two helpers. A competent driller will cooperate with the geologists and is usually enjoyable to work with; an incompetent driller should not be tolerated. We have known a number of exceptionally good drillers who have consistently made a concerted effort to drill holes on a timely basis and, at the same time, have collected good samples for the geologists. A competent driller will—

a. maintain his equipment in good condition;
b. use good judgment in setting up the rig in a level position, with a vertical mast, so that the hole is drilled as vertically as possible;
c. maintain a safe operation, so that he and his helpers and other miscellaneous personnel around the rig are not in danger;
d. maintain a good crew and train his helpers to insure that they do their jobs well;
e. insist that his helpers keep the pit clean so that cuttings are not recirculated when drilling with a steel mud pit (see below), and see that the sample catcher is cleaned carefully after each sample is collected;
f. insure that the samples are laid out where they will not be destroyed when, for example, the mud pit is dumped;
g. select the bit which is best for drill-

ing the hole and producing good samples;
h. change bits at the proper times to assure good samples and also good rates of penetration;
i. cooperate with the geologists in trying to attain a certain objective, watch for a certain interval, anticipate problems, and the like;
j. not try to charge the company fraudulently for bits not used, mud not used, hours not worked, and so forth (see fig. 10-3).

When the field geologist advises the driller that good samples are essential, the driller should take the responsibility to see that the drilling fluid is maintained in good condition in order to recover good samples. Drillers sometimes become negligent in some of their duties, making it necessary for the field geologist to assert control and insist, for example, that the drilling fluid be improved for better sample collection, or that the bit be changed if the driller is attempting to complete a hole with a dull bit.

Mediocre drillers have been known to set up to drill on significantly different locations than those staked and, in addition, without leveling the rig. Others may attempt to drill a hole without conditioning the mud, thereby collecting very poor samples, if any. There are other practices common among drillers which make it impossible for the field geologist to make good interpretations of the geology. The field geologist must assert his control and replace the incorrigibly bad driller.

On the other hand, we must recognize that drilling contractors are not on the job to enjoy the fresh air. They are there to make a profit, and most contractors today are drilling on a *metreage (footage) basis.* This means that the more they drill, the more they get paid. Naturally, if some of the procedures requested by the geologists

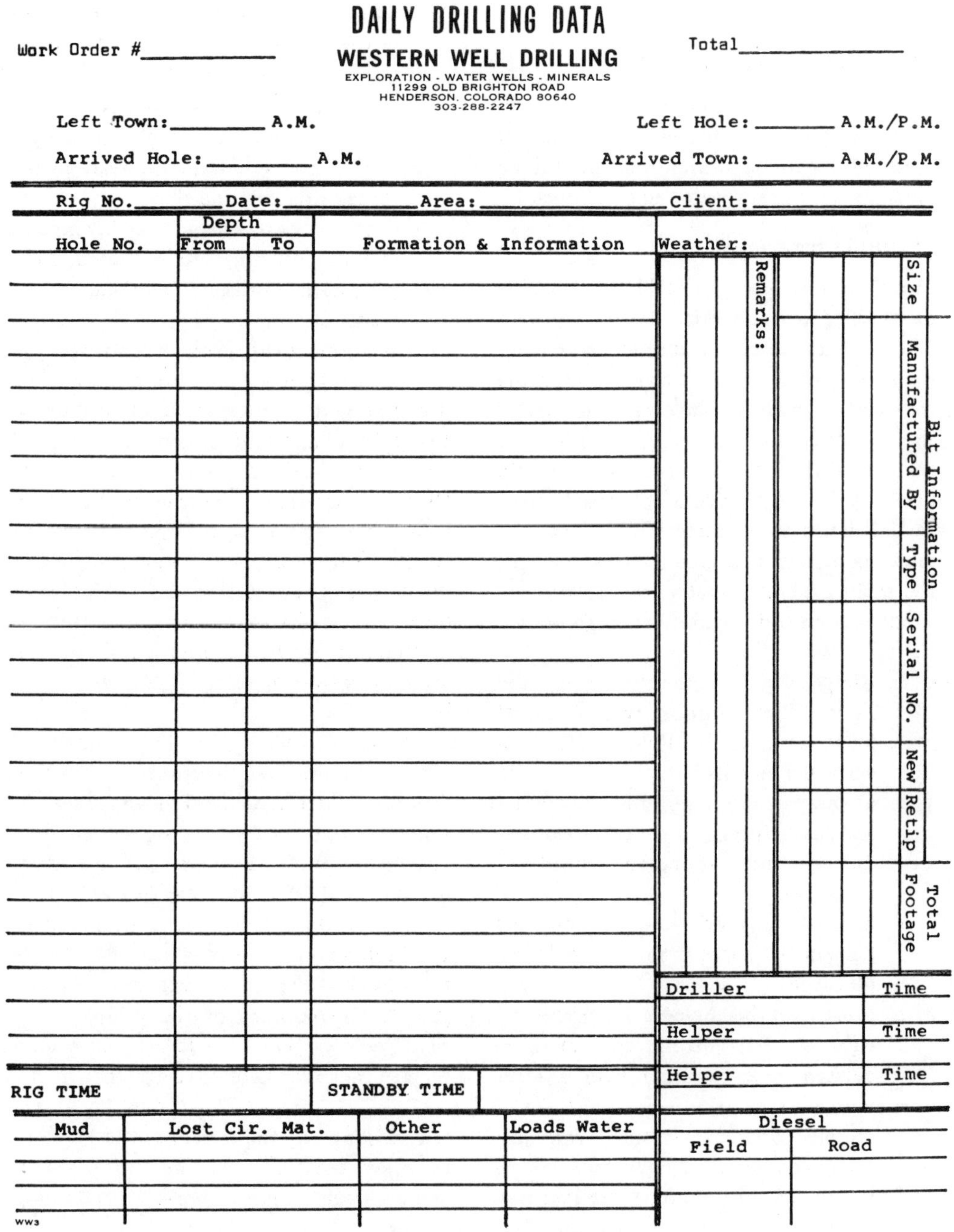

DAILY DRILLING DATA

Work Order #_____

WESTERN WELL DRILLING

EXPLORATION · WATER WELLS · MINERALS
11299 OLD BRIGHTON ROAD
HENDERSON, COLORADO 80640
303-288-2247

Total_____

Left Town:_____ A.M.   Left Hole: _____ A.M./P.M.

Arrived Hole: _____ A.M.   Arrived Town: _____ A.M./P.M.

Rig No._____ Date:_____ Area:_____ Client: _____

Hole No.	Depth From	To	Formation & Information	Weather:	Remarks:					Bit Information — Size / Manufactured By / Type / Serial No. / New Retip / Total Footage

Driller		Time
Helper		Time
Helper		Time

RIG TIME			STANDBY TIME		

Mud	Lost Cir. Mat.	Other	Loads Water	Diesel Field	Road

WW3

Figure 10-3. Type of daily drilling report form used by one drilling contractor. Such forms, when properly prepared, can be of significant value to the explorationist in analyzing drilling conditions and performance of the driller. Such forms are critical for hourly pay contracts (form courtesy of Western Well Drilling).

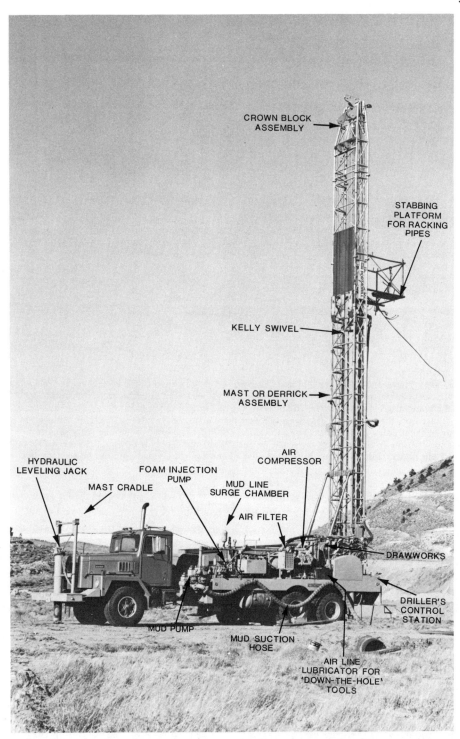

Figure 10-4. Drilling rig showing principal working parts. Courtesy of Portadrill.

Figure 10-5. View of the back end of a drilling rig just before a new drill hole is started, showing the portable steel compartmented mud pit in the foreground, with water gushing into the pit through a hose from the water truck (upper right). The hose on the left side of the mud pit goes to the mud pump centrally located on the rig. The driller is standing on the platform at left where the drill controls are located. One of the driller's helpers is kneeling to thread the drill bit on the end of the kelly. The rotary table is that part of the drilling rig just above the helper's head, and the draw-down chains are visible on both sides of the rotary table. On the upper right side of the rig, the drilling pipe can be seen stacked in the pipe rack.

slow down the drilling rate, the driller will not be happy about it. The geologist must keep this in mind and balance his needs with the necessity to allow the driller to make some good metreage. Good drillers are difficult to find.

Many drillers will get away with everything they can in order to maximize metreage drilled. If the geologists are not observant and do not take corrective action when needed, the poor practices go on. In our experience, there is no question that when poor samples are being collected, or the drill isn't properly leveled, or sample lag is excessive, the problem can be traced to an unobservant or careless geologist. Drillers are often very astute observers, and

if they see that the samples aren't being examined carefully, they won't bother to get good samples. If they see the samples *are* being examined carefully, they will get good samples.

Confidentiality of information must be maintained with drillers and their helpers. The geologist is sometimes deluged with questions about why drilling is being carried out, what the objectives and timing are, and so forth. Most of these come from curious helpers who are eager to learn all they can about something new. The best response we have found to such questions is that such information is dispensed on a "need to know" basis, and if they really need the information, it will be given to them. This will do away with almost all of the questions. Of course, drillers sometimes need to know of drilling plans so they can coordinate their own activities, other contracts, bits and other supplies, and so forth; most drillers restrict their questions to data they need, however, without prying. Drillers and helpers alike should *stay out* of the logging truck.

Drilling Equipment

In recent years, some uranium exploration projects have included drilling targets as deep as 1,200 m (4,000 ft) and greater. These exceptionally deep drilling programs require the use of equipment similar to that which is utilized in the oil industry. Such programs are not common, however, and will not be discussed in detail here. Most uranium exploration programs in sedimentary rocks are directed at targets ranging from the surface down to depths of about 600 to 750 m (2,000 to 2,500 ft). Drilling depths required for a particular project must be determined early in the program in order to select the proper drilling equipment. In most environments where uranium exploration is undertaken, truck-mounted rigs can drill exploration holes to depths as great as 750 m. For shallower drilling, ranging from 30 m (100 ft) to a maximum of 300 m (1,000 ft), a model 1000 drill rig is ideal (figs. 10-4, 10-5, 10-6). This rig has a minimal operating cost, and is generally well suited for shallow drilling. The "1000" designation refers to the depth capacity of the rig—about 1,000 ft (305 m). If most of the drilling to be conducted on a given project is more than 180 m (600 ft) and ranges in depth to 450 m (1,500 ft), a model 1500 drill rig is ideal. Truck-mounted rigs larger than the 1500 are available and sometimes must be called upon for drilling deeper holes in rough, hard-drilling environments (figs. 10-7, 10-8).

Figure 10-6. A good view of the controls and rotary table on a typical 1000 drilling rig. When this picture was taken, the drill hole had been completed and was being logged or probed through the rotary table. The probe truck is out of the picture to the right, and the cable to the rotary table can be seen silhouetted against the bottom row of pipe in the pipe rack. A pulley has been inserted in the rotary table and the cable passes over the pulley and down the drill hole. Drill holes are usually probed through the rotary table (with the driller being paid for by standby time) before taking a sidewall sample or when it is feared the hole might cave in. Drill holes with serious caving problems are often probed through the drill pipe, which is left standing in the hole to protect the probe. The mud pit has not been kept very clear in the drilling of this hole, as evidenced by the sandstone cuttings visible in the lower right corner of the photograph.

Most small, truck-mounted rigs are equipped with dual circulation systems which include both mud pumps for drilling with water or mud, and an air compressor system for drilling with compressed air. Most smaller drill rigs are equipped with a steel mud pit (fig. 10-9). Holes drilled to depths greater than about 150 m should be drilled with pits dug into the ground (dug

Figure 10-7. Drilling is under way as two helpers are kept busy shoveling cuttings from the steel pit. The hose on this side of the mud pit carries water to the mud pump (not visible) on the rig. From the pump, the water travels up the hose left of the driller to a swivel assembly at the top of the kelly, and then through the kelly and drill pipe to the bit. It is not unusual for drillers with small drilling rigs (1500 class and smaller) to complete one hole, move to a nearby hole, and commence drilling in less than 10 minutes. In soft formations with no drilling problems, it is not unusual for one drilling crew to drill three or four 150-m (500-ft) holes in one ten-hour shift.

pits) to give a larger reserve of drilling fluid (fig. 10-10). Large pits dug into the ground also accommodate the greater volume of cuttings which are recovered from deep holes.

When drilling with water, at least one water truck is needed for each drill rig. Some drilling contractors may attempt to operate with only one small water truck with a capacity of about 3,600 l (1,000 gal) assisting each drill rig; however, in most cases, this would not be an efficient operation. It is usually necessary to utilize a water truck capable of hauling in excess of 7,200 l (2,000 gal); otherwise, more than one water truck should assist each rig. Many good drillers use two water trucks

for each rig at all times unless drilling with air.

We have seen old drill rigs, properly maintained, operating very efficiently from day to day when an exceptionally competent driller is in command. On the other hand, we have seen new rigs mistreated and frequently broken down when a negligent driller is running the operation. It is essential that the drilling equipment be kept in good repair and properly operated. This requires a competent driller.

In a normal operation, most exploration holes range in diameter from 108 to 144 mm (4 to 5.5 in.). The average sandstone, mudstone, shale, and coal sequence can be drilled most efficiently with a drag bit

Figure 10-8. Driller (on left) and helper are shown here "making a connection," or adding a joint of drill pipe as they continue to drill a hole. The mud pump must be turned off before any disconnections are made.

(fig. 10-11). This bit has large, thick stationary blades which, when used in conjunction with a high-capacity mud pump on the rig, permit rapid penetration in soft formations. The cutting blades on most drag bits can be replaced with new ones, and these cutting edges are generally fabricated out of tungsten carbide steel, which is a hard alloy. This type of drag bit is frequently referred to as an insert bit.

When a driller encounters rock too hard to penetrate with a drag bit, he may be obliged to utilize a rock bit (fig. 10-12). This bit has teeth mounted on rotary bushings; the teeth rotate against the surface of the rock being cut. Rock bits are more expensive than drag bits, and they cannot penetrate soft sedimentary lithologies as rapidly as drag bits can. Rock bits also grind the cuttings into very small fragments which are not as easily sampled and described as the cuttings resulting from the use of a drag bit.

In igneous and metamorphic terranes, or in hard, well-indurated sedimentary formations, where the objectives are shallow or where shallow (generally less than 60 m) data are required in advance of a deeper drilling program, a wagon drill is frequently used. This is a small truck-mounted percussion drill which generally utilizes a 75-mm (3-in.) tungsten carbide bit and compressed air. Wagon drills are economical, and may be used to drill holes at specified angles to intersect steeply dipping veins or strata (fig. 10-13).

The diamond core drill is often used when a small-diameter core is needed in hard indurated sedimentary, metamorphic, or igneous terranes. This type of drill can core at specific angles to intersect steeply

Figure 10-9. With a bit on the end of the kelly and the back of the steel mud pit filling with water, the driller starts the mud pump and rotates and lowers the kelly to begin digging. The bit descends through a circular opening in the mud pit. In this view, the first water coming into the pit from the drill bit is visible. After the kelly is drilled down, it will be pulled from the hole and the first joint of pipe will be put on and the drag or blade bit (used in soft formations) added. The drag bit and coupler are shown here on the lower right platform.

dipping veins or strata. It is larger than the wagon drill, but smaller than the rotary drill rigs described above. It is capable of cutting a 35-mm (1 ^{3}/8-in.) core (fig. 10-14).

Drilling Contract

Any operator who plans to employ drilling contractors in exploration projects should first prepare a drilling contract. Operators who hire contract drillers on a regular basis usually develop a form for their drilling contract, leaving blanks where drilling rates and other variables may be inserted. An effective drilling contract should include provisions for both metreage rates and hourly

rates. The basis of payment for moving the rig and equipment—either mileage or hourly—should be stated in the contract. Provision should be made for standby time, and there should be a rate for coring.

The contract should also specify the basis for determining depths drilled and metreage payments. In the metreage contract, unless special problems exist, the driller is not paid for holes which do not reach the objective depth. The contract should protect the operator by specifying minimum liability insurance for the drilling contractor. The contract should also provide guidelines concerning hours to be drilled and minimal drilling conditions, to preclude operations which would make drilling too damaging to the surface of the ground.

The drilling contract should make it clear that the field geologist representing the operator will determine when and where drilling shall be undertaken. As a general rule, metreage contracts are better than straight hourly contracts for both the drilling contractor and the operator. A metreage contract with a reputable drilling contractor makes it possible for the operator to prepare reasonably accurate budgets and plan the finances of the drilling program with some reliability. Figure 10-15 is an example of a typical drilling contract.

Some exploration companies assign the responsibility of negotiating drilling contracts to people removed from the field operation (home office administrators). Problems resulting from this procedure could be avoided if the project geologist were given the authority and responsibility of concluding the agreement with the driller. The project geologist is usually in the best position to know working conditions, drilling problems, competence of the driller, going rates in the area, and so forth, and should be able to work out a good price.

Some unqualified drilling contractors will bid unrealistically low in order to be

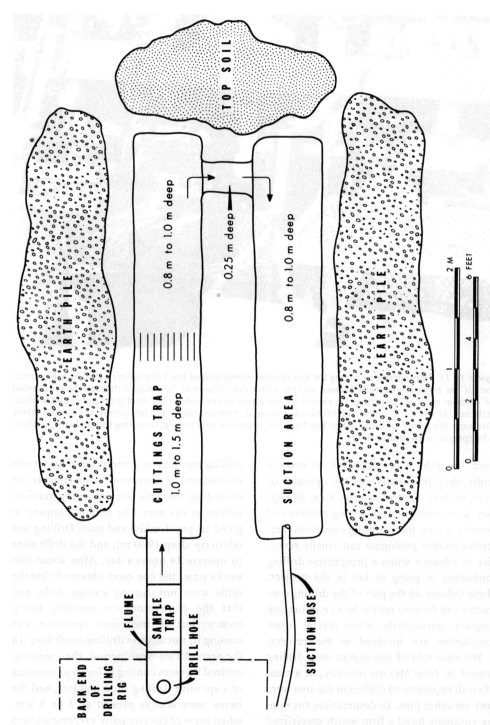

Figure 10-10. Diagram of dug pit, view looking down. Note that topsoil should be saved to be replaced after pits are back-filled.

Figure 10-11. Photograph of a drag bit and coupler on the end of the kelly before a drill hole is started. The bit has holes strategically located so that water can efficiently be pumped through the bit to wash the cuttings away. The cuttings are carried by the water up the hole to the mud pit, where most of them settle out. If the holes in the drill bit become plugged, pressure may build up quickly and cause a safety valve near the pump to blow. When this happens, everyone near the rig, including the geologist, is likely to be sprayed with drilling mud.

given the job. When they fail to make a profit, they frequently ask the operator to increase their metreage rate, even though they were continuously cutting corners and running a very poor drilling operation. Perceptive project geologists can usually recognize in advance when a prospective drilling contractor is going to fail in the project. These failures on the part of the drilling contractor can be very costly to an exploration program, particularly when several other contractors are involved in the project.

We were told of one urgent deep drilling project in New Mexico recently, at a time when all experienced drillers in the area were busy on other jobs. In desperation, the mining company hired a firm which specialized in water wells, but which had some new

drilling equipment. A metreage contract was worked out at a payment scale equal to, or exceeding, that paid to experienced uranium drillers in the area. The mining company agreed to pay for bits and mud. Drilling was relatively deep (900 m), and the drills were to operate 24 hours a day. After about two weeks time, the company observed that the drills were not cutting enough hole, and that the contractor was probably losing money; but an inefficient operation was causing it, not adverse drilling conditions. In the second two-week period, the company noticed invoices coming in for large amounts of expensive drilling mud and bits, and the items were *always* needed at 2 or 3 a.m. when none of the company's representatives were around. An investigation revealed that the drilling contractor was trying to make up losses from drilling by charging for unused mud and bits. Needless to say, this contractor was "run off," as the saying goes here in the western United States.

FIELD PROCEDURES BEFORE DRILLING

Preparation for Access

In areas where the drilling equipment must traverse privately owned land, the exploration operator must make arrangements with the respective landowners for the equipment to pass. If the project involves numerous negotiations of this nature, we recommend hiring a landman to handle the task. If the project involves only a few contacts with landowners, the project geologist or field geologists can usually make the necessary arrangements. If a landman is to be hired, he should report directly to the project geologist and/or the field geologists involved, so that there are no misunderstandings. In some cases, traversing private land necessitates the negotiation of a damage agreement with the landowner. This can be costly and time-consuming and may even require the service of legal counsel.

There are several categories of public domain. Some federal lands are administered by the U.S. Forest Service, and arrangements must be made with the Forest Service to bring drilling equipment onto these lands. Sometimes a bond must be posted to cover possible surface damages, and in most cases, the Forest Service lists stipulations which must be adhered to during the operation. Such stipulations usually include avoiding fire hazards, damage to the surface by traversing with heavy equipment when the land is wet, trenching, and the use of explosives (unless specifically authorized).

Figure 10-12. Two types of three-cone rock bits. The cones are mounted on separate bearings and rotate as the drill pipe is rotated. The drilling medium (mud or water) jets from the center of the bit. The upper bit is for drilling in moderately hard formations such as some sandy shales, red beds, salt, and soft limestone. The lower bit is designed for drilling in hard rocks such as chert, quartzite, pyrite, and granite. These bits are both designed for use on large drilling rigs. Photo courtesy Christensen Diamond Products.

Water Supply

Most drilling operations require a fairly substantial supply of water, which should be located before the operation begins. In areas where water is scarce, it may be necessary to bring in equipment such as a

Figure 10-13. A wagon drill being used for predevelopment drilling in advance of underground uranium mining in the Colorado Plateau area. This drill is not equipped to drill with water; it drills only with compressed air. In areas where substantial water would be encountered, this drill would not be able to function effectively. The sites of four completed drill holes (marked by wooden stakes) are visible in the bladed trail in the right foreground. Neat rows of sandstone samples (light colored) are visible to the right of the line of drill-hole stakes. Photo courtesy John B. Hall.

backhoe and prepare a water hole in advance of the drilling program. Problems such as these are usually the responsibility of the field geologist, and his solutions must not violate the laws of the state or the stipulations set forth by the landowner.

For most drilling operations, where more than one rig is involved, it is necessary that the water hole have the capacity to furnish water for several trucks which may each haul 7,200 to 10,800 l (2,000 to 3,000 gal). It is preferable if 14,500 l of water can be pumped from the water hole within a half-hour's time. The trucks should be able to back up to within 3 m (10 ft) of the water for pumping, without endangering the equipment (fig. 10-16).

In one particularly remote area of Wyo-

ming, where surface water was very scarce, we needed water for a drilling operation and found it necessary to drill and case a hole 120 m (400 ft) deep. A submersible pump was installed and a diesel electric generator brought to the site, and water was pumped into a retaining pond. A field camp was later established there, and we soon learned that the high-sulphate water was an excellent laxative which was impossible to become accustomed to. Coffee made from that water tasted and smelled like sulphur dioxide.

Access Roads and Site Preparation

The field geologist also has the responsibility of scouting for access roads in advance of the drilling program. A typical drilling

Figure 10-14. Diamond core drill making an angle hole at a site in Canada. Drills of this type are available truck mounted, skid mounted, and as components which may be transported to remote locations and assembled. Photo courtesy Longyear Company.

EXPLORATORY DRILLING AGREEMENT

THIS EXPLORATORY DRILLING AGREEMENT made and entered into
this _____ day of   by and between

_____

_____

_____

hereinafter referred to as "Owner"

and

_____

_____

_____

hereinafter referred to as "Contractor";

WITNESSETH:

WHEREAS, Owner desires to conduct an exploratory drilling
program for mineral resources other than oil and gas covering an area in
parts of Larimer, Weld, and Morgan Counties, Colorado which will involve
the drilling of approximately 200,000 feet of hole, consisting of
approximately 700 drill holes with an average depth of approximately 300
feet, with no holes expected to be drilled to a depth greater than 1000 feet.
Some coring or sidewall sampling will be done; and

WHEREAS, Contractor desires to provide the men and equipment

Figure 10-15. Exploratory drilling agreement.

necessary to carry out such drilling program, recognizing that Owner retains the right to discontinue such program at any time;

NOW, THEREFORE, Owner and Contractor mutually agree as follows:

I.   WORK TO BE PERFORMED.

Subject to the right of Owner to abandon the drilling program, Contractor agrees for the consideration hereinafter mentioned in Article VII hereof to commence on the date hereinafter mentioned in Article III hereof and thereafter to diligently prosecute in a workmanlike manner the drilling required by Owner to explore the properties owned or controlled by Owner in certain parts of Larimer, Weld, and Morgan Counties, Colorado at such locations and to such depths as the Owner may specify, provided, however, that Contractor shall not be required to drill any hole to a depth in excess of 1000 feet, and the total footage to be drilled is not expected to exceed 200,000 feet.

II.   CREWS AND EQUIPMENT.

Contractor agrees to furnish at the site where the drilling is to be conducted and maintain in first class operating condition one portable truck-mounted rotary drilling rig, capable of drilling a hole a minimum of four inches in diameter with air or water, to a depth of 1000 feet, equipped with air compressor capable of handling moderate flows of water, together with a crew of three for the drilling rig, and Contractor shall not operate more

Figure 10-15 (continued).

than 40 hours during any work week on a schedule adopted by Owner, unless mutually agreed to in writing. For all non-core drilling, Contractor shall furnish all tools, labor, supervision and fuel. Contractor shall furnish core barrels.

Owner may require the Contractor to discharge any employee deemed by Owner to be incompetent, careless, insubordinate or otherwise objectionable.

III.   COMMENCEMENT AND INSURANCE.

(a) Commencement. Contractor does hereby agree to commence drilling operations within seven (7) days after being notified by Owner at such points as Owner may designate, and continue diligent performance of the work, and to comply with all applicable laws, rules, regulations and orders of the State of Colorado, and of the government of the United States of America, or any sub-division or agency thereof including, but not limited to, the Fair Labor Standard Act and the Walsh-Healy Act, and to maintain and keep the property free and clear of any and all liens, claims or encumbrances of any kind whatsoever arising from or out of the operations of Contractor hereunder.

(b) Workmen's Compensation. Contractor agrees to insure at his sole cost and expense each and every employee and workman employed by the Contractor in the performance of the work, the compensation and benefits provided in the Workmen's Compensation Act of the State of Colorado and any other applicable

Figure 10-15 (continued).

statutory regulation of the State of Colorado and the United States of America relating to Workmen's Compensation, occupational disease, disability and/or employer's liability.

(c) Liability Insurance. The Contractor agrees to secure and maintain at all times during the term of this Agreement and any extension and/or renewals hereof, public liability and automobile liability for injuries to or death of persons and injury to or destruction of property with insurer's liability under the policies representing such insurance to be not less than

(i) bodily injury or death each person, $100,000; each accident, $300,000;

(ii) property damage on each accident, $100,000;

and furnish Owner with certificate establishing that insurance required in this Agreement has been acquired and is being properly maintained and that the premiums therefor are paid and specifying the policy numbers, names of insurers and the expiration date for said policies; all such policies and certificate shall provide that in the event of cancellation thereof that notice of such cancellation shall be given to Owner by the insurer not less than fifteen (15) days prior to the effective date of such cancellation.

IV.   PAYMENT.

Contractor shall invoice Owner at the end of each 30-day drilling

Figure 10-15 (continued).

period for all work performed by Contractor hereunder during the drilling period at the rates specified in Article VII hereof.

Owner agrees to pay Contractor all amounts due under this agreement within 30 calendar days after receipt of Contractor's invoice for such portion of the work, provided Owner is satisfied that Contractor has paid all of his labor and material bills.

V.   LOGS AND PROBING.

(a) Contractor shall maintain accurate and complete written driller's logs in connection with each hole drilled in performance of the work and deliver the original of such logs to Owner. Such driller's logs shall be on forms approved by the Owner and supplied by the Contractor. As and when requested by Owner, Contractor shall gather and retain samples of materials penetrated by the drilling at depth intervals specified by Owner. Samples will be caught in a metal sample trough at footage intervals specified by owner and will be delivered to Owner's representative at drill site in first class condition.

(b) Contractor shall notify Owner verbally not less than one hour in advance of the estimated completion time of each hole drilled in connection with the work to enable Owner to have logging equipment available upon the site of each hole upon completion of the hole. If Contractor should encounter

Figure 10-15 (continued).

adverse ground conditions which might prevent the hole from remaining open, Contractor shall immediately contact Owner's geologist, who shall determine the method to be used to probe the hole.

(c) All information obtained by the Contractor in connection with the materials encountered in performance of the work shall be kept strictly confidential and Contractor shall disclose such information only to Owner or its authorized representative.

(d) Contractor agrees to pay promptly and in full the claims of any and all persons, firms or corporations performing labor upon or furnishing equipment, materials, supplies or power used in or upon or about the performance of or contributed to the work described in this agreement.

VI.   PROTECTION OF ENVIRONMENT AND ECOLOGY.

Contractor agrees to be responsible for its employees, agents or representatives in protecting the environment and ecology of the work area by adhering to, but not restricted to, the following:

(a) All applicable laws and regulations.

(b) All instructions of Owner and appropriate governmental authority.

(c) All applicable provisions of surface agreements communicated to Contractor.

(d) Vehicular travel on areas where there are no existing

Figure 10-15 (continued).

roads shall be kept to a minimum.

(e) Refuse at the drill site, such as oil cans, drilling mud sacks, and food containers shall be buried at the site or shall be transported to an established refuse disposal area.

(f) Open fires shall not be permitted.

(g) Firearms shall not be carried to or on the job site by Contractor, its employees, agents or representatives.

(h) By restoring drill sites in accordance with good conservation practices and as required by applicable law and regulations.

VII.   OWNER'S OBLIGATIONS.

(a) Owner shall pay to contractor as full compensation for the work hereunder a sum determined on a fixed price per foot basis as follows for all drilling performed hereunder:

> From surface to and including 500 feet in depth, at the rate of $ _____ per linear foot;
>
> From 500 feet to and including 800 feet in depth, at the rate of $ _____ per linear foot;
>
> From 800 feet to and including 1000 feet in depth, at the rate of $ _____ per linear foot.

Depth measurements used to determine amounts due Contractor shall be based on footage as measured by Owner's probing of the hole so long as this probing is done not later than two hours after Contractor has completed the drilling of

Figure 10-15 (continued).

the hole, subject to provisions of Article V (b) hereof. If, and solely in the judgment of Owner's geologist, an unusual condition should exist making it impracticable to make depth measurement at the time any hole is probed, or if any hole cannot be probed within two hours after drilling is completed, depth measurements used to calculate the payments shall be made on the basis of drilled footage recorded in the driller's log referred to in Article V (a) hereof, provided however that determination to use or not to use footage recorded in driller's log shall rest solely with Owner's geologist.

(b) Owner shall pay to Contractor at the rate of $_____ per hour for coring or hourly work, requested by Owner's geologist. When working on an hourly basis, Owner will furnish all bits, muds, chemicals and miscellaneous drilling supplies excluding fuels, oils or supplies normally required to operate the drill rig.

(c) Owner shall pay to Contractor at the rate of $_____ per hour for washing out holes, cementing holes, or for sidewall sampling work.

(d) In the event any drill pipe is lost in the hole (through no fault of the Contractor), Owner will pay to Contractor, Contractor's cost less 5% per month depreciation for the lost drill pipe.

(e) If the distance between successive drilling locations exceeds five (5) miles, or if the move consumes more than one (1) hour, Owner shall

Figure 10-15 (continued).

also pay Contractor for the expense of moving drilling rigs either at the rate of $ _____ per hour, or at the rate of $ _____ per mile, whichever Owner shall choose. The construction of all drill location sites and access roads shall be at Owner's expense.

(f) If loss of circulation occurs while normal drilling operations are in progress on a footage rate basis, Contractor shall notify Owner of such loss and shall use all reasonable means to restore the same without compensation for the first _____ hours after the loss. If Owner concurs that such condition exists, operations to restore circulation after said _____ hour period shall be performed on a day work basis as in Article VII (d) hereof provided until normal circulation is restored or the hole abandoned. Determination to abandon the hole shall rest solely with Owner's geologist.

(g) In the event that Contractor is interrupted in the performance of the work solely by the act or omission of Owner, causing Contractor's drilling rigs to be idle at the site of the work, the period of time during which such interruption occurs shall be herein called "standby time", and Owner shall pay Contractor for such standby time at the rate of $ _____ per hour during the period each drilling rig is so idle with crew and $ _____ per hour without crew. On the next working day following the last day during which any such standby time occurs, Contractor shall advise Owner's geologist

Figure 10-15 (continued).

in writing of the duration in hours of such standby time. Contractor shall
make no claim for and Owner shall not be obligated to make payment for
standby time:

     (i)   which has not been approved in writing by the
           Owner's geologist (which approval shall not be
           unreasonably withheld); or
    (ii)   which occurs between the end of any work day and
           the beginning of the next succeeding work day; or
   (iii)   which occurs during any period when drilling has not
           been required by Owner under this agreement; or
    (iv)   which occurs during any period in which Contractor
           is prevented from drilling or is prevented from
           making depth measurements or conducting other
           operations hereunder by reason of any force majeure
           or contingency provided for in Article VIII (a).

In the event that Contractor is interrupted in the performance of
the work by reason of waiting for Owner to complete a logging or depth
measurement, standby time shall not commence sooner than _____ hour after
the Owner has been notified of the estimated completion time of each hole
drilled in accordance with the provisions of Article V (b).

     (h) The unit prices expressed in this contract may be adjusted
from time to time to compensate for the effects of unusual conditions on
Contractor's costs as follows: Either party may at any time notify the
other that it believes an unusual condition may have occurred and a
gross inequity may have resulted. If the unit prices in this agreement

Figure 10-15 (continued).

are less than 75% or more than 125% of prices for comparable work contained in bona fide bids accepted for such work in the area, then an unusual condition will be deemed to have occurred, and the parties agree to renegotiate this agreement in such a manner that the gross inequity is removed. Each party shall, in the case of a gross inequity, furnish the other with whatever documentary evidence it may have in effecting a fair settlement of such gross inequity. If the parties cannot agree upon such settlement within a reasonable time, then this agreement shall be terminated.

VIII.   MUTUAL COVENANTS.

The parties hereto do mutually agree that:

(a) The obligations of the parties under this agreement shall be suspended while, but only so long as and only to the extent that, the parties are prevented from complying with their obligations hereunder either in whole or in part by strikes, order or any court or administrative body, lock-outs, Acts of God, unavoidable accident, state or federal laws, regulations or orders, inclement weather and/or any other matters beyond the reasonable control of the parties, whether similar to the matters herein specifically enumerated or otherwise, provided that performance shall be resumed within a reasonable time after such cause shall have

Figure 10-15 (continued).

been removed.

    (b) In the execution of the work provided herein, the Contractor shall operate as an independent contractor and not as agent of the Owner. Contractor agrees to protect, indemnify, and save Owner harmless from and against all claims, demands and causes of action of every kind and character, including the cost of the defense thereof, arising in favor of Contractor's employees, Owner's employees, or third parties on account of bodily injuries, death, or damage to property in any way resulting from the willful or negligent acts or omissions of Contractor and/or Contractor's agents, employees, representatives, or sub-contractors.

    (c) Any assignment by Contractor of its interest in this agreement without the prior written consent of Owner shall be void, and Contractor shall not subcontract the work or any portion thereof without first obtaining the written consent of Owner.

    (d) Contractor shall indemnify and hold harmless the Owner from any claim (including the cost of investigating and defending same) arising out of the failure of Contractor to comply with all the covenants and conditions of this Agreement or arising out of negligence of Contractor or of Contractor's employees in performing work hereunder.

IX.   RELIANCE BY CONTRACTOR.

    It is understood and agreed by Contractor that this Agreement is made for the consideration herein named and the Contractor has by examination satisfied himself as to the nature and location of the work, the character,

Figure 10-15 (continued).

quantity and kind of materials to be encountered and the character, quantity and equipment needed in the prosecution of the work, and the location, conditions and all other matters which can in any way or manner affect the work under this Agreement. No oral agreement with or statements by any agent or other person whomsoever, either before or after the execution of this Agreement shall affect or modify any of the terms or obligations herein contained, and this contract shall be conclusively considered as containing and expressing all of the terms and conditions agreed upon by the parties hereto. No changes, amendments or modifications of such terms or conditions shall be valid or of any effect unless reduced to writing and signed by the parties hereto.

X.   NOTICES.

Any notice (other than oral notices provided for in Article V (b) of this Agreement, by either party to the other party to be given in writing under this Agreement, shall be considered as having been given by either party to the other party on the mailing thereof to such other party by registered or certified mail at the applicable address set forth below, or at such other address as such other party may from time to time specify in writing.

Notices served upon Contractor shall be addressed to:

and notices to be served upon Owner shall be addressed to:

XI.   TERM.

The term of this Agreement shall commence as of the date first

Figure 10-15 (continued).

above written and shall expire when the drilling program has been completed, provided, however, that Owner may cancel this Agreement at any time upon giving at least seven (7) calendar days' prior written notice, whereupon Owner's only obligation shall be to pay to Contractor the amounts due hereunder up to the time such termination becomes effective. Owner may terminate this Agreement at any time upon written notice to Contractor if:

    (a) Contractor becomes insolvent;

    (b) Any proceedings should be brought seeking any re-organization, arrangement, composition, readjustment, liquidation, dissolution or similar relief of or with respect to Contractor under the present or any future federal bankruptcy act or any other federal, state or any other statute, law or regulation;

    (c) Any proceedings should be brought seeking and appointment of a receiver or similar officer of court with respect to Contractor's business;

    (d) If Contractor should refuse or fail to supply enough properly skilled workmen or proper materials;

    (e) Contractor should fail to make reasonably timely payment for materials or labor;

    (f) If Contractor should disregard governmental laws, ordinances, rules, regulations, or disregard the instructions of the Owner which are not inconsistent with this Agreement; or

    (g) If Contractor should be guilty of violation of any provisions of this Agreement.

    IN WITNESS WHEREOF the parties have caused this Agreement to be executed as of the date and year first above written.

_____   _____

_____   _____

Figure 10-15 (continued).

Figure 10-16. Water truck getting a load of water for use in drilling operations. This tank holds about 7,500 l (2,000 gal) and is equipped with a vacuum pump to draw water into the tank. In fast drilling operations, water trucks capable of hauling no less than 7,500 l are recommended.

rig in a uranium exploration program can negotiate relatively rough terrain; however, sometimes the geologist must consult with the driller in advance of locating some unusually hard-to-reach drill holes.

When actually staking the location for a drill hole, the field geologist should be aware of problems involved in setting up the rig. For example, if the holes to be drilled are deep and require pits to be dug into the ground, this should be taken into consideration in the final decision as to where the holes should be drilled. In areas of rough terrain, it is usually necessary to maintain a bulldozer on the project in order to prepare sites and access roads in advance of the drilling. When such heavy dirt-moving equipment is required, an effort should be made to minimize the damage to the terrain. Topsoil should be preserved and returned to the surface of excavations during reclamation. Approval from the proper governmental authority must be obtained before excavating by bulldozer or other equipment.

Before the drilling equipment is brought into the area, the field geologist should clearly explain to the drillers and their helpers the specific requirements of the area in which they will be drilling. The geologist should specify which roads they must use, which gates to open and which to close, and any particular stipulations of landowners concerning access to property.

DRILLING

Sample Protection and Equipment Orientation

When the drilling equipment arrives at the

Figure 10-17. After the drill hole is completed to desired depth, the drill pipe is withdrawn from the hole, and the driller may drive the rig forward a few metres so that the mud pit can be picked up and flipped over for cleaning. In this photograph, the rig has been moved and the "sand line" attached to the mud pit. The drill hole itself is not visible, but it is just beyond the conical pile of drill cuttings in the center foreground. Some of the samples are visible near the lower left edge of the photograph. Although this drill site is unsightly now, it will be smoothed out and reseeded; within one year's time, the site will be difficult to locate on the ground. Within two years' time, finding the site will be almost impossible.

site of a planned exploration hole, several factors must be considered in orienting and setting up the drill rig. A competent, experienced driller will generally consider these factors and locate his equipment accordingly. However, some drillers will fail to take such factors into consideration, and if they are not properly directed by the field geologist, problems will result.

If a portable mud pit is used, drill cuttings are shoveled out and scattered near the drill site. If, in the course of drilling the hole, the drilling fluid becomes too thick, it is thinned by draining some of the mud out and adding water. Often the drilling fluid is drained out at least once during the time that the hole is being drilled. This

drilling fluid is usually brown, gray, or shades of those colors and is very slippery. If possible, the rig should be set up so that the drilling fluid, when dumped out of the mud pit, will flow away from the rig and from the site where samples are being placed adjacent to the rig. Samples (collection of which is described below) should be placed where they will not be splattered as cuttings are shoveled out of the mud pit; in addition, they should not be covered or splashed with drilling mud when the pit is emptied (fig. 10-17).

If rough terrain or obstacles such as fences or rocks complicate the drill site, caution should be taken when the equipment leaves the site after completion of the hole to avoid damaging samples. The equipment should not be moved through the muddy area that was caused by dumping the mud pit. Usually the drill rig can be oriented when it is set up so that these problems are easily avoided.

When drilling with air on a windy day, the driller should locate the rig so that the samples are laid out upwind from the drilling operation in order to avoid contamination of the samples by wind-blown cuttings material. The field geologists are frequently obliged to remind drillers of this potential hazard to the samples in air drilling.

When a drilling program includes the drilling of closely spaced holes to evaluate mineralization, it is important that the field geologist consistently watch the driller (and the driller's helpers, who usually guide the driller) in setting up the rig to insure that the drill is level and located precisely on the staked spots. Whether the rig is level can be determined by use of a carpenter's level (the longer the better) which many drillers carry, or (more crudely) by suspending the kelly in the rotary table to see if it centers. The geologist can also stand back and sight on the drill pipe with a Brunton compass to see if the pipe is vertical. In random drilling, it is not quite so important that the rig be level, but in closely spaced drilling, it is critical.

Drilling the Hole

How the Drill Functions

The field geologist should have a basic understanding of how the drill functions for several reasons, including: (1) to make certain the driller obtains the best possible samples; (2) to be able to better decide where to drill particular holes with certain equipment; (3) to be better able to make decisions concerning abandoning holes, recovering lost circulation, and the like; (4) to have a better concept of fair rate of pay

for the drillers; and (5) to be able to determine which drillers should be dismissed and do so without hesitancy. When the driller sets up and levels the rig over the hole to be drilled, he should set the jacks to stabilize it during the drilling process. In very shallow, easy drilling, stabilizing with jacks may not be necessary. However, in difficult drilling, the added stability provided by the jacks prevents an excessively crooked hole and hole damage when the driller is raising or lowering the drill pipe, putting a great deal of weight on the bit, or pulling hard on the string of pipe.

After leveling the rig and setting the jacks, the driller should place the mud pit (if a portable steel pit is to be used) and seal around the neck (usually with bentonite) so that the drilling fluid will not leak. When drilling with pits dug into the ground, the driller should use a flume and sample catcher designed for use with dug pits (fig. 10-18). We have seen instances in which the driller simply dug a shallow trench in the ground between the hole to be drilled and the dug pit, and caught samples from the bare ground. This practice is not recommended because the samples are not consistently clean and reliable when collected in this manner.

When beginning a hole, the driller attaches the bit directly onto the kelly bar and drills the first 6.5 m (21 ft). The bit is then removed from the kelly, and the first joint of drill pipe (or drill collar) is screwed on in its place; then the bit is screwed onto the end of the drill pipe (or drill collar). The bit and the drill pipe are then lowered into the hole, and the drill pipe is connected to the kelly bar as the first joint. As the bit rotates and cuts the material in the bottom of the drill hole, fluid is pumped from the mud pit, through the mud pump, through the kelly, and down through the drill pipe and the bit to the cutting surface; the fluid then returns up through the hole around

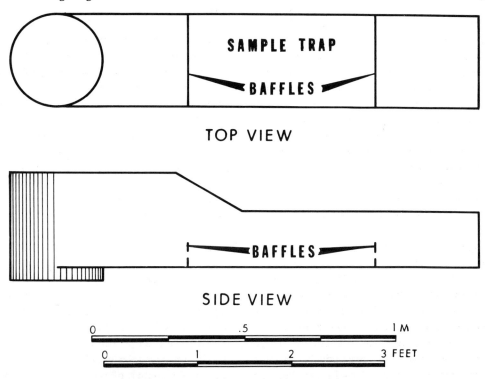

TOP VIEW

SIDE VIEW

Figure 10-18. Diagram of flume and sample catcher for use with dug pits. This device is simple to make and is needed to collect clean samples.

the outside of the drill pipe, carrying the cuttings in suspension. This circulation is possible because the drill bit cuts a hole of larger diameter than the drill pipe.

The drill collars mentioned above are similar to the normal lengths of drill pipe except that they have a larger diameter and, consequently, are much heavier than normal drill pipe. Drill collars are used when more weight is desired on the drilling bit in the bottom of the hole. We do not know the origin of the term "drill collar," but it probably came from the petroleum drilling industry.

Sampling and Sample Lag

Under ordinary drilling conditions, when drilling sandstones, shales, mudstones, siltstones, and coals, the driller can control the rate of penetration by applying varying weights to the drill pipe with the pulldown chain (which essentially transfers the weight of the drill rig onto the drill pipe), and by controlling mud-pump velocity. Because most drilling contractors are paid on a metreage basis, drillers usually attempt to drill holes as fast as is feasible. When the rate of penetration is high and the hole depth in excess of about 30 m (100 ft), there is always an interval between the time when the bit reaches a specific point and the time when the cuttings from that point reach the surface for sampling. The procedure followed by drilling crews is to mark the pulldown chain at 1.5-m (5-ft) intervals (the interval usually sampled in a drilling program) and to collect samples by removing a portion of the drill cuttings from the sample catcher when the pulldown chain indicates that the bit has reached a new 1.5-m increment of depth. After removal of the cuttings for the sample, the sample catcher is then cleaned so that the next 1.5-m sample can be collected. If this procedure is properly followed, the sample for the interval 60- to 61.5-m depth is collected when the bit reaches 61.5 m. If the penetration rate is 3 m per minute, and if it requires two minutes for the cuttings to travel from the bottom of the hole to the surface where they are sampled, the net result is a 6-m "sample lag." In most cases, the rate of circulation remains essentially constant with only slight variations, but the rate of penetration varies considerably and is generally related to the type of lithology encountered by the bit. Shales drill slower than do friable sandstones, and some hard calcareous siltstones or calcite concretions within sandstones drill extremely slowly. These variations result in respectively variable amounts of sample lag. In general, the amount of sample lag varies with the depth of the hole and the rate of penetration.

For those of you who have not seen one of the newer drill rigs drilling the soft sediments in some of the uranium areas, the penetration rate is an amazing sight to behold. The penetration rate can reach 2 to 3 m per minute, which means a 6.2-m drill rod will be drilled down in about two minutes. We have one report of a driller completing 1,000 m (4,000 ft) of hole in six different drill holes in a ten-hour shift (the samples were probably atrocious). Fifteen years ago, such rates were not possible, but the high penetration rates can result in poor samples. We usually allow a driller to drill at the highest rate at which he can get good samples and avoid a large sample lag.

In hard-rock drilling, it is common for the geologist to check on the progress of the drilling only once every few days, or even less frequently. In many situations, the geologist does not go near the drilling equipment at all; instead, he relies on a representative to gather and split the core and to arrange for core transportation to an office or laboratory where the geologist examines it. Drilling of one hole may require several weeks' time. Contrast this with the pace of a "soft-rock" exploration program, where a drill hole to 100 m is completed and probed two and one-half hours after the geologist places a stake designating the drill site. The geologist often must have the interpretation from one hole before spotting the next. We have found it very helpful to have two or more localities in proximity (within 1 km) so that the driller can be deployed at one site while new drilling locations are being staked at the next site. The efficient operation of such a fast-paced program requires skill and speed on the part of the geologists in charge of the operation.

The phenomenon which is most damaging to the quality of samples acquired by drilling with water or mud is best described as "mixing." Mixing results from varying rates of ascension due to different shapes, sizes, and densities of cuttings. Particles which have low density and large surface areas per unit volume generally move more readily (and ascend faster) than particles which are more spherical in shape and more dense. Principles developed by engineers and metallurgists in mill design are applicable in this problem.

Most operators in uranium exploration projects have taken samples at 1.5-m intervals for many years, and when drilling with mud or water, such a procedure is recommended. We have encountered geologists who attempt to make shortcuts by taking 3-m (10-ft) samples; in our opinion, this is a mistake. A typical sample collected from a 1.5-m interval is a small part of the actual volume drilled and removed from the hole. In mineralized and altered sands, it is very easy to miss completely a sample of critical lithology. When the industry converts to the metric system, we strongly recommend that samples be caught at 1-m intervals. The smaller the sample interval, the more likely we are to see all of the lithologic details, but this must be balanced out against excessive sampling in monotonous sequences. When drilling with air, geologists sometimes catch all of the cuttings from .5-m intervals through ore zones to accrue a large volume of material which may later be split for analysis. These large samples can also be used for amenability studies.

Circulation Media

In general, drilling with air is cheaper and faster than drilling with water; as soon as ground water is encountered, however, air drilling must usually be terminated. As discussed later, sometimes wet formations can be drilled with air if the proper amount of detergent is injected with the air. Without detergent, wet clay or slightly wet sand cuttings may form a ball or collect on the side of the hole and cannot be blown out of the hole with air. Also, if unconsolidated sands or gravels are encountered, they tend to cave into the hole. In such cases, water drilling accompanied by a mud program is required in order to drill effectively.

On one occasion, we attempted to drill with air to avoid a 12-km water haul. The drilling was shallow, less than 70 m in depth, and the prospective host rock was a very fine, well-sorted, subrounded quartz sandstone about 30 m in thickness. The first hole we drilled went through shale and encountered the top of the sandstone about 25 m below the surface (the rig we were using had a large-volume air compressor but no equipment for injecting detergent). The drilling was going well until we got about 35 m below the surface, at which point moist sand began to appear in the cuttings at the surface. As we approached

40 m in depth, fewer cuttings were returning to the surface, and by 45 m, *no* cuttings were returning, and very little air. The driller then tried to pick up (lift) the string of drill pipe, and it would move less than a metre. It looked as if we might have some pipe stuck in the hole. The driller then began blowing more air in the hole, and by so doing, and lifting and rotating the pipe, finally got the pipe out. The wet sand had been sticking to the wall of the hole, and we weren't getting a good air return because it was going out into the formation. When we did get the pipe out, the hole blew air for 15 or 20 minutes.

In areas where the water table is low and the formation is dry to considerable depths, it is economical to drill as deep as possible with air before converting to water. In many areas, holes can be drilled with water alone, without mud or other additives. Some formations contain clay beds which contribute to the drilling fluid to form a natural mud. However, when special drilling problems develop, or when conditions in an area produce problems, a mud program must be developed; in some cases, it may be advisable to consult with a mud engineer.

Poorly consolidated sands and gravels commonly cave into the hole when drilling is being done with water or with air, and this may result in the loss of drill pipe; or, if the hole remains open until probing, it can cause the loss of the probe (sonde). The problem may develop if loose rock (gravels are especially bad) falls down the hole and wedges alongside the drill pipe or the probe. When an upward pull is exerted on the pipe or probe, it may become more tightly wedged. In areas of thick, loose surface gravel, it is not uncommon to lose some drill pipe or a probe now and then. Both are expensive losses, but a probe usually costs several thousand dollars. We suggest mixing a good jell-based mud in advance of drilling such unstable lithologies to prevent caving and to keep the hole in gauge. In severe situations, surface casing may have to be placed down through the caving sand and gravel.

Some clays, when exposed to fresh water, dissolve and slough into the hole, thus increasing the diameter of the hole; such out-of-gauge zones result in poor circulation of cuttings to the surface. Cuttings may adhere to the side of the hole where it is widened and out of gauge, and these accumulations often slough in on the probe if it happens to touch them. This can result in a probe being stuck in the hole. Usually there are certain additives which can be used in preparation of the drilling mud to prevent it from reacting with these clays.

Another problem commonly encountered relates to expanding or swelling clays. Clays with a high montmorillonite (bentonite) content expand significantly when exposed to fresh water. Places where swelling clays expand into the hole, reducing the diameter of the hole and inhibiting upward circulation of drilling fluid, are known as "boots." We have seen instances where drillers have repeatedly reamed out boots (drilled through them) and made little daily progress when their mud program (if any) was not properly developed.

Another problem frequently encountered in uranium exploration is the phenomenon known as "lost circulation"—an instance when the drill fluid does not return to the surface. Highly fractured, brittle formations commonly have open fractures capable of carrying off the drilling fluid in the subsurface. In addition, some loose gravels and conglomerates are capable of absorbing the drilling fluid, resulting in lost circulation. In most cases, lost circulation can be controlled by a good mud program, with specific additives included (fig. 10-19). Lost-circulation materials to be added to the mud include cottonseed hulls,

shredded paper, sawdust, and swelling grains or grain products such as bran. In some cases, it is necessary to drill intervals "blind," with no return of cuttings or drill fluid, until the circulation is restored by the plugging action of clays encountered beneath the zone of fracturing. When boots develop in the hole and the pump pressures rise during attempts to maintain circulation, fracturing of porous units beneath the boots sometimes results, causing lost circulation.

In certain situations we have drilled exploration holes blind, even though this is risky and may result in lost pipe. In one area of Wyoming where we were involved, one of the rigs was drilling at a depth of 30 m when a fracture or opening must have been encountered, because all of the drilling fluid in the hole disappeared. Despite our efforts to restore circulation on this important hole, it seemed impossible. We put all kinds of material down that hole: the usual bentonite, bran, sawdust, paper, and hulls, along with cow manure, straw, hay, and old rags off the drilling rig. Nothing seemed to work. We then decided to drill blind. When drilling blind, the water trucks are kept very busy because none of the water pumped into the hole returns to the surface. The water, plus the drill cuttings, goes into whatever fracture or void exists down there. We drilled that hole blind to 120 m, and it appeared that all the cuttings from below 30 m came up the hole to that point and disappeared. We were eventually successful in probing the hole, but it was a relatively expensive one to complete because of the hourly charges by the driller and the additives we put into the hole.

Another problem which requires a specialized mud program for its solution relates to retrieving good samples of fine-grained sandstone in deep holes. If the drilling mud contains too many solids and

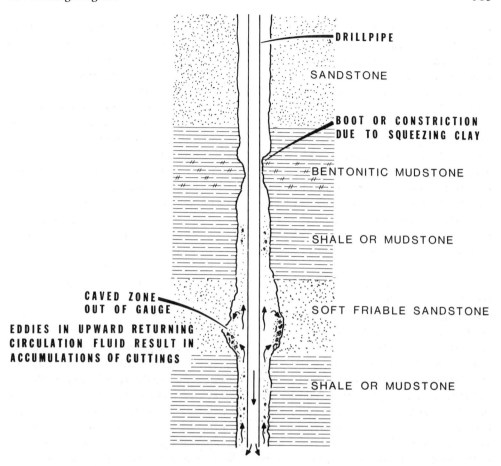

Figure 10-19. Diagram of drill hole illustrating how caved zones or squeezing clays may result in difficulties while drilling and probing. A "boot" or constriction inhibits circulation and might cause additional related damage. Caved zones can also result in problems. These hazards can usually be minimized by developing a good mud program.

is too dense, fine sand will not settle out in the sample catcher; instead, it may remain in suspension throughout the cycle of circulation. If the objectives of deep drilling include fine-grained sandstones which become disaggregated during the drilling process, it may require an exceptional mud program to recover good samples. In such cases, we suggest consulting a mud engineer to develop the best mud program possible. Mud engineers are available from the companies which supply drilling mud and other additives. There is usually no charge for their services.

Uranium exploration companies have encountered serious problems in trying to drill through volcanic rocks in the Grants district in New Mexico. On Mount Taylor, in particular, drilling has been very costly, and some unorthodox drilling techniques have been used. The problem is caused mostly by volcanic rocks (ash and ash flows) and by underclays which consistently result in lost circulation. Drillers will not work by the metre in such situations and must be paid by the hour. Drilling rigs being used are from 2000 up to 7000 class; the latter is an oil rig converted to uranium use. In order to get through the volcanics, aerated mud is usually used, and the companies must rent large-capacity air compressors in order to aerate the mud properly. Air and mud engineers are on hand most of the time.

The volcanic rocks may be 500 m in thickness, and once the drilling penetrates the underlying Cretaceous sediments, casing is set; drilling then may proceed on a metreage basis. Lost circulation may also develop in the Cretaceous beds, however, and so some companies set casing to 600 m or deeper. Aerated mud is usually not needed once the volcanic section is penetrated.

Some explorationists in the Grants district believe that the size of the drilling rig may be quite critical for some of the drilling through volcanics, and they advocate a "delicate balance" between depth of hole and size of rig. A theory has been advanced that the big oil rigs tend to break down the formations and cause extra problems. On the other hand, a 2000 rig may be almost too small and require an expert driller to accomplish the drilling.

In recent years, foam injection has been developed as an efficient drilling method to cope with excessive lost circulation and swelling clays. With this method, foam-generating chemicals (usually detergents) are injected into the circulation system during drilling with compressed air with a two-stage compressor. The foam is dense enough and has sufficiently high surface tension to carry the cuttings out of the hole, yet it is far lighter than either water or mud. The low density reduces the required pressure against the wall of the drill hole, thus preventing loss into the formation. The foam does not react with swelling clays; therefore, the development of boots in the hole is minimized.

It is the driller's responsibility, before leaving a hole, to condition the hole so that

it may be probed. If it is known that a completed hole must remain open for several hours before the probe is available for logging, it is a good practice to condition the mud and circulate this fluid for a sufficient length of time to condition the hole so that it will remain open. An experienced driller can sometimes recognize potential problems which may present hazards for the probe and take preventive steps to avoid costly delays.

Holes drilled with air must be filled with water or mud if electric logs are needed. Usually the operator will rely on the superior samples which can be acquired by air drilling, along with the gamma-ray log, to interpret the geology encountered in air-drilled holes.

Common Problems

Sometimes a swelling clay, a caving gravel, or some other condition of the hole will result in the drill pipe becoming wedged or otherwise immovable. This happens most frequently when the pipe is being pulled out of the hole after the mud pump (or air compressor) and rotation of the pipe have been stopped. The first indication of trouble is evident when it becomes increasingly difficult to pull the pipe upward. When this occurs, the driller should test the nature of the problem by cautiously allowing the weight of the drill pipe to move the pipe downward. In some cases, where poorly consolidated material has sloughed from the wall into the hole and around the drill pipe, some repeated up-and-down movement of the pipe causes the material to move downward, freeing the pipe. However, if this fails to free the pipe, it may be necessary to attach the kelly to the drill pipe (unless it is already attached), rotate, and establish circulation to wash out the material. Usually, if the pipe can be rotated and the circulation to the surface established, the pipe will be recovered from the hole.

If the driller torques up too much in attempting to free the stuck pipe, the pipe sometimes will twist off and break, leaving part of the drill string stuck in the hole. When this happens, it is necessary to employ special tools to recover the pipe; this process of recovery is called "fishing." Fishing is also done in attempts to retrieve other items, such as miscellaneous objects which are sometimes dropped in the hole. One can get quite disturbed when the drill has reached 450 m on a 500-m hole, the driller trips out (brings the pipe and bit out) to put on a new bit and, while the pipe is out of the hole, a clumsy helper drops a pipe wrench in the hole. Drilling through a heavy steel pipe wrench is difficult, so you start fishing. Magnets, overshots, and other gimmicks are used. It is possible to spend $15,000 or more trying to get a $15 wrench out of the hole, and *fail*.

A superior driller will *never* allow an expensive drill hole to remain open at the surface when the pipe is out of the hole, including those times when coring is taking place. A cover, such as a heavy piece of lumber, will be placed over the hole.

In some areas where hard beds or hard rocks are encountered in the drilling, or when the driller puts excessive weight on the bit, or both, crooked holes (those which deviate from vertical) may result. Drill pipe is flexible, and when many drill pipe sections are screwed end to end and extend from the drill (which rotates them, pumps fluid or air through them, and adds or subtracts weight), it is amazing what can happen. As mentioned, the pipe can deviate from vertical and still drill in some preferred direction consistently, or it might corkscrew its way down, or deviate erratically. We sometimes joke about holes deviating so badly that a horseshoe-shaped hole may be formed, with the drilling bit

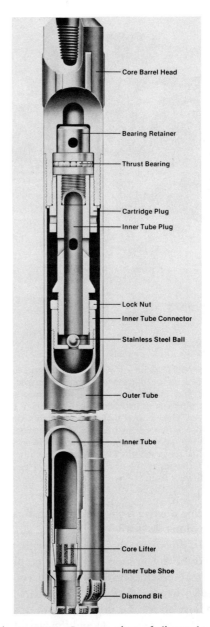

Figure 10-20. Cutaway view of diamond core barrel and bit of a type commonly used in exploration drilling in sedimentary rocks. Also available, but not shown, are "split" core barrels for coring in very soft formations; core barrels which cut an oriented core; and wire line core barrels which allow retrieval of the inner barrel and core through the drill pipe to the surface without "tripping" all of the pipe and barrel out of the hole. Courtesy Christensen Diamond Products.

Figure 10-21. Core of soft, fine sandstone has just been removed from the inner tube (see fig. 10-20) and placed in a core holder made of a plastic pipe cut in half lengthwise, which in turn has been placed in a V-shaped wooden trough. The core reflects light because of a thin coating of drilling mud. This core (and the core barrel) is 3 m (10 ft) in length. The driller is holding one end of the inner tube, and drilling mud can be seen flowing from the near end of the tube.

surfacing not far from the rig. If drilling is being done in hard rock and the driller pushes down on the drill pipe with the weight of the drilling rig, a very crooked hole may result. This is where drill collars come in handy. If several drill collars are used just above the drilling bit, *they* will provide weight on the bit and the weight of the drill rig can be used to a lesser extent. Crooked holes are usually of no real concern to the explorationist in the widely spaced drilling carried out in a preliminary exploration program; however, they are significant when the program involves closely spaced drilling in the evaluation of

mineralization.

Coring and Sidewall Sampling

A core is a cylindrical solid piece of rock which is obtained through the use of a coring bit and tool. Both the bit and the tool have a hollow interior, and the bit cuts the rock away around the hollow so that as the bit and tool advance, a core of the rock is forced up into the tool. When the tool is lifted from the bottom of the hole, a core catcher must function to keep the core from falling out.

When drilling with water or mud, the samples which are collected from the drilling fluid are always mixed; consequently, they are not adequate for chemical analysis, or even for detailed lithologic description. When the explorationist wishes to make a detailed evaluation of a potential uranium host, it is usually necessary to cut cores or take sidewall samples to avoid contamination.

By contamination, we mean from other cuttings and mud, not a biological type of contamination. This may seem like an elementary explanation, yet there is an interesting story to be told. It seems that a young executive in a small uranium exploration corporation (he was untrained in geology or engineering) went to the USGS to ask questions about getting both core and rotary samples from some drilling they planned to do. The Survey obliged, and, as part of the discussion, cautioned the young man about avoiding contaminated samples because they could foul up the reliability of chemical assays. With this advice, the executive promptly went out and bought several cases of sterile plastic gloves, which everyone who worked on the rig and handled the samples or core was required to wear. The fiasco ended when a geologist happened by and asked what they were afraid of getting *from* the samples.

Figure 10-22. Core shown in figure 10-21 being measured so that percentage of recovery may be determined. Measuring tape is usually left near the core for reference when the core is photographed and described mineralogically.

Coring Equipment

Core barrels in common use are capable of retrieving cores 3 m long and ranging from 25.4 mm to 99.3 mm in diameter (fig. 10-20). Recently, core barrels have been designed for recovering oriented cores.

Coring Operation

The obstacles to good core recovery vary from area to area; in general, hard rock and the well-sorted and moderately well indurated sandstones are easier to recover in a core than are the poorly consolidated and poorly sorted sandstones and conglomerates. Some of the sands which host uranium deposits in Wyoming are so unconsolidated and poorly sorted that a coring operation is comparable to attempting to

recover a core from a pebble-strewn beach. The core barrel virtually washes down through these sands, and it can emerge empty if the program is not very carefully conducted.

The mud program must be well controlled, with a low water loss and a low-density mud to avoid heavy pulsating pump pressures. While coring in sandstone, we recommend maintaining a uniformly low pump pressure with the resultant slow and steady circulation of cuttings to the surface. The core bit should be rotated uniformly and at a moderate rate to avoid unnecessary vibration on the cutting surface. Weight on the core bit should be maintained as uniformly as possible and should not be excessive.

Drillers who have consistent success

Figure 10-23. After photographs and descriptions are completed, the core is broken into 1-m sections and wrapped in plastic to avoid contamination.

with coring include those who take extraordinary precautions when "coming out of the hole" with a core. One unusually successful driller in the Shirley Basin district of Wyoming made a practice of padding the slips (devices used in rotary table to keep the drill pipe from falling down the hole) with rags when breaking each connection of the string of pipe to prevent sudden jars from dislodging the core from the core barrel.

When the core barrel is removed from the hole, it should be carried over to where the geologist has made preparations for "laying out the core." (A cover should be placed over the hole!) There are various ways to prepare for laying out a core and, as stated above, soft cores are more difficult to handle than hard cores. If the coring program is extensive enough to justify it, a trough should be built to receive the core (fig. 10-21). For additional protection to the core and to facilitate its handling, it is helpful to utilize a 4.5-m length of plastic casing cut lengthwise in half. This smooth, rounded half-casing can be placed in the V-shaped trough so that the core can slide more freely and with less chance of damage. This is particularly important when coring friable sandstone. For a limited coring operation, when time does not permit the acquisition of the receiving trough, the geologist should lay a heavy sheet of plastic on the ground to accommodate the core.

Core barrels and bits have been improved tremendously in the past 20 years. In the mid-1950s in the Gas Hills district, Wyoming, a great deal of difficulty was encountered when attempts were made to core some of the very coarse, soft sandstone in which much of the uranium ore

Figure 10-24. The cores, wrapped in plastic, are placed in core boxes which are marked for drill-hole number, date, and interval which the core represents.

occurred. One of the techniques used was to drill with diesel fuel and have the mud pit nearly full of pieces of dry ice (solidified CO_2). The idea was to freeze the sandstone so that it would remain coherent and stay in the core barrel. The procedure was very slow, expensive, and only moderately successful.

In the late 1950s in Shirley Basin, Wyoming, core barrels had been improved to the extent where reasonably good core recovery could be obtained *if* a good drilller did the coring. Much of that sandstone was so soft that when it was pushed from the core barrel, it would immediately disintegrate into very soft pieces of sand.

As soon as the core is removed from the core barrel, it should be pieced together as nearly as possible to its original shape. After it is reassembled, it should be measured to determine what percentage of the cored interval was recovered (fig. 10-22). If there has been a loss of core, the core should be examined to determine, as nearly as possible, where the loss occurred. Sometimes the upper part of the cored interval is destroyed by a hard pebble grinding under the core bit for some time before it becomes incorporated with the core. Other times, a hard pebble in the middle of the cored interval may break loose and destroy a portion of the core; also, the lower part of the cored interval may be lost during removal of the core from the hole.

After the core is measured and an effort made to determine where loss, if any, occurs, the core should then be cleaned. The core usually emerges from the core barrel with a thin mudcake surrounding it. This mudcake should be carefully removed, using a knife blade or other sharp, flat instrument. It is sometimes difficult to

Figure 10-25. In the laboratory, core boxes may be opened and the cores removed for detailed mineralogic descriptions through the use of a binocular microscope. A copy of the electric log of the hole should always be on hand when cores are described in the laboratory, as well as a scintillation counter to pinpoint radioactivity sources. The potassium ferrocyanide test (chap. 3) may also be used on cores. Sections of the core may be selected for assay and other tests.

remove the mudcake from weakly indurated sandstone, because this tends to crumble away with the mudcake. The entire core should be cleaned to insure that chemical analyses will not be diluted and that a detailed description may be made of the core. As soon as the core is cleaned, numbered cards should be prepared indicating date, drill-hole number, and depth intervals at about 0.5-m intervals, and these should be placed near the core. A tape measure should also be laid out near the core, and the core should then be photographed. The cards and tape will be of great value in later examinations of the photographs, which, incidentally, should overlap each other (fig. 10-23).

A standard log form can be used for the detailed core description; however, the scale must be expanded to accommodate the amount of detail to be described. Even if the core is to be split for additional description in the lab, it should always be described immediately in the field in as much detail as possible. This is particularly true with unconsolidated or poorly consolidated sandstones, which are often disaggregated after being transported away from the location. Thus, the field description may be the only one made of the sedimentary structures, such as ripple bedding, cross-bedding, carbonaceous laminae in the sand, clay galls, and so forth.

The field description should include an attempt to estimate the porosity and permeability of the sand, as well as the vertical and horizontal fracturing. Color and staining should be accurately described, as well

391

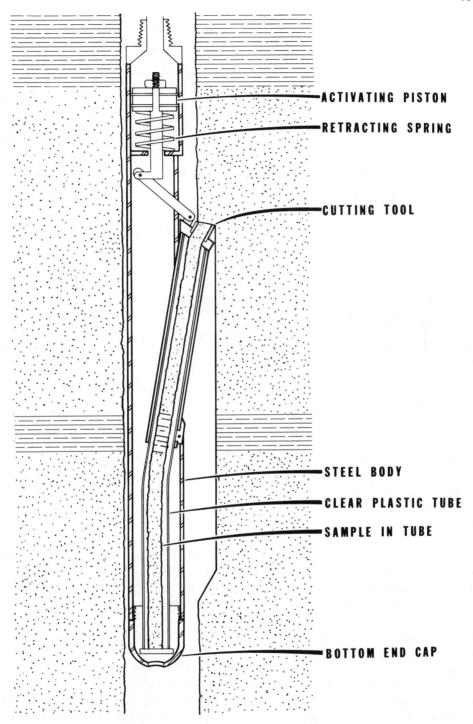

ACTIVATING PISTON

RETRACTING SPRING

CUTTING TOOL

STEEL BODY

CLEAR PLASTIC TUBE

SAMPLE IN TUBE

BOTTOM END CAP

Figure 10-26. Cutaway diagram of a sidewall-sampling device. Courtesy of Mineral Sampling, Riverton, Wyoming.

Figure 10-27. Sidewall-sampling device which has just been removed from the drill hole. Sample is contained in a plastic tube inside the sampler.

as the general mineralogy of the sand and the matrix. Odors associated with the core immediately after removal from the core barrel should also be noted. H_2S is sometimes very pronounced and important. Dilute hydrochloric acid should be used to determine calcareous intervals. A hand lens (14x is good) should be used to examine the core, and a binocular microscope can be valuable when making very detailed examinations or when attempting certain mineral identifications.

After the field description of the core is completed, the core should be placed in a box properly marked with the date, depths, and the number of the core hole (figs. 10-24, 10-25). Boxes should be handled as carefully as possible to minimize damage to friable sandstones or other soft rocks.

When attempting to measure permeabil-

ity and porosity of a weakly consolidated sand which is cored, 15- 20-cm full-diameter samples should be removed from the core. These samples should be representative of the sand to be tested, and they should not be broken or fractured unless that was the condition of the rock in place. If the samples are soft sandstone, they should be dipped in molten wax, giving them a heavy coating all around and over the ends; this should be done without removing the mudcake.

Sidewall Sampling

A sidewall-sampling device has become widely used in recent years to supplement conventional coring (fig. 10-26). One of the chief advantages of the sidewall sampler is that it is cheaper to use than conventional coring; also, an interval may be selected and sampled *after* having run

Figure 10-28. Probe truck operator backs probe truck to a recently completed drill hole for probing, with assistance on hole location by driller's helper, who is holding the probe (sonde) partly in the drill hole. The probe itself is about 3 m in length and is lowered into the hole by means of a steel cable which contains the copper wires which transmit electrical signals from the probe to recording equipment in the truck. Samples taken in the course of drilling the hole are visible in the lower right corner of the photograph. Two drilling rigs are drilling on new locations in the background.

the gamma-ray and electric logs (as opposed to coring, where the interval to be sampled must be selected *before* logs are available). One disadvantage is that the samples collected in a sidewall sampler are usually diluted to some extent with drilling mud.

A typical sidewall-sampling tool used in the industry is about 1.7 m long, and it attaches to the end of the drill pipe for insertion into the drill hole. When the driller has lowered the sidewall sampler to the bottom of the interval to be sampled, he activates the tool by building up the air or water pressure against the piston in the top of the tool; as the piston is pushed downward, a sample cutter is extended into the wall of the hole while the tool is raised

with the draw works of the drilling rig through the interval to be sampled. The sandstone or clay which is sampled moves downward through the cutting tool into a clear plastic tube inside the sampling device. When the interval has been sampled, pressure is released and the spring-loaded sample cutter is retracted and can then be removed from the hole (figs. 10-26, 10-27).

The sample cutter on the sidewall-sampling device is cylindrical, and it projects, when fully extended, completely into the wall, thus recovering a cylindrically shaped sample continuously up the side of the hole through the interval sampled. The tool cuts a core-like sample about 2 cm in diameter, but it will also accommodate

Figure 10-29. Driller and one helper racking the last joint of pipe from a completed drill hole while the other helper holds the probe, which will be lowered into the drill hole through the rotary table.

larger plastic tubing, up to 5 cm in diameter. The expanded sample catcher permits the collection of a continuous sample through thicker zones.

Samples from the sidewall-sampling device can be removed from the plastic tube, described, and then removed for chemical analysis; or they can be left in the plastic tube and taken to the lab for examination, description, and chemical analysis. As stated above, samples caught in the sidewall sampler are usually contaminated to some extent. If uranium mineralization is being sampled, this contamination can be calculated by comparing the equivalent U_3O_8 determination made on the sample with the same interval calculated from the gamma-ray log.

PROBING THE HOLE

General Description

The term "probe" is a holdover from the early days of uranium exploratory drilling, when holes were probed by hand with Geiger-counting equipment. The equipment used to probe (log) holes in the uranium industry has improved substantially during the past several years, and there is an increasing effort to improve it even further. Indeed, it is anticipated that the logging of exploration holes will be substantially more refined over the next few years.

The probing equipment in general use today is mounted on the back of a three-quarter-ton pickup truck or equivalent vehicle (fig. 10-28). The probe, or sonde (which is lowered into the hole), contains a thallium-activated sodium iodide detector, a single-point resistance electrode, and a spontaneous-potential electrode. The probe is suspended on a cable which also houses the leads for the gamma-ray and electrical

Figure 10-30. Probing a drill hole through the rotary table. Probe cable is visible from probe truck (at right) to wheel assembly placed in rotary table. Another loose cable from probe truck to mud near mud pit is the grounding cable and must make good connection with the ground by being placed in water or mud.

logs (figs. 10-29, 10-30). The cable is spooled on a drum that is powered by an electric hoist which can raise and lower the probe at precisely controlled speeds. The rate of movement of the gamma-ray detector must be accurately controlled when logging a uranium ore zone to make possible quantitative calculations. A recorder inside the probe truck draws gamma-ray, resistance, and spontaneous-potential curves simultaneously as the probe is hoisted through the hole (fig. 10-31). Most operators routinely run these three curves.

In areas where the driller has encountered problems with caving holes, swelling clays, or extreme lost circulation in drilling the hole, the probe operator is likely to find similar problems in probing the hole. The most common problem is a "bridge" in the hole; this may be a hard layer which

is out of gauge, a rock which has fallen from loose gravel, or some other obstacle which could cause difficulty for the probe operator in lowering the probe to the bottom of the hole. In some cases, the driller may be obliged to set up over the hole once again and wash out the obstacles in order to allow the probe to be lowered to the bottom. Drillers being paid for completed holes will not be paid for a particular hole if the probe cannot reach bottom.

Perhaps the greatest danger in probing is that a rock might fall into the hole after the probe has passed beyond it; a swelling clay is also hazardous. In such instances, the operator may find the probe has stuck in the hole. If this should happen, and the probe cannot be loosened by pulling and relaxing on the cable, the driller may have to set up on the hole again and wash the

Figure 10-31. View of panel inside probe truck while a drill hole is being logged. Pens are recording on the chart paper (from left to right) gamma-ray log, including 20 K rerun (20,000 counts per second); short (16-in.) normal resistivity; single-point resistance. Holes are probed from bottom to top, and reruns are made wherever the gamma log goes off scale. On the log shown, hole T.D. (total depth) is 280 ft, the gamma rerun peaks at 266 ft, and the pens are recording at 230 ft. Visible below and to the left of the chart paper is the spool which holds the probe cable. Left of and above the cable spool, the meter which shows probe speed (an important factor in grade calculations) can be seen. The coil of paper in the upper panel near the top of the photograph is a digital printout (counts per second and depth printed every 0.2 ft) of the gamma rerun.

probe loose. In going back into the hole to perform such washing, no drill bit should be used on the drill pipe. Just the open-ended pipe should be used, because the bit can cut, twist, kink, or otherwise damage the cable or the probe, or both. If all else fails, there may be cases where a bit will be used, but it must be recognized that there are occasions when washing out the probe cannot be done, and the probe may be lost in the hole. Fortunately, in most areas,

such incidents are extremely rare.

One probe unit can usually handle holes drilled by two or more rigs when the holes are deep, because the drills spend more time on each hole. Conversely, if holes being drilled are shallow and mineralized, then it is extremely difficult for one probe to keep up with two drill rigs.

If a part of the probe truck malfunctions, it can disrupt the entire exploration program and involve other expensive equipment. For this reason, probe units should be well engineered, and the probe operator should carry spare parts for those components which might fail during the operation. The field geologist should become familiar with the probing equipment and have an understanding of the log characteristics in terms of indicated lithologies and mineralization. When malfunctions in the logging equipment are indicated, the field geologist should be able to recognize the problem immediately and see that it is remedied in time for the hole to be logged correctly before being plugged and abandoned. Most problems result from an electrical short circuit in the cable. A competent probe operator should recognize problems and correct them before the geologist becomes involved. It is essential that at least one individual in the field be capable of recognizing and, even better, dealing with such incidents.

When holes are closely spaced to evaluate mineralization, and the mineralized zones occur at depths greater than about 90 m, it is sometimes necessary to run directional surveys to determine the precise locations of the mineralized intercepts. When using a probe truck on contract, it is usually necessary to make advance arrangements for the directional survey tool.

A directional survey tool includes camera equipment to photograph the compass bearing and the angle of inclination of the tool as it is raised out of the drill hole. The data from directional surveys are generally transmitted to a local computer service firm and processed to give the coordinates of the hole at various depths. The geologist can then take the processed information and plot the exact intercepts of the mineralized zone on his maps.

In areas where the mineralized sands tend to wash out during drilling and the mineralized zone is out of gauge, it is sometimes necessary to run a caliper curve to make the appropriate correction when calculating the mineralization. A caliper curve simply logs the variations in hole diameter as the probe is hoisted up through the hole.

When some exploration operators see how expensive drilling and probing rates are, they immediately begin thinking of buying their own drill rigs and their own probing equipment. For long-term development work where the operators and the equipment are going to be stationed at one location for a long period and where drilling will go on 12 months of the year, this is not a bad idea. For *most* programs, however, involving many moves, shutdowns in the winter, and no single stable drilling area, we definitely recommend *against* purchasing drills or probe units. It is a much better choice to contract the work, even if a higher price per metre must be paid to the outside contractors. In the final analysis, the contract work is almost always less expensive, and a great amount of lost time and stress on your own personnel is avoided.

A frequent problem not to be overlooked by the explorationist is the accuracy of the metreing wheel on the probing unit. The metreing wheel measures the amount of cable which is lowered into the drill hole from the spool of cable in the probing unit. The wheel, in a continuing process with use, wears down so that it is turning more revolutions than it should for a given amount of cable. When this happens, mineralized zones appear to be deeper than they actually are. Further, if close attention is not paid to continuing wear of the wheel so that it is replaced on a frequent basis, progressive errors will creep into the logs. For example, if we assume that a wheel wears 0.2 percent per month and it is used on a program nine months, it will be off 1.8 percent at the end of that time. If an ore zone at an actual depth of 350 m is being logged, the depth shown by the probe will be in error by *6.30 m* at 350 m. This will raise havoc with correlations. Further, if such data are used for mine planning, it is obvious that serious and costly mistakes could be made. Underground drifts and scrams would be erroneously planned lower than they should be, thereby disrupting efficient operations.

On the other hand, new metreing wheels might be too large, and thereby show depths shallower than they actually are. For underground mining, this could be even worse than a depth shown too deep because development drifts *above* an ore zone are almost worthless.

It appears that the only good protection the explorationist can have for long-term programs is to have the metreing wheel checked periodically by having a cased drill hole of known depth probed. Such cased holes have been established in many of the existing mining areas by operators in those areas, and such holes are usually available upon request to the appropriate company. ERDA also has a shallow test hole at Grand Junction, Colorado, but it is too shallow to be of significant value in checking metreing-wheel accuracy. More deep cased holes are needed by the industry at various locations around the country. However, in the absence of a test hole, it is recommended that the explorationist insist that the metreing wheel be checked in the field by spooling out cable for a known horizontal distance once monthly in an active drilling

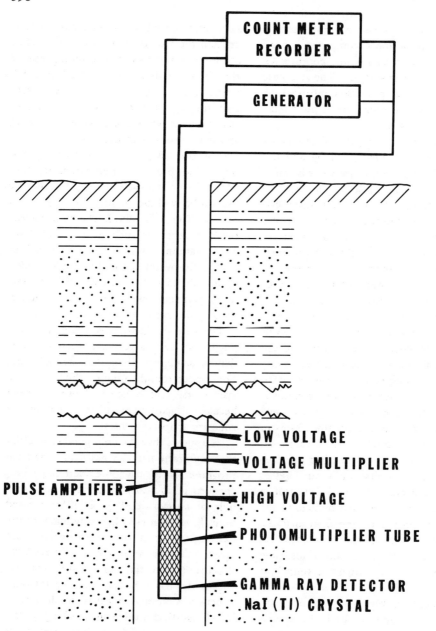

1. NaI(TI) CRYSTAL EMITS MINUTE FLASH OF LIGHT WHEN PENETRATED BY A GAMMA RAY.

2. PHOTOMULTIPLIER TUBE CONTAINS A PHOTO-SENSITIVE SURFACE FUSED TO THE CRYSTAL AND EMITS ELECTRONS. THESE ARE MULTIPLIED IN STAGES WITHIN THE TUBE TO 1,000,000 FOLD INCREASE.

Figure 10-32. Schematic diagram of a gamma-ray logging device.

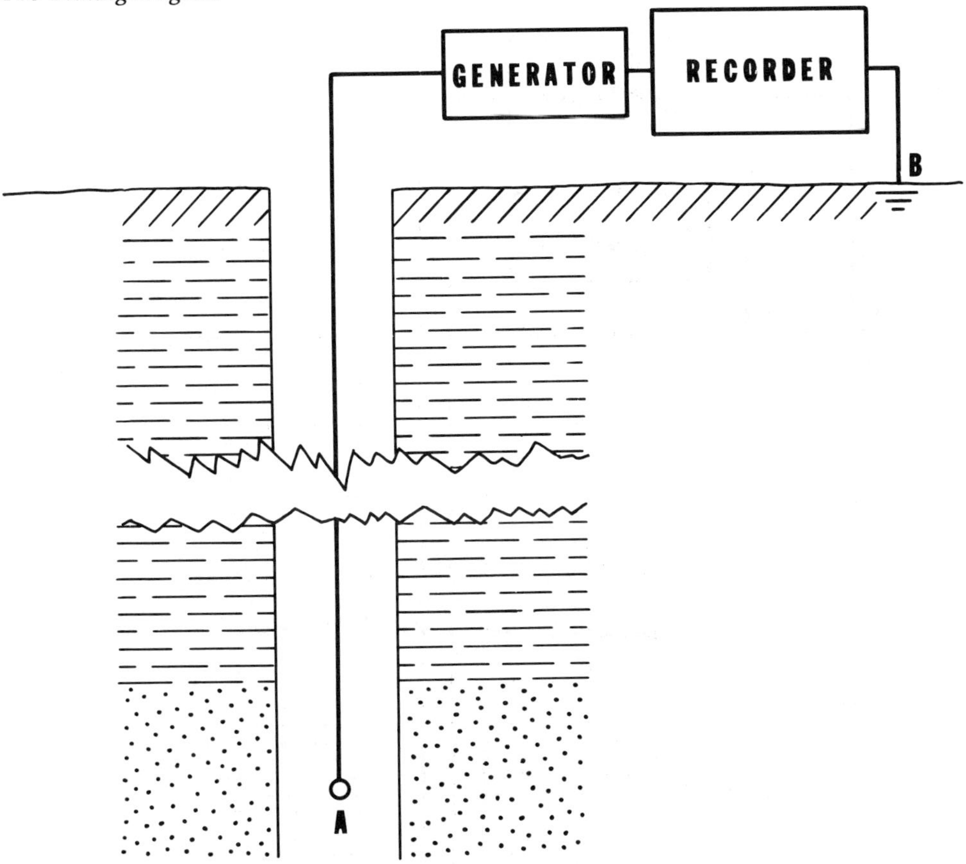

Figure 10-33. Schematic diagram of a single-point resistance logging device. Site of electrode in the probe is marked by A, site of ground electrode by B.

program. A company representative should be on hand to assist and to verify the accuracy.

Gamma-Ray Logging

This log does not require water or mud in the drill hole and can record small variations in gamma activity characteristic of varying lithologies, which may be an aid in sample interpretation and lithologic correlations; however, its main value in uranium exploration is to evaluate mineralization. Gamma-ray logs may be used to interpret lithologies and mineralization behind the casing of cased holes as well as exploratory drill holes, because gamma rays will penetrate steel pipe (although at a reduced intensity).

The equipment used for quantitative gamma-ray logging (fig. 10-32) includes the following parts: (1) a thallium-activated sodium iodide crystal (gamma detector) housed in the probe, (2) power source, (3) cable and hoist, (4) electronic energy multipliers and counting devices, (5) digital instrumentation, and (6) recorder.

When logging exploration holes, the gamma-ray curve is generally run on a sensitive scale to facilitate geologic interpretations. If moderately strong or stronger radioactivity is encountered, the gamma-ray pen goes off scale (and will require a later rerun). When this occurs, the probe operator continues logging up through the

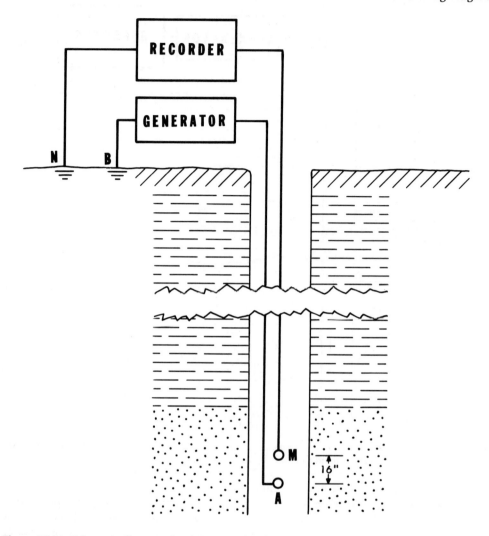

Figure 10-34. Schematic diagram of a short normal resistivity logging device. Site of electrodes in the probe is marked by A and M, site of ground electrodes by B and N.

zone until the gamma-ray pen has returned to background values because of decreased radioactivity. The probe must then be stopped, the recording pens raised, and the gamma recorder reset for a less sensitive scale capable of recording the highest readings encountered. The probe is then lowered down to the base of the radioactive zone and is stopped, the gamma pen is lowered, and the probe is then rerun through the mineralized zone to trace a

gamma curve which can be used to calculate the grade and thickness of the mineralization. If the probe unit is equipped with a digital printout unit, the operator should set this in operation when he begins the rerun. The digital unit prints out depths and the respective counting rates (in counts per second) for small increments of thickness to facilitate mathematical integration of the area under the gamma anomaly.

The probe operator should record the

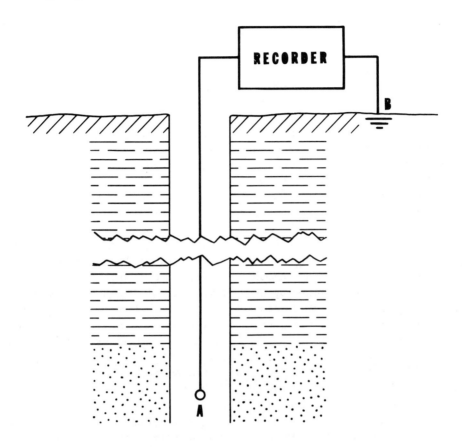

Figure 10-35. Schematic diagram of a spontaneous-potential logging device. Site of electrode in the probe is marked by A, site of ground electrode by B.

factors needed for calculation on each log. The scales for the normal gamma curve and all reruns should be indicated where appropriate on the curves themselves and also on the log heading. The digital printouts should also be properly identified and should include all factors necessary for calculation.

Resistance and Resistivity Logging

Next to the gamma ray, the single-point resistance log is the most useful record made of a typical uranium exploration drill hole. The hole must be filled with water or mud for either of these two logs. An experienced geologist can use the resistance log in conjunction with the samples to identify many ·

subtle variations in lithology. Some explorationists have recently substituted the short normal resistivity log for the single-point resistance to record more accurately the true resistivity of the lithologies; however, the single-point resistance log appears to be more useful in most uranium projects.

The equipment required for single-point resistance logging (fig. 10-33) includes an electrode in the probe, a ground electrode, a direct-current generator, and a galvanometer-recorder. The resistance log is simple to run in most instances, but lithologies with extraordinarily high resistivities, such as coal, limestone, anhydrite, and clean fresh-water sandstones, require an alert operator. The operator should acti-vate and observe the resistance pin (but it should not be recording on paper) as the probe is lowered into the hole; in this manner, he can observe high resistivities and determine the best scale for running the log. Resistance is recorded in ohms.

The equipment necessary for short normal resistivity logging (fig. 10-34) in-cludes two electrodes (usually spaced 16 in. apart) in the probe, a ground electrode, a direct-current generator current meter, and a recorder. Strata or lithologic varia-tions which have thicknesses less than the electrode spacing cannot be discerned on the resistivity log. Shale partings within sandstone units or thin sandstones in shale units, therefore, are not recorded by the resistivity log. The problems encountered in resistivity logging are similar to those en-countered in resistance logging.

Spontaneous-Potential Logging

The spontaneous-potential (SP) log is of questionable value to most uranium explo-ration projects, but it is commonly inclu-ded in the standard suite of logs by most operators. The SP requires the least equip-ment of all the logs; however, its origin and significance are poorly understood. Equip-ment required for the SP is simply an elec-trode in the probe and a ground electrode connected to a common galvanometer-recorder (fig. 10-35). SP is recorded in millivolts.

In shallow exploratory drill holes, the main sources of the SP, known as the "membrane potential" and the "liquid-junction potential," are minimal because of similar salinities between the formation water and the drilling water. For that reason, the probe operator is obliged to record the SP on an extraordinarily sensi-tive scale, which often results in excessive drift of the shale baseline. When this drift results in frequent intersections of the

SP and gamma curves, we recommend that the SP logging be terminated.

USE OF GAMMA–RAY AND ELECTRIC LOGS

General Comments

When a capable geologist is carrying out exploration for uranium deposits in sedi-mentary rocks, he can develop a remark-able facility for comprehending how and why the mineralizing solutions moved as they did, and how to predict the occur-rence of ore based upon his interpretations. In order to develop this facility, the geolo-gist must have at hand excellent data from drilling, such as descriptions of samples from rotary drilling and consistently high quality electric logs. Ideally, the geologist who makes the interpretations also collects the data in the field, but this is seldom the case.

Through the proper use of the sample descriptions and electric logs on working displays, such as cross sections and maps, the geologist can learn to recognize subtle but important variations in gamma, SP, or resistance logs which suggest why miner-alization, or lack thereof, occurs in a cer-tain zone. To accomplish this, however, it is imperative that photocopies or photo-graphically reduced copies of the *original electric logs* be used in the cross sections; otherwise, the slight variations that are so important cannot be observed.

For example, in the Shirley Basin dis-trict of Wyoming, the main uranium-ore-bearing sandstone is about 22 m in thick-ness. Grid drilling had revealed that one area contained ore in the upper one-third of the sandstone, while a nearby area to the north contained ore in the lower one-third and nothing in the upper one-third. Reasons for this phenomenon were not understood until the roll-front concept emerged, and this was coupled with mine

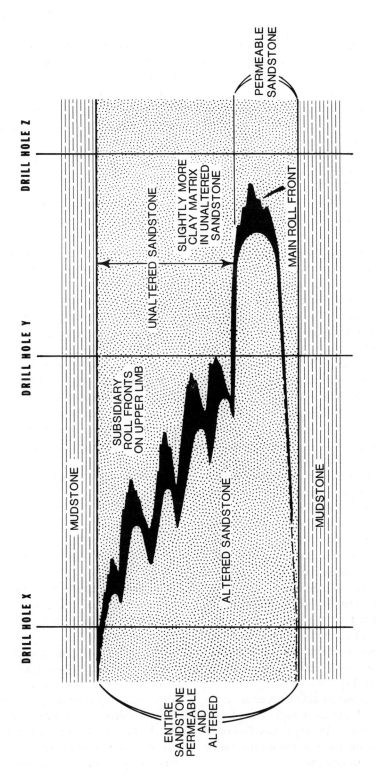

Figure 10-36. Cross section of a roll front showing development of ore on subsidiary roll fronts.

mapping and good geologic cross sections. South of the two ore-bearing areas, the entire thickness of sandstone was altered, and to the north, the upper part of the unit became less permeable. As a result, the upper altered/unaltered contact was developed at a progressively lower level in the sandstone, and this caused ore bodies to form as a series of subsidiary roll fronts on the upper limb. This interpretation is shown in figure 10-36. The question then arose that if ore bodies sometimes resulted in areas where the upper limb lost elevation in a short distance (or the lower limb gained elevation in a short distance) could a subsidiary roll with possible associated ore be predicted on the basis of these elevation changes? The answer was found to be an emphatic yes; such ore occurrences could be predicted. Thus, not only could ore be drilled for between drill holes Y and Z (the main front), but ore might also occur between holes X and Y (see fig. 10-36). We strongly suspect that many ore occurrences are being overlooked by geologists who fail to do good interpretive work from accurate data. It would be simple in the example shown to draw a straight or slightly curved line from the upper limb in X to the upper limb in Y with no suspicion of good mineralization between. Yet we have seen numerous examples of *no* ore on the main front but good ore on a subsidiary, and this applies to both upper and lower limbs.

The predictability of the fronts is directly related to the uniformity of the host rock; and the geologist, with an adequate number of drill holes and through the use of good sample descriptions, good electric logs, and accurate drill-hole elevations, should be able to determine the complexity of the host rock, complexity of the roll fronts (if any), and favorable locations for ore both on the main front or fronts and on subsidiaries.

In one situation we studied, a monofacies roll front had been discovered in a Cretaceous marine sandstone. The occurrence was very unusual in that the front followed an essentially straight line for more than 2 km (1.25 mi). The lack of sinuosity was directly related to the uniformity of the host rock, but when detailed cross sections were assembled from the data gathered in fence drilling across the front, a second phenomenon was discovered. There were *two* roll fronts in the sandstone, essentially paralleling each other for virtually the entire 2 km. Seldom were they separated laterally by more than 20 m, and vertically they were separated by only 3 to 4 m of clean sandstone. The lower front consistently had the highest grade of associated uranium mineralization. We concluded that these two fronts were reflecting subtle differences in the porosity and permeability of the sandstone, and an even greater effort had to be made to interpret these subtle differences from electric logs because there was *no way* to see the difference in the samples, or even in the cores. In such situations, the use of multiple electric logs in cross sections can provide a more useable interpretation for ore-finding than any other technique we are aware of.

Interpretation of Logs

Gamma-Ray Logs

The gamma-ray log can be used on a sensitive scale as an aid in identifying lithologic units. In general, clay and mica minerals are more radioactive than are quartz, feldspar, chert, calcite, coal, and anhydrite. This results in low gamma readings in clean sandstone, limestone, coal, and anhydrite (evaporite) beds, and higher gamma readings in shale, mudstone, siltstone, and other clay-rich beds. The gamma-ray log may be used in conjunction with the resistance log to estimate clay content in impure sandstone,

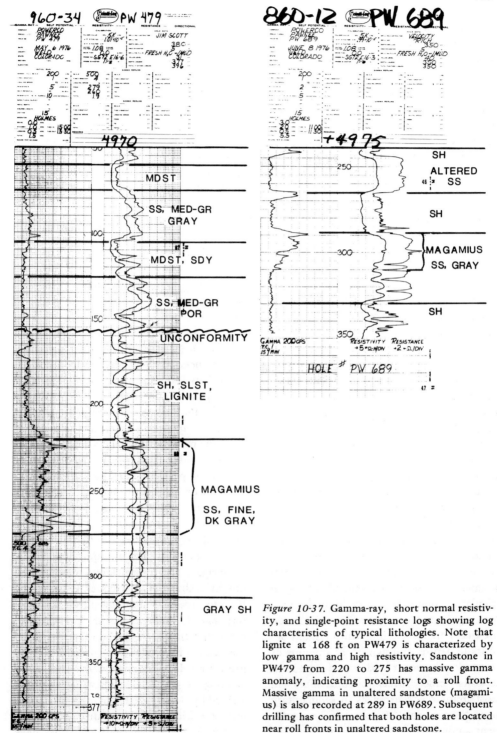

Figure 10-37. Gamma-ray, short normal resistivity, and single-point resistance logs showing log characteristics of typical lithologies. Note that lignite at 168 ft on PW479 is characterized by low gamma and high resistivity. Sandstone in PW479 from 220 to 275 has massive gamma anomaly, indicating proximity to a roll front. Massive gamma in unaltered sandstone (magamius) is also recorded at 289 in PW689. Subsequent drilling has confirmed that both holes are located near roll fronts in unaltered sandstone.

coals, and other lithologies (fig. 10-37).

The predominant use of the gamma-ray log in uranium exploration is to detect radioactive minerals and, with certain assumptions, to calculate the thickness and average grade of uranium mineralization in drill holes.

As the uranium exploration industry developed in the western United States, the U.S. Atomic Energy Commission (AEC) established standards for quantitative use of gamma-ray logs in calculating uranium values in drill holes. The AEC established standard model holes in their facility at Grand Junction, Colorado, which could be used to calibrate gamma-ray logging equipment. Some companies maintain cased holes through mineralization in some areas where mining is being carried on (such as at Shirley Basin), and these cased holes are probed periodically to verify the accuracy of the probe unit. The AEC and others also established theoretical and empirical methods for calculating equivalent U_3O_8 (eU_3O_8) from the gamma-ray logs (Dodd and Eschliman, 1972; Scott, 1961). With improvements in the electronics, in the recorders, and in the detectors, the available gamma-ray logging equipment has become quite reliable, and the results are reproducible.

The gamma rays are emitted by certain radioactive isotopes in the uranium (and thorium) decay series as well as by the conversion of ^{40}K to ^{40}A (see fig. 3-4). When thorium or ^{40}K is present in significant amounts, the use of the gamma-ray log in calculating uranium is very limited. However, significant concentrations of ^{40}K do not occur in association with uranium deposits in nature, and significant amounts of thorium only occur with certain uranium deposits under special circumstances. In most uranium occurrences in sedimentary hosts, thorium and ^{40}K do not occur in sufficient amounts to confuse calculations. The gamma-ray log is the basis for calculating

equivalent U_3O_8 (eU_3O_8), which is the U_3O_8 that would be present if the uranium and all of its decay products were present in the deposits and fully developed (in equilibrium). It requires about 300,000 yr to fully establish the chain of decay products which emit gamma rays after uranium has been transported in ionic form and concentrated (Rosholt, 1959, p. 10). In addition, due to differing solubilities, uranium or one of its decay products might be differentially leached out of a deposit. If the uranium deposit is too young, or if chemical leaching has removed uranium or one of its decay products, *radioactive disequilibrium* is the result. If there is a geologically recent net addition of uranium relative to the gamma-ray emitters (or net loss of gamma-ray emitters) of the deposit, it is out of equilibrium in favor of chemical analysis ($U_3O_8 > eU_3O_8$). Conversely, if uranium has been recently leached from the deposit, it will be out of equilibrium in favor of the gamma-ray log ($U_3O_8 < eU_3O_8$). Disequilibrium is common in uranium deposits and must always be taken into account in any quantitative considerations. This is more fully discussed in chapter 11.

The empirical model for standard calculation of equivalent grade of U_3O_8 from the gamma-ray log has been expressed as

$$e\bar{G}T = KA,$$

in which $e\bar{G}$ is the average radiometric grade of uranium in percent equivalent U_3O_8 by weight, T is the thickness of the uranium mineralization, K (frequently called the K factor) is a constant of proportionality determined by calibration of the probe equipment in standard model holes, and A is the corrected area under the anomaly on the gamma-ray log. The manual method of calculating eU_3O_8 from gamma logs was described by Scott (1961), and this procedure has been widely used by industry field personnel. Many exploration companies have used standard log calculation forms (fig.

DATE __2/24/76__ LOGGING CO. _____ HOLE NO. __PW 226__

LOCATION __1062-24 Weld County, Colorado__

K FACTOR __2.4 x 10⁻⁵__ DEAD TIME __0.0__ μ SEC.

INTERPRETER __J. R. Nies__

BASE	218.5		213.0	
TOP	205.5		211.5	
T	13.0		1.5	
E_1	1000		4000	
E_2	640		3000	
$1.38 (E_1 + E_2)$	2260		9660	
I_1, I_{14}	2400	15600	10000	
$I_2,$	1600	8000	15600	
$I_3,$	800	3000	8000	
$I_4,$	700	1200		
	700	500		
	800	500		
I_7, I_{20}	1100	800		
	1800	1000		
	1800	1200		
$I_{10},$	1800	1100		
	2000	1100		
	4000	1100		
I_{13}, I_{26}	10000	1000		
AREA	67863		43260	
X MUD FACTOR	73224		46720	
KA=GT	1.76		1.12	
G	0.14		0.75	
T-G-D	13'-0.14% eU_3O_8 - 205.5'		1.5'-0.75% eU_3O_8-211.5'	

HOLE NO. __PW 226__

Figure 10-38. Gamma-ray calculation form with a calculation of the mineralized zone shown on the log in figure 10-39. Note that the entire zone is calculated and that the high-grade zone is calculated separately.

10-38) for field calculation of logs in an effort to further simplify the method set forth by Scott (1961).

The first step in calculating a gamma-ray log manually (with one-half-ft determinations) is to mark the upper and lower boundaries of apparent mineralization to be calculated. The boundaries are usually picked at points near the half-amplitude of the gamma-ray anomaly, but if the anomaly includes a zone of low-grade mineralization which brings the average grade below the economic cutoff level, a separate calculation excluding the low-grade mineralization can be made. An effort should be made to calculate the zones which have the most potential for mining, in addition to the entire mineralized zone. When a log calculation form is used, the depths to the top and bottom of the zone to be calculated are entered first and then the thickness is entered (fig. 10-39).

The area under the anomalous part of the gamma-ray curve is subdivided into two tail areas and a central area. The value of the sum of the two tail areas is approximated by adding together the counting-rate values (with dead-time corrections if necessary) at E_1 and E_2 and then multiplying by a "tail factor." For most logging equipment, when using one-half-ft intervals, this tail factor is 1.38. When E_1 and E_2 are read at or below the half-amplitude points on the curve, the error in calculating the value of the tail areas is less than 2 percent. E_1 is usually picked at the top of the mineralized zone and E_2 about one-half ft below the bottom of the mineralized zone. Next, the successive intermediate values I_1, I_2, etc., are determined from the gamma-ray log at one-half-ft depth intervals and entered on the form. If dead-time corrections are necessary, they are made before adding the values. Dead time is more completely discussed below.

The corrected intermediate values are added to the calculated tail areas to arrive at a total value for the area under the anomalous part of the gamma-ray curve. Because most K factors are determined for air-filled holes, the total area must be multiplied by a water factor if the logged hole is filled with water or drilling mud. The water factor is usually about 1.08 for holes drilled about 11.5 to 14 cm (4.5 to 5.5 in.) in diameter. Next, the corrected total area is multiplied by the K factor, and this product is equal to the average grade multiplied by the thickness (called the grade-thickness product or GT). Finally, GT is divided by T to get the average grade in percent eU_3O_8. It is customary to express the results of the calculation in terms of thickness, grade, and depth to the top of the zone.

Much of the logging equipment in use today has the capability to print out digital counting rates on 0.2-ft intervals (table 10-1). Many of these units are set up to record in centimetres, but the industry has not fully embraced the metric system at this writing. With these digital printout sheets, it is only necessary to add the counts within the mineralized zone plus six or seven counts above and six or seven below the boundaries and multiply by the water factor and K factor to calculate GT.

Computer programs have been set up for calculating gamma-ray logs (Scott, 1963; Dodd and Eschliman, 1972). Commercial computer services now available are also capable of digitizing old gamma-ray logs and making calculations rapidly and accurately.

Dead time is the time which is required by the logging equipment to detect a gamma ray and to process the pulse. Corrections for dead time are necessary unless the logging system makes this correction internally. Indeed, many old logs from logging systems used extensively during the 1950s and 1960s require significant corrections for dead time even in low-grade ore. Most modern logging systems have internal

TABLE 10-1. DIGITAL PRINTOUT FOR GAMMA-RAY LOG

HOLE NO: PW-226
UNIT NO: 4 DATE: 2-24-76
2 FT. K 9.6 x 10^{-6}
5 FT. K
WATER FACTOR 1.08

Depth (ft)	Counts per second	Depth (ft)	Counts per second	Depth (ft)	Counts per second
2 0 4 6	0 0 3 0 0	2 0 7 6	0 0 7 0 0	2 1 0 6	0 1 9 0 0
2 0 4 8	0 0 4 0 0	2 0 7 8	0 0 7 0 0	2 1 0 8	0 2 0 0 0
2 0 5 0	0 0 5 0 0	2 0 8 0	0 0 7 0 0	2 1 1 0	0 2 5 0 0
2 0 5 2	0 0 8 0 0	2 0 8 2	0 0 8 0 0	2 1 1 2	0 3 2 0 0
Top 2 0 5 4	0 1 0 0 0	2 0 8 4	0 0 9 0 0	2 1 1 4	0 4 5 0 0
2 0 5 6	0 1 5 0 0	2 0 8 6	0 0 9 0 0	2 1 1 6	0 6 7 0 0
2 0 5 8	0 2 3 0 0	2 0 8 8	0 1 1 0 0	2 1 1 8	0 9 7 0 0
2 0 6 0	0 2 6 0 0	2 0 9 0	0 1 3 0 0	2 1 2 0	1 3 5 0 0
2 0 6 2	0 2 0 0 0	2 0 9 2	0 1 6 0 0	2 1 2 2	1 5 5 0 0
2 0 6 4	0 1 4 0 0	2 0 9 4	0 1 8 0 0	2 1 2 4	1 4 5 0 0
2 0 6 6	0 1 0 0 0	2 0 9 6	0 1 9 0 0	2 1 2 6	1 1 0 0 0
2 0 6 8	0 0 9 0 0	2 0 9 8	0 1 8 0 0	2 1 2 8	0 8 1 0 0
2 0 7 0	0 0 7 0 0	2 1 0 0	0 1 8 0 0	2 1 3 0	0 5 8 0 0
2 0 7 2	0 0 7 0 0	2 1 0 2	0 1 8 0 0	2 1 3 2	0 3 9 0 0
2 0 7 4	0 0 7 0 0	2 1 0 4	0 1 9 0 0	2 1 3 4	0 2 6 0 0
2 1 3 6	0 1 7 0 0	2 1 5 8	0 1 0 0 0	2 1 7 8	0 1 1 0 0
2 1 3 8	0 1 2 0 0	2 1 6 0	0 1 1 0 0	2 1 8 0	0 1 1 0 0
2 1 4 0	0 0 8 0 0	2 1 6 2	0 1 2 0 0	2 1 8 2	0 1 0 0 0
2 1 4 2	0 0 6 0 0	2 1 6 4	0 1 1 0 0	Base 2 1 8 4	0 0 9 0 0
2 1 4 4	0 0 5 0 0	2 1 6 6	0 1 2 0 0	2 1 8 6	0 0 8 0 0
2 1 4 6	0 0 5 0 0	2 1 6 8	0 1 1 0 0	2 1 8 8	0 0 6 0 0
2 1 4 8	0 0 6 0 0	2 1 7 0	•0 1 1 0 0	2 1 9 0	0 0 6 0 0
2 1 5 0	0 0 6 0 0	2 1 7 2	0 1 1 0 0	2 1 9 2	0 0 5 0 0
2 1 5 2	0 0 8 0 0	2 1 7 4	0 1 1 0 0	2 1 9 4	0 0 3 0 0
2 1 5 4	0 0 9 0 0	2 1 7 6	0 1 1 0 0	2 1 9 6	0 0 2 0 0
2 1 5 6	0 1 0 0 0				

Base: 218.4 ft
Top: 205.4 ft
$T = 13$ ft
$GT = 1.755$
$G = 0.14$
13 ft of 0.14% eU_3O_8 at 205.4 ft

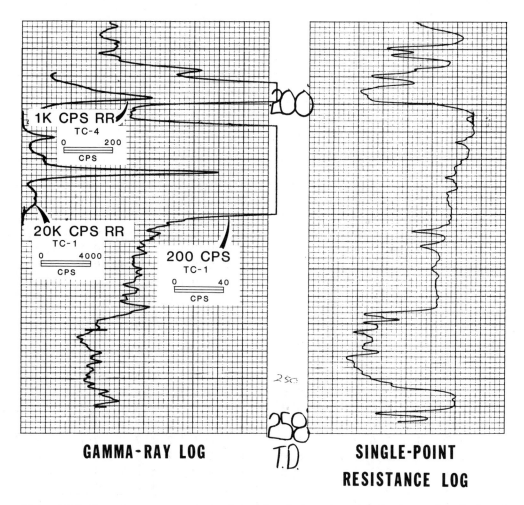

GAMMA-RAY LOG   T.D.   SINGLE-POINT
                                             RESISTANCE LOG

Figure 10-39. Part of a gamma-ray and resistance log (hole no. PW 266) showing a mineralized sandstone. Equivalent U_3O_8 is calculated in figure 10-38. Digital equivalent is illustrated and calculated in table 10-1.

corrections for dead time. The statistical correction for dead time is made by the equation

$$N = \frac{n}{1-nt},$$

where N is the true rate of gamma emission, n is the observed count rate, and t is dead time expressed in microseconds. Most operators have prepared graphs to make rapid corrections for dead time (fig. 10-40).

Large-diameter holes filled with water or drilling mud can result in significant reduction of the observed count rate due to a lowering of the energy of gamma rays and due to absorption of low-energy gamma rays. Probing holes with casing and probing through drill pipe also result in significant reduction of count rates. Correction factors or curves can be developed for varying diameters of drill holes filled with water or mud and for varying thicknesses of casing or drill pipe. They

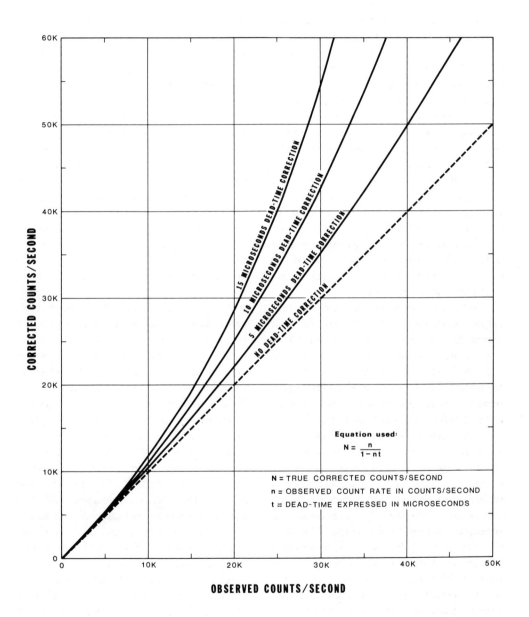

Figure 10-40. Dead-time correction chart.

will vary with individual probes (Dodd and Eschliman, 1972).

One of the most important interpretations for which the gamma-ray log is used in exploration for roll-front deposits lies in its use in distinguishing altered from unaltered sandstones. The classic example, often illustrated in the literature, shows altered sandstone with low gamma readings bounded above and below by thin zones (limbs) with high gamma readings. In practice, however, the limb mineralization is often weak or nonexistent, but a thin carbonaceous clay might be mineralized. If the gamma reading is highest at a clay-shale contact and low in clean, porous parts of the sandstone, the porous zones are often interpreted to be altered; conversely, however, if a high gamma anomaly occurs in the most permeable part of the sand, it is interpreted to be unaltered but probably near a roll front.

Thick zones of weak uranium mineralization commonly extend beyond roll fronts for distances ranging from 10 to 100 m into the more permeable layers of the unaltered sandstone. The gamma anomalies on logs of drill holes which penetrate these gradational zones of weak mineralization are sometimes called "protore" or "remote seepage" (Rubin, 1970). Both of these terms should be tossed out of the literature. The term "protore" should be reserved for mineralization which is nearly ore and which is characteristically enriched by weathering or supergene action to ore grade. The classic examples of this are copper deposits of sub-ore grade below a zone of secondary enrichment. The implication is that if the weathering process continues, the sub-ore–grade material will become ore. The use of the term "protore" in a description of weakly mineralized sandstone near a roll front implies that if advancing weathering continues, the sandstone will become ore. We dispute the suggestion that weathering is contributing to the migration of a roll front. If the hypothesis we advanced in chapter 2 has any credibility, most of the sandstone uranium mineralization we see today was deposited shortly after the sands themselves were deposited. Certainly, local adjustments to changing conditions have been and are now taking place, but evidence is lacking that the process which emplaced the ore bodies is now active, at least in the United States.

To illustrate the inappropriateness of the term "protore" in connection with sandstone uranium deposits, Shirley Basin is a good example. Ground-water flow at the present time is *opposite* to the direction of ground-water flow when mineralization took place. If we are going to use the term "protore" there, it would have to be used for the altered sandstone. Let's avoid the term "protore" in connection with roll-front uranium deposits; it has a useful place in connection with supergene enrichment, and let's keep it there.

The term "remote seepage" is misleading, because the weak mineralization in unaltered sandstone typically occurs immediately adjacent to roll-front deposits and therefore is not "remote." "Seepage" implies that a certain amount of the mineralizing solution was able to find a path through the media to deposit uranium and that no other part of the solution could get through. This is ridiculous, and it is not the process almost unanimously agreed to by those who have published papers on roll-front geology. These authors envision a ground-water solution migrating (not seeping) through the sandstone, with certain chemical phenomena (such as Eh and pH change) taking place in both the host rock and the solution as the feature called a roll front advances in the direction of ground-water movement. Not all of the uranium (or other migrant elements) precipitates at the roll front, and some is carried forward to precipitate "downstream" from the front itself. This is not a "seepage" phenomenon.

A gamma anomaly in unaltered sandstone is a favorable indicator of a *nearby* roll front. A term is needed for this important and characteristic gamma-ray–log feature; we prefer to call it a "gamius" (*gam*ma *i*n *u*naltered *s*andstone). The term "magamius" is useful for describing massive gamma in unaltered sandstone.

Resistance and Resistivity Logs

The single-point resistance log is the most widely used supplement to the gamma-ray log. It may be used in conjunction with the samples to identify and pinpoint sandstone, siltstone, shale, coal, sandy mudstone, clayey sandstone, and other lithologies (fig. 10-37). The single-point resistance does not penetrate far beyond the borehole, and it cannot be used for calculating resistivities of formation fluids. It is a very useful but qualitative tool capable of identifying thin beds of interest to uranium explorationists. When a clean sandstone is present for comparison, it is usually possible to recognize relatively clayey sandstones. Often a thin (5 cm or less) mudstone parting or calcite concentration in a sandstone is important to the uranium explorationist, and this is usally seen on the single-point resistance log. Resistivity logs do not distinguish thin beds of about 30 cm or less in thickness.

The 41-cm (16-in.) resistivity is being used by some operators in an effort to get more quantitative data on sandstones, but until now, it has been difficult to see any advantage over the single-point resistance. The most important uses for the resistance or resistivity logs are interpreting lithologies and making correlations between drill holes. The resistivity log might in some cases give a more reliable basis for evaluating the porosity and permeability of sandstones due to its greater penetrating capability, but we have not seen convincing evidence of this.

Spontaneous-Potential Logs

The spontaneous-potential (SP) log is included in the basic logging program of many exploration projects, as a supplement to the resistance curves. The SP log indicates porous and permeable beds in deep sandstone and shale sequences, where interstitial water is more saline than the drilling fluid. In this environment, the Na^+ cations and the Cl^- anions will move down the chemical gradients caused by fresh drilling water invading the relatively saline environments. Shales have a layered clay texture with charges on the clay layers to permit the passage of Na^+ cations, but are impervious to Cl^- anions. This results in selective movement of ions through the shales to the mud in the drill hole, and the resultant potential across the sandstone-shale contact is known as the "membrane potential." Other similar causes of voltage potential are related to differences in salinity of drilling fluid and formation water. However, these sources of potential are not well developed in most environments that are less than about 400 m in depth.

Unfortunately, the SP log is poorly understood, has been of very limited value in most uranium exploration programs, and appears to be sorely in need of technical improvement. Some advocates suggest that the SP is useful in interpreting redox potentials associated with altered sands in areas where roll fronts are being investigated; however, we have not seen this use of the SP illustrated to our satisfaction.

The SP logs are sometimes characterized by wide fluctuations (both positive and negative) in sediments where altered complexes and roll fronts are developed, but we have seen no systematic relationship to the geochemical interfaces. Indeed, in shallow, fresh-water sands lacking alteration and roll fronts, similar positive and negative fluctuations are often logged on the SP.

Induced-Polarization Logs

Several members of the uranium exploration industry, as well as the Uranium and Thorium Branch of the U.S. Geological Survey, haved studied induced-polarization (IP) logs obtained in drill holes, in an effort to determine if differences in concentration of sulphide minerals (particularly pyrite) or clay minerals with high ion-exchange capacities can be measured across roll fronts. It is known that pyrite is sometimes concentrated along the reduced sides of roll fronts; also, there is evidence that clay minerals are altered by mineralizing solutions in such a way that the ion-exchange capacities of the altered clays are reduced (Scott and Daniels, 1976; Robert Rodriguez, unpub. data). The U.S. Geological Survey is also studying the usefulness of hole-to-hole IP measurements in determining if a potential host sand is barren or mineralized between two drill holes 100 to 300 m apart.

In some instances, the chargeability is distinctly higher in the unaltered sand than it is in the equivalent altered sand, but in other instances, the differences are subtle and inconclusive. Use of the IP log is still in the experimental stage, but it might be developed as a useful tool to differentiate between oxidized (altered) and reduced (unaltered) sandstone equivalents.

Magnetic-Susceptibility Logs

The magnetic property of a sandstone which is dependent upon the direction and strength of an externally applied magnetic field is known as "induced magnetization." The magnetic susceptibility of the sandstone is a measure of its ease of magnetization, and this is directly related to the content of magnetite and maghemite (Ellis and others, 1968; Scott and Daniels, 1976). Where alteration complexes and roll fronts are developed, mineralizing and oxidizing solutions convert maghemite and magnetite into hydrous iron oxides and hematite, thus reducing the magnetic susceptibility of the altered sandstone relative to its unaltered equivalent.

Industry has fielded several probes to measure magnetic susceptibility, but these proved to be too unstable with varying temperatures to make the required measurements. The U.S. Geological Survey has developed a prototype probe which has the capability of making measurements to the desired accuracy, but its usefulness remains unknown. The magnetite and maghemite contents of most host sandstones are not high; thus, the theoretical differences in magnetic susceptibility to be found in association with roll fronts are not substantial.

It has not been determined to what extent magnetite is converted to less susceptible minerals during the typical alteration process associated with roll-front development. Reported measurements show slight decreases of magnetic susceptibility in close association with uranium concentrations, but not in altered sandstone removed from mineralization. Indeed, Ellis and others (1968) reported a negative correlation between magnetic susceptibility and equivalent U_3O_8. Magnetic susceptibility is a tool in the experimental stage with remote potential for the uranium explorationist.

Delayed-Neutron Activation Analysis

Kerr-McGee Corporation first demonstrated the feasibility of *in situ* assaying of uranium, using Californium-252 (^{252}Cf) for neutron-induced fission and delayed-neutron detection. This original device utilized a large ^{252}Cf source and a very heavy shielding to protect personnel from harmful radiation while it was above the surface of the ground.

Under contract with the U.S. Energy Research and Development Administration, IRT Corporation has developed an improved ^{252}Cf-based logging system for *in situ* assaying of uranium (D. K. Steinman, D. G. Costello, C. S. Pepper and others, unpub. data). Mobil Oil Company developed a similar device which utilizes a pulse neutron source, and they have used this system for about two years (Robert Rodriguez, unpub. data).

Correlations

Uranium exploration programs which are designed to locate strata-controlled uranium deposits usually require detailed correlations of lithologic units. Such correlations are the first step necessary to working out stratigraphic relationships which may control uranium mineralization. The resistance log is usually more useful than other logs in correlation of units, as well as identification of lithologies; however, distinctive beds are occasionally identifiable on the gamma-ray log, making it a fair tool for correlation as well. Shale, siltstone, and sandstone sequences frequently contain zones with distinctive variations on the resistance logs which facilitate correlations (fig. 10-41). These variations are only rarely to be seen in the samples.

LOGGING SAMPLES

The Need for Good Sample Logs

Accurate and detailed sample descriptions, prepared by the geologists in the form of logs, are essential to any uranium exploration program. Samples often provide the most reliable basis for distinguish-

ing altered sand from unaltered sand, favorable host rock, bleaching, humate staining, and so forth, and in many cases this is the most valuable geologic determination to be made. Stratigraphic relationships such as facies changes, unconformities, formation boundaries, and correlations frequently require accurate sample logs for their reliable definition. In most cases, understanding these stratigraphic relationships is essential to a determination of the controls on mineralization.

Electric and gamma-ray logs measure the exact depths of bedded lithologies penetrated in a drill hole, but these mechanical logs can only be used to make an approximate interpretation of the lithologies in most instances. If good samples are obtained and accurately described with the aid of the electric logs (also called mechanical logs), usually a fairly accurate interpretation of the lithologies encountered in drilling can be made. The sample log is an interpretation, because there is always some mixing of cuttings. Also, variable rates of penetration result in variations in the amount of "sample lag."

There are a few situations where it is not necessary to prepare sample logs of *every* drill hole, such as in closely spaced development or advanced exploration drilling. For example, if the objective is clearly defined and limited to a sand in which the altered and unaltered parts cannot be differentiated by sample examination, sample logs might be less useful than mechanical logs. Good judgment on the part of a project geologist might, in other cases where drilling is in an advanced stage, eliminate selectively the description of part of the section so that the geologist's attention can be focused on the objectives, but it should be emphasized that too often the logging of samples is treated with almost casual abandon, and insufficient data are

416

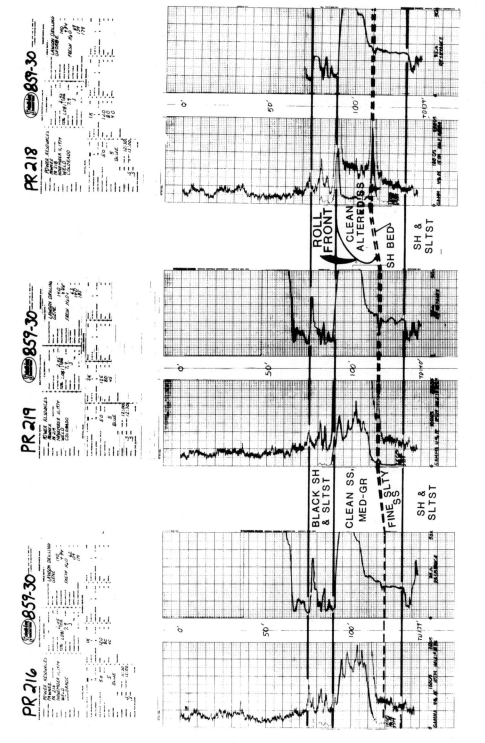

Figure 10-41. Correlation diagram illustrating the use of electric and gamma-ray logs for correlation. Thin shale and siltstone units are clearly recognized on the resistance curve. This diagram also shows a typical roll front, with the limb mineralization on the right associated with altered sandstone.

gained from expensive drilling programs. This is a real tragedy of widespread importance in the uranium industry today.

Acquiring Good Samples

Inexperienced or poorly trained geologists are frequently inclined to depend too much on the judgment, expertise, and initiative of drillers to collect good samples. Some drillers are conscientious and will try to collect good samples, but they have no way of knowing what the objectives are or what peculiar characteristics of the sands in question might cause problems in recovering good samples. The field geologist who supervises a drilling operation should be adequately trained to take steps necessary to assure that best efforts are made to obtain good samples.

Each situation requires a unique solution, and there are too many variables to be inclusively discussed herein. One common example is a fine-grained sandstone objective which occurs stratigraphically below mudstones which tend to break down and create a heavy silty drilling mud. Fine sand grains will not settle out of heavy, viscous mud, and it is possible to drill through a good host sandstone and see *nothing* in the samples to suggest that a sand even exists there. In such situations, the geologist should instruct the driller to condition the drilling mud before reaching such an objective. If fresh water entering the drill hole will cause caving or other problems, it might be necessary to work out a mud program with special additives to get good sample recovery. It is recommended that mud engineers be consulted on such occasions. The important thing to remember here is that the geologist is the only person in the field who knows where the primary objectives are in the hole and what their characteristics are; therefore, he is the only one who can really define the problem and adequately seek a solution.

Obtaining good samples is almost always the responsibility of the geologist, and it is he who must take the blame for poor samples.

Preparing and Describing Samples

It is our experience that samples are best described right on the spot shortly after the driller places them on the ground (electric log of hole must be on hand), and before they are placed in sacks or containers, if this is done at all (fig. 10-42). Many sandstone hosts are poorly consolidated and easily disaggregated when drilled in the conventional manner. Clay or other soft matrix material is often lost to the drilling mud except for small amounts which are more firmly stuck to sand grains. The black stain which indicates the presence of humic substances, or the red or brownish-yellow stains of a hematite, limonite, or goethite are probably the most important details which support an interpretation as to whether a sand is altered or unaltered. These stains can be masked or even obliterated when samples are sacked, transported, and perhaps stored for a period of time.

Some operators have set up programs in which a person without geologic training works in the field with the drillers and collects samples. The samples are sacked or put in some sort of sample containers, the containers are identified, and they are then transported to the office or laboratory where geologists examine them (usually under binocular microscopes). Those who favor this procedure mistakenly believe that a more rigorous examination is likely to be had in the comfort of an office or lab than in the field. In practice, this procedure has many pitfalls, and we strongly recommend against it. First, it is important to have a geologist supervising the drillers to insure that good samples are acquired (especially of the critical zones). Excessive sample lag or missed samples

Figure 10-42. With a 14X hand lens, a geologist takes a close look at mineral composition of sandstone from samples near bottom of hole, collected by driller while hole was being drilled. Sample piles each represent 5 ft (1.52 m) of interval in the drill hole, and they are laid out 20 to a row or 100 ft (30.5 m) per row. Samples from the top of this drill hole are in the left row at the bottom of the photograph, the rock hammer is at 160 ft (48.8 m), and the geologist's right foot is at the 260-ft (79-m) sample. Total depth of this hole is 280 ft (85 m). The probe record for this hole, along with the sample logging form, is in the vehicle, safely out of the wind.

will soon be apparent. Further, the geologist is the only person with the data and competence to know where the objectives are and what particular characteristics of the objectives might cause problems in acquiring samples. Second, the geologist should be able to instruct, supervise, or, if necessary, carry out any special sample preparations which may be required. Third, most efficient drilling programs should be constantly modified, depending on the results of each hole. The data from each

hole must be evaluated promptly in order to be utilized in planning and modifying the program. When the samples are collected and transported for examination in the office, a critical and intolerable delay is built into the data processing. Fourth, the observant geologist promptly examining drilling samples will acquire a firsthand working knowledge of mineralization habits, appearance of certain units on the electric log, and details of other mineralogic evidence such as staining, corrosion, anomalous increases in contents of sulphides, and so forth. *The geologist can learn what to look for.* This working knowledge will usually increase as the samples from each hole are described in a given area and, coupled with some good cross sections, a powerful tool for the forecasting of ore occurrences will be available.

When drilling is being done in moderately friable sandstones, siltstones, and even mudstones, we have found that one of the valuable sample-examination techniques the geologist should use is to select unbroken fragments of the material from the sample pile and break the fragments open to reveal undisturbed material inside. Typically, these pieces will range from the size of a pea to the size of a large bean, and the sands and silts will be rounded to subrounded due to abrasion in traveling up the drill hole. Clays will be less rounded. Very coarse (and coarser) friable sands with no clay matrix and no cement will arrive at the surface as loose sand and not as fragments.

The geologist should squeeze and crumble the samples between the fingers to feel for clay content, friability, texture, and so forth while the samples are wet. If the samples have dried, water should be available to wet some of the material. A small bottle of dilute hydrochloric acid with a dropper bulb should be on hand so that the sample may be tested for effervescence to determine if it is calcareous.

Some operators have established a routine with their drillers whereby drilling mud on each sample is washed away by placing the sample in a coarse sieve about 15 cm in diameter and dunking it vigorously in a bucket of water before placing it on the ground. To us, this procedure is almost always *unacceptable.* It is acceptable only if the lithologies are such that important components will not be lost through the sieve, and in 99 situations out of 100, important components will be lost, and the finer grains will be lost consistently. To verify this, if the water is poured out of the driller's bucket after washing a few samples, a thick layer of fine sand and other components will be found in the bottom together with silt and mud.

On occasion, we have used extra-fine sieves for cleaning mud from sand samples and for concentrating heavy materials such as pyrite, but even when carefully using a fine sieve, fine sampled material is almost always lost. Given this precaution, however, the fine sieve is a quick and effective way to clean sand samples and concentrate the heavy minerals in a small part of the sample collected by the driller. The fine sieve may be filled with sand and gently dunked in water with a slight sideways movement to permit the partially suspended grains to sort themselves by size and density. The washed sample should be turned upside-down near its parent pile, on the palm of a hand, on a sample tray, or into a sample container (fig. 10-43). Certain coarse heavy minerals, often including pyrite, will then be concentrated on the top of the rounded and washed sample.

In most cases where heavy drilling mud tends to coat the cuttings and prevent good description, it is possible to carefully pour some water on the fresh sample pile and clean the upper part of it. This is a quick and easy method which, to be effective, must be done while the samples are still wet; once the mud dries, this will not work. Best results are obtained on coarse material, because fine sand tends to be washed away with the mud.

If it is important to concentrate the heavy materials in a sand sample, the old panning method used in gold panning is valuable. Even though this is more time consuming than the methods referred to above, it is comforting to know that you are going to get *all* the heavy materials, coarse and fine, and this may help in making critical decisions. Many geologists logging "no visible pyrite" for a certain stratigraphic section or mineralized interval have then panned a sample of the material and have been surprised to discover abundant fine pyrite. We recommend that the gold pan be used much more frequently than it is at the present time. Some explorationists carry two gold pans as standard equipment in drilling for uranium in sediments: one is a small (14-cm diameter) pan, and the other a standard (30-cm diameter) pan. The smaller one is usually used to pan samples from rotary drill holes.

Obviously there are more sophisticated methods than those outlined above for cleaning samples and concentrating heavy minerals, but these are the quick and easy field methods which have been applied successfully to large numbers of samples.

Potentially important samples (if you have a *good* idea of which ones may fall in this category) should be collected and preserved for more detailed preparation and study in the event that they might provide data needed in possible later investigations. However, again we caution that collection of samples is often used as an excuse for poor initial work in the field. Good field logging eliminates the need for most samples (except core or sidewall samples for assay, of course). In a program of drilling 200 to 300 holes averaging

Figure 10-43. Concentrating heavy minerals and cleaning a sand sample, utilizing a fine sieve. The sample was washed carefully in a bucket of water and then turned upside-down in the palm of the geologist's hand. Sometimes the pan is used to concentrate heavy minerals.

100 m in depth, it is doubtful that we would collect more than three or four samples unless something very unusual showed up. Some of this philosophy relates to the low cost of redrilling if it were ever necessary. If we were coring a deep, expensive hole, we would retain every sample regardless of its value for assay data.

Regarding the matter of the driller (or helper) collecting samples from the rotary drilling of a hole, in most situations it is expedient, and the results are acceptable, when the samples are laid out on the ground for examination. At the present time, most operators in the United States have set up a standard procedure whereby samples are laid out in rows representing 30.48 m (100 ft) so that they can be examined much as one reads standard script from left to right and top to bottom. It is also helpful to the geologist if spaces are left between 6.09 m (20 ft) intervals in

the rows. These procedures will reduce the possibility of errors in depth being made by the geologist who describes the samples.

When drilling is carried on in subfreezing weather, the wet samples will usually freeze quickly after they are placed on the ground. If they are rinsed off with water before they freeze, so that the surface of the sample is not covered with drilling mud, the geologist can have a reasonably good look at the mineral constituents even though the sample may be frozen solid. Frozen samples may have to be broken up with a rock hammer.

Obtaining and describing samples requires desire and effort on the part of the geologist if the job is to be done right. Some geologists make a strong effort and do obtain very good sample descriptions in the field even under poor conditions. On the other hand, there are those who cannot or do not get sample descriptions completed in the field; some have been known to make fictitious sample descriptions in the motel room at night when they really had no idea what the samples looked like. Such descriptions can make later interpretations extremely difficult and hazardous. Recognizing that weather conditions occasionally can become so severe that writing is impossible because of numb fingers, we still opt for the honest log, with a warning to the person who might later use it that the log was made under certain adverse weather conditions. For example, in Wyoming, strong winds and sub-zero temperatures are common in winter storms, resulting in extreme chill factors. Such conditions should be written directly on the sample log form.

Under almost all operating conditions, we prefer to delay preparation of the sample log until the mechanical log is available. We have found that, in most circumstances, people (including ourselves) fail to see many subtle differences or changes in the samples unless they are specifically looking for them. The mechanical log will usually indicate where lithologic changes take place, and it will give clues as to where sands are altered and unaltered. Major distinctive lithologic contacts such as sandstone, mudstone, and coal beds can be plotted on the sample log using the mechanical log before examining the samples in detail. Gamma anomalies should also be indicated on the sample log at this time.

Each log should include an accurate location of the hole relative to nearby holes or a description of how the hole was located relative to a section corner or other recognizable and permanent landmark. Sample logs should allow enough space to print a thorough description of significant lithologic units (fig. 10-44). For most drilling operations, a scale of 1:240 is good. Core logs should, of course, be

expanded to accommodate a more detailed description.

In our opinion, sample log forms which consist primarily of check-off columns for color, grain size, sorting, rounding, and accessories may have weaknesses. First, it is easy to make errors in checking long, thin columns. We have often found it difficult to tell if the geologist was in error when colors were checked on a listing, but if it is written out, we at least know what color the geologist thought was prevalent. Also, because of mixing, or if there are several colors in the particular interval, there is always some interpretation involved in concluding what color belongs to the zone being described. A written description permits the geologist to express his impressions and best judgments.

A typical sample of disaggregated sand taken as a whole will commonly have one tint of color, but individual grains seen through the hand lens might be stained with distinct colors which provide the best clue as to whether the sand has been altered. The "check the box" type of sample logging is stifling to a thinking geologist, and we contend that it is important that the geologist be given the opportunity and incentive to focus on gathering information which appears to be most useful in interpreting the potential of the rocks under study as hosts for ore bodies. If a geologist cannot describe samples by any method other than the "check the box" method, we wonder if he is a geological technician.

In most sandstone-mudstone sequences, it is adequate to describe most of the samples in the field with a good 14-power hand lens. In some fine-grained sandstone sequences, however, it might be better to use a 20-power lens. It is always good to have a binocular microscope available for more detailed examination of critical samples. Even the microscope work, however,

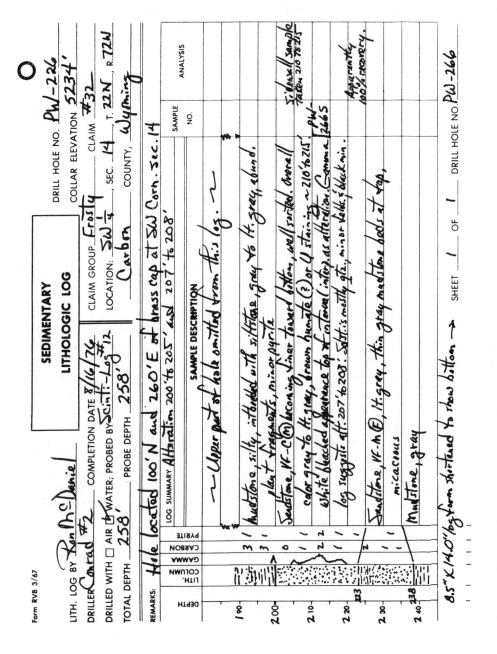

Figure 10-44. Sample log of sandstone illustrated in figure 10-39.

should be done without unnecessary delay.

Normally it is best to describe the samples from the surface downward, so that the depths of first appearance of critical lithologies can be noted. If the host rock sought is a sandstone, then the sandstones are most closely scrutinized. In the initial phases of exploration in an area, it is important to identify accessory minerals in the sands and fully classify the sands by describing mineral content, ratio of feldspar to quartz, grain size, rounding, and so forth. Careful classification of sands has sometimes been helpful in working out patterns of deposition, ultimately shedding light on the origin of mineralization.

In most uranium exploration, color of the sands is generally considered to be important, because the degree of oxidation is reflected by the colors of the interstitial iron. Some colors are significantly different when wet or dry, and, when necessary for accuracy, the geologist should indicate wet or dry when describing the colors. Iron oxide coatings on the grains are usually yellow, rusty brown, or red. Sometimes the sand grains appear clean, but the clay matrix will be discolored. Unoxidized host sands often have visible amounts of carbon or humate. Humate is sometimes evidenced by black staining on the grains.

Some years ago, the Geological Society of America arranged for the printing of color charts that assigned letters and numbers to colors to facilitate descriptions of rocks and minerals. These charts have not found general acceptance in many segments of the industry, including uranium exploration, because standard color terminology has proven to be very adequate and not cumbersome. Further, it can be read and understood by almost anyone, and it is unlikely that the geologist will list every color observed anyway. For example,

an altered sandstone in samples might be described as ranging in color from pale rusty yellow to dark rusty brown, or from bleached white to pale pink to old rose. These are good descriptive terms for colors. For some individuals, there may be an important link between carefully observing colors and then describing those colors satisfactorily in writing, a process which requires a personal or artistic touch; such a touch could easily be lost in translating observed colors to cold, impersonal letter-number designations such as those on color charts.

There is a further hazard in sample logging forms which allow only a small space for color designation, and that is that the geologist is forced to shorten color descriptions (and to abbreviate) to fit the space. This is a serious shortcoming, because colors are so important. We will not accept the argument that additional color descriptions can be satisfactorily placed under a "general comments" column.

The mechanical log can often be used to aid in interpreting how much clay matrix a sand contains, but sands which disaggregate during drilling usually lose their matrix to the drilling mud. The clay matrix, if any, should be mentioned in the sample log, including the color if it is different from the sand as a whole.

PLUGGING DRILL HOLES

Past Practices

During the early development of the uranium exploration industry, many operators abandoned exploration holes with no effort to reclaim the surface or plug the hole. In some areas where drilling was widespread and extensive, this practice led to significant changes in the wildlife. An example is the Gas Hills area of Wyoming, where drilling was very extensive during the 1950s and early 1960s and many of the drill holes were abandoned. Rattlesnakes were quite common in the Gas Hills area when the uranium exploration got under way, but they are very scarce in the area now. Rattlesnakes have a tendency to enter small holes for protection from the hot sun or when seeking rodents as food, and the drill holes appear to have been very effective traps for these snakes. It is also likely that these open holes served as deadly traps for other small animals, including rodents and perhaps even small rabbits.

Also, when a hole is left open, it may tend to cave in over a period of years and leave a sharp, deep pit in the ground, hazardous to people or livestock traversing the area. During the 1960s, most conscientious operators in the industry voluntarily set up procedures for plugging holes, which would insure against caving and protect the wildlife in the area. The most common procedure, first employed during the 1960s, was to insert a plastic and metal plug in the hole 3 to 6 ft below the surface and pack soil and dirt above the plug, mounding it up at the surface. We have visited drill sites eight to ten years after this procedure was used, and the holes cannot be recognized unless a permanent marker was put there to indicate where the hole was drilled.

State and Federal Requirements

Several states have passed legislation which requires specific procedures for plugging exploration drill holes. In most cases, these state laws require the use of concrete beneath the surface of the ground as part of the plugging procedure. On Indian lands and federal lands supervised by the Conservation Branch of the U.S. Geological Survey, concrete is also generally required as a part of the plugging procedure. In addition, the U.S. Geological Survey requires the cementing of

aquifers encountered during drilling (fig. 10-45).

In New Mexico, the state engineer supervises the plugging of drill holes; his office published the following procedure for holes drilled in the San Juan Basin which do not penetrate formations older than Jurassic:

Drill holes that are not plugged with cement shall be plugged to the land surface at the time of abandonment with drilling fluids which meet the following specifications:

(a) ten minute gel strength of at least 20 lbs./100 sq. ft.; and

(b) filtrate volume not to exceed 13.5 cc.

The above properties shall be determined in accordance with RP 13-B, Sections 2 and 3 (low temperature test), respectively, Standard Procedure for Testing Drilling Fluids, Third Edition, February 1971, American Petroleum Institute. The tests shall be conducted on a drilling fluid sample taken at the hole collar after the total depth of the hole has been reached and all circulation has been completed. A cement plug may be used at the surface for a top cap.

The weight of the drilling fluid left in the drill hole at the time of abandonment shall be sufficient to prevent flow of water into the hole from any aquifer penetrated. In the alternative drill holes may be plugged bottom to top with a neat cement slurry weighing not less than 15 lbs. per gallon; the weight of the neat cement shall be sufficient to prevent flow of water into the hole from any aquifer penetrated.

Drilling-mud additives have been developed which safely prevent damage to aquifers (figs. 10-46, 10-47).

425

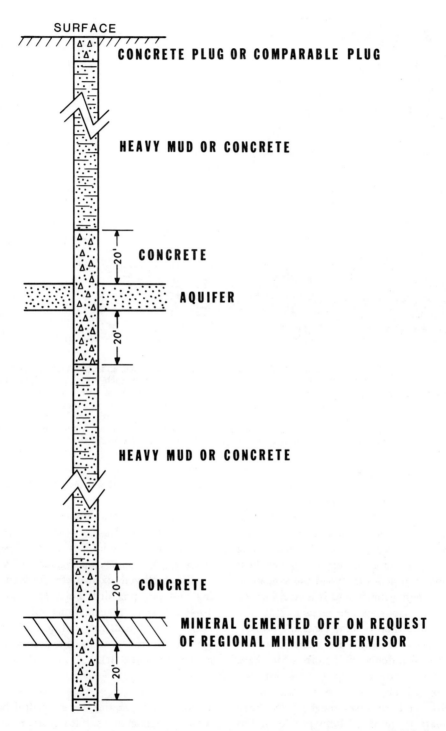

SURFACE

CONCRETE PLUG OR COMPARABLE PLUG

HEAVY MUD OR CONCRETE

20' CONCRETE

AQUIFER

20'

HEAVY MUD OR CONCRETE

20' CONCRETE

MINERAL CEMENTED OFF ON REQUEST
OF REGIONAL MINING SUPERVISOR

20'

Figure 10-45. Diagram illustrating federal specifications for plugging exploration drill holes.

Figure 10-46. Drilling mud in mud pit after total depth has been drilled and mud has been circulated with the drill pipe and bit still on the bottom of the hole. In order to reach the proper weight and viscosity, it is often necessary for the driller to mix drilling mud (such as bentonite) and other additives with the mud in the pit and circulate the mixture through the hole. In shallow holes where lost circulation is not a problem, and where low aquifer pressures prevail, such mud mixtures effectively prevent communication between aquifers. The drill hole is usually probed after the mud is mixed and the drill has left the hole. "Mudding up" the hole also helps prevent caving of the walls of the hole.

Special Procedures

Since 1975, solution mining (*in situ* leaching) of uranium has become important to the industry, and, in some of these operations, it is necessary to thoroughly plug with cement those sandstones which host the uranium deposits. When the sandstone has low permeability and it is necessary to inject solutions under pressure, it has been found that in some cases the pressurized solutions will enter and move up through abandoned drill holes which have not been carefully plugged. It has not been determined at this time if standard or special mud additives might have been adequate to prevent these problems, but they probably would. It is difficult to imagine a column of thick bentonite being pushed out of a hole by low nearby pressure.

When substantial amounts of cement are used in the plugging of drill holes, the cost of exploration goes up very significantly. In some operations, a small rig is employed exclusively for cementing and plugging drill holes (figs. 10-48, 10-49). A small rig cementing holes behind two standard drill rigs will add 25 to 30 percent to the cost of the program (including cement cost).

Hole Markers

It is a good practice to mark drill holes with a permanent surface marker unless the area is a cultivated field. A 5 x 5-

Figure 10-47. Drilling-mud sample is taken from the mud pit and poured into a testing funnel so that viscosity may be measured. As a second step, the weight of the mud will be determined on a special balance designed for that purpose. If the viscosity and/or weights are too low, additives will be mixed with the mud, recirculated through the hole, and then retested.

cm (2 x 2-in.) wooden stake with a stainless steel metal tag embedded in the cement plug is a fairly long-lasting marker for a drill hole. In cultivated areas, it is a good idea to mark the hole with a short length of heavy steel reinforcing rod (rebar) placed in the upper part of the hole but below the soil zone so that it will not interfere with cultivation. If a resurvey is needed, this metal marker can be located with a metal detector. It is always a good practice to survey the holes accurately soon after they are drilled for accurate map preparation and easy location of the holes at a later time.

RECLAMATION

Most exploration drilling causes some temporary damage to the surface of the ground. The use of a portable steel mud pit results in the least surface damage when drill holes are less than 100 m in depth. After such a hole is completed, a small cuttings pile remains, which should be scattered out by shovel around the drill site. If the hole is located in a cultivated field, the surface will be reclaimed as part of the cultivation process. On rangeland, it is usually good practice to seed (with recommended seed) any location where significant grass disturbance occurs. We have had occasion to return to some drill sites just one year after the holes were drilled and have been unable to find them. This has happened in arid Colorado and Wyoming during periods of low rainfall.

When pits are dug into the ground in conjunction with drilling holes deeper than 100 m, it is important to set the topsoil aside for reclamation purposes after

Figure 10-48. Drilling rig being used to cement drill holes after the holes are probed. In this photograph, dry cement is being dumped into a mixing tank. The kelly and the pump hose of the drilling rig are also in the tank preparatory to mixing the cement. Note that the water truck in use here has a small-capacity tank.

the pits are refilled. Dug pits can result in damage which lasts for years if the topsoil is not properly replaced.

In areas where topographic problems require the use of a bulldozer for leveling drill sites and constructing access roads, it is important to minimize loss of topsoil. Grass will usually grow over an abandoned road if topsoil is not completely removed. If topsoil must be removed, it should be set aside and replaced over the road or site later.

When roads are constructed on slopes, water bars should be carefully installed. These are horizontal, shallow, and rounded ridges of earth which extend across the inclined road at intervals to divert surface runoff to the side of the road during rains or spring thaw. Water bars, when properly installed, effectively reduce erosion.

Until most vegetation has been restored, an area of recent concentrated drilling activity should be periodically examined and steps taken to assure successful reclamation. It may be necessary to seed grass in the spring, or to fertilize the grass where soil has been disturbed. It is the responsibility of the explorationist to restore the surface in areas where activities under his direction have damaged it. Funds for reclamation should be part of every exploration budget involving land disturbance.

HAZARDS

The exploration geologist should be alert to hazards in drilling projects. Drillers and their helpers have been seriously hurt and even killed by raising the mast of the drilling rig into overhead power lines. Even more common is the act of driving a rig to the next location with the

Figure 10-49. After the cement is mixed, open-ended drill pipe is placed to the bottom of the hole and the cement is pumped into the hole, displacing mud already in the hole. In some cases, columns of cement will be placed at certain intervals in a drill hole. At the locality pictured here, closely spaced drilling is being carried out in fences across a roll front, and the drill holes are being cemented in anticipation of solution mining. The rig at left is cementing a hole, and the rig at right is drilling a new hole. In between the two rigs, a small stake may be seen which marks the location of a previously drilled hole.

mast raised, to save time; in such a procedure, there is a risk of running into power lines (fig. 10-50). If this happens, the driller should back away or *stay in the truck*. If he gets out and steps to the ground he will most likely be instantly electrocuted.

Drillers have also drilled into buried pipelines that transmit gasoline or natural gas. These accidents can usually be avoided if the geologist observes the dangers and modifies the drilling pattern. Some areas have shallow pockets of flammable natural gas which can blow out during drilling operations; for-

tunately such occurrences are rare.

Drill collars and other equipment around a drill rig are heavy and can be hazardous if a driller or his helpers are careless. The geologist should be alert to hazards and encourage careful procedures. One of the authors was nearly killed when a driller carelessly released a drill collar which had been stacked vertically in a rack against the mast of the drill rig. The collar crashed into the rear end of the probe truck very close to where the author was standing, and lifted the front end of the truck about 1 m off the ground.

Figure 10-50. Drilling rig and water truck moving to a new drilling site with the mast raised. Such a procedure should not be followed where hazards such as overhead power lines exist. Several drillers have been electrocuted by stepping out of the truck while the mast was in contact with a power line.

11
Using Geology in Planning and Executing Drilling

GENERAL COMMENTS

Most uranium exploration drilling is directed at potentially mineable targets, but there are situations where other information is sought—in stratigraphic test drill holes (strat holes), for example. It is the responsibility of the explorationist to use all available geological, geophysical, and geochemical evidence in planning drilling, whether one hole is planned or one thousand. As the program progresses, each new increment of data must be integrated, interpretations modified, and the drilling program appropriately adjusted to the new interpretations. Unfortunately, many uranium exploration programs have been understaffed or staffed with inexperienced geologists, with the result that the programs suffer from inadequate planning.

In most situations, the large majority of data gathered from drilling will be negative. That is, it will not indicate that environments in which ore bodies may have formed are, or were ever, present. Such negative information nevertheless has an important value to the geologist because it helps sharpen the focus of later work on areas which may possess potential. Without negative information, we would not know how to weed out areas of low or no potential.

The skill of the geologist comes into play as data are gathered and interpreted and decisions are made. Good basic information must be gathered in the course of the program, or decisions made on the basis of that information will not be good. We have all heard the comment made about computer-processed information—"garbage in, garbage out"—and the same may be said of data gathered and used in drilling programs. With good basic information, a skillful geologist can make deductions pertaining to the direction a drilling program should take that will optimize expenditures and result in conclusions upon which corporate management can rely.

PHASED DRILLING CONCEPT

Most drilling programs with which we are familiar and which are well planned and executed can best be described as an application of a phased drilling concept (fig. 11-1).

Initial Phase

The main objective of the initial drilling phase should be to determine if the prospect area has the basic qualities necessary to host mineable uranium deposits. In sedimentary basins, where strata-controlled deposits are anticipated, the initial drilling usually consists of widely spaced holes drilled to depths which are believed to include all of the reasonably potential zones. In contrast, the initial phase of drilling in metamorphic and igneous terranes, where structure- or intrusive-controlled deposits are anticipated, frequently consists of shallow holes. These shallow holes are sometimes drilled in more closely spaced patterns than those that characterize the later stages of deeper drilling, as we shall discuss later in this chapter.

Intermediate Phases

This drilling is designed to progressively define and evaluate controls on mineralization. In sedimentary basins, these intermediate phases usually consist of more closely spaced holes directed at smaller and more distinct targets.

Predevelopment Phase

The last phase of exploration drilling is directed at those targets which have been enhanced by intermediate drilling and appear to contain mineable deposits, or for which additional information is needed. This phase is sometimes called "predevelopment drilling," and it usually provides enough data to make some preliminary ore-reserve estimates. The predevelopment drilling program should include sufficient coring or sidewall sampling for assaying to determine if other metals of value occur with the uranium and to determine if a disequilibrium problem exists in the uranium occurrences.

Disequilibrium

Uranium in nature includes two isotopes: ^{238}U (99.3 percent) and ^{235}U (0.7 percent). The ratio of these two uranium isotopes is constant in nature except for a rare occurrence at Oklo in the Gabon Republic (Africa), where "reactor zones" within a uranium-ore body contain as low as 0.44 percent ^{235}U (Cowan, 1976). Because of the dominance of ^{238}U and its decay products, ^{235}U and its daughters are not included in the chart of isotopes (see fig. 3-4), and ^{238}U is the isotope with which the explorationist deals.

Most drilling programs depend heavily on gamma-ray logs from inception to completion for recognizing uranium mineralization as well as calculating the grades and thicknesses of uranium occurrences encountered by drill holes. The calculation of eU_3O_8 from gamma-ray logs has been discussed in chapter 10, and disequilibrium was briefly described. During predevelopment drilling, it is important that any disequilibrium be identified and evaluated so that better direction of drilling will be possible and so that corrections may be applied to reserve estimates.

After a uranium deposit has been formed by precipitation from solution, it requires about 3×10^5 years to develop a complete chain of radionuclides, all of which have reached a steady state or equilibrium concentration. Approximately 98 percent of the gamma emission is produced by the radium group (radium and its daughters), particularly ^{214}Pb and ^{214}Bi.

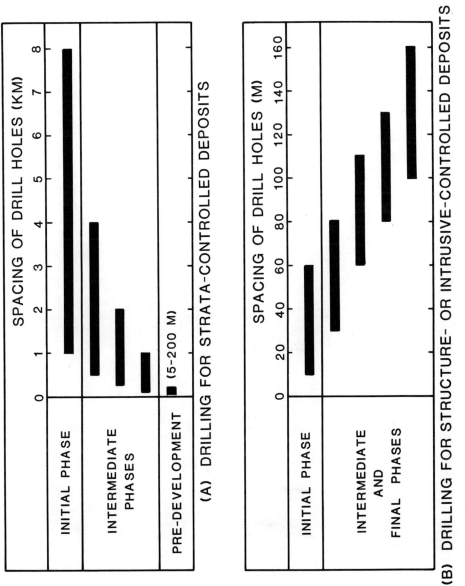

Figure 11-1. Chart showing spacing of initial, intermediate, and predevelopment drilling in strata-controlled deposits contrasted with phased drilling program typical of structure- or intrusive-controlled deposits.

434

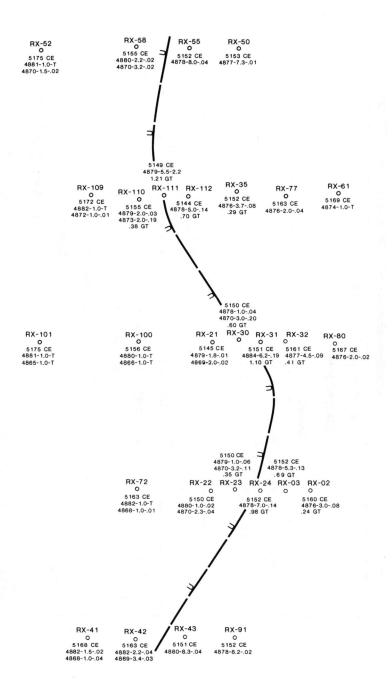

Figure 11-2. Intercept map of the RB ore body. *Note:* these maps, which illustrate the methods of preliminary evaluation of an ore body, were originally drawn at a scale of 1:600. They have been reduced for publication.

435

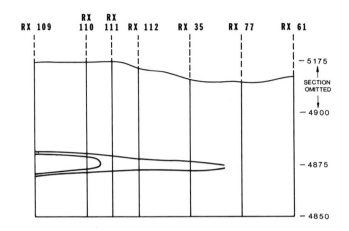

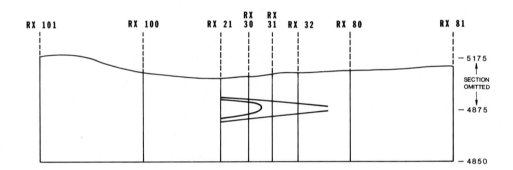

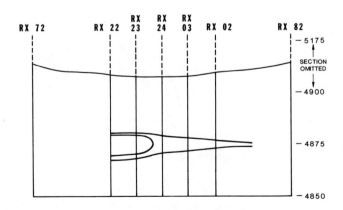

Figure 11-3. Cross sections showing profile configuration of the RB ore body.

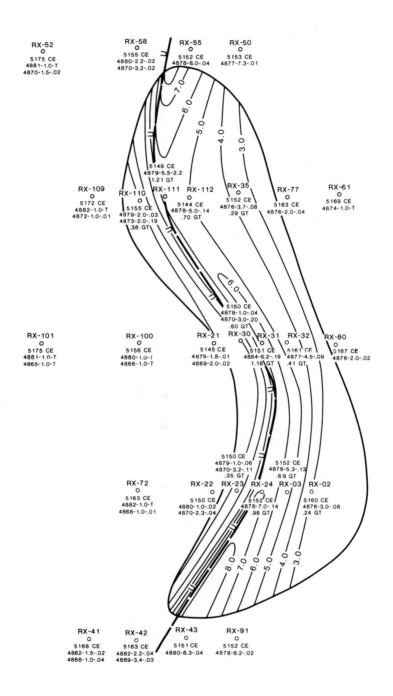

Figure 11-4. Isopachous map of the RB ore body.

The gamma-ray log is a measure of the amount of ^{214}Pb and ^{214}Bi present in the ore. If the ore is not in equilibrium between ^{238}U and ^{214}Bi, calculations based on gamma-ray logs will be in error. The value calculated from a gamma-ray log is called eU_3O_8 or gamma-*equivalent* U_3O_8. The calculated eU_3O_8 might be either higher or lower than the actual U_3O_8 content of the ore, depending on the nature of the disequilibrium.

A common form of disequilibrium is caused by radon loss, which can be measured by the so-called closed-can gamma assay. The closed-can assay method partially regenerates equilibrium conditions by sealing an ore sample in an airtight container for several days to permit radon-222 and its immediate short-lived daughters to accumulate.

Disequilibrium may develop when oxidizing ground water passes through an ore zone, dissolves uranium from one part of the ore body, becomes reduced, and precipitates the uranium in another part. In such circumstances, there might be a net addition or loss of uranium relative to established gamma-emitting daughters. In theory, it is possible to remove all of the uranium from an ore body, leaving the gamma-emitting daughters in place. Because of the ease of movement of water through sandstone, disequilibrium is common in shallow roll-front deposits which are exposed to weathering or oxidizing ground water. Corrections can be made for disequilibrium by comparing total gamma-count calculations with either chemical analyses or beta-gamma analyses.

The beta-gamma analysis is based on two measurements: (1) high-energy beta, which includes the beta emissions of ^{234}Pa, ^{214}Pb, and ^{214}Bi, and (2) high-energy gamma, which includes ^{214}Pb and ^{214}Bi. If the high-energy gamma measurement is subtracted from the high-energy beta measurement, the result is a measurement of ^{234}Pa concentration in the sample. Protactinium-234 is always in equilibrium with ^{238}U, because of the short half-life (24.1 days) of its predecessor ^{234}Th and its own short half-life of 2.18 minutes (see fig. 3-4); hence a determination of ^{234}Pa concentration can be converted to ^{238}U.

Projected Reserve Calculations

One of the most important objectives of the predevelopment drilling is to gather data for reserve estimation. The approximate shape of the ore body must be determined, and the volume must be estimated. Some generalities about the distribution of grade within the ore body must be formulated. Later on, when uniformly dense development drilling has been completed, the calculation of reserves will often be clear and straightforward. However, calculations using predevelopment drilling information usually require a certain amount of interpretation based on apparent ore controls and other geologic concepts in order to come up with reasonably reliable reserve estimates. The geometry of an ore body projected from predevelopment drilling should conform to the geologic controls which are known or believed to exist. An isopachous map which shows the thickness and planimetric form of the ore body should be prepared with the cross sections as guides (figs. 11-2, 11-3, 11-4). Once an ore body has been illustrated on numerous cross sections, a grade-thickness product (GT) map may be prepared for calculating the projected reserves. This, in effect, is a contour map showing the weighted average grades and thicknesses of the ore body (fig. 11-5). The GT map should resemble the isopachous map, unless the grades vary inversely with the thicknesses. In most cases, there is a positive correlation between

438

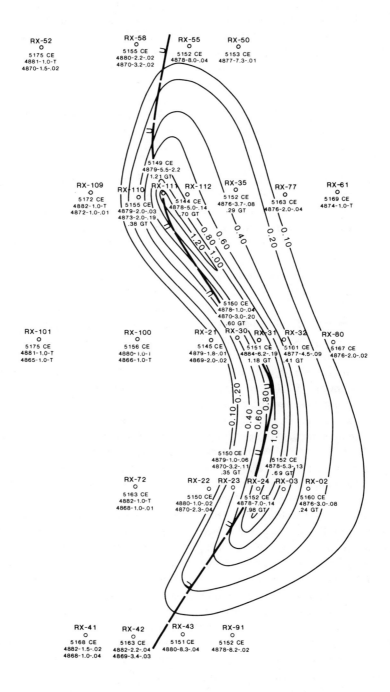

Figure 11-5. Grade-thickness product (GT) map of the RB ore body. Note contrast with the isopachous map (fig. 11-4). This map integrates grade variations with volumetric map.

TABLE 11.1. CALCULATION OF PROJECTED RESERVES
BASED ON GRADE-THICKNESS MAP

GT increment	$\overline{GT}$*	Map area (sq. in.)[†]	Area (A) (sq. ft)[*]	$(\overline{GT})(A)$[#]	U_3O_8 (lb)[**]
0.10-0.20	0.15	7.96	19,900	2985.0	3,852
0.20-0.40	0.30	8.97	22,425	6727.5	8,681
0.40-0.60	0.50	4.75	11,875	5937.5	7,661
0.60-0.80	0.70	3.79	9,475	6632.5	8,558
0.80-1.00	0.90	1.79	4,475	4027.5	5,197
1.00-1.20	1.10	1.73	4,325	4757.5	6,139
1.20 $\longrightarrow$	1.20	0.17	425	510.0	658
					40,745

$\overline{GT}$ = averaged grade-thickness product. The grade-thickness product map is a method of integrating grade and thickness values for the ore body. G is expressed without the percentage. For example, if $G = 0.2$, it represents a grade of 0.2 percent U_3O_8. G x 20 = pounds per ton.

[†]These values are usually obtained by planimeter.

[*]Values converted from the map scale.

[#] $\overline{GT}(A)$ is an integrated value, equivalent to $\overline{GV}$ where the grade and volume are averaged. The ore-density conversion factor used in these calculations is 15.5 cu. ft = 1 short ton (2,000 lb).

[**]$\dfrac{(\overline{GT})(A)(20)}{15.5}$ = lb U_3O_8 per GT increment.

TABLE 11-2. CALCULATION OF PROJECTED RESERVES
BASED ON BLOCK METHOD

Block	Map area (sq. in.)	Area (sq. ft)	Thickness (ft)	Volume (cu. ft)	Tons ore	Grade	U_3O_8 lb
I	1.83	4,575	2.0	9,150	590	.19	2,243
II	1.30	3,250	5.5	17,875	1,153	.22	5,074
III	2.20	5,500	5.0	27,500	1,774	.14	4,968
IV	3.03	7,575	3.7	28,027	1,808	.08	2,893
V	1.85	4,625	3.0	13,875	895	.20	3,581
VI	0.96	2,400	6.2	14,880	960	.19	3,648
VII	2.08	5,200	4.5	23,400	1,510	.09	2,717
VIII	1.54	3,850	3.2	12,320	795	.11	1,749
IX	1.12	2,800	7.0	19,600	1,265	.14	3,541
X	1.14	2,850	5.3	15,105	975	.13	2,534
XI	2.66	6,650	3.0	19,950	1,287	.08	2,059
							35,007

grades and thicknesses.

As illustrated in table 11-1, the projected reserves may be calculated from the GT map. The reserves calculated by this method are expressed in terms of contained U_3O_8. The actual volume (or weight) of the ore body may be determined by planimeter from the ore isopachous map. If this is done, the average grade might also be calculated.

We frequently use the GT map method for calculating projected reserves from predevelopment drilling because it appears to be one of the most realistic methods of integrating and extrapolating the data obtained during a typical predevelopment phase of drilling. The GT map permits the interpretive use of ore controls and other geologic information. It can be easily adapted to ore bodies with complex configurations.

The profile projection method, also called the block method (fig. 11-6), is a simple way of making a rough calculation of projected ore reserves when the configuration of the ore body is not complex. When profiles have been drilled across an ore body, cross-sectional areas and grades may be approximated by projecting ore thicknesses and grades halfway between adjacent holes and halfway between profiles. The explorationist is limited in his use of geologic parameters when employing this method, because it requires simple geometry and arbitrary but uniform boundaries. The method has the advantage that it is simple, but the areas still must be measured with a planimeter in many cases.

The methods presented here for the RB ore body differ by 14 percent in the total reserve estimate, and this might be anticipated if one examines the two maps carefully, noting the differences in extrapolation of data. The GT map probably gives a more accurate result.

DRILLING FOR STRATA-CONTROLLED DEPOSITS

Trend Deposits

The ore bodies in trend deposits range from a few tons to several hundred thousand tons, and they have been found at depths ranging from a few metres to more than a thousand metres. These variations in size and depth impose respective variations in the drilling programs. Exploration drilling for trend deposits includes evaluation of regional distribution of favorable host rocks, and provides information on more specific prospects as well. Our initial phase of exploration drilling is partly completed in programs designed to extend the limits of known mineralization.

Initial Drilling Phase

The initial drilling should provide information concerning the existence of potential host rocks and, if they are present, their extent and quality. On occasion, this phase of the drilling program results in the discovery of ore-grade mineralization.

In new prospect areas far removed from known occurrences, the basis of the initial drilling usually depends largely on the depth of the potential host rocks. For example, the Westwater Canyon occurrences in the San Juan Basin north of the Ambrosia Lake trend are too deep to be detected by surface techniques (at least those now in use). The initial drilling for these deep deposits is necessarily based on regional geological interpretations (with heavy dependence on a *working hypothesis*). Our hypothesis is that most trend deposits occur in the major trunks of paleodrainage systems; thus, if the regional facies can be mapped, the basis for an initial drilling phase might result (fig. 11-7).

If the potential host rocks are exposed in outcrops near the prospect area, mapping

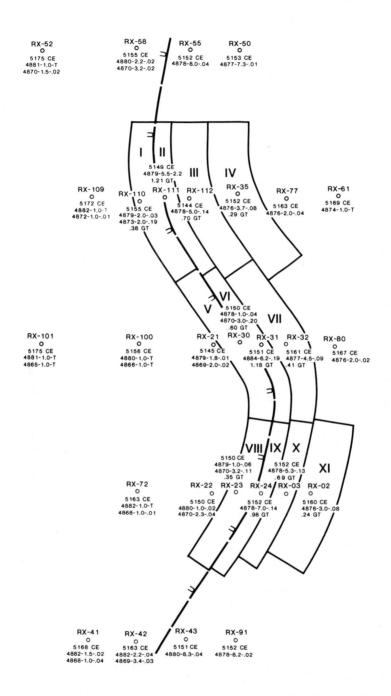

Figure 11-6. Block reserve map of the RB ore body. Note comparison with the GT map (fig. 11-5). This map provides a rough method of calculating ore reserves, but it is not recommended for complex ore bodies.

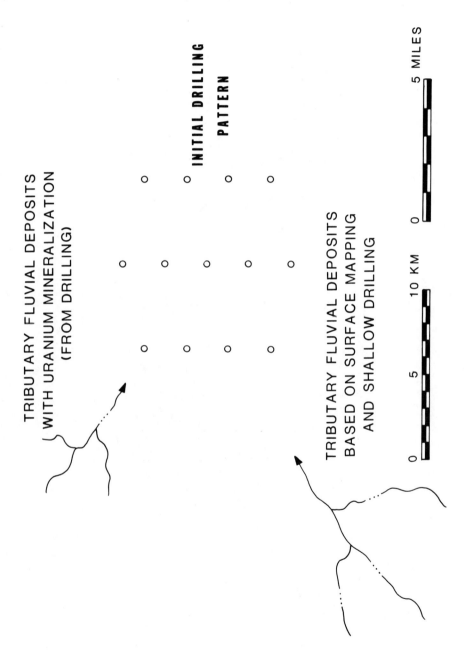

TRIBUTARY FLUVIAL DEPOSITS
WITH URANIUM MINERALIZATION
(FROM DRILLING)

INITIAL DRILLING
PATTERN

TRIBUTARY FLUVIAL DEPOSITS
BASED ON SURFACE MAPPING
AND SHALLOW DRILLING

0   5   10 KM

0   5 MILES

Figure 11-7. Idealized map showing paleodrainage pattern developed by surface work and shallow drilling. This might form the basis for an initial drilling program in a new prospect for trend depths.

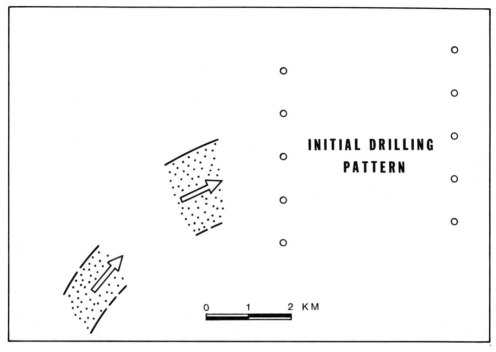

EXPLANATION

 FLUVIAL SANDSTONE

 PALEOCURRENT DIRECTION

Figure 11-8. Idealized map illustrating the use of surface data in projecting favorable trends for initial drilling. Surface exposures often reveal the reasonable size of the targets to establish optimum spacing within rows of holes.

in detail may enhance the regional mapping. From this information, projections may be made into the subsurface where drilling is planned. Mapping which may be of value includes current directional data, reduced sandstones, or even mineralization. Numerous Shinarump, Mossback, and Salt Wash outcrops in the Colorado Plateau include mineralized and reduced sandstones. In that area, it is frequently useful for drill planning to prepare maps which show the exposed channel sandstone in the setting of the regional paleodrainage pattern (fig. 11-8).

As discussed in chapter 6, geochemical data from ground water, including uranium, radon, and helium, can be useful in planning the initial drilling program. The first patterns may be designed to intercept projected sandstone trends or mineralization trends, as well as to evaluate geochemical anomalies.

Mapping and planning for the initial drilling phase is generally done on topographic base maps (when available) at a scale of 1:24,000. If these maps are not available, it is advisable to acquire aerial photographs and prepare a photomosaic. Some exploration projects include the preparation of topographic base maps from

Figure 11-9a. Geologist working on cross-section correlations of sandstones from drill hole to drill hole in an area of closely spaced drilling. The probe records, which have been taped to cross-section paper, have been reduced from an original scale of 1:120 to a scale of 1:480. Such cross sections are invaluable in the interpretation of geology.

aerial photographs and control surveys.

Depending upon the location of properties under control, drill-hole patterns for the initial drilling phase are usually designed to intercept and delimit channel-sandstone hosts. Initial drill-hole patterns consist of rows (commonly called "fences" in the industry) oriented normal to the projected trends of the fluvial or channel sandstones. Holes in these fences or rows are usually spaced 1 to 2 km apart, and the rows are spaced 4 to 8 km apart. The object of the initial drilling phase is to delineate sandstones which are potential hosts for uranium deposits and to establish the depths of the targets.

Holes in the initial drilling phase should be drilled to sufficient depth to completely penetrate any potential target zones. Some inexperienced but conscientious geologists have a tendency to try, early in a program, to predict the target zones precisely, and sometimes this leads to the unfortunate situation where some holes are not quite deep enough to reach and evaluate all of the targets. We have seen numerous cases where holes drilled early in an exploration program fell short of mineralization discovered by later drilling. Another reason for drilling the initial holes to depths which might seem generous is that early correlations based on widely spaced holes are often difficult, and deeper holes make these correlations more feasible.

As discussed in chapter 2, local ore controls in trend deposits are often intraformational disconformities such as channel scour surfaces, carbonaceous mudstone layers, clay-pebble conglomerate lenses, and other similar stratigraphic features. It

Figure 11-9b. Cross sections which have been prepared by the geologists through the use of reduced probe records may be assembled into groups, fastened together at one end, and conveniently suspended from overhead supports, as shown in this photograph. Such cross sections, and the correlations made from them, are indispensable in properly interpreting the geology of most areas where stratigraphy played an important role in the mineralization process.

is important that these features be recognized and correlated in the initial drilling program. To accomplish this, the drill-hole data should be described and correlated daily to systematically refine the geological interpretations. Therefore, it is important that field geologists have full-scale copies of all the logs of drill holes in the area where they are working. When the geologists are attempting to make interpretations of the mineralization and stratigraphic relationships in the field where drilling is being carried out, they will naturally be more alert to evidence in the samples. During the initial drilling phase in most uranium prospects, it is a continuing struggle to establish reliable correlations and interpretations of the controls on mineralization.

As the drilling progresses during the first initial drilling phase, it is important that maps and cross sections be prepared and updated as the data are collected. Usually it is desirable to photographically reduce the gamma-ray electric logs to a convenient scale for cross sections prepared during this phase. Service firms with the capacity of reducing logs are present in most of the larger cities. In the reduction, we prefer a scale of 1:480 (1 ft = 40 ft) when the original logs were 1:120 (1 ft = 10 ft). Details of mineralization and lithology may still be seen on the logs when they are reduced to this scale, but they are small enough to include a thick section and numerous drill holes on individual cross sections. As part of the geologic program in support of drilling, numerous cross sections should be prepared to illustrate all significant geologic features, including the relationship of mineralization to individual sandstone bodies and complexes of sandstone bodies. The cross sections should form the basis for maps which show mineralized trends, sand trends, and favorable facies patterns.

To minimize delay in the preparation of the cross sections, it is recommended that a simple procedure be adopted for use by the project geologist and the field geologist whereby the reduced prints of logs are taped on ordinary cross-section paper (figs. 11-9a, 11-9b). As additional holes are drilled, reduced logs from them can readily be inserted into the cross sections, thus providing needed space and flexibility. When time is not so critical, additional cross sections can be prepared for reproduction in the normal manner.

During the initial drilling phase, it is recommended that posting of drill-hole locations and preparation of geologic maps be done at a scale of 1:24,000. In areas where topographic maps are available, it is ideal to use them for both field operations and for planning the drilling program. One map should be prepared and maintained which shows *all* of the mineralization in one area encountered by the drilling program, while other maps might be prepared to show certain structural or stratigraphic features, or to show interpreted mineralization controls and inferred loci of mineralization. Such loci, of course, constitute drilling targets, and as the initial drilling phase progresses, these targets should be defined systematically. The mineralization intercepts should be systematically shown as illustrated in figure 11-10. As soon as drilling has progressed to the point where interpretations can be made, they should be included in the cross sections. Controls on mineralization should also be indicated

and highlighted.

Uranium-ore controls in sedimentary rocks vary from one environment to another, and in fact the total control on mineralization almost always consists of a composite group of geologic factors. In some environments, sand trends or channel systems are the most important controls on mineralization. Once drilling has delimited the sand fairways, the next step is to define additional controls therein. This might consist of locating reduced environments within a generally oxidized system (fig. 11-11). A specific sandstone unit might be found to host the significant uranium mineralization. When this is the case, it is important to establish good correlations which enable the geologist to isolate and map the controlling unit.

As the geologic picture begins to emerge, appropriate maps should be maintained and continually modified as subsurface data are gathered. Surface data, of course, should not be forgotten as the drilling program progresses; in many cases, surface data can be integrated with the drill-hole data to enhance the mapping of sand trends and other factors which constitute controls on mineralization.

In most cases, the initial phase of the drilling program must serve as a guide to a complementary land-leasing or land-acquisition program. As the factors controlling mineralization become better known and more clearly defined, the area of interest is sometimes modified, and this requires a change in the acquisition program. Indeed, as the drilling program defines new areas for acquisition, it is likely to define areas for abandonment as well. In extreme cases, the initial drilling phase provides sufficient data to abandon the entire project. A decision to abandon all or part of a prospect area is often a very difficult one to make on the basis of widely spaced drill holes. If, within the

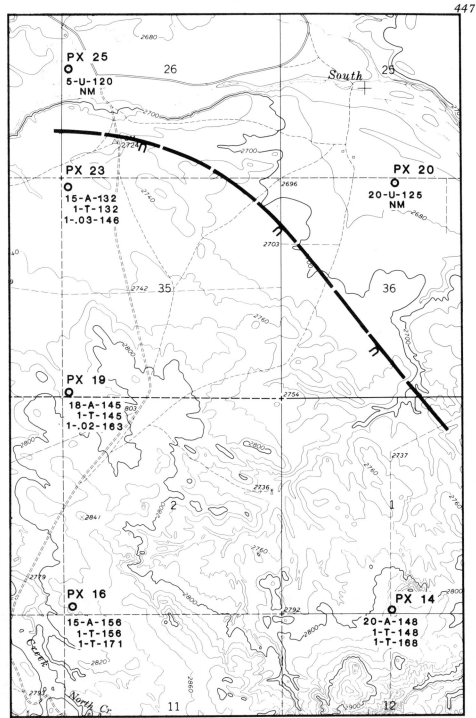

Figure 11-10. Map illustrating a method of plotting the results of an initial drilling phase on a topographic map. The first line beneath each hole gives the thickness of the sandstone, shows whether it is altered (A) or unaltered (U), and indicates the depth to the top. The following lines give the mineralized intercepts, if any. NM = no mineralization, T = trace mineralization.

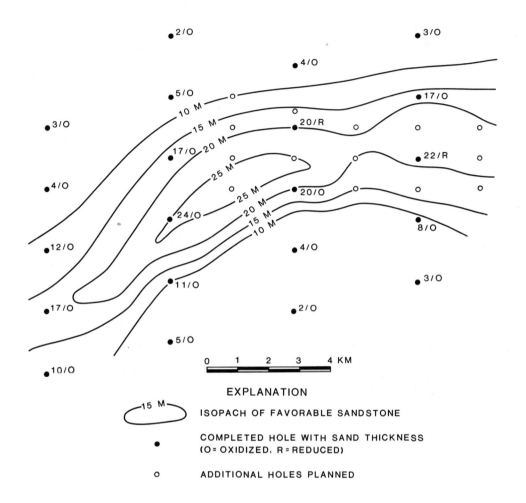

Figure 11-11. Idealized map showing initial drilling patterns with additional holes planned to evaluate secondary controls (reduced sandstone).

area under consideration, all of the holes encounter poor sand development and no mineralization, the decision to abandon can be easily reached. In many cases, however, some holes will encounter fair or even good sand development, while other drill holes log poor sand development. At an early stage of drilling, correlations are not always reliable. Some holes may encounter weak mineralization, but the widely spaced holes of this phase usually leave abundant room for improvement for both hosts and mineralization in untested areas. These are the cases where the decision to abandon is a very difficult one. In most projects, when there is some encouraging mineralization and some favorable sandstone development, it is prudent to go into an intermediate drilling phase before making a final decision to abandon the property.

Intermediate Drilling Phases

Intermediate drilling is a term used to describe drilling carried out between the initial drilling phase and the so-called predevelopment drilling. In most cases, intermediate drilling includes several phases which are gradational between the initial widely spaced drilling and the more closely spaced predevelopment drilling. Many areas thought earlier to be targets are abandoned during some phase of intermediate drilling. Typically, each phase of drilling will diminish the size of the area of potential and increase the definition of surviving targets.

Intermediate drilling phases are designed to locate and define favorable areas, where the sandstones are well developed and reduced. The sandstone bodies should be at least 10 m thick, gray, and associated with green and gray mudstones. Carbonaceous material, especially in the form of humates, is a very favorable accessory.

Successive intermediate phases of drilling are directed at areas that appear favorable. The first intermediate phase generally consists of drilling holes within preexisting rows at 300 to 500 m spacing and starting additional rows of holes to better delineate favorable areas. Rows may be spaced as closely as 500 m when the intermediate drilling is completed.

Typical intermediate drilling data can be plotted on topographic base maps at a scale of 1:24,000, but larger-scale maps should be prepared for favorable areas at some phase of the intermediate drilling. It is usually advisable to accurately survey all of the holes in promising areas, including surface elevations, so that accurate maps and cross sections can be prepared. We also specify that the surveyors locate and make note of drainages, fences, roads, power lines, and other important features when they are in an area. This provides the necessary data for preparation of useful base maps. A good scale for intermediate base maps is 1:6,000 if the favorable areas are large, or 1:2,000 if they are small.

Some detailed (closely spaced) drilling (that is, 50- to 100-m spacing) might be carried out during the intermediate phases to evaluate an unusually favorable area of mineralization. It is sometimes desirable to get cores for chemical assay and to test the continuity of mineralization. However, we wish to emphasize that the explorationist must be careful not to get too deeply committed to detailed drilling until sufficient drilling has been accomplished to allow selection of the most favorable targets. Even a relatively small area of detailed drilling may consume a large part of an exploration budget.

As stated earlier, intermediate drilling provides the subsurface control which is necessary to establish reliable correlations with enough detail to identify and evaluate controls on mineralization. If the specific sandstone units in which mineralization has taken place can be isolated and correlated throughout a favorable area, isopach maps of such units may be useful in defining the targets more precisely.

As the intermediate drilling progresses, each hole should be correlated and integrated into the interpretation of the geology and mineralization immediately after it is drilled and before an offset is drilled. In situations where easy, shallow drilling is being done or where for some other reason the explorationists lack adequate time to gather and integrate drilling information into locating additional drill holes, a technique of *dual drilling areas* may be employed. In this technique, the explorationists have two separate but nearby areas where drilling is being carried out. The drill, or drills, are deployed to drill no two consecutive holes in the same area, and they move back and forth between the two areas as each drill hole is completed. This technique almost always provides time for the explorationist to have a drill hole probed, complete the sample description, integrate the new data with the old, and stake a new drilling location before the drill returns from the nearby area where another hole has been drilled. The explorationist then goes to the nearby area and repeats the data-gathering and integration procedure before the rig returns there. The drillers do not mind such moves if the distance between areas is short and easily traversed. It would be unwise to implement this procedure in rough terrain, but where it is useable, standby time will be minimized.

Concerning supplementary information, a mudstone or siltstone unit sometimes is determined to be a partial control on mineralization as a result of influence on mineralizing solutions, and an isopach map of such a unit might help define the target. Detailed cross sections are indispensable, and they should be prepared for each row of holes. Longitudinal sections should also be prepared to tie the cross sections together. The cross sections, together with maps, give a three-dimensional representation of the mineralized zones.

As a general rule, it is best to evaluate all of the favorable targets with intermediate-spaced drilling before selecting the most favorable places to carry out the initial closely-spaced drilling. Uniformly distributed intermediate-spaced drilling also gives the explorationist the necessary data to make a broad evaluation of the prospect so that unfavorable areas can be abandoned. This drilling approach also makes it possible to put a limited amount of detailed drilling into better perspective in evaluating the overall prospect.

After sufficient intermediate-spaced drilling has been carried out to provide data from which a preliminary evaluation of the favorable areas in the prospect may be made, it is usually possible to select a few of the most favorable sites for some detailed drilling to make an evaluation of the mineralization. This is the last phase of intermediate drilling, and if it indicates that the mineralization is sufficiently well developed, it will lead naturally to predevelopment drilling. When this detailed drilling is carried out, significant mineralization should be selectively cored or samples taken with a sidewall sampler for determination of disequilibrium. Chemical data are also necessary at this point to determine if any metals such as vanadium occur in significant quantities with the uranium. This detailed drilling at or near the end of the intermediate drilling phase may provide sufficient information so that a decision may be made to abandon the entire project or to begin predevelopment drilling.

Predevelopment Drilling

Uranium occurrences which appear to have favorable potential on the basis of results of the intermediate drilling must be further evaluated in a predevelopment drilling program. One of the primary objectives of this program is to determine the size, grade, and shape of the potential ore bodies. This is usually accomplished by drilling a grid pattern in the potential area, as determined from the results of earlier intermediate drilling. Holes in the initial grid may be spaced, depending on the anticipated size of the mineralized area, from 60 to 200 m apart. The grid drilling program should be laid out so that each hole can be evaluated and the grid modified accordingly before moving the rig on to an offsetting hole which expands the grid pattern, and the dual drilling technique may also be used, as previously described. For a continuous drilling program

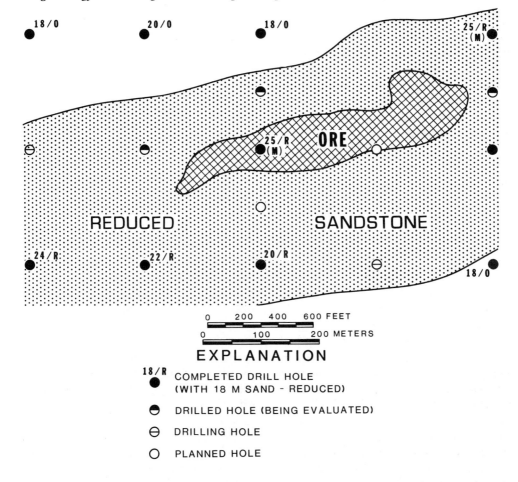

0   200   400   600 FEET

0   100   200 METERS

EXPLANATION

18/R
●   COMPLETED DRILL HOLE
      (WITH 18 M SAND - REDUCED)

⊖   DRILLED HOLE (BEING EVALUATED)

⊖   DRILLING HOLE

○   PLANNED HOLE

Figure 11-12. Idealized map illustrating how a continuous drilling program with two rigs can be planned and executed utilizing a progressively modified grid.

with more than one rig, this grid procedure requires a pattern with at least four lines of expansion (fig. 11-12). When a drilling program is planned and executed so that the data from each hole are evaluated and the interpretations modified before drilling offsets, we call the program a *progressively modified grid*. Holes are offset when the interpretations indicate that additional data are needed. The concept of dual drilling areas can be readily used in a progressively-modified-grid program.

As the drilling progresses, an effort should be made to establish all possible ore controls to minimize the drilling necessary to make an evaluation and estimate the potential (or projected) reserves. This is facilitated by preparing cross sections oriented east-west and north-south (or other directions) through all of the holes. These cross sections should show the details of the mineralized zones, and they should also show continuity of mineralization from hole to hole. For the preparation of such cross sections, we have found that it is usually desirable to reduce the logs to a scale of 1:240 (one-half of the original scale). Wherever appropriate, calculated grades, thicknesses, and depths should be posted on the cross sections. Known ore controls and details which appear to represent ore controls should be properly indicated on the sections. The geologists themselves should do the interpretive work on the cross sections, *not* assistants or draftsmen.

One map at a scale of 1:600 should be prepared to show all of the mineralized intercepts, and the interpretation of areas where ore is developed. Ore controls may be plotted on this map, or on another map of the same scale if the data become too cluttered and confusing. The maps and cross sections should be updated and modified as the drilling progresses, and they should be used in planning the modifications to the drilling program.

Predevelopment drilling for trend deposits might involve a substantial amount of coring to confirm gamma-log calculations and to determine if significant amounts of accessory elements (such as vanadium) are present. Selected parts of potential ore bodies may be drilled in relatively dense patterns to confirm continuity and to provide details of mineralization which may be extrapolated for the overall evaluation.

Trend deposits have a tabular habit and are frequently elongate in plan. However, their planimetric outlines are often complex with irregular shapes. We recommend that the GT map method be used for preliminary calculations of projected reserves based on predevelopment drilling.

Roll-Front Deposits

Drilling for roll-front deposits usually includes the phases outlined above, but certain useful controls which guide the explorationist are unique to these deposits. Most important of these controls are the roll fronts themselves and the associated alteration complexes. Indeed, the explorationist who is drilling for roll-front deposits

in a new area must devote a great deal of time and effort to the task of developing criteria for distinguishing altered and unaltered sandstones. Later in the program, the explorationist must attempt to determine where the best conditions for ore deposits occur along the roll fronts. It is not uncommon for certain roll fronts to contain no ore-grade mineralization along their entire length; others may contain ore bodies erratically spaced along their length.

Initial Drilling Phase

In areas far removed from known uranium occurrences, the initial drilling phase may be planned on the basis of mapped surface occurrences of altered and unaltered sandstones. The sandstone trends may be mapped on the surface and projected into the subsurface where drilling is planned; the initial drilling should then be designed to intercept these projected sandstone trends. The spacing and drill patterns used will depend, to some extent, on the degree of control at the surface which can be projected into the subsurface. In some instances, a mineralized roll front can actually be mapped at the surface and the initial drill patterns may simply be set up to drill at right angles across the front and thereby extend the mapping into the subsurface from the surface. In such situations, we recommend drilling the first pattern very near the front, depending on topography and dip, so that the direction of front trend may be quickly established.

In most cases, roll fronts will be poorly exposed or not exposed at all, and initial drilling will consist of strat tests designed primarily to locate and define the sandstone trends, if any, and to determine where alteration, if any, exists in the sandstone trends. In situations where sandstone trends can be determined from surface work, or after early drill-hole data have suggested sandstone trends, the following drill holes may be drilled in rows or fences oriented normal to the projection of the trends. Such drill holes may be spaced initially 2 km apart within rows, and the rows may be spaced 5 to 7 km apart, depending, of course, on the size of the target.

From the very outset of the exploration program, it is important to establish criteria for recognizing altered sandstone, if at all possible. In some areas, alteration is easily recognized in cuttings from drill holes by the distinctive coloration produced by iron oxides and hydrous iron oxides. It is frequently difficult to distinguish alteration colors from colors produced by recent near-surface chemical weathering (oxidation) of sandstone outcrops, but in these same areas, drill cuttings from below the water table often may easily be identified as either altered or unaltered. In the reduced facies where roll-front deposits occur, unaltered sandstones are normally gray below the zone of weathering. If the alteration is distinguished by any iron oxide or hydrous iron oxide staining, it will contrast sharply with the unaltered and unweathered sandstone color. Alteration typically is easier to recognize in the argillaceous to fine-sand-matrix part of a sandstone, as opposed to coatings on the coarser sand grains.

There are sandstone hosts, however, in which the alteration is not easily recognized in the samples by coloration because iron oxides are not present to any substantial degree. Indeed, there are also cases where iron minerals in the altered sandstones are actually reduced (assumed to be postalteration reduction), with the result that there is no visible evidence of alteration in the sandstones. In addition, as described in chapter 2, many alteration complexes are zoned with respect to color. In the zoned complexes, it is com-

mon for the altered sandstone near roll fronts (within 30 m or so) to lack iron staining, and bright colors are farther back in the altered sandstone from the roll front. In some cases, as described earlier, substantial volumes of altered sand have been reduced subsequent to ore development. This may have resulted from reduction by migrating hydrocarbons or from reduction by migrating volatile components of organic carbonaceous material.

When alteration cannot be recognized in the samples, it is necessary to establish other criteria for alteration. The most widely used criterion is the characteristic gamma-ray peaks (sometimes called "limbs") marginal to the altered zones on gamma-ray logs. Limb mineralization is usually distinguished by its occurrence marginal to porous and permeable altered sandstone. Such mineralization frequently occurs in shaly zones within the altered sandstone or at shale or mudstone contacts with altered sandstones (fig. 11-13). Limb mineralization is typically recognized on the gamma-ray log as thin radioactive zones, and usually is less than a metre in thickness. Frequently, carbonaceous silty or argillaceous zones in contact with altered sandstones will have retained some uranium or daughters, and thus they will be radioactive and easily recognized on the gamma-ray log. The most permeable part of the altered sandstone often will be characterized by low gamma-ray activity, which will contrast with the radioactive shaly and carbonaceous zones marginal to the sandstones. Such log characteristics are usually exhibited best in holes drilled in altered sandstone near a roll front. The mineralization (and radioactivity) on the limbs frequently increases as the roll front is approached. However, there are numerous exceptions to this. We have seen some cases where virtually no limb mineralization occurs from ore-bearing roll fronts 30 m away from the front, whereas

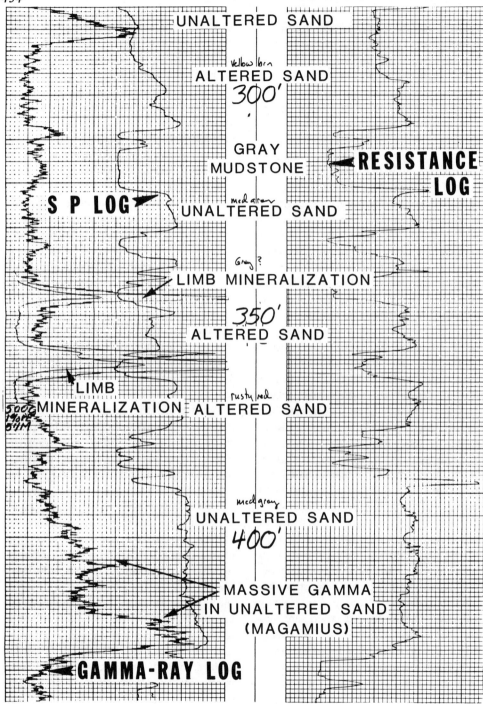

Figure 11-13. Gamma-ray and electric log illustrating log characteristics of altered and unaltered sand-stones. Note the reversed SP throughout the zone. Gamma peaks are well developed marginal to several altered sands. Massive sand at base of log is unaltered but near a roll front. Drilling within 30 m of this hole encountered ore-grade mineralization more than 3 m thick.

radioactive limbs may be logged in holes farther removed from the front.

The important and *consistent* characteristic of limb mineralization is that it does *not* occur in the *permeable* part of the sandstone which is altered. Also the gamma-ray log (set on a sensitive scale for exploration) will typically show limb mineralization as sharp peaks (frequently called "kicks," "tails," "spikes," or "rabbit ears" by some personnel in the industry).

Gamma-ray logs of unaltered sandstone usually do not record radioactivity in the shaly or silty margins above and below permeable sandstones. If any anomalous gamma-ray readings are logged, they are most commonly in the permeable sandstone. Gamma-ray anomalies in unaltered sandstones are usually relatively thick and have low amplitude readings (fig. 11-13).

Roll-front uranium deposits are usually asymmetric in cross section, with a sharp interface between high-grade ore and altered sandstone on one side and gradually diminishing uranium content away from the roll front on the other (unaltered) side. In some cases, a characteristic gamma-ray anomaly (the gamius) can be recorded on sensitive exploration logs as far as 100 to 150 m away from a roll front in permeable unaltered sandstones. Usually such an anomaly appears as a weak but thick trace of radioactivity. Next to drilling a hole in ore, the gamius gives the best evidence on the unaltered side that a hole is located near a roll front. On the altered side, strongly mineralized limbs are the best evidence that a hole is located near a roll front, and the better the grade of mineralization on the limbs, the better the chance of finding ore on the front.

During the initial phase of drilling for roll-front deposits, it is important to attempt correlation of sand zones so that the sand trends can be mapped. If the section drilled and logged includes marine or lacustrine elements, the correlations are usually straightforward and can be made with confidence during the initial drilling phase. However, if the section is entirely fluvial, correlations will be more difficult; indeed, usually only gross zonal correlations can be accomplished during the initial drilling stage. In most cases, even in entirely fluvial sections, certain zones have a fair degree of continuity. As soon as a zone of sands can be correlated with some degree of confidence, it should be given a letter or number designation so that the units or zones may be mapped. If the correlations are good enough at this point, the initial map should be prepared at a scale of 1:24,000, and isolith maps of the sand should be prepared. Data relating to alteration and interpretations of altered and unaltered sandstones should be superimposed on these isolith maps. As soon as possible, roll-front projections should also be plotted on these maps.

Some effort has been made to use induced-polarization (IP) logs and magnetic-susceptibility logs to differentiate altered sandstones from unaltered sandstones. What is known about the alteration of clay minerals suggests that they should have lower capacities for induced polarization than the more chargeable clays found in unaltered sandstones. In addition, iron sulphides in the unaltered sandstones are more chargeable than their oxidized equivalents found in altered sandstones. IP logs seem to support the theory in some cases, and they may be useful in helping with some difficult determinations of altered and unaltered sandstones. Geologists and geophysicists with the Uranium and Thorium Branch of the U.S. Geological Survey are conducting research in the use of the IP log in uranium exploration, and this is discussed more fully in chapter 13.

Magnetic-susceptibility logs have been

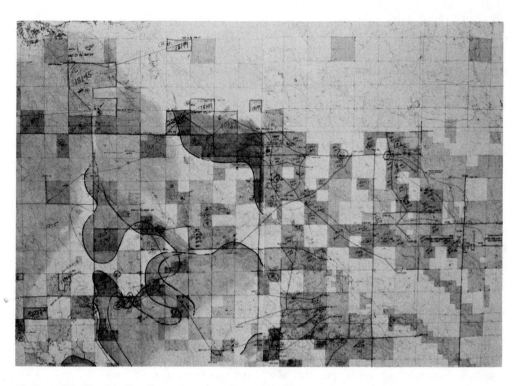

Figure 11-14. Photograph of a work map used by geologists to show mineralized features, such as roll fronts, superimposed on properties leased, staked, or otherwise controlled. The land or drilling activity of competitors may also be shown. Actual scale of the work map is 1:24,000, which is not adequate for advanced stages of intermediate drilling.

used without success to date. The first attempts were made with inadequate instruments, but the USGS is developing better instrumentation which should be adequate to the task. The few measurements of magnetic susceptibility that have been made on cores of altered and unaltered sandstones do not support the theory that measurable amounts of magnetite and other minerals with relatively high magnetic susceptibility are destroyed by the alteration processes. There frequently appears to be a measurable low *in* the uranium-ore bodies, but values in altered or unaltered sandstones do not appear to be distinguishable.

Intermediate Drilling Phases

After the initial drilling is completed and sandstone trends are approximately determined, the first phase of intermediate drilling should begin with more closely spaced holes within the sand trends and within the initial rows, providing adequate control to strengthen and confirm correlations. This drilling should be focused between areas mapped as altered and unaltered from earlier drill holes so that roll-front mapping may begin (fig. 11-14). Additional rows of holes on trend with the roll fronts and within areas of better sandstone development should be drilled to confirm correlations between rows of holes and to improve the reliability of the roll-front mapping. At this intermediate stage, the holes should be spaced within rows close enough to have firm correlations, particularly where the roll fronts occur.

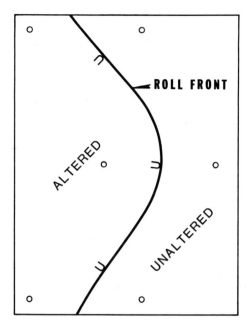
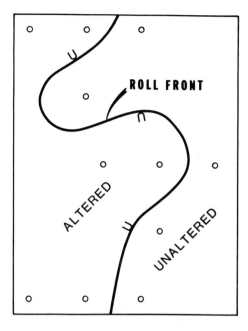
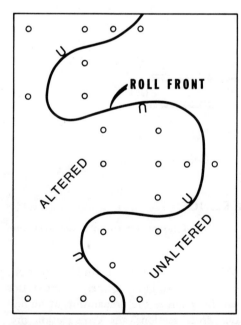
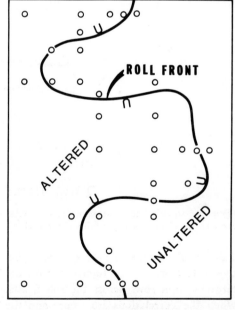

Figure 11-15. Series of idealized maps illustrating four stages in the development of a progressively modified grid (PMG). Note how the drilling is concentrated along the projected roll front and how this projection is modified as drilling progresses. The PMG method is simple when the drilling program is properly planned and executed.

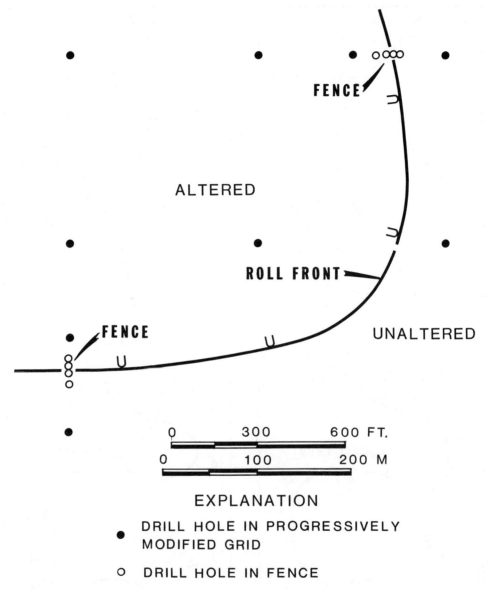

Figure 11-16. Idealized map showing how fences, or profiles of closely spaced holes, are sometimes drilled after a progressively modified grid is completed.

This usually results in hole spacing of between 150 and 30 m (depending on the target) within rows across the roll fronts. When the correlations are secure and the roll fronts are bracketed on 150- to 30-m centers within rows of holes, it is time to prepare detailed cross sections for each profile, as well as longitudinal sections tying the profiles together. Because this usually requires the preparation of numerous cross sections, the cross-section preparation method should be kept simple to facilitate modification of the sections as new data come in. One good approach is to reduce the gamma-ray electric logs to an appropriate scale and tape these reduced logs onto cross-section paper, using a correlation datum or using elevation as a datum. When the cross sections are prepared, the geologist should work with both the map and cross sections to determine roll-front locations. The map and cross sections showing roll fronts for each zone of sandstone or each sandstone thus can be kept updated.

After profiles have been drilled at approximately 2-km intervals along the roll fronts and correlations are reasonably reliable within profiles and between profiles, favorable areas may be selected in which to begin drilling progressively modified grids (fig. 11-15). This modified grid drilling should be planned on maps prepared at a scale of 1:6,000 or some other carefully selected but suitable scale. As each hole is drilled, correlations and interpretations should be made, and, when necessary, the roll front should be modified to be consistent with the new data. Drill-hole spacing in the grid should be condensed in this fashion sufficiently to determine the degree of sinuosity of the roll fronts. In some cases, this may require holes as close together as 60 m in the portions of the grid along the roll front.

After the configuration of the roll front has been well defined by the modified grid drilling, several fences should be drilled across the fronts in promising places to evaluate the mineralization (fig. 11-16). If possible, the sand should be isopached and the front plotted with notations, including such information as abundance of pyrite and organic carbonaceous material in the sandstone, apparent permeability on both sides of the front in the sandstone, and *all* mineralization and calculated intercepts. Some geologists make the mistake of not plotting or posting weak mineralization (called traces).

Favorable areas for initial profiling usually occur where the roll front is developed within thick permeable sand. Pyrite and organic carbonaceous material are favorable accessories. Of course, strong mineralization is particularly encouraging, and places where limb mineralization is strong near the roll front are favorable for profile drilling. In a few cases, the modified grid drilling may be used to determine that intervals along a roll front do not have potential for mineable ore. An example of this might be where the altered sand gradually becomes very thin and, at the same locality, has weak limb mineralization. In this instance, a hole in the grid pattern might encounter such a thin altered sandstone with weak limb mineralization, indicating the drill hole is near the front but that mineralization is weak (fig. 11-17). In such interpretations, the skill of the geologist comes into play.

The profiles, of course, should be oriented nearly perpendicular to the trend of the roll front itself. The initial drilling profiles should be closely spaced to show the details of mineralization across the front. At depths less than 120 m, the front should be drilled to 10-m centers (fig. 11-18). Deeper fronts may be fenced with hole spacing as close as 10 to 15 m, if the holes can be drilled at low cost and if directional surveys are run on the holes.

It is important during the intermediate drilling phases to try to determine which fronts contain the best mineralization. It is also important to try to determine what stratigraphic or geochemical controls appear to influence mineralization along the roll fronts, so that the "best" areas may be found quickly. The mineralization may be concentrated in the thicker and more permeable sands, particularly where humate or other carbonaceous organic matter is abundant. In some cases, a reasonable interpretation of the

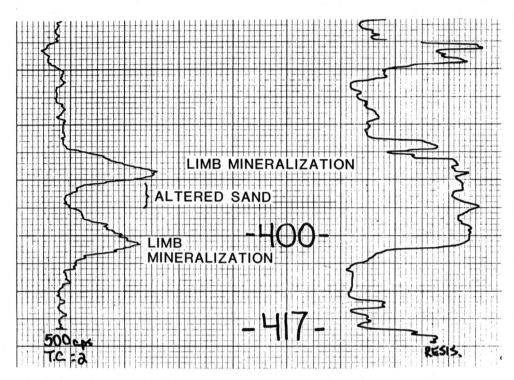

Figure 11-17. Gammay-ray and resistance log of hole drilled near a roll front where only weak mineralization is anticipated. The gamma peaks are close together and within the sand, indicating that the hole is located on the altered side of the roll front. The interpretation was confirmed by sample description. Such a hole drilled as part of a progressively modified grid would be strong evidence that the roll front is weakly mineralized in that immediate area.

paleohydrodynamics which controlled the movement of mineralizing solutions might be useful in determining where uranium is most concentrated along a roll front.

Geophysics and geochemistry may become useful guides in some intermediate drilling programs. Negative magnetic anomalies have been recorded over shallow roll fronts (Smith and others, 1976), and negative magnetic-susceptibility anomalies have been measured in uranium-ore bodies. Preliminary studies also indicate that ^{4}He anomalies may be detected in detailed soil-gas surveys over ore bodies. Radon anomalies in soil-gas surveys might be useful where mineralization is shallow. These geochemical and geophysical methods may be very useful supplements to intermediate-phase drilling programs.

Predevelopment Drilling

Roll fronts vary in geometry as well as in their distribution of mineralization. Some roll fronts are simple linear features with virtually uniform mineral distribution. Others are very sinuous in plan, with numerous diverging subsidiary rolls along which uranium mineralization may be distributed erratically. We have seen simple linear fronts which were evaluated by profiles spaced at 180-m intervals. In that program, holes were drilled as close as 8 m within profiles at right angles to the front trend to determine the grade-thickness distribution. Preliminary reserve calculations based on the 180-m profile spacing

Figure 11-18. Stakes marking the sites of previously drilled holes in front of the drilling rig indicate that the mineralized interval is not deep; otherwise, deviation of the drill holes from vertical would cause interpretive problems, not to mention excessive cost. In this instance, the target horizon is about 79 m (260 ft) below the surface, and drilling costs are low. The hole being drilled is not in line with the others because it is on a line between a drill hole in the fence and a previously drilled hole which was not on line with the fence of holes. The area of drilling will be carefully smoothed and reseeded when drilling is completed.

proved to be reliable. In some roll-front complexes, specific subsidiary rolls or roll fronts may control large reserves; some of these features may be simple enough to evaluate with widely spaced profiles (100 to 140 m), whereas others might require more detailed profiling, or even dense grid drilling. When a roll front is very sinuous and complex in geometry, the uranium distribution along the front is usually irregular and difficult to evaluate with drilling.

If a uranium roll-front deposit occurs in permeable sandstone below the water table where no unusual ground-water problems (such as fast water movement) exist, the occurrence should be evaluated for solu-

tion mining. Conventional cores should be taken for porosity and permeability testing and identification of clay minerals and other accessories of the host sandstones. This preliminary information may be integrated with other data, such as electric logs, to determine if the deposits have characteristics favorable for solution mining.

Other Strata-controlled Deposits

In any drilling program, it is important to attempt identification of the various ore controls during the initial phase. This approach should result in increased efficiency as the program develops. It requires careful correlations, numerous cross

462

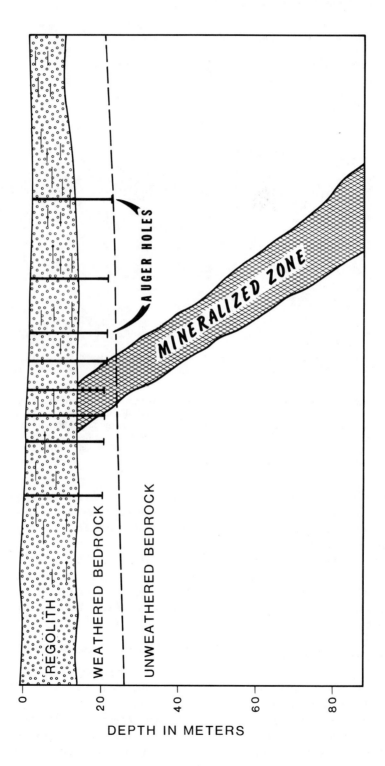

Figure 11-19. Idealized vertical cross section through a structure-controlled uranium deposit showing how auger drilling to 20 m can determine shallow projection of mineralization and dip. This method sometimes provides low-cost data necessary to planning an efficient deep diamond drilling program.

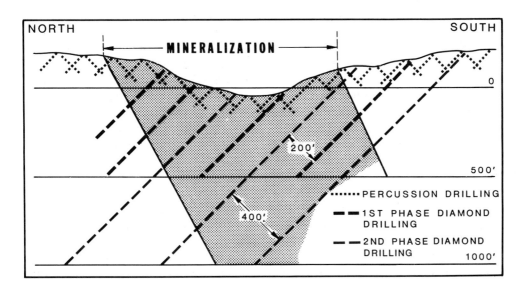

Figure 11-20. Simplified cross section through the Rössing uranium deposit showing drilling patterns in one profile. Shallow percussion drilling with a wagon drill established the near-surface configuration and projected dip. Successively deeper phases of diamond drilling were efficiently planned; the relatively low-cost data were fully interpreted. After Berning and others, 1976.

sections, innovative maps, and meticulous scrutiny of all the data.

The Proterozoic (Basal Aphebian) heavy-mineral deposits, which contain detrital uraninite together with sulphides of detrital origin, are primarily stratigraphically controlled. These deposits appear to have a placer origin, and they occur in quartz-pebble conglomerates and sandstones which were deposited under reducing conditions. Besides the obvious stratigraphic control, these ancient placer deposits have additional sedimentologic control. Extremely fine detrital heavy minerals (including uraninite) tend to concentrate in specific parts of the ancient stream channels or along the ancient beaches. Drilling programs may be designed to first establish stratigraphic correlations and identify the more favorable host sandstones and conglomerates. Next, the programs may be designed to identify the favorable facies within the host units. As the drilling program pro-

gresses, the controls may be continuously refined and the targets thereby systematically restricted and finally pinpointed.

DRILLING FOR STRUCTURE-CONTROLLED AND INTRUSIVE-CONTROLLED DEPOSITS

Uranium deposits which are controlled by structure, or which occur disseminated in intrusive rocks, frequently extend to very near the surface (perhaps being covered by regolith) and dip at high angles into the subsurface. The country rocks usually are hard, making deep drilling costly. It is common practice to initiate the drilling of these deposits with densely spaced (that is, 3 to 10 m apart), shallow, relatively low-cost holes. Auger drills have been used to drill through superficial deposits of alluvium or regolith to provide information from which the bedrock geology may be mapped, and to prospect for near-surface extensions of mineralization. Auger drills capable of drilling to depths of 25 m are available. In some cases, the explorationist can determine the near-surface extent of the ore zone (or altered fracture-control system) and its dip from the auger data (fig. 11-19). Such information is important in planning an efficient deep diamond drilling program.

In bedrock which is too hard for auger drilling, the shallow data can be acquired with wagon drills (scout percussion). This is also relatively low-cost drilling which may include a dense grid of inclined or vertical holes (fig. 11-20). In most rocks, the percussion drilling can evaluate an ore deposit (or altered fracture zone) to depths as great as 60 m. This initial drilling may form the basis for projecting the dip of

the mineralized zone to facilitate the planning of the diamond drilling program.

The intermediate and final drilling phases in exploration for structure- and intrusive-controlled deposits are usually accomplished with a diamond core drill. If possible, and on the basis of the configuration and dip of a mineralized zone to be evaluated, diamond drill holes should be collared downdip to drill through the hanging wall and intersect the mineralized zone at high angles to its long dimensions (fig. 11-20). This drilling might be continued in phases which include more widely spaced but deeper final drilling to confirm reserves sufficient to support a mining and milling facility.

12

Transition from Exploration to Development

DISCUSSION

The terms "exploration" and "development" are used in varying contexts by different organizations and persons in the industry. Generally speaking, "exploration" is understood to mean geological, geochemical, and geophysical field work and other activities, including drilling, performed in an area where properties may have been acquired but no commercial discovery has been made. "Development" refers to the work which is directed toward putting the property into production, including the opening of mines, construction of mills, or other similar activities. The terms "development drilling" or "predevelopment drilling" are used by some exploration companies in referring to situations where a discovery has been made; they are satisfied that commercial deposits exist, and they then perform additional drilling for further confirmation of reserves (or perhaps to provide

greater definition for mine planning).

In this chapter, we discuss the transition that usually takes place after exploration has resulted in a discovery and the property is to be turned over to a mining department or division for development. The specific stage at which the development operators (miners) take over from the explorationists is frequently dictated by the size and location of ore bodies discovered. For example, discovery of a 240,000-kg (500,000-lb) "ore body" in a remote, rugged location might not lead to immediate development because of the economics of mining and hauling ore to a mill site. According to the precise definition, such a mineral occurrence would not be considered ore; however, most explorationists refer to such an occurrence as an ore body if it meets the standard definition for the region, even though it may be uneconomical at the time of discovery because of location, depth, and so forth. Further, an occurrence which, under normal circumstances, may be uneconomical because of distance from a mill could well be economical (and thus considered an ore body) if it is mineable by *in situ* leaching, a process which does not require a mill. A uranium occurrence when concentrate sells for $22.00 per kilogram may be an ore body when concentrate sells for $88.00 per kilogram.

The transition from exploration to development may sometimes take place abruptly, in a matter of a few weeks; in other instances, it can be eight to ten years after a discovery is made before development and subsequent mining take place. This may be attributed to several conditions, such as the market for the particular commodity, the availability of funds to finance drilling, the character and depth of the ore bodies, the availability of milling facilities, and the organization and objectives of the companies or individuals concerned with the project.

Regardless of the reason for or the timing of the transition from exploration to development, it is imperative that the explorationist adequately define the overall geology of the deposits before turning the project over to the miners (distinguished from laborers). It must be remembered that most miners are engineers; as such, their philosophies and viewpoints differ from those of explorationists and geologists. We should also point out that while explorationists may have greatly varying backgrounds (as we explained in chapter 1), the competent interpretation and drilling out of ore bodies and mineral occurrences requires specific technical training and experience. This task is usually delegated to geologists or their staff members, although mining engineers are occasionally involved in working on the geology of a deposit. Geologists, however, are usually best qualified for this work because of their specialized training.

Adequately defining the geology of a deposit is sometimes difficult to accomplish because of insufficient time and funds. For example, some companies may find it necessary to send their geologists to new locations before they have had an opportunity to perform adequate drilling or conduct other studies needed for a comprehensive report to be written. In such a situation, monthly reports must usually be depended upon for much of the information which is relayed to the miners.

THE WRITTEN REPORT AND SUPPORTING DATA

The Report

It is essential that an adequate report, together with supporting maps and cross sections, be prepared by the geologists when a property is to be transferred from exploration to mining and development. This report conveys to the miners all pertinent information about the project. It should be prepared in the usual format of a regular geologic report, usually including an abstract, introduction, acknowledgments, regional geology, local geology, orebody description (emphasizing unusual characteristics), summary of exploration work to date (including drilling costs), and so forth. It should also include maps, cross sections, reserve estimates, and other information which may be of value to the personnel who will plan and carry out the mining.

The report should include references, and perhaps several photographs to illustrate pertinent geologic or topographic features of the project. A voluminous report containing perhaps more information than necessary is preferable to a brief one, which may risk omitting critical information.

Preparing and assembling the reports,

including the photographs and other illustrations, can be very time consuming. A careful count should be made to determine the number of copies needed; in addition, several extra copies should be prepared. To make 10 or 12 copies is not unusual; 20 or 25 would probably be excessive. Aside from distribution to personnel at the project site (where mine managers and superintendents, mill superintendents, project geologists, and others have been known to collect project geological reports as personal mementos), copies of the report should also be distributed to the appropriate offices of all participating companies. Each company can then distribute two or three copies of the report to the various departments involved with the project.

Maps and Cross Sections

Quantity, Proper Scale, and Adequate Detail

Accompanying the report there should be a regional map, topographic maps of the area, and such additional maps as are required to show in detail the geology of the ore deposits and their precise locations. Some of the maps may be on a scale of 1:600, or perhaps even 1:240, if such a large scale is necessary to show the details of ore-body configuration. Cross sections should also be prepared to show the regional geology (including, perhaps, data from outcrops and from drilling) and the local geology, with particular emphasis on cross sections through the ore deposits. It is the responsibility of the geologists, through competent preparation of maps and cross sections, to explain and display adequately to the miners the important geological concepts for the ore bodies under study, including structure, stratigraphy, fluid movement, and placement of the ore bodies. Even though many details

of the geology may not be resolved at the time the maps and cross sections are prepared, such interpretations as are available must be properly shown.

Ore Controls and Ore-Body Configuration

Geologists must also attempt to explain to the miners, as clearly as possible, the mineralization controls of a deposit. Some deposits (as, for example, in the Crooks Gap area of Wyoming) are so complex that it is very difficult to show the ore controls and their relationship to ore-body configuration. However, when controls on mineralization can be clearly defined, as in some typical Wyoming deposits, where roll fronts can often be followed for hundreds or thousands of metres (in some cases, even for kilometres), geologists should indicate these phenomena on their maps and cross sections to further assist the miners in their planning.

The uranium-ore bodies in the Shirley Basin district of Wyoming were discovered in 1955. By 1959, Utah International was sinking a shaft near some of the ore bodies, and numerous holes had been drilled on a grid on the properties that were controlled by that company (fig. 12-1). In areas where mineralization had been encountered with the initial spacing of 300 to 600 m (1,000 to 2,000 ft) between drill holes, additional drilling was carried out to close in the grid to 30 m (100 ft) between drill holes. On the basis of the results of this drilling pattern, reserve estimates were made utilizing the area-of-interest concept. Plans for an underground mine were then formulated. The roll-front concept was not recognized in the industry at that time, and geologists working on the project could describe the deposits only as "pods" and "lenses." Many maps and cross sections were drafted in an effort to decipher the apparently complex mineralization in the thick sandstones; the correlations and configurations,

Figure 12-1. Utah International Shirley Basin uranium mine in early spring, 1976. Two drill rigs utilized for drilling ore-evaluation holes just ahead of mining are visible near center of photograph. Top of boom of electric shovel can be seen just to the left of the rigs. This mine was originally developed as an underground mine, but high mining costs, together with a better recovery factor for open pit, resulted in a change to an open-pit operation.

however, did not seem to explain the mineral emplacement.

In 1960, underground workings in Shirley Basin reached the first ore, and detailed mapping was undertaken on a scale of 1:240 and 1:60. All mine walls were mapped in intricate detail. At the same time, exploration and development drilling was being carried out on properties where mining was tentatively planned, and also in areas where additional exploration data was needed. The late Philip N. Shockey, a geologist working on the job logging samples from rotary drilling, noted that the uranium mineralization in drill holes consistently occurred where pale, rusty-green sandstone came into contact with pale gray to gray sand. Shockey's observation accelerated the recognition of one of the

most important ore guides ever to be utilized in exploration for minerals. Soon, mine mapping and interpretation of drilling at Utah International's Shirley Basin operation evolved into a skillful application of prediction of ore at an oxidized/reduced interface. The Colorado Plateau terminology of uranium "rolls" was adopted and was later combined with the term "front" to become "roll front," although many geologists still prefer the more cumbersome, but technically accurate, "geochemical interface."

The geologists at the Utah International mine in Shirley Basin were then able to begin accurately forecasting ore occurrences, even between "barren" drill holes. They found that their previous terms, "pods" and "lenses," were no longer needed, and that mineralized intervals previously misunderstood tied in remarkably well with the new concept. Reserve estimation procedures were modified to fit the geology, and mine planning came to rely heavily on roll-front interpretations. Geologists employed by the U.S. Geological Survey and the U.S. Atomic Energy Commission were invited to witness these mapping and interpretation techniques, as were geologists with other companies. Word of the results soon spread throughout the region, then nationwide, and subsequently worldwide.

Continuity of Mineralization

When planning a mining operation, knowledge of the configuration of the mineralization in a restricted locality is of the utmost importance. Awareness of the continuity of the mineralization along or within a particular geologic feature, such as a trend, stack, or roll front, is another vital factor. As explained earlier, ore deposits along any of these three features typically occur wherever the fluid movement and geochemical conditions were favorable for the precipitation of larger concentrations of uranium. However, these areas of higher-grade mineralization often are connected by areas of low-grade mineralization. All mineralization and continuity should be appropriately indicated on the maps and cross sections, to be taken into account when preparing plans for mining and extracting the ore.

Further, it is often impossible, or simply uneconomical, to carry out drilling along the total length and breadth of a particular feature. Take, for example, the case of a roll front where fences may have been drilled at 30-m (100-ft) intervals, with drill holes in the fences spaced 7.5 m (25 ft) apart. If the depth of this front is 240 to 270 m (800 or 900 ft) below the surface, and if the drilling is reasonably difficult, it may not be prudent to drill fences at 15-m (50-ft) intervals.

The project manager may decide to accept reserve estimates calculated from the more widely spaced drilling, and perhaps to carry out additional exploratory long-hole drilling underground once the mining operation is under way. This is one reason why it is important that the mining staff have maps showing the projected locality of the roll front or other mineralized feature. If there is a need for additional work to be carried out in order to define the precise reserves or exact configuration between the fences of drill holes or other drilling control, the miners can plan their work accordingly.

Knowledge of the continuity of mineralization is most important from the point of view of calculating reserves. Given a specific selling price for uranium concentrate, the reserve estimate cutoff might be established at a grade of 0.05 percent U_3O_8; if the price should increase, the cutoff may be lowered to 0.03 percent. Consequently, mineralization which might not have been included in reserves estimated from one set of calculations may be included in reserves estimated through different calculations. Further, in the case of open-pit mining, low-grade mineralization is often passed through, and thus removed, as the mining operation proceeds down to a higher-grade horizon. If the development operators are aware that the low-grade mineralization exists, or if they are instructed to be on the alert as excavation proceeds, the low-grade ore can be segregated and taken to a stockpile.

Property Situation

As we stressed earlier, if the explorationist does not control the property in a particular area, then he cannot proceed with exploration or development because he has nothing to work on. A very clear understanding must exist about the property situation at the time the explorationist turns the project over to the development operators. Typically, one company will not control all the properties in a particular area where exploration and development are being carried out. Negotiations between property owners frequently extend over many years before agreements are reached concerning which parties are to operate on the various properties. Geologists should therefore prepare comprehensive maps for the development operators, clearly indicating those properties held by the company, other properties which might be optioned by the company, nearby properties (whether favorably located or otherwise), and their ownership. Information about approaches to other owners and the results of such approaches should be given. This could also include some suggestions for approaching other landowners with desirable properties, and for possible deals which might be arranged.

In a given area, geologists frequently drill out and estimate reserves on just a portion of the company's properties to the point where such properties can be turned over to the development operators; yet many nearby or outlying properties may still remain in the exploration category. In such cases, the geologists may continue working on the other properties; therefore, it is unnecessary for them to furnish the miners with data concerning these outlying properties. The geologists remain responsible for such properties, which can include making deals for new outlying properties.

The land- or lease-maintenance records of any exploration group are extremely important, as we described in chapter 5. When properties are turned over to development operators by geologists, complete land records should accompany them. In addition to the basic records, development operators should be given written descrip-

tions of the property status, on a tract-by-tract basis. For example, in a situation where corner and side center monuments must be maintained for lode-mining claims, the geologists might recommend replacement with better posts, or perhaps suggest a procedure for maintaining a vigil for "claim jumpers." Many companies hire security guards to patrol mining claims in areas where competition is intense.

Reserve Estimates

Coordinate Procedure with Miners

After a discovery has been made on a property, the geologists proceed with determining the size and grade of the occurrence and prepare preliminary reserve estimates from the drilling and sampling information. At a later date, such estimates are turned over to the miners to facilitate preliminary discussions concerning development of the properties. However, at a relatively early stage, when explorationists are developing procedures for estimating reserves, they should hold discussions with the mining operators concerning a mutually agreeable method of estimating reserves. Geologists frequently tend to estimate reserves based on the ore-body configuration, as we explained above. On the other hand, many mining engineers have a tendency to use a block-type (modified or otherwise) reserve estimation method.

Tabulate Reserves

Reserve estimates should be tabulated in such a manner that they are readily accessible and can be utilized by both geologists and development operators. Drill-hole designations (such as numbers) should be listed, and coordinates should be given; surface elevation, depth to the mineralized horizon, thickness of the mineralized horizon, and grade should be indicated, and there should also be a notation as to whether the analysis is chemical or radiometric. The area of interest assigned to a given hole at a particular horizon should be tabulated. Grade and grade-thickness cutoffs should be included among the categories.

The reserves should be tabulated and totaled by area, either within a certain part of a coordinate system on the properties, or within a certain horizon, or both. This depends upon the wishes and the requirements of the explorationist and the development operators for a clearer understanding of the precise location of the reserves and how they might best be extracted.

When to Computerize

A computer is designed to store data, retrieve it on demand, and process it according to specific instructions at a given time. These instructions are known as computer programs. Programs for use in reserve estimates are written by professional programmers who have a keen understanding of the mathematical and geologic concepts involved in reserve estimates. Some computer programming companies (known as software firms) maintain programs which are available for immediate use; other programs may be modified as required. A *new* program, designed for a specific situation, could be an expensive undertaking, requiring the services of a specialized person, or a geologist working closely with a competent programmer. In all cases, data has to be prepared according to a specified format for entry into computer storage. Although the format may vary from one program to another, any given program requires the data to be stored in a specific manner.

Computer programs for block-type estimates are simple and relatively inexpensive. The results are as dependable as block

estimates ever are, and they can be quickly obtained. If more sophisticated techniques are to be utilized, such as estimating reserves along uranium roll fronts where fences have been drilled across the fronts, the computer program becomes a more complex matter. Even firms with several years of experience with this type of program express concern about computer-generated estimates for roll-front deposits, because feeling and artistry, which are incorporated in reserve work to some degree by a good geologist, cannot be programmed into the computer.

Computers have proven to be especially valuable aids in the rapid calculation of reserves of differing grades, differing grade-thickness cutoffs, and at different depths. For example, a uranium reserve estimate might be made for a certain open-pit mine, using a grade cut-off of 0.05 percent U_3O_8 and a grade-thickness product of 6 cm or .06 m (0.20 ft) percent U_3O_8. Later, the cut-off might be lowered to 0.03 percent and the grade thickness to 4.5 cm or .045 m (0.15 ft) percent, and calculation of new reserve estimates would be required. In such an event, so long as the data are properly compiled and fed into the computer, it is a relatively simple matter for the computer to recalculate the reserves by using the new cutoff percentages. A diligent effort must be made to feed into the computer data from every drill hole so that as differing cutoffs arise, possibly through economic changes, the task of obtaining new reserve estimates is greatly simplified.

There are many divergent opinions as to when a company should computerize its findings. Indeed, unless programs are written to use computers in processing exploration data, it is hardly worthwhile to consider placing drill-hole data in the computer until there appears a reasonable chance that mineralization found can be developed into an ore body. Current practice in the uranium industry, at least among those who use computers, is to computerize data at the stage where a project is progressing toward development, and there is reasonably good assurance that the mineralization will prove to be an ore body. At this point, the data could be assembled and put onto tape or cards for use in the computer applications. Computers can be used in conjunction with plotters, a valuable method for plotting certain types of data in map form.

In our opinion, there is a major problem inherent in plotting and calculating uranium reserves in sandstones by computer: a certain creative effort is required to interpret the form of the ore occurrences. This is especially true in the case of roll fronts, which, as we described earlier, are sinuous, curving features with ore of differing grades, thicknesses, and widths deposited in response to chemical and stratigraphic phenomena. We have yet to see a computer program which can adequately plot a roll-front configuration, or adequately estimate reserves on a roll front.

When Is Drilling Density Adequate?

The drilling density necessary to determine the amount of mineralization and ore present at a given locality is directly related to the complexity of the mineralization and the ore bodies, if any. Depth and cost of the drilling are also factors to be considered.

In an area of simple, straightforward geology, with simple geometry of the mineralization, drilling can usually be carried out to the point where there is no question that the ore body has been correctly defined, and few doubts remain about the quantity of reserves present, their depth, and their grade. However, excessive drilling costs may cause the geologist to sacrifice a "proven" reserve estimate for a combination proven, indicated, and inferred estimate. In other instances, ore-body config-

uration may be complicated to such an extent that closely spaced drilling provides no data advantage over widely spaced drilling, and the additional cost is generally not justified in terms of the amount of significant information gathered.

In the Crook's Gap area in Wyoming, much of the ore lies under Green Mountain, just east of the topographic feature known as Crook's Gap. Drilling holes to depths of 450 to 600 m (1,500 to 2,000 ft), the operators found costs to be excessively high compared with drilling costs in other parts of Wyoming. In 1970, costs were about $11.50 per metre ($3.50 per foot), chiefly because of difficult drilling through boulder conglomerate beds near the surface. A change from rotary drilling with mud to hammer drilling with foam injection reduced the costs to about $6.50 per metre ($2.00 per foot). The engineers and geologists decided that the best approach for evaluating reserves in the area was to drill a basic grid on 150-m (500-ft) centers, followed by a denser grid of holes on 60-m (200-ft) spacing. Virtually no drilling was carried out in an attempt to intersect roll fronts, even though the mineralization controls at Crook's Gap–Green Mountain are roll-front related. The small, erratic nature of fronts at Crook's Gap, together with the deep, difficult drilling, renders drilling for projected fronts infeasible. On the basis of their experience in the area, the engineers and geologists concluded that a drilling grid on 60-m (200-ft) centers provided adequate data for a decision to proceed with sinking a shaft more than 300 m deep and to plan the mining operations. In this situation, the geologists were unable to provide reliable maps showing roll-front locations—not because this task was impossible, but because the economics of data gathering were prohibitive.

Several recently discovered areas, now

in various stages of development in the Grants uranium region, require drilling to depths as great as 1,500 m. Such holes are very costly, and it is impossible to drill a dense pattern of individual holes to determine grades and thicknesses of uranium mineralization which may be present. This has led to utilization of a development drilling technique originated by the petroleum industry called "whipstocking." The technique—an attempt to reduce costs of development drilling for oil or gas—allows two to six wells to be drilled from one drilling platform. A single, usually vertical hole is drilled through the objective horizon. This hole is logged and cased; then a deflection device is set in the hole several hundred to a few thousand feet above the objective. A second hole is then drilled, angled away from the first hole, to penetrate the objective at a different location. As many as seven development wells can then be drilled from one original vertical hole. In such an operation, every hole requires a directional survey to determine its location at various depths. The whipstock technique is especially useful when the upper part of the drilled section is hard or otherwise difficult to drill, as in the Mount Taylor district of the Grants region. Many uranium explorationists in the Grants region now drill no closer than 60 m (200 ft) between drill holes in areas where depths exceed 1,050 m (3,500 ft). In such situations, however, certain assumptions must be made about the grade and continuity of ore, and mining must be planned accordingly.

In the Colorado Plateau area, it was proven to be uneconomical to drill and block out all the reserves ahead of mining. In this area, an exploratory underground drift is frequently driven out to the ore-grade mineralization found in a drill hole. From that point on, the configuration of the ore body is deciphered. This often requires the drilling of so-called "long holes" from underground workings, rather than attempting to carry out the drilling from the surface. This type of drilling is exploration drilling associated with a mining operation.

In the Gas Hills and Powder River Basin areas of Wyoming, many uranium operators consider drilling on 30-m (100-ft) centers to be adequate if mineralization appears to be continuous and readily predictable. Others drill a 30-m grid and later close this in to a 15-m grid; still others prefer working with a 30- to 60-m grid which they modify by drilling fences of closely spaced holes across the predicted locations of roll fronts. In Texas, 30-m grids are commonly drilled, but fences are also drilled across the trend of the roll fronts to gain information about the grade and continuity of ore.

Thus it can be seen that the density of drilling varies considerably from area to area; the personal philosophies of individual operators together with economics are determining factors.

Ore and Host-Rock Characteristics

Mineralogy, Geochemistry, and Theory of Origin

The geologist's report to the development operators should include a discussion of the mineralogy of the host rock and of the ore itself and also a discussion of geochemical theories which have been developed by the geologists to explain the occurrence. This can often be tied in with the discussion of ore controls and ore-body configuration and should be covered in considerable detail. The discussion should include descriptions of the cores taken, assays, petrographic studies, relationship of the various constituents of the ore and host rock (such as the relationship of calcite or humate to the ore mineral), and identification of the ore mineral(s). If possible, a discussion of the relationship, or apparent relationship, among the various minerals and their role in the formation of the ore body is of value.

A working hypothesis or model which explains the origin of the host rock and the origin of the ore itself can be very useful to mining engineers and geologists in making projections and interpretations based on available drill-hole information. This is particularly true when solution mining or underground mining is being considered.

Amenability Tests

The geologists should give the development operators an adequate description of amenability tests which may have been carried out on the ore. Such tests should be conducted shortly after an ore body is found. The tests are performed on material taken from cores, although they may also be performed on bulk samples which have been collected from underground workings, test pits, or outcrops. The results of these tests should be included in the geologists' report to the development operators, together with any comments made by the lab which conducted the tests. The purpose of amenability tests is to determine (1) how the ore reacts to various treatment methods for extracting the elements or minerals of value, such as uranium, and associated elements, such as vanadium, and (2) what recovery resulted from the various tests which were performed. If only a few standard treatment methods were tried, then the geologists might discuss other methods which could be feasible, but which have not been tried.

If the physical or chemical characteristics of the ore or of the host rock vary from one location to another on the properties, this fact should be brought to the attention of the development operators. Occasionally the character of an ore body changes so significantly from one area to another on a given property that a mining and/or milling procedure which was satisfactory in one area is not satisfactory in another. For example, ore from one particular area might be deficient in calcite, but in an adjacent area, the amount of calcite in the ore could be extremely high, with consequent excessive acid consumption in the treatment of the ore (if the mill is the acid-leach type). Such hazards, if they are known, or suspected, should be discussed in the report.

Disequilibrium

The nature of disequilibrium has been discussed earlier in this text. When geologists are preparing to turn over a property to the miners, careful descriptions should be given of disequilibrium studies made and any apparent problems with disequilibrium which the geologists may have encountered in the course of their work. Geologists should also include in the report a discussion of procedures used in determining equilibrium, together with a description of the laboratories which made the analyses.

In some cases, movement of uranium from one part of an ore body to another in recent geological time is systematic, so that disequilibrium can be projected reliably with limited data. An attempt should be made to work out such relationships, and such studies should be incorporated in the report to the miners. Failure to recognize a disequilibrium problem can result in serious misinterpretations in the course of mining.

Competence of Host Rock

The development operators should be made aware of the competence of the host rock; this can be a very important factor in developing a mining plan. For example, if the host rock is known to be competent, and will hold up well in operations such as stoping, then mining costs per ton for underground mining will be far lower than with incompetent host rock which caves easily. Mining of an ore body in incompetent rock may require substantial timbering and roof support. On the other hand, competent rock in an open-pit operation may require extensive drilling and blasting, resulting in increased mining costs. Ideally, for underground mining the host rock should be competent, whereas for open-pit mining the host rock should be soft enough for easy excavation while the overburden should be sufficiently competent that high walls will hold up well. For uranium solution mining, it is essential that mineralization occur in a low-clay, poorly cemented, permeable part of a sandstone horizon, regardless of the competence of the host rock. The competence of the host rock may vary from one location to another within an ore body. This can be directly related to the amount of cement in the host rock. Generally speaking, if the rock is well cemented, it will be competent; if cement is absent, the rock will be relatively incompetent and will cave easily. However, in well-cemented rock, fractures can result in hazards.

In Utah International's underground mining operations in the Shirley Basin district of Wyoming, it was found that the sand and gravel in the Wind River Formation, which composed the host rock, were extremely soft and would spall and "flow" into the workings if openings were left unsupported for even a few hours. Consequently, the mine was heavily timbered, with resultant high costs. This particular mine also suffered from being situated in an extremely permeable water-saturated sand; as much as 14,500 l (4,000 gal) per minute were pumped from the mine and nearby wells to keep the water level from rising into the mine workings.

Chiefly because of the high costs of the underground mining, Utah International ultimately decided to mine by open pit at this operation. Ironically, soft rock continued to plague the operation, but this time it was not the host sandstone which caused problems. They resulted instead from a soft, water-saturated organic clay bed which occurred about 45 m (150 ft) below the ground surface. When the pit had been excavated to a depth of about 75 m (250 ft), one wall unexpectedly failed due to movement on the clay bed, and several million cubic metres of highwall slumped into the pit.

The underground operations in the Colorado Plateau area are in older rocks of Mesozoic age. As a general rule, these rocks are more competent than many of the younger host rocks, such as the Wind River Formation of Wyoming. Underground mining costs in the Colorado Plateau area can be kept to a minimum because the rock holds up well.

We do not wish to imply here that geologists are responsible for determining host-rock competence. On the contrary, this responsibility belongs to the development operators. However, since the geologists will be reviewing and studying many rotary samples, sidewall samples, and cores, some study and thought could well be given to rock competence at the same time, to the possible benefit of the entire mining operation.

Additional Information

Description of Sedimentary Column

Aside from a knowledge of the characteristics of host rock, it is important that the miners know something about the characteristics of the rocks which overlie an ore body. This information is essential from the standpoint of mine planning, regardless of whether an underground mine, an open pit, or even solution mining is planned.

Geologists typically gather considerable information about the overlying rocks in the course of drilling exploration holes. This information should be digested into a comprehensive discussion of the characteristics of the overlying rocks. It can then be of significant value to the miners as they review alternative mining plans. In planning an open-pit mine, for example, it is desirable to know whether any blasting will be required as the pit is excavated. In the case of a shaft, it is important to know whether any significant aquifers might be encountered in sinking the shaft. With solution (*in situ*) mining, it is important to identify the character of the rocks because of additional costs involved with drilling and casing through hard rock, squeezing shales, soft surface gravels, and so forth.

Suggestions of Areas for Additional Geologic Study

If the geologists have encountered areas where they recognized a problem or a situation requiring additional study but were unable to give the necessary attention themselves, such areas should be brought to the miners' attention. For example, perhaps the explorationists noted a lignitic shale or lignite that overlay a host sandstone and that they suspected to be water saturated. Such a condition conceivably could cause the pit walls to cave if the lignite were not adequately drained prior to mining. If the hazard were considered serious, some benches might be planned in the pit. Problems such as these should be pointed out to the miners, so that they may devote additional study to prospective hazards as they plan for mining.

Another problem which might be encountered in deep mining operations relates to rock temperature. In some opera-

tions planned in the Grants district, where depths of 1,200 m (4,000 ft) below the surface may be reached in some underground workings, high temperatures may present an acute problem for miners planning to operate at those depths. When a discovery is made in a deep zone, down-the-hole temperature measurements should be made in order to formulate ideas about special problems which may exist. In other situations, noxious or explosive gases may exist, permeating the workings and creating a hazardous or even a deadly environment. About 1962, Union Carbide sank a shaft on the Palangana salt dome in Texas, attempting to mine by underground methods some of the uranium ore which had been discovered overlying this dome. Two major problems were encountered during the operation: first, the temperature was over 40°C (100°F), which was attributed to the heat conductivity of the salt; second, the workings could not be kept free of toxic hydrogen sulphide gas. Union Carbide found these problems to be insurmountable, and the workings had to be closed. Attempts are now being made to mine the ore at Palangana dome by solution-mining methods.

Another matter which could require additional study is the occurrence of densely cemented zones within an overlying rock, or perhaps within the host rock. The geologists might find it impossible to precisely define such zones because of inadequate data, but their existence should be pointed out to the development operators in order to alert them to a potential problem. Development operators frequently take over a project before the geologists are satisfied that they have answered most of the geological questions in the area; indeed, some of the questions may not have been raised. The development operators should be fully aware that the geologists' studies may be incomplete.

*Suggestions of Possible
Cost-saving Techniques*

Geologists should operate the drilling programs and other exploration activities at reasonable costs, and the development operators should be fully informed of the procedures, including cost details. The geologists may also make suggestions for additional drilling and related cost-saving techniques which could be implemented, or experimented with, by the miners.

It is the miners' responsibility to determine which mining techniques will work best in a given area under certain conditions; however, the geologists should be encouraged to make suggestions about mining procedures which might be used. For example, the explorationists may recognize that one ore body is particularly amenable to solution mining, whereas others in the area might require a different mining method because of low permeability resulting from dense calcite cementation or other problems. Geologists should volunteer such views so that the miners can develop their mine plan accordingly. Some uranium companies have a program of requiring inexperienced exploration geologists to participate for a time in mine planning.

Activities of Others in the Area

Reports to the miners by the geologists should contain a summary of the activities of other companies in the area. This should include discussions of both reported and rumored discoveries by others in the area, with details of the location, depth, grade, reserves, and so forth. In addition, land acquisition efforts should be described, as well as ongoing negotiations, drilling, or any other type of activities by parties who may be competing for uranium properties in the area. As it may be desirable for the miners or the geologists to attempt to ac-

quire certain properties, a report concerning activities of others in the area might be of considerable value. This is referred to in the oil business as "scouting." Many oil companies have been able to acquire oil and gas leases on properties which ultimately were productive through prompt action on scouting reports.

Environmental Considerations

*Surface Disturbance
and Restoration Practices*

In the course of their exploration work, explorationists generally become acquainted with various drilling techniques which cause minimal surface disturbance. They also become knowledgeable about those restoration methods which produce the best results, such as sowing a particular type of grass seed, and utilizing straw or fertilizer on drill sites to cover surface disturbances which might result from an exploration program. Explorationists sometimes find it desirable, in the course of their work, to confer with agronomists or with appropriate federal government agencies to determine which grasses might grow best in a given area or to research other questions related to land disturbance (fig. 12-2). They should then pass this information on to the development operators.

Surface and Ground Water

A certain amount of water is used in almost all exploration and mining operations, and the geologists should prepare for the miners a discussion of the characteristics and availability of both surface and ground water in the area to be developed. The geologists frequently purchase the surface water, and if this is the case, they should inform the development operators of the terms of such purchase, indicating whether they encountered any problems relating thereto.

Figure 12-2. Previous site of open-pit mining for uraniferous lignite in North Dakota has been stocked with trout and is now a favorite fishing spot for local ranchers and farmers. The pond is fed by ground water and has no inlet or outlet. The water level in the pond accurately reflects the water-table depth at this location and is about 10 m below the surface. Such beneficial ponds can often be developed in the course of open-pit mining and should be included in many reclamation plans instead of the all-inclusive "restore to original contour" concept.

Many mining operations, with the exception of solution mining, can cause a significant disturbance of the ground water in the immediate area surrounding the project. The geologists should discuss such disturbances, provided they have sufficient information to do so at the time. For example, let's assume that a uranium-ore body exists in a particular area at a depth of 105 m (350 ft), the water table is at 30 m (100 ft), and there are some stock water wells in the immediate vicinity, drilled and cased to a depth of 45 m (150 ft). If the stock water wells are close enough to the excavation, then a cone of depression will almost certainly cause the water table to drop below the depth of the water wells as the mine excavation

progresses and water is pumped out. In such a case, the geologists might suggest to the miners that an advance agreement be made with the owners of the water wells whereby the wells could be moved to a new location or deepened to another aquifer. As an alternative, water from the mining operation could be pumped to those pastures where it is used by livestock, thereby satisfying the rancher's water requirements for his stock. The latter course is less desirable, however, because the specter of uranium and daughter products in the water may be raised.

In the case of solution mining, geologists should inform the development operators of the regional direction of ground-water movement, and also the rate of movement, if such information is available. Such data could be of great significance to solution miners in determining how many patterns, or what type of patterns, they might use in the solution-mining process (figs. 12-3, 12-4).

Drill-Hole Plugging Procedures

As we have discussed, most states have enacted statutes pertaining to the plugging of wells of all types, including exploration drill holes. Such laws should be discussed or even quoted in geologists' reports to the miners, together with a description of plugging procedures which they have used in the course of their program. The costs for plugging should be given, as well as suggestions for modifying procedures to reduce costs but still comply with the law.

Federal and State Environmental Agencies

Explorationists, geologists, and miners are required to have ever-increasing dealings with the personnel of various federal and state agencies. As with any individuals, in or out of government, some are reasonable and cooperative and others are not. Information concerning individual bureaucrats can be valuable to development operators, and the geologists should relate these facts, including suggestions for expediting certain applications or procedures through the necessary agencies. They should also point out any particular problem areas which may exist in dealing with the agencies.

CONTINUING GEOLOGIC MAPPING AND INTERPRETATION DURING MINING

Discussion

It is extremely important that geologic mapping and interpretation be continued during advanced exploration and through the development program of any mining operation. There are two basic reasons.

First, if the geologic information is properly gathered, it can be used to significant advantage by the development operators in understanding the geology of the mine, thereby facilitating greater ore recovery. With accurate geology, the operators will be able to predict with greater reliability where the ore is going, and where it is likely or unlikely to occur in the mine. Therefore, it is extremely important that persons with geologic competence be assigned the task of carrying out geologic mapping and interpretation in the course of the mining operation. Second, such data can be used to help guide exploration in new areas, either in the same general district or in more distant areas where the company might wish to conduct exploration or mineral investigations. After all, new ore deposits are usually found through attempts to understand those which have already been discovered; the knowledge of known deposits is extremely valuable in seeking new occurrences. Yet despite this fact, many development operators require little or no geologic investigation to be carried out on a project during the course of mining.

Stuart R. Wallace, a highly respected geologist, addressed some of these problems in an excellent paper delivered in 1974. Excerpts from this paper (p. 216–227) follow:

> In spite of continuing technical advances by the mining industry in the more efficient handling of muck in the mine and its more effective treatment in various types of plants, new discoveries of ore grade material, or even interesting mineralization, are just plain hard to come by. It is increasingly difficult, expensive, and time consuming to find something that qualifies as an ore body. . . . if the mining companies are to fulfill their responsibility of supplying our raw

Figure 12-3. Solution-mining test site in Texas. Drilling pattern, visible in background near top of photo, is evidenced by light-colored patches on ground where drill cuttings have been scattered on the surface. Such evidence usually disappears within a short time. Courtesy Wyoming Mineral Corporation.

mineral needs, they must work harder and do better than they have in the past at the business of finding ore. . . .

My basic premise is that the intelligent search for additional ore reserves must be based on a thorough understanding of known deposits and that this understanding can come only from really detailed, careful, and painstaking study.

I would like to make a general comment on the geometry of ore bodies. An ore body drawn with geologic feeling or character is much more informative than one drawn by a computer or one drawn with straight lines or by the polygonal method. I know this cannot always be done—it depends upon the amount of information available—but if you can, do it. The ore boundary and the grade-zone boundaries are the most

important geologic contacts drawn by the economic geologist, not just for reserve calculation or mining control but because they may tell him a great deal about the origin of the deposit. There are obvious difficulties with those deposits that have been strongly affected by supergene processes, but the ore zone, properly treated, can still be a tremendously useful tool.

. . . I want to comment . . . on the general problem of when to quit and when to keep going. There are always reasons why a program should be terminated or a property dropped—not just political or economic or geographic reasons, but geologic ones. Negative geologic factors are always easy to come up with and they are difficult to counter.

I am not saying that negative features should be ignored, but I think that

Figure 12-4. Solution-mining plant in Texas. Uranium-ore body exists under the elongate area in the center of the photograph. Solutions which dissolve the uranium are circulated from the plant to injection wells, through the ore body to production wells, and back to the plant. When mining is complete, the wells are cemented and the surface is restored and planted. Courtesy Wyoming Mineral Corporation.

scientific knowledge must be tempered with a good deal of common sense, particularly in the business of exploration. I also think that a certain amount of persistence is a virtue. I have known geologists who were unquestionably very intelligent men, but who for some reason lacked what I would term a real geologic sense. There are times when the dumbest thing you can do is to be too smart. Mother Nature is extremely capricious; never assume that you can successfully predict her infinite vagaries beyond a certain general point. Unfortunately for us, specifics and not generalities are what determine the exact location of most ore bodies.

The judgment to continue or to quit is difficult to make and can never be completely objective. If you persist and discover something, you may be con-sidered brilliant or perhaps just lucky; but in any event a hero of sorts. If you fail, you are apt to be labeled quite a few other things; this is a hazard of the business. If you are not willing to stick your neck out once in awhile, I don't really think you should expect to find an ore body.

There is another vital factor in the discovery equation, and that is management. There is no substitute for a management that is sympathetic to your problems and understands the odds of mineral exploration.

I would like now to summarize and make a few recommendations.

1) Ore bodies are complex geologic features and result from very special sets of conditions. If this were not so, ore bodies would be commonplace—clearly they are not. The complexities will be at

various scales and of differing kinds— from the oxidation state of an element in a mineral to the structural development of a mountain range. A roll-front uranium ore body is not the same as a bauxite deposit, and both are different from a molybdenite stockwork. Thus, the complexities will be different, but they are there somewhere. Discover these complexities, decipher them, and apply this knowledge to the search for additional ore.

2) It takes time—a long time—to develop a model—and only then if you are persistent in your attack. Many ore bodies are less complicated than Climax, but I have no doubt that others are more so. Most ore bodies are probably far more complicated than has been recognized. Those complications that are part of the ore body itself must be investigated while the ore body is being opened up and mined. After it is gone, it's too late. There are no shortcuts— there is no panacea—there is no black-box magic that will give you the answer. Laboratory studies are invaluable for total understanding; they can tell us much, but only if the geologist provides an accurate model into which the numbers that are developed can be properly fitted.

There is no substitute for the geologic map and section—absolutely none. There never has been and there never will be. The basic geology still must come first—and if it is wrong, everything that follows from it will probably be wrong.

3) No two ore bodies are the same. A geologic model of a deposit therefore should be not only descriptive, it should provide as much genetic information as possible, because it is only by understanding the origin that the geologist is able to interpret the differences be-

tween ore bodies and thereby intelligently apply a model in varied geologic settings.

. . . I have stressed the development and use of geologic models, and this emphasis reflects my own experience. I realize that there is far more to exploration than the modeling of individual deposits, but doing so does not in itself preclude any other exploration approach. Ideally, it should complement larger scale analysis. Understanding details is an integral part of understanding generalities. Understanding today's ore bodies does not necessarily limit our approach to exploration or confine our thinking. It simply gives us some tools to think with.

. . . The fact is, however, that we do not do these seemingly obvious things. By "we" I mean the mining industry and the geologic profession. Both are very much remiss in that many ore bodies are being mined, and geologically destroyed, and there is little or no real effort made to determine the model.

Geologists by nature are a hypercritical lot, especially of their own kind, but even taking this into account, I find that I come away from perhaps half of the mines that I visit with a sense of disappointment and frustration and the conviction that the job will never be done.

I have visited mines, and you have too, where the geologists spend most of their time surveying, controlling mining grade, or doing some other type of nongeologic or border-line geologic work. We all expect some of this, and a certain amount of it is good; but it's the kind of thing that easily gets out of hand. Sometimes this is the manager's fault and sometimes the blame lies with the geologist himself. At other mines, an engineer will do whatever geology is thought to be required, and at still others, including some pretty fair-sized operations, there is absolutely no attempt to do any geology at all.

To those of you who are managers, or who will be shortly, I say this—you have a chance to learn some things about the ore bodies that are being mined today that may help in the search for those that will surely be needed tomorrow. Take this opportunity, it will never come again. Make certain that you have a qualified geologic staff and see that they have the opportunity to do some research geology as well as operational geology. If they have this opportunity and do not take it—get some who will.

To mining geologists I say—do the operational geology, and do it well, but don't neglect the generic modeling part of the job either. It will serve you well in the search for ore, both at your own mine and elsewhere.

We have observed that geologists working on mining projects often spend their time on matters other than mapping, such as development drilling or making ore-reserve studies. These are also important functions, but meanwhile, mining goes on, with new exposures appearing and being mined away every working day with no detailed mapping carried out. Geologists are used extensively in underground work in the Grants district, but primarily they are ore-control engineers. While ore control is a needed function, the importance of good mine mapping and interpretation should not be ignored. Likewise, there appears to be essentially no geologic mapping being done in any of the major uranium areas of Wyoming. One can only conclude that the mine operators see little or no value in the mapping, and consequently are not willing to have geologists on the payroll unless they are directly connected with ore production.

Organizational Structure

The structure of an organization is often a stumbling block in attempting to obtain accurate geologic information from a mining operation. Generally the mine superintendent or the project manager, or both, are mining engineers. Because of their training and experience, they are accustomed to dealing with relatively firm numbers regarding ore reserves, and they usually have definite opinions relating to procedures for obtaining reserve estimates. From an engineering standpoint, they believe they know the kind of information that is required and the amount of dollars which should be spent on a particular project to arrive at the lowest dollar cost per ton of ore produced. Unfortunately, the geological viewpoint is too often overlooked, many times because firm numbers cannot be attached to the data.

Only the most open-minded project managers or superintendents who are not geologists seem to be able to recognize the significant value in a geologist's mapping of a deposit which has already been found and is now being mined. Unfortunately, regardless of the original intention, when geologists are assigned to a mining project, they frequently are obliged to assume the roles of ore-control engineers, reserve-estimate engineers, or surveyors (such as for the laying of sewer lines for the town site), or to perform other menial tasks which are totally unrelated to geology. When this happens, the whole operation ultimately suffers.

The value of mapping is sometimes difficult to recognize, and often mapping must go on for months before helpful results begin to appear. Even though the geologist is gathering data which could result in lower mining costs, and perhaps additional ore recovered, it may still be difficult to pinpoint where a significant contribution took place. Consequently, geology is often neglected as mine managers go about the day-by-day procedures to produce ore, disregarding geological mapping.

In order to accomplish accurate, worthwhile mapping in an active mine, we recommend having the geologists on the project report directly to an exploration office (probably at a different location), rather than to a manager at the mine. This type of organization can result in some friction on the job site, because project managers usually prefer that all personnel working on the project be under their control, but we maintain that it is counterproductive for the project manager to control or direct the activities of geologists to other chores when they have the responsibility of mapping the mine. Another solution might be to have the geologists' responsibilities and work assignments spelled out in a job description from which no deviation would be permitted, but it is doubtful that this would be workable in many situations.

If the mine geologists are under the direction of an exploration office, they proceed to gather useful basic information concerning the overall geology, occurrences of ore in the mine, and ore-body configuration and control, unhampered by any attempts of the project manager to interfere. Although copies of most information gathered and recommendations made by the geologists would be given to the project or mine manager on the site, the geologists would not be required to report to these people. Drilling costs can be minimized, additional ore can often be recovered in the mine, driving useless drifts can be avoided, and the overall operation can become more efficient through good geologic mapping and interpretation.

Gathering Data, Including Photographs

Geologists assigned to geologic mapping and interpretation during mining operations should be prepared for long working hours and extensive data gathering. Mine geologists also usually spend considerable time working on the drilling program, either surface or long-hole drilling underground. We recommend that they carry a camera and take numerous photographs, both of the geology in the mine workings and of mining and surface activities. Such photographs will prove valuable in correspondence and company reports, and they could also be used in publications which may result from the work.

We suggest using both black-and-white and color photography. Color slides and prints are valuable tools for oral presentations, and are also useful to accompany certain reports with limited distribution. Black-and-white photography is used in most publications; utilizing black-and-white original film produces sharper prints than those converted from color negatives or slides. We recommend taking a *large* number of photographs; in that way, you can be assured of at least some good ones. Prints should be marked promptly before storing, and we suggest using card files, with a key number/letter code on the photograph and the corresponding card, for permanently recording the subject matter.

Data gathering, particularly in a mine, must be done on a continual basis. Whether it be an open-pit or an underground mine, a new face is exposed virtually every working day, and new material becomes available for inspection. We have found it is far easier, more interesting, and usually safer for the geologist to map in an underground mine for the following reasons:

1. It is often possible to map floor, walls, and roof in an underground

mine.

2. Advancement usually takes place into the ore body in drifts, raises, crosscuts, or winzes a few metres at a time, and so the details can readily be mapped, photographed, or sampled.

3. There are usually two or three miners working at a face or in a stope, and the conditions in the workings are such that the examination of walls, roof, or floor is facilitated.

4. Miners and geologists are usually on friendly terms, and the miners can be very helpful in aiding the geologists in their observations and interpretations, particularly if the geologists converse with the miners about the mine geology.

5. Both miners and geologists have a closer working relationship with the ore bodies and other rocks than is the case in an open-pit mine.

6. Areas away from active workings are usually quiet.

7. Miners usually are willing to turn off their equipment in order for the geologists to observe the walls and other areas.

8. The climate in an underground mine is almost always comfortable, with no extreme temperatures (most mines are cool). There is no wind (except in air conduits or corridors), and no rain or snow. In wet mines, of course, water will be running in here and there, but this usually presents no problem. All of these conditions considerably facilitate the mapping activities.

In an underground operation, the geologist may be required to enter the mine during unconventional working hours if he wishes to observe a particular face or feature exposed. Regardless of whether it is midnight or on a weekend, if a geologist is interested in seeing a certain face being mined, he has to be available at that specific time; otherwise, he must miss it altogether.

Drawbacks for the geologist working in an open-pit mine, on the other hand, include the following:

1. Mining usually consists of removing overburden down to the top of the ore body; certain parts of the ore body are then mined by a combination of vertical and lateral excavations, often involving ripping, scraping, and/or dozing. It is therefore extremely difficult, most of the time, to find a clean face to map.

2. Advancement of the mining operation usually takes place rapidly. Generally, the ore is dug out with a power shovel and loaded into trucks which are constantly coming and going, making it almost impossible for the geologist to observe the advancing ore face (or waste).

3. Relationships between geologists and equipment operators in the pit are usually far more distant than those in an underground mine.

4. The climate in an open pit can be severe, ranging from temperatures above 40°C (100°F) to –40°C (–40°F), with rain, snow, wind, and dust causing frequent problems.

5. Heavy equipment is noisy and dangerous and can be an extreme hazard for the geologist who becomes engrossed in mapping, photographing, or sampling.

If geologic information at a mine site is properly gathered, geologists can assemble an accurate picture of what transpired during the mineralization process. They can gain an understanding of how and in which direction the solutions moved, why precipitation took place where it did, the role of the various minerals and humic substances (if any) in the precipitation process, and perhaps the role of other minor elements, such as selenium and molybdenum, as they appear at various locations in the mine workings.

Publishing or Otherwise Using Data

Information gathered by geologists on a mining project can be extremely valuable to other departments in the company or to other company geologists in guiding exploration plans and activities for deposits in the same or different regions. Although most companies are aware of the value of such information, they all have different approaches and policies concerning handling the information. Some guard it jealously, and under no circumstances will they permit any of their employees to present oral or published papers regarding observations which have been made in the course of exploration, or in mapping and gathering data at a mining project. We believe this attitude is understandable for the short term; however, after some time has passed—say, two or three years—the company should have had ample opportunity to utilize the data, and it should then permit its employees to publish major parts of the information. Data or interpretations which are very critical can be (and usually are) withheld. Students of geology should be aware that many papers written by industry geologists often do not contain information which would aid competitors in the race for ore deposit discovery. Yet, the policy of publication is a good one for the industry in general; it gives the individual recognition within the industry, and it offers the company an opportunity to make a contribution which may further the exploration efforts of all those involved in the search for minerals.

In our opinion, geologists who are employed by a company should be encouraged (and even helped) to publish

papers. It causes them to consider and review their ideas more carefully and to pay greater attention as they gather geologic information in their programs, whether exploration or mining. Further, it provides an additional form of compensation to the employee in the form of recognition throughout the industry for original or inventive work on a particular project. New ideas may have been developed or old ideas may be resubstantiated, but the approval and support of geological papers by an employer is an extremely important factor for everyone involved in the exploration industry. As pointed out by Wallace (1974) in the quotation above, our search for new deposits is based largely on our knowledge of previously discovered deposits.

Several substantial companies have been involved in uranium exploration and mining since the early 1950s and have employed literally hundreds of geologists; yet the number of papers published by their geologists can be counted on one hand. It is apparent, and very unfortunate, that such companies have maintained a policy of not permitting their employees to publish papers concerning uranium geology. Their common excuse for such a policy is that the information gathered may one day be of use to them in a competitive situation. They use the same argument for confidential logs from exploration drill holes. If petroleum exploration companies behaved in a similarly secretive manner, the United States would be even more sadly lacking in petroleum resources than it is.

It is our firm belief that companies have no justifiable reason for keeping most data confidential after a two- or three-year period. In any case, the secretive employer cannot prevent the resignation of employees; they may become dissatisfied with certain aspects of the job or the employer, or perhaps a more favorable situation presents itself. Consequently, any truly important information ultimately becomes disseminated throughout the industry, but in this case, neither the company which originally financed the gathering of the information nor the particular employee who worked on the project receives any credit.

We are not proposing that employees initiate a conflict-of-interest situation in which they resign from a company and then utilize information which they have gathered while in that company's employ against the best interests of the previous employer. In fact, many companies require that an employee sign an agreement stating that when he leaves the employ of that company he will not operate in areas where

the company has properties, or may be acquiring properties, for a one-year period. Naturally, the previous employer cannot require an employee to obliterate from his mind all that he may have learned while on the job. Consequently, when the employee leaves, he takes with him certain information which has been mentally retained.

We again stress that, after a reasonable time has passed, employers should encourage their employees to publish papers on data gathered, information assembled, and interpretations made. Such a procedure is beneficial to everyone: the employer, the employee, and the industry as a whole.

13
A Forward Look

OVERVIEW

International interest in uranium has increased during the last decade, with concomitant increases in research aimed at expanding and improving exploration techniques. The International Atomic Energy Agency has sponsored several meetings to coordinate research activities and encourage communication between research and exploration groups. Within the United States, the U.S. Energy Research and Development Administration has accelerated its program of supporting research to improve technologies and procedures in uranium exploration. The Uranium and Thorium Branch of the U.S. Geological Survey has accelerated and expanded its programs which are aimed at improving exploration techniques as well

as developing new geophysical and geochemical tools. Some groups within the industry have also taken steps to carry out research to develop new methods as well as new and improved tools for exploration. Significant advances in technology have resulted from projects which were financed and executed by groups of companies and by joint efforts between companies and government. Some research has been accomplished by universities with financial aid from government. This increased research activity is resulting in numerous improvements affecting exploration. The usefulness of some of the developing technologies is difficult to assess at this time because of inadequate field testing. The most promising avenues of research which are being followed are discussed in the following pages.

GEOLOGICAL RESEARCH

Regional Studies

Integrated studies of mineralized regions are becoming feasible as a result of accumulated masses of drill-hole information and other data which have been acquired by various exploration programs. In addition to describing an ore deposit, its geological setting and interpreted origin and history, we may be able to describe entire mineralized regions in detail, relating the various mineral occurrences to one another.

To illustrate, let us consider the roll-front deposits of Wyoming and surrounding areas. Many published reports have described individual ore deposits; several reports have described individual districts; and a few reports have outlined theories of origin (or partial theories) for the roll-front deposits in general. Drilling, however, has been carried out within all of the known uranium districts as well as in large areas surrounding the districts. Perhaps 30 companies have most of the drill-hole information and other data in their files, but no single company has sufficient information to develop a totally integrated picture of the entire region. At this point, no effort has been made to present an integrated description of the entire mineralized region that relates one mineralized occurrence to another or the various districts to one another.

Broad regional controls on mineralization may be recognized when an integrated study is carried out. When the host rocks of the various occurrences are correlated regionally, it may become apparent that bifacies roll-front deposits occur in reduced facies deposited by large trunk streams of certain major paleodrainage systems. These regional studies may also show that specific identifiable zones are preferentially mineralized within particular parts of the

region. We believe that the southern Powder River Basin deposits are among the oldest bifacies deposits in the Wyoming mineralized region. These deposits are probably earliest Eocene in age. The bifacies deposits in the southern part of the Great Divide Basin are probably contemporary with those of the southern Powder River Basin. The Gas Hills, Shirley Basin, and Crook's Gap deposits occur in the upper part of the lower Eocene strata within 50 to 100 m of the contact with the overlying Wagon Bed Formation. We believe that these deposits were formed in late early Eocene time.

Monofacies deposits, which occur beneath the pre-Oligocene unconformity in northeastern Colorado, eastern Wyoming, western South Dakota, southeastern Montana, and southwestern North Dakota, were probably formed during late Eocene or earliest Oligocene time. These deposits appear to be related to paleodrainage systems which can be mapped by regional studies of the unconformity.

Regional studies of mineralized regions will prove to be extremely useful in developing new prospect areas, but these cannot be carried out in sufficient detail unless the exploration companies develop a cooperative program to make the data available. Companies should adopt the attitude that when a project within a portion of the region is reasonably complete, the data therefrom should be made available for use by the industry at large. This is the only way that comprehensive regional studies will be made. Yet, realistically, it seems unlikely that the companies will cooperate to this extent unless legislation is passed either at the state or local level which requires that drilling data be filed with a state agency within a reasonable period of time.

We predict that in the next few years an increasing number of states will begin requiring that drilling information be filed with an agency such as the state survey, and that after two or three years, the information will be available to the public. Will this result in more ore being found? Of course it will; the petroleum industry has already demonstrated as much. The difference is that the petroleum companies make the information available voluntarily; the mining community is not inclined to do so.

Aerial Photographs and Multispectral Scanning Techniques

Remote-sensing technology has been improved in recent years. Airborne geophysical methods have been improved and expanded as discussed below; also, aerial photographs and multispectral scanning techniques for recording reflectance data have been improved. It is difficult to predict the future usefulness of aerial photography. Perhaps increased use of films which are more sensitive to longer visible wavelengths will permit better resolution when the improved methods of photography begin to be used on high-altitude flights.

Some research is being carried out by the U.S. Geological Survey to improve and expand the use of multispectral-scanner reflectance data. This work includes on-the-ground studies of altered rocks to determine which parts of the spectral reflectance data characterize such alteration. Most of the multispectral-scanner reflectance data have been collected by Landsat orbital systems at an altitude of about 910 km (Spirakis and Condit, 1975). These data can be enhanced by computer programs to distinguish large exposed areas of alteration, but the resolution is inadequate for mapping detailed geology.

The technology of multispectral reflectance scanning and data processing is being developed and adapted to mineral explora-tion. If the costs can be reduced so that the technology might be used for high resolution at moderate flight altitudes, we believe it will be a very useful tool for future improved geological mapping.

Mapping Lithologies and Facies

Sedimentary rocks have always been subdivided into members, formations, groups, and supergroups which in turn have been measured, described, and mapped. Until the present time, most field geologists have been concerned with identifying rock outcrops in terms of the correct and formal unit to which they belong. A formation, or even a member, frequently includes more than 100 m of diverse lithologies. These lithologies may change laterally as well as vertically, due to facies variations within a formation.

Most geologic maps of sedimentary rocks fail to show lithologic details within the formal map units, and it is a rare exception which gives any indication that rock outcrops appear to be altered. Even igneous and metamorphic rocks are sometimes subdivided into formations and mapped with only limited specific mention of lithologic detail. Fortunately, however, these "hard" rocks are usually mapped as specific lithologies. We believe that the next generation of geologic maps in the sedimentary environments should show specific lithologies. Significant facies should all be mapped and described. Such mapping is facilitated by good color photography.

Hydrodynamics

In sedimentary basins, past movement of ground water often played an important role in processes of mineralization. Present ground-water movement and analyses of the water are important to exploration programs seeking hidden mineral deposits. Most geochemical surveys must take into account ground-water movement. Helium and radon anomalies in soil gas frequently appear to be displaced by active ground-water movement. When ground water itself is sampled in the course of geochemical prospecting, the movement becomes particularly critical. We believe that much can be gained by explorationists if more knowledge is acquired about hydrodynamics in sedimentary environments. Exploration programs can be more efficiently planned, and data can be better interpreted. A better understanding of hydrodynamics should result in improved concepts of the origin of strata-controlled uranium deposits.

Deciphering the History of Mineral Deposits

Exploration geologists are always on the alert for information which might shed light on the origin and history of mineral deposits. Their interest has a very practical basis, because knowledge of the origin and history of a deposit is important in establishing controls on mineralization which guide exploration.

Rapid and low-cost identification and computerized analytical techniques which have been developed for clay minerals and a variety of other minerals may facilitate the detailed and extensive studies which are necessary to determine what mineralogical differences distinguish altered rocks from their unaltered equivalents, or a good host rock from a poor one. The limited studies which have been reported in the literature suggest that alteration of clay minerals, in particular, might be significant. For example, the work of Lee (1975) in New Mexico suggested that Rb-Sr age determination on vanadium-enriched clays may date the alteration and mineralization much more reliably than previous attempts utilizing uranium-ore minerals.

As discussed in chapter 2, geologists have, in some cases, attempted to use fragmentary data in support of theories explaining part of the history and origin of uranium deposits. Their ideas seem to be plausible regarding the use of the data, but we have questioned the validity of such fragmentary data because of the limited scope, coupled with the variations in the mineralogy of the rocks. An example would be the theories advanced that uranium originated outside the present host rocks in Wyoming roll-front deposits because a few analyses of altered sandstone show an enrichment of uranium. With better coring and sampling techniques and improved low-cost analyses, it should be possible to acquire the body of data necessary to contrast altered and unaltered rocks.

Research on stable isotopes of sulphur, oxygen, and other elements might help to decipher the history of some uranium deposits. Recent work by the U.S. Geological Survey has indicated that some calcite concentrations near ore deposits in New Mexico derived carbon from organic sources. The research also shows that the organic carbonaceous matter, which appears to have been transported as soluble humic acids during the time of uranium mineralization, is now almost pure carbon. We believe that a better understanding of these organic carbonaceous materials and their chemical evolution will help to explain other aspects of the uranium deposits and will possibly define additional controls on mineralization.

Studies of phytogenic materials and their chemical modification with time in the subsurface environments should be useful in understanding the formation of uranium deposits. We also believe that research on the role of climates and climatic change in controlling chemical weathering and transport of organic material, metals, and other elements in solution might help in deciphering the history and origin of uranium deposits. There is evidence that climates and climatic changes played important roles in the formation of many strata-controlled uranium deposits. Some evidence also suggests that climatic conditions influenced the formation of structure-controlled uranium deposits in northern Saskatchewan.

When the origin and evolution of known uranium occurrences are better understood, exploration will be enhanced. The explorationist will have better models on which to base his programs.

GEOCHEMISTRY

Hydrogeochemical Surveys

In sedimentary environments, ground water can yield valuable data to reconnaissance exploration programs, and new technologies are beginning to make it possible to acquire more of these data. Most hydrogeochemical surveys in uranium exploration have focused on uranium. Radon analyses of surface waters have been included in hydrogeochemical surveys in Canadian Shield areas (Dyck, 1975), but few surveys utilizing ground water have included analyses of radon contents.

The technology for rapid, high-precision, low-cost analysis of helium in water (or soil gas) has only recently been developed. Field testing of the helium-analysis technology is now under way. We believe that hydrogeochemical surveys which utilize ground water should include analyses of uranium, radon, and helium. Uranium is very soluble in oxidizing ground water, particularly when carbonate or sulphate ions are present to form uranyl complexes. In many sedimentary basins, uranium may form fairly widespread anomalies in oxidized aquifers, but uranium contents are dependent on the oxidation potential of the ground water.

When the ground water moves into reducing environments (where the aquifer contains humates or other reductants) the uranium is precipitated, and the water analyses will probably indicate less than 5 ppb uranium. A water sample taken from a well near a uranium deposit, but in the reduced aquifer, might contain less uranium than a sample taken far from any uranium deposit but in an oxidized part of the aquifer. At present, uranium analyses are obviously useful in establishing the presence of uranium in an area, and in addition, they may shed some light on the distribution of altered sandstone in an area where roll fronts are anticipated.

Radon and helium are both soluble in water, and this solubility is not dependent on the oxidation potential or the presence or absence of any complexing ions. Radon is restricted in its migration by its half-life of 3.82 days and by its low rate of diffusion. In many cases, radon anomalies in ground water should give sharper focus on uranium deposits than uranium anomalies. Helium will migrate much farther from a uranium source than radon, because it is a stable isotope and very diffusible. Helium anomalies in ground water may be detected at distances of several kilometres from a uranium source.

The detection of the combination of uranium, radon, and helium in ground water should produce good results in regional exploration programs in sedimentary basins where wells are available for sampling.

Soil-Gas Surveys

The recent development of helium-detection technology has added a new dimension to soil-gas surveys. Until now, soil-gas surveys have simply consisted of radon analyses. As discussed in chapter 6, the quantitative data available indicate that radon has a very limited range of migration in sedimentary rocks, soil, or water. We have not seen any conclusive evidence of radon migration upward through soils, rock, or water involving distances of more than 10 to 15 m unless open fractures or other conduits are present. Gabelman (1972) measured radon over the Starks salt dome in Louisiana; he concluded that radon migrates greater distances than those suggested by Tanner (1964). However, Gabelman was unable to acquire good background data around the dome due to ubiquitous swamps. It is also likely that salt domes have anomalous characteristics favorable to radon migration, such as numerous tension fractures. Uncased drill holes on the dome may permit gases to migrate rapidly upward to disperse in shallow, permeable sandy layers.

Helium has a much greater capacity for long-range vertical migration through rock, soils, and water than does radon. However, due to its high rate of diffusion, helium disperses through the soil, and the anomalies usually are of low amplitude. It requires extreme care in sampling and analysis to confirm a helium anomaly in soil gas. The technology is not adequately field tested at this time, but we believe that helium analysis of soil gas will give greater dimension to soil-gas surveys.

Pathfinder Elements

Some modest encouragement has been reported from the analysis of pathfinder elements other than uranium, radon, and helium in water, bottom-sediment, and soil-sampling programs. We have not seen any results which show these methods to be conclusively successful, however. As discussed in chapter 6, soil sampling and bottom-sediment sampling are fraught with uncertainties when analysis is for uranium, but other elements might be used successfully in such surveys. Some research along

this line is being carried out by the U.S. Geological Survey.

NONRADIOMETRIC SURFACE GEOPHYSICS
(by Bruce D. Smith, Geophysicist, U.S. Geological Survey)

Introduction

The success of any mineral exploration program depends upon careful integration of all information available to the explorationist. This information is gathered using geological, geochemical, and geophysical tools. The conceptual or physical model of the particular exploration target of interest indicates which of the many geophysical tools are likely to be most effective. In uranium exploration, the construction of such a conceptual or physical model must be somewhat innovative, since many subtle aspects of even the more typical targets are not well known. The current ignorance of the subtle aspects of uranium targets is partly due to the relative youth of uranium exploration in comparison to exploration for many other mineral deposits. A consequence of this relative youth is that neither the conceptual models nor many of the geological, geochemical, and geophysical tools have been fully developed. This is particularly true for nonradiometric geophysical methods.

Nonradiometric geophysical methods have not been extensively used to date in uranium exploration primarily because uranium mineral deposits, unlike sulphide or iron mineral deposits, do not seem to have a distinctive physical property which can lead to a characteristic geophysical anomaly. Thus state-of-the-art nonradiometric geophysical methods can at best only indirectly indicate favorable areas for uranium deposits. The following discussion of applications of these methods demonstrates that they can be used very effectively in detecting and defining certain aspects of the geochemical and geological environment of uranium deposits. In addition, rapidly increasing sensitivities of geophysical instrumentation and improved methods in studying the physical properties of rocks may one day lead to more direct application of nonradiometric geophysical methods to uranium exploration.

A complete description of all nonradiometric geophysical methods which could be applied to uranium exploration would require a text as long as this book. The following discussion is limited to the most commonly used methods which at this time appear most promising. Consequently, some of the more exotic methods such as heat flow, and some of the less well understood methods such as spontaneous potential, are not discussed. No one geophysical method will prove to be the best for all types of uranium deposits because the geological and geochemical settings of different deposits are highly variable. The selection of particular or complementary geophysical methods must be based on geologic and geochemical information about the nature of the target. Consequently, for each general type of geophysical method, the more important physical parameters which control the response are briefly described. A summary of reported applications and an estimate of the potential applications for each method are given in the following discussion.

Seismic Methods

A comprehensive discussion of seismic methods was given by Telford and others (1976). Dix (1952), Dobrin (1960), Griffiths and King (1965), and Zohdy and others (1974) gave good general descriptions of the seismic method. Generally, seismic methods can be divided into reflection and refraction methods. Refraction work detects increases in acoustic velocity of rocks. Reflection work detects varia-

tions in the acoustic impedance (the product of seismic velocity and density of rocks). Refraction surveys are generally cheaper, but they detect high-velocity units only, and they can detect such units only to a depth of about one-third the aperture (the distance over the ground of the geophone or sensor spread). Reflection methods, while more expensive, can detect both low and high seismic velocity units and usually see to depths many times the aperture. Applications discussed below illustrate the types of seismically detectable variations in rock properties which may prove useful in uranium exploration.

The seismic refraction method has been used by Black and others (1962) to locate channels cut into the Triassic Moenkopi Formation in the Monument Valley area. These channels are filled with sedimentary rocks of the Triassic Shinarump Member of the Chinle Formation. The seismic survey was successful in locating the buried channels, several of which were defined by drilling subsequent to the survey. The success of the survey can be attributed to a velocity contrast on the order of a factor of 2 between the Shinarump (2,380 m per sec) and the Moenkopi (4,270 m per sec). Using the refraction method, the depth to the contact between the Moenkopi and Shinarump could be mapped.

The seismic reflection method, the major geophysical tool in oil exploration, is by far more commonly used than the refraction method. High-resolution seismic reflection methods have a potential application to uranium exploration where the target is not deeper than approximately 300 to 400 m. We will confine our discussion to this method, as it is likely to be important in exploration in the future. Other versions of seismic reflection methods are described in the references mentioned at the beginning of this discussion.

The high-resolution seismic method is a

type of reflection method in which the frequency of the received seismic signal is much higher than that commonly used in oil exploration. Routine seismic refraction surveys use a low-frequency range (8 to 20 Hz or cycles per sec) because the higher-frequency seismic signals are attenuated too much to reach the depths of many oil exploration targets (several kilometres). The high-resolution seismic method makes use of signals in the frequency range of 50 to several hundred hertz. Use of high frequencies in seismic exploration enables mapping of much smaller structures at shallower depths than can be mapped with lower frequencies. Therefore, the high-resolution seismic method may prove to be useful in mapping stratigraphic features which are important in uranium exploration.

The high-resolution seismic method must at present be viewed as somewhat experimental in comparison to more routine reflection surveys. The data acquisition and processing must be carried out with greater care than in more routine surveys. The sources used to generate the high-frequency signals must be carefully tailored to the particular geological environment. Attenuation of high-frequency signals in weathered strata above the water table can be a problem if the seismic source (an explosion) cannot be placed at or near the water table.

There are some general considerations in the implementation of many seismic surveys. A major limitation is the high cost of conducting a survey and processing the data. A second consideration in seismic methods is that explosive seismic sources may have an adverse environmental impact. Fortunately, considerable progress has recently been made in the use of truck-mounted vibrating or impacting sources, some of which can be used even in residential areas. In the case of high-resolution surveys, some experimentation is currently

being conducted on sources consisting of prima-cord (similar in size to a telephone cable) buried in a narrow trench. By varying the length (3 to 4.5 m) and the number of segments (greater than two), an effective seismic source with minimal environmental impact can be used. A third important consideration in designing and interpreting data from seismic surveys is that very little is known about the seismic velocities of many uranium-bearing sediments. This is particularly true of Tertiary sedimentary rocks such as in the Powder River Basin of Wyoming.

The single reported use of high-resolution seismic data in uranium exploration is from the Sweetwater uplift area of Wyoming (White, 1975). Here the basic geological problem is to define paleotopographic features which occur at depths as great as 400 m. White (1975) reported that the success factor in drilling increased by a factor of 3 to 4 when the drill holes were located on the basis of seismic data. Results from high-resolution seismic surveys have also been reported by Western Geophysical Co. (1975, written commun.) in the Texas coastal region. The survey defined faults with a throw as small as 4 to 5 m to depths on the order of 60 to 80 m. Such subtle features could provide important control in localization of uranium mineralization.

In summary, seismic methods will probably enjoy an increasing application in future uranium exploration. The growth of the seismic applications will be limited by the high cost of conducting both research and surveys. Fortunately, much of this cost is being borne by major oil companies whose seismic programs produce technical advances which can be applied to uranium exploration problems.

Gravity Methods

General aspects of gravity measurements

and their interpretation have been given by Telford and others (1976), Grant and West (1965), Griffiths and King (1965), and Zohdy and others (1974). Gravity measurements record the relative difference in the acceleration of gravity at various points on the earth's surface. The variations in the measured values of gravity acceleration are due to variations in the density of rocks. A well-known limitation in the interpretation of gravity measurements is that a shallow body can create the same gravity anomaly as a deeper body with a greater density contrast. Therefore, in order to interpret gravity anomalies adequately, something must be known about the density of rocks within the survey area. Other aspects of the interpretation of gravity measurements are discussed below in particular applications.

The gravity method is one of the cheaper geophysical tools which can be used in mineral exploration; consequently, gravity surveys are often used in regional surveys. Fortunately, within the United States many gravity studies have been made on a regional scale. These gravity data are available as maps or open-file reports from the U.S. Geological Survey or from Department of Defense files.

A good example of potential use of gravity measurements on a regional scale was given by Cady (1976), who took the available gravity data for the state of Wyoming (some 9,000 measurements) and performed a fairly routine computer processing of the data. The processed data showed that the Sweetwater uplift is associated with a broad, east-west-trending gravity high. The east-west trend is anomalous in a setting of the northwest-trending anomalies caused by Upper Cretaceous uplifts. A tentative hypothesis could be formed that east-west–trending Precambrian uplifts accompanied by broad gravity highs are favorable source environments for uranium. This hypothesis then leads to the conclusion that the sedimentary rocks flanking the Owl Creek and Uinta Mountains constitute potential target areas for uranium exploration. The fact that sedimentary rocks flanking the Owl Creek Mountains have proven uranium mineralization and that the area of the Uinta Mountains is currently receiving greater interest in uranium exploration suggests that similar regional gravity studies may be useful elsewhere.

Gravity surveys can also be used on a somewhat smaller scale to aid in the analysis of sedimentary basins. Some good examples of basin analysis by gravity methods were given by Zohdy and others (1974) in terms of hydrologic problems. In basin analysis, the gravity method might be applied to such problems as the determination of sedimentary-rock thickness and examination of basement structural controls.

Small-scale gravity surveys can aid in geological mapping where density contrasts are associated with different lithologic units. Cady (1976) gave an example of both regional and local applications of gravity measurements in the Spokane, Washington, area. The local and regional surveys were successful in locating areas of metasedimentary rocks adjacent to zoned plutons which constitute target areas for uranium exploration. Generally, the metasedimentary rocks are more dense than the plutonic rocks. Consequently, plutons are associated with gravity lows, and metasedimentary rocks are associated with gravity highs.

In sedimentary terranes, small-scale gravity surveys can only be used in relatively simple geologic problems such as mapping buried channels or permeability barriers to uranium-bearing fluids. Black and others (1962) calculated that a buried channel in the Monument Valley area would yield a very small (.04 mgal) gravity anomaly. This small gravity anomaly tended to be masked by gravity anomalies due to variations in the depth of the near-surface alluvium. Consequently, gravity measurements may not always be suitable to channel-location problems. Painstaking work would be required to see anomalies of this magnitude, for typical gravity meters have a sensitivity on the order of .01 mgal.

In summary, the gravity method can be used in both regional and local surveys in uranium exploration. Generally, gravity methods can be used to aid in geologic mapping where different rock types have sufficiently contrasting densities. The use of gravity measurements in this context is most suitable to uranium exploration in hard-rock terranes. The method will be most successful when combined with good basic geological models.

D.C. Apparent Resistivity

General references for the d.c. resistivity method are Zohdy and others (1974), Telford and others (1976), Parasnis (1962), and Dobrin (1960). More technical references are Keller and Frischknecht (1966) and Grant and West (1965). In the d.c. resistivity method, a direct current (such as from a battery) is put into the ground through metal stakes or electrodes. A voltage is then measured across two other electrodes. An apparent resistivity of the earth is computed by dividing the voltage by the current and multiplying by a constant. The constant accounts for the geometry between the current electrodes and the voltage electrodes. The resistivity value is calculated as if the earth were homogeneous—hence the term, "apparent resistivity."

Variations in the electrical resistivity of the earth produce variations in values of apparent resistivity. There are several factors which are important in governing the resistivity of rocks. The two most important factors are the porosity and the amount and nature of the water contained in the rocks. Keller and Frischknecht (1966) and Olhoeft (1976) have given good descriptions of the factors influencing rock resistivity.

Resistivity logs can be used to evaluate potential applications of surface d.c. resistivity surveys in a given geological environment. However, there are important differences between surface d.c. resistivity surveys and borehole logging measurements. For example, the most commonly used borehole method, the single-point resistance log, measures resistance (units of ohms), but surface surveys measure resistivity (units of ohms times metres). An additional complication in single-point resistance logs is that the measurements are not absolute, but relative, values of earth resistance. The measured values of resistance may vary according to the size of the borehole and the resistivity of the fluid in the borehole. Consequently, the single-point resistance may vary from absolute values within or between boreholes. If surface-resistivity measurements are expected to be used in a given area, we highly recommend that normal borehole resistivity logs be used. This is because the normal resistivity-log measurements can be used to calculate the absolute resistivity and resistance of the earth. These logs are much more easily correlated with surface-resistivity measurements.

The two general types of resistivity surveys are horizontal profiling and vertical sounding. Horizontal profiling surveys are usually designed to map variations of earth resistivity to a particular depth. Electrical vertical sounding surveys are designed to determine the variation of resistivity of the earth with depth at a single location. Most resistivity surveys in uranium exploration which have been described in the literature have been of the horizontal profiling type.

There are many different geometrical arrangements (arrays) of current and voltage electrodes used to make profiling and sounding resistivity surveys. Two of the commonly used arrays are shown in figure 13-1. The Schlumberger array is employed in vertical electrical soundings. A vertical electrical sounding is accomplished by making readings at incrementally increasing separations of the current electrodes (AB). The distance between the voltage electrodes (MN) is less than one-fifth of the distance AB and is usually increased only when the voltage between MN becomes too small to read accurately. Routine resistivity surveys use a maximum separation of about 6 km between current electrodes (AB), though the exact distance between AB depends on the depth to the layer of interest. The exact depth to which a given vertical electrical sounding will sense is determined by the electrical structure of the earth, but a gross estimate of the depth of exploration is one-third of the distance AB. The dipole-dipole array (fig. 13-1) is typical of horizontal profiling methods used in resistivity and induced-polarization surveys discussed in a following section. In a dipole-dipole survey, successive readings are made with separations between dipoles which are integer (N) multiples of the dipole length (a). Most surveys are conducted with a maximum N of 4 or 5. The transmitter dipole is then moved a unit dipole length, and successive measurements are made as above. In this manner, a horizontal profile of earth resistivity is obtained. The depth of exploration for a dipole-dipole survey depends upon the resistivity structure of the earth and the value of N and of a. Generally the depth of exploration increases logarithmically as a function of increasing N. A gross estimate of the depth of exploration is that at N = 1 the exploration depth is 1/2 Na, at N = 2

the exploration depth is 1/3 Na, and so on to the maximum value of N used in the survey.

Horizontal resistivity-profiling surveys have been reported by Stahl (1974). These surveys were conducted in a hard-rock environment to define geologic features associated with vein-type uranium deposits. The survey performed by Stahl (1974) successfully defined faults at the Schwarzwalder uranium mine near Denver, Colorado. Faults can be detected by horizontal profiling methods under the following general conditions. Brecciation of country rock associated with faulting can make the fault zone less resistive than the country rock because it contains more water or clay-alteration products. However, some fault zones may be less permeable than the country rock (due to silicification) and more resistive. A fault can often be detected when it separates rocks of different resistivities.

Results of combined profiling and sounding resistivity surveys have been described by Collin and others (1958). The surveys were conducted in a hard-rock environment at the Escarpière mine in France. The profiling surveys were used to define fault zones and to map the contact between granites and mica schists. In addition, Collin and others reported that areas of anomalously low resistivity are almost always associated with uranium mineralization. Resistivity-sounding surveys were conducted to aid in the design of underground mining operations. The integration of the application of this geophysical survey to both exploration for, and development of, a mineral deposit is highly recommended but is more advanced than typical practices in the United States.

Davis (1951) reported the application of combined profiling and sounding resistivity measurements to exploration for uranium deposits in the eastern Colorado Plateau

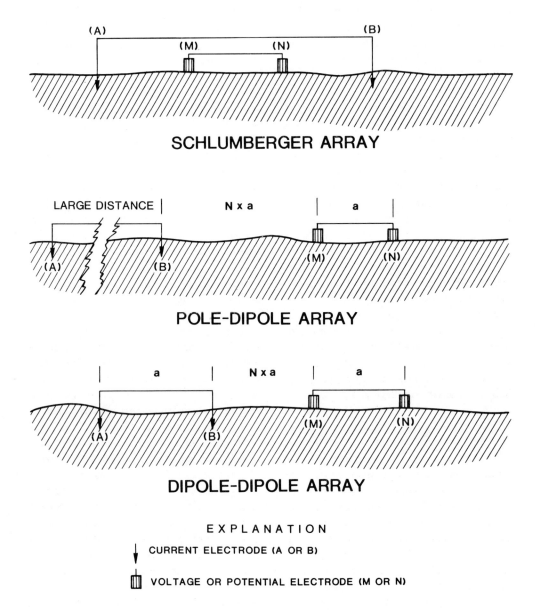

SCHLUMBERGER ARRAY

POLE-DIPOLE ARRAY

DIPOLE-DIPOLE ARRAY

E X P L A N A T I O N

CURRENT ELECTRODE (A OR B)

VOLTAGE OR POTENTIAL ELECTRODE (M OR N)

Figure 13-1. Electrode configurations commonly used in d.c. resistivity measurements.

area. In contrast to the survey reported from France (Collin and others, 1958), these surveys demonstrated that the uranium deposits were associated with resistivity highs. The resistivity highs can be correlated with variations in resistivity of the Shinarump Member of the Chinle Formation. Keller (1959) followed up this survey by making directional resistivity surveys. In this type of survey, one current electrode is placed in a borehole, and the other electrode is placed at a great distance from the borehole. The voltage electrodes are then moved around in the area near the borehole. Keller (1959) converted measured values of apparent resistivity to a favorability index. This index was based on a correlation of numerous resistivity measurements and uranium mineralization in the general region. Generally, the areas of high favorability tended to correlate with trends of known uranium mineralization.

Besides the previously mentioned exploration situation, one situation that could be effectively examined by resistivity methods is related to solution mining. The hydrological setting of the area to be mined is a very important aspect of a solution-mining operation. Since d.c. resistivity sounding is one of the major geophysical tools in ground-water exploration, it has obvious potential application toward examination of the hydrologic setting of a uranium-ore deposit.

The past successes of the resistivity method should serve at least to increase the interest of uranium explorationists. Since the time that many of the previously discussed surveys were conducted, both the method of making resistivity measurements and the methods of interpretation have advanced greatly.

Electromagnetic Methods

Telford and others (1976) and Parasnis (1962) have given good summaries of the various electromagentic (EM) geophysical methods, and more technical discussions have been contributed by Grant and West (1965), Keller and Frischknecht (1966) and Hansen and others (1967). The fundamental principle of EM methods is as follows. A source of controlled EM waves, which may be a loop of wire placed on the surface of the earth, is required. The loop is energized by an a.c. current source (alternating current generated by a motor generator). Electrical and magnetic fields are generated by the current flowing in the loop of wire. These fields are modified by the presence of the conductive structure of the earth. By measuring the modified electrical and/or magnetic fields and comparing these resultant fields with analytical or physical models, the electrical structure of the earth can be interpreted.

There are so many different types of EM methods that even a brief description would require as much space as all the other geophysical methods combined. A few guideposts for those not familiar with EM methods are as follows.

First, electromagnetic methods can be divided into those using natural electromagnetic fields (passive sources) and those using artificially generated fields (controlled sources). Second, various frequencies of EM waves are used from around 20 kHz (kilohertz) to around 10 Hz or lower. Figure 13-2 gives the frequency band for some commonly used EM methods discussed below. Just as with seismic waves, the lower-frequency techniques will penetrate deeper within the earth with less resolution than the high-frequency methods. Third, some techniques measure electric fields only, some measure magnetic fields only, and some measure both.

Unlike the d.c. resistivity methods, EM surveys can be made from the air as well as on the ground. EM surveys made on the ground can be either horizontal profiling or

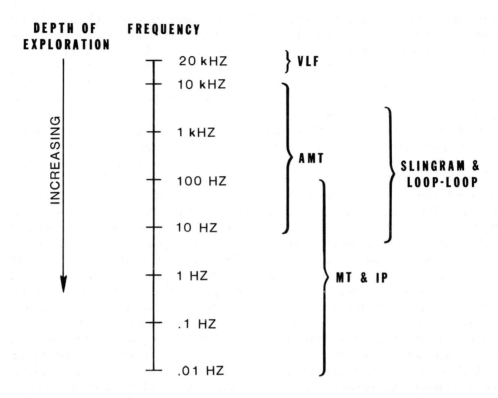

Figure 13-2. Generalized relative depths of exploration for some electromagnetic methods.

vertical sounding surveys. Vertical sounding surveys are accomplished by measuring the electrical and/or magnetic fields at several frequencies or by varying the distance between source and receiver. A fundamental aspect of all EM methods is that the depth of penetration of a signal decreases as the frequency increases. In a sounding survey with an artificial source, several frequencies (say more than ten) may be used with measurements being made at one or more receiver sites. In horizontal profiling surveys, measurements are made at many receiver sites, usually at only one or two frequencies. The frequencies used should be chosen to give the required depth of penetration in a given geologic setting.

The electrical properties of rocks which are important in the implementation and interpretation of EM surveys are described by Keller and Frischknecht (1966) and Olhoeft (1976). Conductivity is the most fundamental physical property of rocks when EM surveys are interpreted. For the purpose of most interpretation of EM data, conductivity is considered to be the reciprocal of resistivity.

There have been no documented cases where EM surveys have been used on a regional scale in uranium exploration. However, EM surveys could have several possible applications in regional exploration problems. Oil companies have used magnetotelluric (MT) and audiomagnetotelluric (AMT) methods to study the structures of sedimentary basins. These EM methods use natural electromagnetic signals generated in the ionosphere and by worldwide thunderstorm activity. The information gained from MT and AMT surveys of sedimentary basins could prove useful in initial stages of a uranium exploration program. However, the structures of most sedimentary basins in the United States are fairly well defined as a result of intensive exploration for oil. It is likely, therefore, that MT and AMT measurements for the definition of basin structure probably will receive relatively little attention in uranium exploration in the United States.

Low-frequency airborne EM methods could be used to define sedimentary channels if (1) they are large enough (more than 15 to 30 m in width), (2) they contain rocks which have a contrasting conductivity with surrounding rocks, and (3) they are within 100 m of the earth's surface. Low frequencies, no more than a few hundred hertz (cycles per second), should be used in this application. The reason low frequencies are needed is that higher frequencies are more likely to be attenuated too greatly by the sedimentary rocks to provide the useful information. Higher-frequency airborne EM surveys might be used to supplement geologic mapping.

Ground EM methods are generally suited to aid in subsurface geologic mapping on a smaller scale than airborne methods. The Slingram method is a typical example of controlled-source, horizontal profiling; this is a class of EM methods described by Keller and Frischknecht (1966) and Hansen and others (1967). The apparatus for making Slingram measurements consists of two loops, one acting as a transmitter and the other as a receiver. Each loop, usually in a horizontal plane, is carried by one person. A wire connects the two loops, which are maintained at a fixed separation. A single fixed frequency, chosen from the band, of a few hundred hertz to tens of kilohertz, depending on expected conductivities in the region, is

usually used to make measurements. The separation between the loops can range from 8 to 200 m. A rule of thumb is that the depth of exploration for this sytem is about one-third the distance between the two loops. However, as with all EM methods, the depth of exploration depends on the transmitted frequency and the conductivity of the target and the host rock. The measurements are the in-phase (real) and out-of-phase (quadrature) components of the received signal as compared to the transmitted signal. These measurements can usually be made by one member of a two-person team in a matter of seconds.

Two examples of the application of the Slingram method are illustrated by surveys conducted by the U.S. Geological Survey. In one application, reported by Black and others (1962), the Slingram method was used to define buried channels containing uranium mineralization. The EM surveys were made to supplement resistivity surveys. The Slingram method was found to be effective in locating and delineating buried sedimentary channels. The EM anomalies are thought to be due more to conductive mudstones (15 to 30 m deep) than to the sandstone in the channels.

In an application reported by Flanigan (1976), the Slingram method was used to aid in mapping structural features which could control mineralization. The EM survey was conducted in a uranium-bearing alaskite in the area of Mount Spokane, near Spokane, Washington. The alaskite is uniform and fracture zones are difficult to detect, yet the EM survey demonstrated that fractures were marked by anomalies indicating low-resistivity rocks associated with alteration. Zones of alteration can be anomalously conductive in comparison to the less-altered host rocks, if the alteration has produced clay minerals and increased the rock porosity.

The VLF method (very low frequencies,

20 kHz, in comparison to radio frequencies of hundreds of kilohertz) is an example of EM methods which make use of a controlled source that is distant from the survey area. The source of VLF electromagnetic signals can be any one of a number of powerful VLF transmitters (used primarily in naval communications) located in the Western Hemisphere. The magnetic field measurements are very simply and rapidly made with the VLF receiver. This is the principal advantage of using the VLF method in comparison to many other EM methods. There are two principal disadvantages of the VLF method. (1) The high frequencies used in the VLF method (on the order of 20 kHz) do not have a very great depth of penetration (on the order of 60 m in unaltered, fresh basalts, and 6 m in conductive sediments). (2) The orientation of the VLF station, which is fixed, may not produce maximum anomalies in a given survey area (the strike of the target should be in line with the VLF station). Paterson and Ronka (1971) gave a detailed discussion of the VLF method and its application to metallic mineral exploration.

The VLF survey involves one or two persons. The VLF receiver is very light and simple to operate. In the case that only magnetic fields are measured, one person can perform the survey. In some cases, both electrical and magnetic fields are measured to yield values of the apparent resistivity of the earth. Surveys which yield values of the earth's apparent resistivity require two persons for maximum efficiency.

VLF surveys used in hard-rock uranium exploration problems have been reported by Flanigan (1976) and Campbell and Flanigan (1975). The survey conducted by Flanigan (1976) supplemented the Slingram survey previously described. The differences between the depth of exploration for the VLF and Slingram surveys pro-

vided some control in determining possible sources for the EM anomalies. In addition, the measured VLF apparent resistivities aided in interpreting the background and geologic noise for the Slingram survey.

The survey described by Campbell and Flanigan (1975), conducted at the Midnite Mine near Spokane, Washington, was supplemented by a total field magnetic survey. Contoured maps of the VLF and magnetic data show many features in common with the geologic map as well as a few unique features. The major unique feature suggested by the geophysical data is a small normal fault traversing the area. The presence of this fault had not previously been suspected because soil cover made outcrops scarce. A target area for further uranium mineralization north of the fault was suggested by the quick and inexpensive geophysical survey.

VLF surveys made to aid in uranium exploration in sedimentary host rocks have not been described in the geophysical literature. In part, this reflects the limited depth of penetration of VLF signals in conductive sedimentary terranes. A pilot study of the application of the VLF method in sedimentary environments was conducted by the U.S. Geological Survey in the south Texas coastal plain uranium district. One of the many hypotheses relating to the controls of uranium deposition in this area is that reductants effecting the precipitation of the uranium were introduced through permeable fault zones. Consequently, the trend and relative permeability of fault zones could be important factors in uranium exploration in this area. Two types of faults which are exposed at or near the earth's surface have different conductivity contrasts with the surrounding rocks. The permeable fault zones contain more water than the unfaulted rocks. Consequently, they are more conductive. Impermeable fault zones have been silicified and contain less water than the surrounding rocks; they are therefore less conductive. The VLF magnetic field and apparent resistivity measurements showed distinctive differences between these two types of faults. Though only a few profiles were made over the two types of faults, the implications of the pilot study are: (1) the VLF method can be used to map the faults where exposures are poor, as is the case in the south Texas coastal plain, and (2) the type of fault zone may be interpreted from the VLF data.

There have been no reported applications of EM sounding methods to uranium exploration problems. In general, the EM sounding method can be very effectively applied to supplement EM profiling surveys. An example of a potential application is in the determination of the probable electrical nature of buried channels or other sedimentary structures which can have an important spatial relationship to uranium mineralization. For example, the Slingram survey in the Monument Valley area of Arizona (Black and others, 1962) produced anomalous responses over buried channels. However, the character of the anomalies suggested that their source was not related to the actual channel geometry but perhaps to mudstones associated with the channel. A more detailed set of EM sounding measurements could be used to aid in identification of the conductive rocks causing the Slingram anomalies.

There are two important factors in implementing EM surveys. The first is an estimate of the nature of the possible source of anomalies. The important general features of the potential target are (1) predicted contrast of electrical properties between the target and surrounding rocks, (2) size of target, and (3) depth to target. When these aspects are defined, then one or more of the EM techniques can be selected as most appropriate for the definition of the target.

A second important consideration in EM measurements is the general source of noise. For each type of EM survey, the importance of a given source of noise will vary. The two general sources of noise are geologic and cultural. Geologic noise is created by geologic features other than the potential target. One common source of geologic noise in EM measurements arises from rough terrain. For example, in VLF surveys where the surface sediments are conductive, the measurements can be an image of the topography. EM methods that use lower frequencies are somewhat less affected by the terrain. Cultural noise consists of EM signals generated by or induced in manmade structures. The EM signals that are generated by buried pipes (as in cased drill holes), by metallic fences, and by power lines are sometimes strong enough to mask the signals from the target of the survey. Cultural noise was found to be a handicap in applying EM methods in parts of the south Texas coastal plain.

This discussion of EM methods has been very brief. During the last five years, the state of the art of instruments used in electromagnetic surveys and the methods used to interpret the survey data have been improved dramatically. As EM surveying instruments get better, they will tend to be used to define more subtle geologic features associated with uranium deposits.

Induced-Polarization Methods

The induced-polarization (IP) method has been described generally by Telford and others (1975) and in more detail by Hansen and others (1967) and Keller and Frischknecht (1966). In practice, the IP method is much like the d.c. resistivity method. The IP method can be visualized in the following fashion. Consider a four-electrode array, such as that shown in figure 13-1, where a d.c. current is transmitted into two electrodes and a voltage is measured across two other electrodes. When the d.c. current is turned off, the receiver voltage does not instantly drop to zero; rather it instantly drops somewhat and then decays to zero. The time required for the voltage to decay to zero may vary from a fraction of a second to as long as minutes, though long decay times are relatively rare.

The reason that the received voltage does not drop instantaneously to zero is that the earth acts much like a capacitor. (A condenser or capacitor in an electrical circuit will charge up as a current is applied and discharge when the current is removed.) The electrical properties of the earth which give rise to IP effects have been described by Keller and Frischknecht (1966), Hansen and others (1967), and Olhoeft (1976). A very simplified discussion of the IP mechanism follows. The decay of the received voltage to zero is most directly a result of the storage of part of the energy of the transmitted current in the form of chemical energy in the ground. The chemical energy storage is the result of: (1) variations in the mobility of ions in fluids or solids and (2) the variation in the mechanisms of current conduction between metallic and most nonmetallic substances. The first effect is termed "membrane polarization," and the second is termed "electrode polarization" or "overvoltage." Both of these contribute to the IP effect that is measured in IP surveys.

There are two types of IP surveys: time domain and frequency domain. There are proponents of each method, and quite frequently, one is termed superior to the other. In principle, there is no difference between the two types of measurements, though in their implementation, differences may sometimes be seen. Time-domain IP measurements detect the variation of voltage as a function of time. Frequency-domain measurements detect the change in the apparent resistivity at two different frequencies, usually below 10 Hz. The frequency-domain approach to IP measurements is not as intuitively obvious to many people—hence much of the early work involved the time-domain approach. The frequency-domain principle can be simply explained as follows. When the transmitter is suddenly turned on, the voltage measured at the receiver will rise very rapidly, then more gradually increase until it reaches the steady-state d.c. value. The rapid rise represents high frequencies, and the steady state represents low frequencies. The variation of voltage with frequency produces a low value of apparent resistivity at high frequencies and a relatively high value at low frequencies when IP effects are present. The time-domain measurement of the IP effect is termed "chargeability," and the frequency-domain measurement is termed a "percent frequency effect" (PFE).

Most of the laboratory measurements of the IP effects of rock samples are made in the frequency domain (see, for example, Olhoeft, 1976). Here a sine wave of a.c. current is transmitted into a rock sample. The measured wave form is shifted in phase from the transmitted wave form, because in a gross sense, the sample acts like a capacitor. Consequently, both amplitude and phase measurements are used to describe the sample's IP effect. The measurement of amplitude and phase is made at many frequencies between .001 Hz and several kilohertz. Measurements made over this wide frequency range allow an interpretation to be made of the sources of the IP effect in the sample. When the measurements are made for a number of samples of known petrology, then predictions can be made as to the variation in field IP measurements. The field-observed variations then can be related to the mapping of possible petrologic variations that are meaningful in exploration. An example of this application to metallic mineral exploration has been described by Zong and Wynn (1976). Their measurements suggest that the IP method can be used to differentiate between IP responses related to mineralized ground containing copper-bearing minerals and those related to nonmineralized ground.

The factors that give rise to the polarization or chargeability of sediments are not well known. An extensive laboratory measurement program to isolate some of the more important factors should provide important information which would be relevant to uranium exploration programs. In lieu of such a program, only a few general statements can be made about the probable parameters which may cause variations in IP measurements that would be meaningful in uranium exploration. There are three major petrologic variations which may be relevant. (1) A variation of less than 1 percent in the concentration of sulphide minerals associated with a uranium deposit can yield a detectable IP response. (2) A variation from montmorillonite to kaolin clays might produce a variation from high to low IP response. (3) A change in the nature of fluids within capillary fractures would produce a change in IP response.

Although the IP method is capable of detecting and mapping subtle variations in the geochemical environment of uranium ore, there are several important considerations in designing an IP survey. The method in its most general sense is an electromagnetic method. The sources of noise discussed for EM methods are the same for IP measurements. An important difference, though, is that the IP measurement is somewhat less affected by

topography than most EM surveys. A fundamental difference between EM surveys and IP surveys is that the latter are affected by EM coupling. In an EM survey, only EM coupling is measured, whereas in an IP survey, both induced polarization and EM coupling are present in the measurements. EM coupling is described as follows: The electromagnetic signal created by switching off or turning on the transmitter creates a decay in the received voltage or frequency variation similar to that produced by polarization of the earth. Consequently, the effects of EM coupling can be confused with the IP effect in IP surveys. This problem has been a particularly important consideration in exploration for sulphide minerals buried beneath conductive sediments in the southwestern part of the United States.

There have been several methods used to reduce the effect of EM coupling in IP surveys. The first method is to use low-frequency signals. In time-domain measurements, a longer time is used between times that the transmitter current is switched on and off. A simple method to reduce the effects of EM coupling is to reduce the distance between transmitting and receiving electrodes. Several methods have been proposed to analytically remove coupling effects from the field measurements. Wynn and Zong (1975) discussed the effectiveness of these methods in various geologic settings. The removal of EM coupling from IP measurements is important in many uranium exploration applications, since variations in the IP measurements can be very subtle.

To date, the induced-polarization method has been used mainly in hard-rock uranium environments. Ketola and Sarikkola (1973) described results from an IP survey conducted in Finland to define sulphides associated with uranium mineralization. They implied that IP surveys are a standard part of uranium exploration in Finland. Smith and Daniels (1976) described an IP survey performed at the Midnite Mine near Spokane, Washington. This survey was also designed to detect sulphides associated with uranium ore. An analysis of the survey data indicated that the uranium mineralization is not directly associated with the IP anomalies but is consistently located at the margin of the anomaly. The reason for the indirect association is that the highest concentration of sulphide minerals is not coincident with the highest concentration of uranium. Two important general conclusions are illustrated by this survey. (1) Although geophysical methods can provide meaningful data in uranium exploration, they do not directly indicate uranium mineralization. (2) Drilling geophysical anomalies is not always productive because of the indirect association between uranium mineralization and the source of the anomaly.

The only reported application of IP methods in sedimentary uranium environments has been described by Smith and Daniels (1976). They demonstrated that the success or failure of the application of IP methods to this uranium environment may be dependent on the nature and distribution of clay minerals and the variation in concentration of pyrite. The combination of both pyrite and polarizable clay minerals (montmorillonites) can lead to larger anomalies than would be predicted just from the occurrence of montmorillonite or pyrite alone.

Detectable variations in the concentrations of pyrite and polarizable clay minerals are associated with some uranium deposits, but such variations are not detectable with IP methods in all cases. It is suggested that the application of IP surveys will be most successful when estimates of the probable geochemical

environment of the uranium can be made. Massive concentrations of pyrite occur marginal to some Wyoming roll fronts. Pyrite concentration along roll fronts is particularly favored when large volumes of altered sandstone are white, indicating that iron was leached as the geochemical interface advanced.

There are certain factors other than uranium mineralization (and its associated pyrite or clay minerals) which influence the distribution of polarizable minerals in sedimentary rocks. Fluvial sediments may contain local concentrations of montmorillonite or other polarizable materials which have no relationship to uranium deposits. Pyrite is not a common detrital mineral in many sedimentary environments. However, authigenic pyrite is particularly common in paleosedimentary environments which contain organic material. The geochemical processes which lead to the formation of authigenic pyrite may at best be only indirectly associated spatially with uranium mineralization.

IP methods will probably receive greater use in uranium exploration because of increasing knowledge of the geochemical setting of uranium deposits and because of advances in the implementation and interpretation of IP data. One important application will be the use of IP methods to map subtle variations in the alteration of rocks associated with uranium deposits. This approach has been discussed by Zong and Wynn (1976) in exploration for porphyry copper deposits. As more laboratory measurements are made, for example of the variation of the polarizability of clay minerals, the interpretation of IP surveys in the geologic setting of sedimentary uranium deposits should be greatly enhanced.

Magnetic-Field Measurements

Descriptions of magnetic methods can be

found in the texts written by Telford and others (1976), Parasnis (1962), Grant and West (1965), and Dobrin (1960). The magnetic method is one of the simplest and cheapest geophysical methods. For this reason and because magnetic-field measurements have a wide variety of applications, magnetic surveys are a common component of many metallic-mineral exploration programs.

In magnetic surveys, measurements can be made of components, the gradient, or the total magnetic field. Most commonly today, the total magnetic field is measured. Variations in the magnetic-field measurements reflect differences in the magnetization of various rock types and the diurnal variation of the earth's magnetic field. Sometimes strong magnetic storms, generated by sunspot activity, cause very erratic and large magnetic-field variations. At these times, meaningful magnetic-field surveys cannot be made.

The magnetization of rocks can be either induced or permanent. A material which acquires a certain intensity of magnetization when placed in a magnetic field exhibits induced magnetization. Some substances such as magnetite, pyrrhotite, and chromite are magnetized even when removed from the magnetic field. These materials are said to have permanent or remanent magnetization.

The ability of materials to acquire induced magnetization is measured by their volume susceptibility. The susceptibilities of magnetite, pyrrhotite, magnetic ilmenite, and some chromites are large in comparison to pyrites and hematites. A solid solution series exists between ilmenite and hematite. Although both end members have low susceptibilities, the intermediate members can have large susceptibilities. In the sedimentary uranium environment, magnetite and magnetic ilmenite are commonly the most important magnetic

minerals. In crystalline terranes, the distribution of pyrrhotite may also bear on the distribution of uranium mineralization.

Permanent or remanent magnetization can be acquired by most of the common minerals which have high susceptibilities. Rocks containing these minerals acquire a remanent magnetization by a variety of processes. The three most common types of permanent magnetization are thermoremanent magnetization (TRM), detrital remanent magnetization (DRM), and chemical remanent magnetization (CRM). The remanent magnetization of igneous rocks is primarily a TRM and is acquired when a rock initially cools through the Curie temperature—the temperature above which materials cannot maintain a permanent magnetization—of the magnetic grains in the rock. DRM is acquired by sediments in which fine-grained magnetite particles have slowly settled out of suspension into alignment with the earth's magnetic field. The magnetization of all rocks can be modified or changed at a later time by the acquisition of CRM resulting from the replacement of existing magnetic grains by other magnetic phases or by the growth of authigenic magnetic grains. Remanent magnetization in rocks is capable of recording the direction of the earth's magnetic field at the time magnetization is acquired. The field is known to have changed both polarity and location of the north and south magnetic poles through geologic time. The remanent magnetization in rocks records the history of the earth's magnetic fields.

The paleomagnetic features of rocks associated with uranium deposits could indicate their age relative to surrounding sediments and the relative age of magnetic minerals associated with uranium deposition. However, this application will, at best, be used only in research, for the following reasons. (1) The time history of the earth's magnetic field may not be known to high enough precision to establish a meaningful difference between the age of host rocks and uranium mineralization. (2) Oriented samples needed to establish paleomagnetic dates are very difficult to obtain from the sedimentary rocks which host a majority of uranium deposits in the United States. (3) The special geochemical conditions which lead to a difference in the remanent magnetization of the host rock and uranium-mineralized rock may not be common. However unworkable the specific applications of paleomagnetic studies may be to practical uranium exploration problems, they are nevertheless an important part of understanding the overall magnetization of rocks associated with uranium deposits.

Both local and regional magnetic surveys can be applied to uranium exploration problems. Airborne magnetic surveys are useful in regional exploration problems. Aeromagnetic surveys may be flown separately or concurrently with radiometric surveys as an additional option. This practice is particularly common in Canadian uranium exploration surveys. As in the case of gravity data, the U.S. Geological Survey has published the results of airborne magnetic surveys over much of the United States. A good example of the application of airborne magnetic-field measurements is the regional analysis of the Midnite Mine, discussed above in the section on gravity methods. The magnetic map of the Midnite Mine region can be used to indicate the presence of zoned plutons (Cady, 1976), which could be important in the orogeneses of uranium deposits in the area. Cady (1976) concluded that the magnetic highs in some cases are coincident with gravity lows. The correlation of these two types of anomalies could indicate the presence of foundered roof pendants or

anatectic centers of zoned plutons. Since the contact between such a pluton and surrounding metasedimentary rocks is associated with the uranium mineralization at the Midnite Mine, several similar, potentially mineralized areas are identified by the regional magnetic and gravity data.

There are several possible applications of local magnetic surveys in detailed uranium exploration problems. In crystalline terranes, the magnetic method is most effectively used to aid in geological mapping in areas of poor exposures. Campbell and Flanigan (1975) described an application of magnetic methods to aid in geologic mapping in the Midnite Mine area. The magnetic survey, in conjunction with a VLF survey, succeeded in defining a structural feature, probably a fault, which could prove important in uranium deposition. This structural feature offsets a diabase dike, which is interpreted to have a dip—hence the offset of the dike by the fault indicates the fault movement. Since thickness of the metasedimentary host rock is an important control for uranium mineralization in this area, the downthrown side of the fault constitutes a more favorable area of mineralization.

The electromagnetic surveys conducted by Flanigan (1976), previously described, were supplemented by ground magnetic surveys. The areas of high conductivity generally correlated with magnetic lows. It seems reasonable to infer that areas of magnetic lows and high conductivity indicate altered areas where water has moved through fractures in the alaskite, altering magnetic minerals. Such fracture zones appear to be favorable areas for uranium mineralization.

Targets of magnetic-field surveys in sedimentary environments tend to have smaller magnetic susceptibility contrasts than the targets in hard-rock terranes. Consequently, some modification of normal magnetic-field survey methods is often required. The modifications are directed toward increasing the precision of the magnetic-field measurements. In high-precision surveys, two matched magnetometers with high sensitivity are used. One magnetometer is used to conduct the magnetic survey. The other is used as a base station to continuously record time and magnetic field. The base-station readings are used to correct the survey magnetic-field measurements for diurnal time variations of the earth's magnetic field. In surveys conducted by the U.S. Geological Survey, we have found that recording the base-station field every 60 seconds is sufficient.

Factors which control the differences in magnetization of sediments in a uranium environment are not well documented. A few general observations can be made about possible variations of the nature and concentration of magnetic minerals. Each of these predicted variations can produce a magnetic anomaly which, under favorable circumstances, can be detected by surface magnetic-field measurements.

One type of variation found in sediments is a low magnetization in the area of uranium mineralization. Petrologic and drill-core susceptibility measurements made on two different deposits in the Texas coastal plain suggest the low magnetization could be caused by destruction of magnetic minerals by geochemical conditions favoring uranium deposition (Richard Reynolds, personal commun.). The susceptibility measurements demonstrated that a halo of low-susceptibility rocks exists around the uranium deposit in the mineralized horizon. The expected magnetic anomaly from a decrease in magnetic ilmenite would be a magnetic low. A surface survey conducted on one of the properties (Smith and others, 1976) has a steep magnetic gradient over the zone of mineralization. Unfortunately, the

survey could not be extended far enough to determine whether the gradient was part of a magnetic low or a change in level of magnetic field across the mineralized zone. Another magnetic survey conducted over a shallow uranium deposit in northern Colorado (Smith and others, 1976) did show a pronounced magnetic low directly over the roll front, which is persistent for about 2 km. Unfortunately, core samples are not available for the survey area. Analysis of such samples would greatly aid in determining the nature of the source of these anomalies. Computer models suggest that a 0.1 percent decrease in magnetite could account for the magnitude of the anomalies. Additional evidence for destruction of magnetic minerals associated with uranium mineralization has been reported by Adams and others (1974), who demonstrated an antipathetic relationship between detrital ilmenite and magnetite and uranium mineralization.

Uranium mineralization is not the sole source of variation of the magnetization of sediments. The distribution of magnetic minerals can be strongly controlled by paleosedimentary features. For example, magnetic minerals can be concentrated in beach sands or at the margins of stream channels. Magnetic anomalies from these features can be indirectly related to uranium deposits (for example, uranium in channels) or they might mask a weaker anomaly from the mineralized horizon.

Magnetization may change between altered and unaltered lithologies. A change in magnetic minerals could be caused by destruction of magnetic grains either by oxidation or reduction. For example, magnetite can be replaced by hematite in oxidizing environments or replaced by sulphide minerals under reducing conditions (Richard Reynolds, personal commun.). A detailed study of the variation of magnetization of altered and unaltered lithologies

TABLE 13-1. GENERALIZED COSTS FOR GEOPHYSICAL SURVEYS

Method	Generalized cost per line-mile* (in terms of seismic method)
Seismic	
High resolution	
Reflection	1
Refraction	1/5
D.C. resistivity	1/20
Gravity	
Regional	1/60
Detail-local	1/30
Electromagnetic (ground)	1/15
VLF	≈1/40
Induced polarization	1/10
Magnetic	
(ground)	1/40
(airborne)	1/70

*From worldwide average costs (*Geophysics,* 1975, p. 791-793). Costs are expressed as fractions of the reflection seismic method, which is assigned an arbitrary value of 1. For example, if reflection seismic costs $2,000 per line-mile, then d.c. resistivity costs $100 per line-mile.

has not been reported in the literature. Pending a more detailed study, the magnetic-field anomaly which would be associated with this type of variation trends from high magnetic-field values over the highly magnetized ground to a low value over less magnetized ground.

The geochemical process which forms uranium mineralization in sediments may have a profound influence on the nature and distribution of magnetic minerals. Where this occurs, magnetic surveys may provide an easy and cheap method to aid in uranium exploration problems.

Conclusions

One of the more important considerations in the implementation of geophysical methods is their relative costs. Unfortunately, the specific costs of a contracted survey are not easy to generalize, because the cost depends upon how much work is to be done, amount of data interpretation

that is required, and the mobilization costs for the survey crew. Comparison of costs of geophysical surveys is frequently made on the basis of dollars per line-mile of survey. Although this is a convenient basis of comparison of costs, the actual numbers may be somewhat misleading. For example, a regional gravity survey involves stations spread over a large region, but an IP survey involves many stations in a limited region. Table 13-1 gives the line-mile costs of some geophysical surveys in terms of a unit line-mile cost for a high-resolution seismic survey. This type of comparison demonstrates a general trend of costs of the different methods. In order to judge the general costs of a given geophysical survey, the cost of a high-resolution seismic survey might be on the order of $3,000 per line-mile. This figure, though, may be somewhat misleading, because the minimum cost to conduct a high-resolution survey is about $50,000, exclusive of special problems that may be present in data acquisition and processing. On the other hand, a routine IP survey might be conducted for a minimum of $3,000. The selection of a particular type of geophysical survey should never be made solely on the basis of cost, but also in terms of the physical properties, size and depth of the target, and the general exploration objectives. In a given situation, the most costly survey may not always be the most effective.

The other important consideration in implementation of a particular geophysical method is an estimate of what the work can accomplish. This has been discussed in the previous sections for each of the various types of geophysical methods. These methods should never be treated as a black box which will produce a "typical" anomaly. The application of many of the methods to uranium exploration requires much more care in designing and carrying out measurements than their now-routine use in oil and metallic-mineral exploration might indicate. Some of the "failures" which have been reported by various uranium explorationists have been in part a result of a lack of care in implementing geophysical surveys. Other geophysical surveys are termed failures because the interpretation of the geophysical data has not been very innovative. Not all geophysical surveys will yield a pronounced and simple anomaly associated with the ore deposit. However, a well-designed geophysical program, when properly performed and interpreted, can almost always contribute to an understanding of the geological and/or geochemical setting of a uranium deposit.

General References

Adams, J. A. S., and Gasparini, P., 1970, Gamma-ray spectrometry of rocks: Elsevier, Amsterdam, 295 p.

Adams, S. S., Curtis, H. S., and Hafen, P. L., 1974, Alteration of detrital magnetite-ilmenite in continental sandstones of the Morrison Formation, New Mexico, *in* Formation of uranium ore deposits: Vienna, Internat. Atomic Energy Agency, p. 219-252.

Adler, H. H., 1974, Concepts of uranium-ore formation in reducing environments in sandstones and other sediments, *in* Formation of uranium ore deposits: Vienna, Internat. Atomic Energy Agency, Proc., p. 141-166.

_____1975, Geological aspects of foreign and domestic uranium deposits and their bearing on exploration: U.S. Energy Research and Development Administration open-file rept., 19 p.

Altschuler, Z. S., Jaffe, E. B., and Cuttitta, F., 1956, The aluminum phosphate zone of the Bone Valley Formation, Florida, and its uranium deposits, *in* Page, L. R., Stocking, H. E., and Smith, H. B., comps., Contributions to the geology of uranium and thorium by the United States Geological Survey and Atomic Energy Commission for the United Nations International Conference on Peaceful Uses of Atomic Energy, Geneva, Switzerland, 1955: U.S. Geol. Survey Prof. Paper 300, p. 483-487.

Anderson, C. C., 1969, Uranium deposits of the Gas Hills: Wyoming Univ. Contr. Geology, v. 8, p. 93-103.

Armstrong, F. C., 1970, Geologic factors controlling uranium resources in the Gas Hills District, Wyoming: Wyoming Geol. Assoc. Guidebook, p. 31-44.

Bailey, R. V., 1965, Applied geology in the Shirley Basin uranium district, Wyoming: Wyoming Univ. Contr. Geology, v. 4, no. 1, p. 27-35.

_____1969, Uranium deposits in the Great Divide Basin-Crooks Gap area, Fremont and Sweetwater Counties, Wyoming: Wyoming Univ. Contr. Geology, v. 8, no. 2, p. 105-120.

Bailly, P. A., 1976, The problems of converting resources to reserves: Mining Eng., v. 28, no. 1, p. 27-37.

Becraft, G. W., and Weiss, P. L., 1963, Geology and mineral deposits of the Turtle Lake Quadrangle, Washington: U.S. Geol. Survey Bull. 1131.

Berning, J., Cooke, R., Heimstra, S. A., and Hoffman, U., 1976, The Rössing uranium deposit, South West Africa: Econ. Geology, v. 71, p. 351-368.

Black, R. A., Frischknecht, F. C., Hazelwood, R. M., and Jackson, W. U., 1962, Geophysical methods of exploring for buried channels in the Monument Valley Area, Arizona and Utah: U.S. Geol. Survey Bull. 1083-F.

Bloomenthal, Harold S., 1976, The evolution of the uranium joint venture, *in* Uranium exploration and development institute: Denver, Rocky Mountain Mineral Law Foundation, p. 8-1–8-20.

Bowie, S. H. U., and Cameron, J., 1976, Existing and new techniques in uranium exploration, *in* Symposium on exploration of uranium ore deposits: Vienna, Internat. Atomic Energy Agency (in press).

Brock, B. B., and Pretorius, D. A., 1964, An introduction to the stratigraphy and structure of the Rand goldfield, *in* Haughton, S. H., ed., The geology of some ore deposits in southern Africa: Geol. Soc. South Africa, v. 1, p. 25-62.

Cady, J. W., 1976, Regional gravity and aeromagnetic studies applied to uranium exploration in northeast Washington and Wyoming: U.S. Geol. Survey open-file rept. 76-317, 21 p.

Campbell, D. L., and Flanigan, V., 1975, Ground magnetic and VLF studies at Midnite uranium mine, Stevens County, Washington: U.S. Geol. Survey open-file rept.

Caneer, W. T., and Saum, N. M., 1974, Radon emanometry in uranium exploration: Mining Eng., v. 26, no. 5, p. 26-29.

Chenowith, W. L., and Malan, R. C., 1973, The uranium deposits of northeastern Arizona: U.S. Atomic Energy Commission rept., 23 p.

Childers, M. O., 1970, Uranium geology of the Kaycee area, Johnson County, Wyoming: Wyoming Geol. Assoc. Guidebook, p. 13-20.

_____ 1974, Uranium occurrences in Upper Cretaceous and Tertiary strata of Wyoming and northern Colorado: Mtn. Geologist, v. 11, no. 4, p. 131-147.

Clark, D. S., and Havenstrite, S. R., 1963, Geology and ore deposits of the Cliffside mine, Ambrosia Lake area, *in* Kelley, V. C., chm., Geology and technology of the Grants uranium region: New Mexico Bur. Mines and Mineral Resources Mem. 15, p. 108-116.

Collin, C. R., Sanselme, A., and Huot, G., 1958, An example of the use of geophysical exploration methods in metalliferous mining, the Escarpière uranium mine: Geophysical Surveys in Mining, Hydrological and Engineering Projects: The Hague, The Netherlands, European Assoc. Exploration Geophysicists.

Corbett, R. G., 1963, Uranium and vanadium minerals occurring in Section 22 mine, Ambrosia Lake area, *in* Kelley, V. C., chm., Geology and technology of Grants uranium region: New Mexico Bur. Mines and Mineral Resources Mem. 15, p. 80-81.

Cowan, G. A., 1976, A natural fission reactor: Sci. American, v. 235, no. 1, p. 36-47.

Darnley, A. G., 1972, Airborne gamma-ray survey techniques, *in* Bowie, S.H.U., Davis, M., and Ostle, D., eds., Uranium prospecting handbook: London, Inst.

Mining and Metallurgy, p. 174-211.

———1975, Geophysics in uranium exploration: Canada, Geol. Survey Paper 75-26, p. 21-31.

Davis, Dudley L., 1973, Uranium deposits of the Great Plateau: unpub. rept., 9 p.

Davis, J. F., 1969, Uranium deposits of the Powder River Basin: Wyoming Univ. Contr. Geology, v. 8, no. 2, p. 131-142.

Davis, W. E., 1951, Electrical resistivity investigations of carnotite deposits in the Colorado Plateau: U.S. Geol. Survey TEM-232, issued by U.S. Atomic Energy Commission Tech. Inf. Service, Oak Ridge, Tenn.

Denson, N. M., and Gill, J. R., 1965, Uranium-bearing lignite and carbonaceous shale in the southwestern part of the Williston Basin—A regional study: U.S. Geol. Survey Prof. Paper 463, 75 p.

Denson, N. M., Bachman, G. O., and Zeller, H. D., 1959, Uranium-bearing lignite in northwestern South Dakota and adjacent states: U.S. Geol. Survey Bull. 1055-B, p. 11-58.

Dix, C. H., 1952, Seismic prospecting for oil: New York, Harper Bros., 414 p.

Dobrin, M. B., 1960, Introduction to geophysical prospecting (2d ed.): New York, McGraw-Hill Book Co., Inc., 446 p.

Dodd, P. H., and Eschliman, D. H., 1972, Borehole logging techniques for uranium exploration and evaluation, *in* Bowie, S. H. U., Davis, M., and Ostle, D., eds., Uranium prospecting handbook: London, Inst. Mining and Metallurgy, p. 244-276.

Dodson, R. G., 1972, Some environments of formation of uranium deposits, *in* Bowie, S. H. U., Davis, M., and Ostle, D., eds., Uranium prospecting handbook: London, Inst. Mining and Metallurgy, p. 33-44.

Dodson, R. G., Needham, R. S., Wilkes, P. G., Page, R. W., Smart, P. G., and

Watchman, A. L., 1974, Uranium mineralization in the Rum Jungle-Alligator Rivers Province, Northern Territory, Australia, *in* Formation of uranium ore deposits: Vienna, Internat. Atomic Energy Agency, p. 551-568.

Dooley, J. R., Jr., Harshman, E. N., and Rosholt, J. N., 1974, Uranium-lead ages of the uranium deposits of the Gas Hills and Shirley Basin, Wyoming: Econ. Geology, v. 69, p. 527-531.

Dooley, J. R., Jr., Tatsumoto, M., and Rosholt, J. N., 1964, Radioactive disequilibrium studies of roll features, Shirley Basin, Wyoming: Econ. Geology, v. 59, no. 4, p. 586-595.

Droullard, R. F., and Dodd, P. H., 1958, Gamma-ray logging techniques in uranium evaluation, *in* Second United Nations International Conference on Peaceful Uses of Atomic Energy, Proc.: Geneva, United Nations, v. 3, p. 46-53.

Dunbar and Rodgers, 1957, Principles of stratigraphy: New York, John Wiley & Sons, Inc., p. 356.

Dyck, Willy, 1972, Radon methods of prospecting in Canada, *in* Bowie, S. H. U., Davis, M., and Ostle, D., eds, Uranium prospecting handbook: London, Inst. Mining and Metallurgy, p. 212-241.

———1974, Geochemical studies in the surficial environment of the Beaverlodge area, Saskatchewan: Canada Geol. Survey Paper 74-32, 30 p.

———1975, Geochemistry applied to uranium exploration: Canada Geol. Survey Paper 75-26, p. 33-47.

Eargle, D. H., 1959a, Sedimentation and structure, Jackson Group, south-central Texas: Gulf Coast Assoc. Geol. Socs. Trans., v. 9, p. 31-39.

———1959b, Stratigraphy of Jackson Group (Eocene), south-central Texas: Am. Assoc. Petroleum Geologists Bull. v. 43, no. 11, p. 2623-2635.

Eargle, D. H., and Weeks, A. D., 1961, Possible relation between hydrogen sulfide-bearing hydrocarbons in fault-line oil fields and uranium deposits in the southeast Texas coastal plain: U.S. Geol. Survey Prof. Paper 424-D, p. D7-D9.

Eargle, D. H., Dickinson, K. A., and Davis, B. O., 1975, South Texas uranium deposits: Am. Assoc. Petroleum Geologists Bull., v. 59, p. 766-779.

Ellis, J. R., Austin, S. R., and Droullard, R. F., 1968, Magnetic susceptibility and geochemical relationships as uranium prospecting guides: U. S. Atomic Energy Commission, RID-4, 21 p.

Fischer, R. P., 1942, Vanadium deposits of Colorado and Utah, a preliminary report: U.S. Geol. Survey Bull. 936-P, p. 363-394 (1943).

———— 1968, The uranium and vanadium deposits of the Colorado Plateau region, *in* Ore deposits of the United States, 1933-1967 (Graton-Sales Vol. 1): Am. Inst. Mining, Metallurgical, and Petroleum Engineers, Inc., p. 735-746.

Fisher, W. L., Proctor, C. V., Jr., Galloway, W. E., and Nagle, J. S., 1970, Depositional systems in the Jackson Group of Texas—Their relationship to oil, gas, and uranium: Gulf Coast Assoc. Geol. Socs. Trans., v. 20, p. 234-261.

Flanigan, V. J., 1976, Geophysical survey of uranium mineralization in alaskitic rocks, eastern Washington: U.S. Geol. Survey open-file rept. 76-679, 25 p.

Frarey, M. J., and Roscoe, S. M., 1970, The Huronian Supergroup north of Lake Huron: Canada Geol. Survey Paper 70-40, p. 143-157.

Gabelman, J. W., 1972, Radon emanometry of Starks salt dome, Calcasien Parish, Louisiana: U.S. Atomic Energy Commission, RME-4114, UC-51, 75 p.

Garrels, R. M., and Christ, C. L., 1959, Behavior of uranium minerals during oxidation, *in* Garrels, R. M., and Larsen, E. S., Geochemistry and mineralogy of the Colorado Plateau uranium ores: U.S. Geol. Survey Prof. Paper 320, p. 81-90.

Gingrich, J. E., and Fisher, J. C., 1976, Uranium exploration using the track etch method: Vienna, Internat. Atomic Energy Agency (unpub. preprint).

Gott, G. B., Wolcott, D. E., and Bowles, C. G., 1974, Stratigraphy of the Inyan Kara Group and localization of uranium deposits, southern Black Hills, South Dakota and Wyoming: U.S. Geol. Survey Prof. Paper 763, 57 p.

Gould, W., Smith, R. B., Metzger, S. P., and Melancon, P. E., 1963, Geology of the Homestake-Sapin uranium deposits, Ambrosia Lake area, *in* Kelley, V. C., chm., Geology and technology of the Grants uranium region: New Mexico Bur. Mines and Mineral Resources Mem. 15, p. 66-71.

Granger, H. C., 1963, Mineralogy, *in* Kelley, V. C., chm., Geology and technology of the Grants uranium region: New Mexico Bur. Mines and Mineral Resources Mem. 15, p. 21-37.

————1968, Localization and control of uranium deposits in the southern San Juan mineral belt, New Mexico—An hypothesis: U.S. Geol. Survey Prof. Paper 600-B, p. 60-70.

Granger, H. C., and Raup, R. B., 1969, Geology of uranium deposits in the Dripping Spring Quartzite, Gila County, Arizona: U.S. Geol. Survey Prof. Paper 595, 108 p.

Granger, H. C., Santos, E. S., Dean, B. G., and Moore, F. B., 1961, Sandstone-type uranium deposits at Ambrosia Lake, New Mexico—An interim report: Econ. Geology, v. 56, no. 7, p. 1179-1210.

Grant, F. S., and West, G. F., 1965, Interpretation theory in applied geophysics: New York, McGraw-Hill Book Co., Inc., 581 p.

Gregory, A. F., 1960, Geological interpretation of aeroradiometric data: Canada Geol. Survey Bull. 66, 29 p.

Griffiths, D. H., and King, R. F., 1965, Applied geophysics for engineers and geologists: London, Pergamon Press, 222 p.

Groth, F. A., 1970, New sandstone uranium deposit in the Battle Spring Formation, Lost Soldier–Green Mountain area, Sweetwater County, Wyoming: Wyoming Geol. Assoc. Guidebook, p. 9–12.

Hansen, D. A., Heinrichs, W. E., Jr., Holmer, R. C., MacDougall, R. E., Rodgers, G. R., Sumner, J. S., and Ward, S. H., eds., 1967, Mining geophysics, vol. 2: Tulsa, Soc. Exploration Geophysicists, 708 p.

Harshman, E. N., 1961, Paleotopographic control of a uranium mineral belt, Shirley Basin, Wyoming, *in* Short papers in the geologic and hydrologic sciences: U.S. Geol. Survey Prof. Paper 424-C, p. C4–C6.

_____ 1962, Alteration as a guide to uranium ore, Shirley Basin, Wyoming, *in* Short papers in geology, hydrology, and topography: U.S. Geol. Survey Prof. Paper 450-D, p. D8–D10.

_____ 1966, Genetic implications of some elements associated with uranium deposits, Shirley Basin, Wyoming, *in* Geological Survey research 1966: U.S. Geol. Survey Prof. Paper 550-C, p. C167–C173.

_____ 1968, Uranium deposits of the Shirley Basin, Wyoming, *in* Ridge, J. D., ed., Ore deposits of the United States, 1933–1967 (Graton-Sales Vol. 1): New York, Am. Inst. Mining, Metallurgical, and Petroleum Engineers, p. 849–856.

_____ 1972, Geology and uranium deposits, Shirley Basin area, Wyoming: U.S. Geol. Survey Prof. Paper 745, p. 82.

Hart, O. M., 1968, Uranium in the Black Hills, *in* Ridge, J. D., ed., Ore deposits of the United States, 1933–1967: New York, Am. Inst. Mining and Metallurgical Engineers, p. 832–837.

Hazlett, G. W., and Kreek, J., 1963, Geology and ore deposits of the southeastern part of the Ambrosia Lake area, *in* Kelley, V. C., chm., Geology and technology of the Grants uranium region: New Mexico Bur. Mines and Mineral Resources Mem. 15, p. 82–89.

Hiemstra, S. A., 1968, The mineralogy and petrology of the uraniferous conglomerate of the Dominion Reefs mine, Klerksdorp area: Geol. Soc. South Africa Trans., v. 71, p. 1–65.

Hilpert, L. S., 1963, Regional and local stratigraphy of uranium-bearing rocks, *in* Kelley, V. C., chm., Geology and technology of the Grants uranium region: New Mexico Bur. Mines and Mineral Resources Mem. 15, p. 6–18.

_____ 1969, Uranium resources of northwestern New Mexico: U.S. Geol. Survey Prof. Paper 603, 166 p.

Hilpert, L. S., and Moench, R. H., 1960, Uranium deposits of the southern part of the San Juan Basin, New Mexico: Econ. Geology, v. 55, no. 3, p. 429–464.

Hoskins, W. G., 1963, Geology of the Black Jack No. 2 mine, Smith Lake area, *in* Kelley, V. C., chm., Geology and technology of the Grants uranium region: New Mexico Bur. Mines and Mineral Resources Mem. 15, p. 49–52.

Hostetler, P. B., and Garrels, R. M., 1962, The transportation and precipitation of uranium and vanadium at low temperatures, with special reference to sandstone-type uranium deposits: Econ. Geology, v. 57, p. 137–167.

Houston, R. S., 1969, Aspects of the geologic history of Wyoming related to the formation of uranium deposits: Wyoming Univ. Contr. Geology, v. 8, no. 2, p. 67–79.

International Atomic Energy Agency, 1974, Formation of uranium ore deposits: Vienna, International Atomic Energy Agency, 728 p.

Joralemon, P., 1975, The ore finders: Mining Eng., v. 27, p. 32-35.

Keller, G. V., 1959, Directional resistivity measurements in exploration for uranium deposits on the Colorado Plateau: U.S. Geol. Survey Bull. 1083-B.

Keller, G. V., and Frischknecht, F. C., 1966, Electrical methods in geophysical prospecting: Oxford, Pergamon Press, 517 p.

Kelley, S. F., 1958, Geological studies of uranium-vanadium deposits by geophysical exploration methods: Geophysical Surveys in Mining, Hydrological and Engineering Projects: The Hague, The Netherlands, European Assoc. Exploration Geophysicists.

Ketola, M., and Sarikkola, R., 1973, Some aspects concerning the feasibility of radiometric methods for uranium exploration in Finland, *in* Uranium exploration methods: Vienna, Internat. Atomic Energy Agency, p. 31.

King, J. W., and Austin, S. R., 1966, Some characteristics of roll-type uranium deposits at Gas Hills, Wyoming: Mining Eng., v. 18, p. 73-80.

Kittel, D. F., 1963, Geology of the Jackpile mine area, *in* Kelley, V. C., chm., Geology and technology of the Grants uranium region: New Mexico Bur. Mines and Mineral Resources Mem. 15, p. 167-176.

Klohn, M. L., and Pickens, W. R., 1970, Geology of the Felder uranium deposit, Live Oak County, Texas: Paper presented at A.I.M.E. Meeting, Denver, Colorado, Feb. 15-19, 1970: Mining Engineers Soc. preprint no. 70-1-38, 19 p.

Knipping, H. D., 1974, The concepts of supergene versus hypogene emplacement of uranium at Rabbit Lake, Saskatchewan, Canada, *in* Formation of uranium ore deposits: Vienna, Internat. Atomic Energy Agency, p. 531-549.

Kraner, H. W., Schroeder, G. L., and Evans, R. D., 1964, Measurements of the effects of atmospheric variables on radon-222 flux and soil-gas concentrations, *in* Adams, J. A. S., and Lowder, W. M., eds., The natural radiation environment: Chicago, Univ. Chicago Press, p. 191-215.

Krasnikov, V. I., and Sharkov, Yu. V., 1962, Spacial and genetic relation between the exogenetic and metamorphic uranium deposits and the arid zones of the geologic past: Akad. Nauk SSSR, Doklady, Earth Sci. Sec. 1964, v. 144, (p. 165-167) trans. from v. 144, 1962 (p. 1359-1362).

Langen, R. A., and Kidwell, A. L., 1974, Geology and geochemistry of the Highland uranium deposit, Converse County, Wyoming: Mtn. Geologist, v. 11, p. 85-93.

Langford, F. F., 1974, A supergene origin for vein-type uranium ores in the light of the western Australian calcrete-carnotite deposits: Econ. Geology, v. 69, p. 516-526.

Laverty, R. A., Ashwill, W. R., Chenowith, W. L., and Norton, D. L., 1963, Ore processes, *in* Kelley, V. C., chm., Geology and technology of the Grants uranium region: New Mexico Bur. Mines and Mineral Resources Mem. 15, p. 191-204.

Lee, M. J., 1975, Rb-Sr geochronological study of the Westwater Canyon Member, Morrison Formation (late Jurassic), Grants Mineral Belt, New Mexico: Geol. Soc. America abstracts for 1975, p. 1164.

Lintott, K. G., Pyke, M. W., and Netolitzky, R. K., 1976, Uranium—Recent developments in Saskatchewan: Paper presented at the Prospectors and Developers Assoc. Ann. Mtg., March 7–10, 1976, 18 p.

Little, H. W., 1970, Distribution of types of uranium deposits and favourable environments for uranium exploration, *in* Uranium exploration geology: Internat. Atomic Energy Agency, Proc. 1970, p. 35–46.

Locke, Augustus, 1921, The profession of ore-hunting: Econ. Geology, v. 16, p. 243–278.

Love, J. D., 1952, Preliminary report on uranium deposits in the Pumpkin Buttes area, Powder River Basin, Wyoming: U.S. Geol. Survey Circ. 176, p. 37.

———1964, Uraniferous phosphatic lake beds of Eocene age in intermontane basin of Wyoming and Utah: U.S. Geol. Survey Prof. Paper 474-E, p. E1–E66.

———1970, Cenozoic geology of the Granite Mountains area, central Wyoming: U.S. Geol. Survey Prof. Paper 495-C, p. C1–C154.

Low, J. W., 1952, Plane table mapping: New York, Harper Bros., 365 p.

Malan, R. C., 1968, The uranium mining industry and geology of the Monument Valley and White Canyon districts, Arizona and Utah, *in* Ridge, J. D., ed., Ore deposits in the United States, 1933–1967: New York, Am. Inst. Mining, Metallurgical and Petroleum Engineers, p. 790–804.

Mason, B., 1952, Principles of geochemistry: New York, John Wiley & Sons, Inc., 276 p.

Masursky, Harold, 1962, Uranium-bearing coal in the eastern part of the Red Desert area, Wyoming: U.S. Geol. Survey Bull, 1099-B, p. 152.

McKelvey, V. E., and Carswell, L. D., 1956, Uranium in the Phosphoria Formation, *in* Page, L. R., Stocking, H. E., and Smith, H. B., comps., Contributions to the geology of uranium and thorium by the United States Geological Survey and Atomic Energy Commission for the United Nations International Conference on Peaceful Uses of Atomic Energy, Geneva, Switzerland, 1955: U.S. Geol. Survey Prof. Paper 300, p. 483–487.

McLaughlin, E. D., Jr., 1963, Uranium deposits in the Todilto Limestone of the Grants district, *in* Kelley, V. C., chm., Geology and technology of the Grants uranium region: New Mexico Bur. Mines and Mineral Resources Mem. 15, p. 136–149.

Miller, D. S., and Kulp, J. L., 1963, Isotopic evidence on the origin of the Colorado Plateau uranium ores: Geol. Soc. America Bull., v. 74, p. 609–630.

Miller, L. J., 1976, Corporations, ore discovery, and the geologist: Econ. Geology, v. 71, p. 836–847.

Minter, W. E. L., 1976, Detrital gold, uranium, and pyrite concentrations related to sedimentology in the Precambrian Vaal Reef Placer, Witwatersrand, South Africa: Econ. Geology, v. 71, p. 157–176.

Moench, R. H., 1963, Geologic limitations on the age of uranium deposits in the Laguna district, *in* Kelley, V. C., chm., Geology and technology of the Grants uranium region: New Mexico Bur. Mines and Mineral Resources Mem. 15, p. 157–166.

Morley, L. W., ed., 1967, Mining and groundwater geophysics: Canada Geol. Survey, Econ. Geology Rept. no. 26, 722 p.

Motica, J. E., 1968, Geology and uranium-vanadium deposits in the Uravan mineral belt, southwestern Colorado, *in* Ridge, J. D., ed., Ore deposits in the United States 1933–1967: New York, Am. Inst. Mining, Metallurgical, and Petroleum Engineers, p. 805–813.

Mrak, V. A., 1968, Uranium deposits in

the Eocene sandstones of the Powder River Basin, Wyoming, *in* Ridge, J. D., ed., Ore deposits of the United States, 1933–1967 (Graton-Sales Vol. 1): New York, Am. Inst. Mining, Metallurgical, and Petroleum Engineers, Inc., p. 838–848.

Nash, J. T., and Lehrmann, N. J., 1975, Geology of the Midnite uranium mine, Stevens County, Washington—a preliminary report: U.S. Geol. Survey open-file rept. 75-402.

Newman, Joseph, directing ed., 1975, What everyone needs to know about law: Washington, D. C., U.S. News & World Report Books, p. 225–229.

Offield, T. W., 1976, Remote sensing in uranium exploration: Internat. Atomic Energy Agency, Proc., SM/208-15 (in press).

Olhoeft, G. R., 1976, Electrical properties of rocks, *in* Strens, R. G. J., ed., The physics and chemistry of minerals and rocks: New York, John Wiley & Sons, Inc., p. 261–278.

Parasnis, D. S., 1962, Principles of applied geophysics: London, Methuen.

Paterson, N. R., and Ronka, V., 1971, Five years of surveying with the very-low-frequency electromagnetic method: Geoexploration, v. 9, p. 7–26.

Perry, B. L., 1963, Limestone reefs as an ore control in the Jurassic Todilto Limestone of the Grants district, *in* Kelley, V. C., chm., Geology and technology of the Grants uranium region: New Mexico Bur. Mines and Mineral Resources Mem. 15, p. 150–156.

Pipiringos, G. N., 1955, Tertiary rocks in the central part of the Great Divide Basin, Sweetwater County, Wyoming: Wyoming Geol. Assoc. Guidebook, p. 100–104.

———1961, Uranium-bearing coal in the central part of the Great Divide Basin: U.S. Geol. Survey Bull. 1099-A, p. 104.

Pipiringos, G. N., and Denson, N. M., 1970, The Battle Spring Formation in south-central Wyoming: Wyoming Geol. Assoc. Guidebook, p. 161–168.

Pretorius, D. A., 1966, Conceptual geological models in the exploration for gold mineralisation in the Witwatersrand Basin: Univ. Witwatersrand Econ. Geology Research Unit Inf. Circ. 33, 38 p.

Rackley, R. I., and others, 1968, Concepts and methods of uranium exploration: Wyoming Geol. Assoc. Guidebook, p. 115–124.

Rankama, K., 1963, Progress in isotope geology: New York, Interscience Publishers, div. of John Wiley & Sons, Inc., p. 705.

Reimer, G. M., 1976a, Helium detection as a guide for uranium exploration: U.S. Geol. Survey open-file rept. 76-240, 14 p.

———1976b, Design and assembly of a portable helium detector for evaluation as a uranium exploration instrument: U.S. Geol. Survey open-file rept. 76-398, 18 p.

———1976c, Helium in soil gas and well water in the vicinity of a uranium deposit, Weld County, Colorado: U.S. Geol. Survey open-file rept. 76-699, 10 p.

Renfro, A. R., 1969, Uranium deposits in the Lower Cretaceous of the Black Hills: Univ. Wyoming Contr. Geology, v. 8, p. 87–92.

Rich, R. A., Holland, H. D., and Petersen, U., 1975, Vein-type uranium deposits: U.S. Energy Research and Development Administration, GJO-1640, 383 p.

Robertson, D. S., 1974, Basal Proterozoic units as fossil time markers and their use in uranium prospection, *in* Formation of uranium ore deposits: Vienna, Internat. Atomic Energy Agency, p. 495–512.

Rosholt, J. N., Jr., 1959, Natural radioactive disequilibrium of the uranium series: U.S. Geol. Survey Bull. 1084-A, 30 p.

Rowntree, J. C., and Mosher, D. V., 1976, A case history of the discovery of the Jabiluka uranium deposits, East Alligator River region, Northern Territory of Australia, *in* Symposium on exploration of uranium ore deposits: Vienna, Internat. Atomic Energy Agency (in press).

Rubin, B., 1970, Uranium roll front zonation in the southern Powder River Basin, Wyoming: Wyoming Geol. Assoc. Earth Sci. Bull., v. 3, p. 5–12.

Ruzicka, V., 1975, New sources of uranium? Types of uranium deposits presently unknown in Canada: Canada Geol. Survey Paper 75-26. p. 13–20.

Santos, E. S., 1963, Relation of ore deposits to the stratigraphy of the Ambrosia Lake area, *in* Kelley, V. C., chm., Geology and technology of the Grants uranium region: New Mexico Bur. Mines and Mineral Resources Mem. 15, p. 53–59.

Scott, J. H., 1961, Quantitative interpretation of gamma-ray log: Jour. Geophysics, v. 26, p. 182–191.

———1963, Computer analysis of gamma-ray logs: Jour. Geophysics, v. 28, p. 457-465.

Scott, J. H., and Daniels, J. J., 1976, Nonradiometric borehole geophysical detection of geochemical halos surrounding sedimentary uranium deposits, *in* Symposium on exploration of uranium ore deposits: Vienna, Internat. Atomic Energy Agency (in press).

Sharp, W. N., and Gibbons, H. B., 1964, Geology and uranium deposits of the southern part of the Powder River Basin, Wyoming: U.S. Geol. Survey Bull. 1147-D, 60 p.

Shockey, P. N., Rackley, R. I., and Dahill, M. P., 1967, Source beds and solution fronts: Paper given at A.I.M.E. meeting, Casper, Wyoming, December 1967, 7 p.

Smith, B., Cady, J. W., Campbell, D. L., Daniels, J. J., and Flanigan, V. J., 1976, A case for "other" geophysical methods in exploration for uranium deposits, *in* Symposium on exploration of uranium ore deposits: Vienna, Internat. Atomic Energy Agency (in press).

Smith, B. D., and Daniels, J. J., 1976, Induced polarization surveys applied to exploration for roll-front uranium deposits: U.S. Geol. Survey open-file rept. 76-73, 21 p.

Soister, P. E., 1966, Puddle Springs Arkose Member of Wind River Formation, *in* Cohee, G. V., and West, W. S., Changes in stratigraphic nomenclature by the U.S. Geological Survey, 1965: U.S. Geol. Survey Bull. 1244-A, p. 42–46.

———1967, Geology of the Puddle Springs quadrangle, Fremont County, Wyoming: U.S. Geol. Survey Bull. 1242-C, p. C1–C36.

———1968, Stratigraphy of the Wind River Formation in south-central Wind River Basin, Wyoming: U. S. Geol. Survey Prof. Paper 594-A, p. 50.

Sørensen, H., 1974, Low-grade uranium deposits in agpaitic nepheline syenites, south Greenland, *in* Uranium exploration geology: Vienna, Internat. Atomic Energy Agency, p. 151–159.

Spirakis, C. S., and Condit, C. D., 1975, Preliminary report on the use of Landsat-1 (ERTS-1) reflectance data in locating alteration zones associated with uranium mineralization near Cameron, Arizona: U.S. Geol. Survey open-file rept. 75-416, 20 p.

Squires, J. B., 1972, Uranium deposits of the Grants region, New Mexico: Wyoming Geol. Assoc. Earth Sci. Bull., v. 5, no. 3, p. 3–12.

Stahl, R. L., 1974, Detection and delineation of faults by surface resistivity measurements, Schwartzwalder mine, Jefferson County, Colorado: U.S. Bur. Mines Rept. Inv. 7975.

Svenke, Erik, 1956, The occurrence of uranium and thorium in Sweden, *in* United Nations, Geology of uranium and thorium: Geneva, Internat. Conference on Peaceful Uses of Atomic Energy, 1955, Proc., v. 6, p. 198-199.

———1975, Potential and limitations for beneficiation of low grade uranium resources: Paper presented at Atomic Industrial Forum Internat. Conf. on Nuclear Fuel Cycle, 28-31 October 1975.

Swanson, V. E., 1961, Geology and geochemistry of uranium in marine black shales, a review: U.S. Geol. Survey Prof. Paper 356-C, p. 67-112.

Szalay, A., 1966, The significance of humus in the geochemical enrichment of uranium: Geneva, Second United Nations Conference on the Peaceful Uses of Atomic Energy, p. 182-186.

Tanner, A. B., 1964, Radon migration in the ground: A review, *in* Adams, J.A.S., and Lowder, W. M., eds., The natural radiation environment: Chicago, Univ. Chicago Press, 1069 p.

Telford, W. M., Geldart, L. P., Sheriff, R. E., and Keys, D. A., 1976, Applied geophysics: New York, Cambridge Univ. Press, 860 p.

Tipper, D. B., and Lawrence, G., 1972, The Nabarlek area, Arnhemland, Australia: A case history, *in* Bowie, S.H.U., Davis, M., and Ostle, D., eds., Uranium prospecting handbook: London, Inst. Mining and Metallurgy, p. 301-304.

Tremblay, L. P., 1972, Geology of the Beaverlodge mining area, Saskatchewan (rev. ed.): Canada Geol. Survey Mem. 367, 265 p.

Van Houten, R. B., 1948, Origin of red-banded early Cenozoic deposits in the Rocky Mountain region: Am. Assoc. Petroleum Geologists Bull., v. 32, p.

2083-2126.

Vickers, R. C., 1957, Alteration of sandstone as a guide to uranium deposits and their origin, northern Black Hills, South Dakota: Econ. Geology, v. 52, p. 599-611.

Vine, J. D., 1972, Geology of uranium in coaly and carbonaceous rocks: U.S. Geol. Survey Prof. Paper 356-D, p. 113-170.

Vine, J. D., Swanson, V. E., and Bell, K. G., 1966, The role of humic acids in the geochemistry of uranium: Geneva, Second United Nations Conference on Peaceful Uses of Atomic Energy, p. 187-191.

von Backstrom, J. W., 1974, Other uranium deposits (I.A.E.A-SM-183/27, Review Paper), *in* Formation of uranium ore deposits: Vienna, Internat. Atomic Energy Agency, p. 605-624.

Wallace, Stuart R., 1974, The Henderson ore body—Elements of discovery, reflections, *in* Am. Inst. Mining, Metallurgical, and Petroleum Engineers, Trans., v. 256, p. 216-227.

Walthier, T. N., 1976, The shrinking world of exploration: Mining Eng., v. 28, no. 4, p. 27-31; and v. 28, no. 5, p. 46-50.

Wayland, T. E., 1976, Remote sensing continues to open new horizons for ore seekers: Eng. and Mining Jour., v. 177, no. 7, p. 67-73.

Weeks, A. D., and Eargle, D. H., 1960, Uranium at Palangana salt dome, Duval County, Texas: U.S. Geol. Survey Prof. Paper 400-B, p. B48-B52.

———1963, Relation of diagenetic alteration and soil-forming processes to the uranium deposits of the southeast Texas coastal plain, *in* Clays and clay minerals, v. 10—10th Natl. Conf. Clays and Clay Minerals, 1961, Proc.: New York, Macmillan and Co., p. 23-41.

Wenrich-Verbeek, K. J., 1976, Water and stream-sediment sampling techniques for use in uranium exploration: U.S. Geol. Survey open-file rept. 76-77, 30 p.

White, L., 1975, Wyoming uranium miners set sights on higher production: Eng. and Mining Jour., Dec., p. 61–71.

Wood, H. B., 1968, Geology and exploitation of uranium deposits in the Lisbon Valley area, Utah, *in* Ridge, J. D., ed., Ore deposits in the United States, 1933–1967: New York, Am. Inst. Mining, Metallurgical, and Petroleum Engineers, p. 770–789.

Wyant, D. G., Sharp, W. N., and Sheridan, D. M., 1956, Reconnaissance study of uranium deposits in the Red Desert, Sweetwater County, Wyoming: U.S. Geol. Survey Bull. 1030-I, p. 237–308.

Wynn, J. C., and Zong, K. L., 1975, EM coupling, its intrinsic value, its removal and the cultural coupling problem: Geophysics, p. 831–850.

Young, R. G., and Ealy, G. K., 1956, Uranium occurrences in the Ambrosia Lake area, McKinley County, New Mexico: Washington, D. C., U.S. Atomic Energy Commission, RME-86, Tech. Ser., Dept. Commerce.

Zohdy, A.A.R., Eaton, G. P., and Mabey, D. R., 1974, Application of surface geophysics to ground water investigations, Ch. D1, Techniques of water-resources investigations of the United States Geological Survey: Washington, D. C., U.S. Govt. Printing Office.

Zong, K. L., and Wynn, J. C., 1976, Recent advances and applications in complex resistivity measurements: Geophysics, p. 851–864.

Index

Page numbers in italics refer to illustrations.

Abstract companies, 161-162
Access roads, 360
Accounting departments, 337
Accounting procedures, 314, 315, 337-343
Acquired lands, federal, 143, 182; mineral entry, 146
Administration, 8, 11, 346; exploration manager, 19, 346; project manager, 21
Advance royalties, 281, 308, 310
Aerial photography, 2, 98, 100-101, 102, 491-493; alteration complexes, 100; black and white, 100; controlled mosaics, 220; geological research, 491; hematite staining, 100; infrared, 101; Landsat, 101, 491; natural color, 100, 221; scale, 100; surface mapping, 220; U.S. Department of Agriculture, 100; U.S. Forest Service, 101; U.S. Geological Survey, 100
Aerial radioactivity surveys, 224
Aeroradiometric anomalies, 124
Aeroradiometry, 105, 108, *110*, 219; aircraft, 219; equipment, 108; flight-line markers, 219; importance of, 228; preliminary ground surveys, 219; Saskatchewan, 228; structure-controlled deposits, 225
Affidavit of assessment, 201
Age determination. *See* Uranium deposits, age determination
Agreement in principle (AIP), 303
Airborne electromagnetic (EM) methods, 503
Airborne gamma-ray surveys. *See* Aeroradiometry
Airborne radiometric surveys, 2; flight patterns, *218*, 219
Airborne reconnaissance, 101, *112*; aircraft, 101, 105; field notes, 102, 105; gamma-ray spectrometry, 113; magnetic-field surveys, 510; maps, 102-103, *103*, *104*, 105; purpose, 103; scintillation counter, 105, 108, 109; strategic, 101; structure-controlled deposits, 124; total gamma-radiation surveys, 109-110, 221, *221*
Alabama, Chattanooga Shale, 28, 85

Alaska, hard-mineral exploration, 251
Alkaline water, 73; Red Desert deposits, 82; Shirley Basin, 73; spontaneous-potential logs, 413
Alpha-sensitive film method, 234, *235*, 235; disadvantages, 235; Jabiluka deposits, 239; procedures, 234, 235; radon detection, 234-235; suppliers, 234
Alteration complexes: aerial photography, 100; Black Hills area, 65; Brule Formation, 75; Fox Hills Formation, 58; Gas Hills district, 55-56; geochemical cell, *48*, 76; geological reconnaissance, 114; Lance deposits, 75; origins, 73, 74; Powder River Basin, 65; Puddle Springs, 74; Rabbit Lake, 92; roll-front deposits, 53, 54, 117-118, 452-453; Shirley Basin, 56
Alteration tongues: geochemical cell, 76; roll-front deposits, 53, 54
Ambrosia Lake, 35, *36*; Cliffside deposit, 47; cross section, 78; new reserves, 95; Nose Rock occurrence, 95; paleodrainage patterns, 33
Amenability tests, 474
Aquifers, 317, 477-478; plugging, 424
Arikaree Formation, 58
Arizona: Monument Valley trend, 27, 495; strata-controlled deposits, 26; Triassic Chinle Formation, 27, 28, 39; White Canyon Trend, 27
Army Map Service, 102
Assessment work, 153-154; affidavits, 154; holding costs, 201; mineral development agreements, 312; recording, 166; resumption, 154; waivers, 154
Assessor's office, 161
Athabasca Basin, 90, *93*
Athabasca deposits, *96*
Atomic Energy Commission: gamma-ray log standards, 406; land withdrawal, 182
Atomic Industrial Forum (AIF), 286
Audiomagnetotelluric (AMT) geophysical method, 502
Australia, Northern Territory: Jabiluka deposit,